Molecular Genetics of Axial Patterning, Growth and Disease in the *Drosophila* Eye

Drosophila eye during different stages of development. (Contributed by Oorvashi Roy Puli, Meghana Tare and Amit Singh)

Amit Singh • Madhuri Kango-Singh
Editors

Molecular Genetics of Axial Patterning, Growth and Disease in the *Drosophila* Eye

Springer

Editors
Amit Singh
Department of Biology
Center for Tissue Regeneration
and Engineering at Dayton (TREND)
Premedical Program
University of Dayton
Dayton
Ohio 45469
USA

Madhuri Kango-Singh
Department of Biology
Center for Tissue Regeneration
and Engineering at Dayton (TREND)
Premedical Program
University of Dayton
Dayton
Ohio 45469
USA

ISBN 978-1-4939-5303-5 ISBN 978-1-4614-8232-1 (eBook)
DOI 10.1007/978-1-4614-8232-1
Springer New York Heidelberg Dordrecht London

Printed on acid-free paper

Springer is part of Springer Science+Business Media (www.springer.com)

To

OurParents
Late Prof. Hari Om Singh
Mrs. Dinesh Kumari Singh
Late Mr. D.M. Kango
Prof. Mangala Kango

&

Our Daughters
Aditi Singh
Manasi Singh

Introduction

The fly sat upon the axle tree of the chariot-wheel and said, what a dust do I raise!

—Aesop

An important question in developmental biology is how a single-celled embryo gets transformed into a multicellular three-dimensional organism with complex structure and functions. The quest to understand this important facet of development resembles the search for the holy grail of modern day biology. Patterning and development of an organism require production of specific number of cells whose fate is determined by a genetic circuitry. Any perturbation in this finely tuned process results in defects. Therefore, the basic cell biological process of cell proliferation, cell differentiation, and cell death play important roles in sculpting an organ during organogenesis. In developmental biology, it is important to unravel the mechanism of fate assignment and differentiation.

The time tested *Drosophila melanogaster* (fruit fly) model has played a central role in developmental biology during the twentieth century. The *Drosophila* model has a long genetic legacy, beginning with Thomas Hunt Morgan in early 1900 (Morgan 1911). A judicious blend of molecular and developmental genetics has proved beyond doubt that *Drosophila* is a valuable model for addressing important questions of modern day biology. There are several thousand people whose work/lives center around the little fruit fly *Drosophila melanogaster*. In recent years, the emphasis of their studies has shifted from inheritance to development and disease. In the hands of a small number of particularly imaginative scientists, traditional genetics, experimental embryology, and new molecular genetic techniques have been combined to build a picture of developmental mechanisms. To date, *Drosophila* has maintained its status as a trusted and highly versatile model to study patterning, growth, and disease. Among all the adult body structures, the *Drosophila* eye, because of its simple structure, and easy amenability to mutations and genome-wide screens has become an important tool in the hands of Drosophilists.

The study of developing eye from a two-dimensional eye primordium to a three-dimensional adult eye and visual system, and use of eye model to study patterning, growth, development, evolution, and disease is the topic of the current book. The *Drosophila* eye has been intensively studied to explore cell biological processes like cell fate specification, patterning, growth, and cell signaling, etc. Understanding the generation and functioning of eye as an organ, our primary sensory modality, is important. We are curious to know how the visual system assembles.

It is now almost 37 years since the seminal paper from Ready et al. (1976) described the development and structure of *Drosophila* compound eye. The discovery of morphogenetic furrow (MF), a wave of differentiation, which is initiated from the posterior margin of the eye imaginal disc and sweeps in the anterior direction (Ready et al. 1976), is considered to be a major milestone in *Drosophila* eye field. It results in differentiation of retinal precursor cells to photoreceptor neurons. It was known that adult appendage develops from a group of cells set aside during embryonic development, which grows during larval stages and then metamorphose into adult appendages. Tomlinson provided the electron microscopic view of cellular events that follow the formation of morphogenetic furrow (Tomlinson 1985). Generation of monoclonal antibodies to detect early cell differentiation was another major landmark (Fujita et al. 1982). Enhancer trap technique using P element-mediated transgenesis proved to be an important tool that still remains an asset in the arsenal of modern day fly geneticist's tool kit (Bellen et al. 1989; Grossniklaus et al. 1989; Wilson et al. 1989). Another important milestone was demonstration of structural and functional similarity in the genetic circuitry involved in eye development in flies and humans (Halder et al. 1995; Quiring et al. 1994). These studies completely changed the outlook of the eye field. Halder et al. (1995) reported the master selector gene concept in the eye where they demonstrated that *eyeless (ey) Drosophila* homolog of *PAX-6* gene could reprogram other tissues and generate ectopic eyes in the wing, leg, and antenna. These studies provided a great impetus to the *Drosophila* eye model, which, by then, was also used to address questions for human disease. The evolution of *Drosophila* eye research cannot be complete without mentioning the contributions of Seymour Benzer, Walter Gehring, and Gerald M Rubin. The hard work of Gerald Rubin and his collaborators came to fruition when fly genome was published in the year 2000 (Adams et al. 2000; Myers et al. 2000; Rubin et al. 2000). It was instrumental in validating the observation of Gehring's group that there is a strong conservation in the genetic circuitry of flies with that of humans and other vertebrates. It completely changed the field and put the fly model on the forefront among all other animal models. These discoveries led to generation of new genetic and molecular technology, and put *Drosophila* eye model system on the forefront of biological research to address important questions related to human diseases like retinal diseases, neurodegenerative disorders, cancers, etc. Furthermore, the *Drosophila* eye model provided more versatility to study basic cell biological processes of patterning, growth, cell proliferation, and cell death and to carry out genome-wide screens.

This picture is new and exciting, although far from complete. It represents the beginnings of a real understanding of how one animal is designed and built. This book, which is written for the students as well as the specialists, aims to give an up-to-date glimpse of that picture. However, the field is developing so fast that some of the things may change; therefore, we have tried to use well-established material. We have made an attempt to provide an overview of approaches used in the fly eye model. We have dealt with the basic question of patterning of how eye develops starting from early events of specification to molecular mechanisms involved in transition of eye from a monolayer epithelium to a three-dimensional structure. During this transition, one of the hallmark events is formation of the morphogenetic furrow (MF). This book also highlights events of morphogenesis, cell polarity, cell adhesion, and negative regulation of neural patterning in developing *Drosophila* eye. Other areas discussed in this book are use of *Drosophila* eye model to understand protein homeostasis network, organ size control mechanism, and genetic basis of neurodegeneration. The book also encompasses an important aspect of development and evolution during early eye development as well as larval eye or Bolwig's organ.

The collection of chapters in this book helps us celebrate hundred plus years of research using *Drosophila* eye model, and provides a blueprint of future research directions and frontiers in this field. We hope you enjoy reading this book as much as we did. We would like to end with a quotation (Dryden J (1696) from: The epilogue to The Husband his own Cuckold, lines 35–37):

Fools change in England, and new fools arise'
For, tho' th' immortal species never dies,
Yet ev'ry year new maggots make new flies. . . . '

References Cited

Adams, M.D., Celniker, S.E., Holt, R.A., Evans, C.A., Gocayne, J.D., Amanatides, P.G., Scherer, S.E., Li, P.W., Hoskins, R.A., Galle, R.F., *et al.* (2000). The genome sequence of Drosophila melanogaster. Science *287*, 2185-2195.

Bellen, H.J., O'Kane, C.J., Wilson, C., Grossniklaus, U., Pearson, R.K., and Gehring, W.J. (1989). P-element-mediated enhancer detection: a versatile method to study development in Drosophila. Genes Dev *3*, 1288-1300.

Fujita, S.C., Zipursky, S.L., Benzer, S., Ferrus, A., and Shotwell, S.L. (1982). Monoclonal antibodies against the Drosophila nervous system. Proc Natl Acad Sci U S A *79*, 7929-7933.

Grossniklaus, U., Bellen, H.J., Wilson, C., and Gehring, W.J. (1989). P-element-mediated enhancer detection applied to the study of oogenesis in Drosophila. Development *107*, 189-200.

Halder, G., Callaerts, P., and Gehring, W.J. (1995). Induction of ectopic eyes by targeted expression of the eyeless gene in Drosophila. Science *267*, 1788-1792.

Morgan, T.H. (1911). The Origin of Five Mutations in Eye Color in Drosophila and Their Modes of Inheritance. Science *33*, 534-537.

Myers, E.W., Sutton, G.G., Delcher, A.L., Dew, I.M., Fasulo, D.P., Flanigan, M.J., Kravitz, S.A., Mobarry, C.M., Reinert, K.H., Remington, K.A., *et al.* (2000). A whole-genome assembly of Drosophila. Science *287*, 2196-2204.

Quiring, R., Walldorf, U., Kloter, U., and Gehring, W.J. (1994). Homology of the eyeless gene of Drosophila to the Small eye gene in mice and Aniridia in humans. Science *265*, 785-789.

Ready, D.F., Hanson, T.E., and Benzer, S. (1976). Development of the Drosophila retina, a neurocrystalline lattice. Dev Biol *53*, 217-240.

Rubin, G.M., Hong, L., Brokstein, P., Evans-Holm, M., Frise, E., Stapleton, M., and Harvey, D.A. (2000). A Drosophila complementary DNA resource. Science *287*, 2222-2224.

Tomlinson, A. (1985). The cellular dynamics of pattern formation in the eye of Drosophila. J Embryol Exp Morphol *89*, 313-331.

Wilson, C., Pearson, R.K., Bellen, H.J., O'Kane, C.J., Grossniklaus, U., and Gehring, W.J. (1989). P-element-mediated enhancer detection: an efficient method for isolating and characterizing developmentally regulated genes in Drosophila. Genes Dev *3*, 1301-1313.

Acknowledgment

At the outset, we would like to extend our gratitude to the fly community and their support for this venture. There is not enough space to mention the names of all the researchers whose contribution in the *Drosophila* eye field has been instrumental in making this book a reality. The *Drosophila* eye model has been exceptionally lucky to get attention from a strong group of highly accomplished scientists. It has been a great pleasure for us to work or interact with many of them, and to hear them at symposia and meetings. We would like to thank all the fly pushers who have been kind enough to contribute to this book.

We are grateful to our mentors Henry Sun, Kwang-Wook Choi, Georg Halder, and Pradip Sinha for introducing us to this "Golden Bug" or "Cinderella of Modern Genetics" and to the field of patterning and growth in the developing eye and other organs.

This book would not have been possible without the excellent support from Ms. Melanie Tucker, Editor, Genetics and Systems Biology at Springer, and from Ms. Meredith Clinton, Assistant Editor, who worked patiently and diligently to help keep the process of writing streamlined, and manageable.

The encouragement and support for initiating this project was provided by my mother Dinesh Kumari Singh. I would like to thank my brother Rohit Singh who never doubted our ability to take on this project. We would like to thank our daughters Aditi and Manasi and brother Rohit Singh who never doubted our ability even though have been alongside during this endeavor, and can not believe that any sane person can be fascinated by flies to this extent. Their perception of the fruit fly is enclosed in the accompanying figure.

Contents

Early Eye Development: Specification and Determination

Emmi Bürgy-Roukala, Sara Miellet, Abhishek K. Mishra and Simon G. Sprecher

Introduction

The compound eyes of insects are typically composed of a large array of unit eyes termed ommatidia (Fig. 1a). The number of ommatidia and the size of the eyes are variable within the group of insects. The compound eye of the fruit fly *Drosophila melanogaster* is composed of 750–800 ommatidia, forming a highly stereotypically organized, virtually crystalline lattice. In turn, each ommatidium is composed of photoreceptor (PR), cone, and pigment cells (reviewed in Wolff and Ready 1993). Most adult structures develop from larval epithelial structures called imaginal discs. The adult compound eye of *Drosophila* originates from the eye-antennal imaginal disc. The eye-antennal disc develops into the adult eye, antenna, head capsule, and the ocelli, a group of extra-retinal photoreceptors (Kenyon et al. 2003). The discs' precursors are specified during embryogenesis and the imaginal discs keep proliferating throughout the three larval instar stages. During third instar, retinal differentiation is initiated in the eye disc from posterior to anterior by the dynamic progression of an epithelial groove called the morphogenetic furrow (MF). After metamorphosis, the pair of eye-antennal discs has transformed into the whole head capsule of the adult fly.

The eye disc is specified and determined during embryogenesis and larval stages by a genetic network called the retinal determination network (RDN). One of the earliest genes expressed in the presumptive eye field is called *eyeless* (*ey*). Mildred Hoge described the *ey* mutant in *Drosophila* almost 100 years ago (Hoge 1915), and genetically mapped the gene causing this phenotype to the fourth chromosome of the fly. Much later, the *ey* gene was cloned and sequenced, leading to the astonishing observation that this gene is a homolog of the vertebrate *Pax6* gene, which upon mutation causes a developmental syndrome of retina called aniridia in humans and a similar disorder caused by the *small eyes* mutation in mice (Quiring et al. 1994;

S. G. Sprecher (✉) · E. Bürgy-Roukala · S. Miellet · A. K. Mishra
Institute of Cell and Developmental Biology, Department of Biology,
University of Fribourg, Chemin du Musée 10, 1700 Fribourg, Switzerland
e-mail: simon.sprecher@unifr.ch

A. Singh, M. Kango-Singh (eds.), *Molecular Genetics of Axial Patterning, Growth and Disease in the Drosophila Eye*, DOI 10.1007/978-1-4614-8232-1_1,

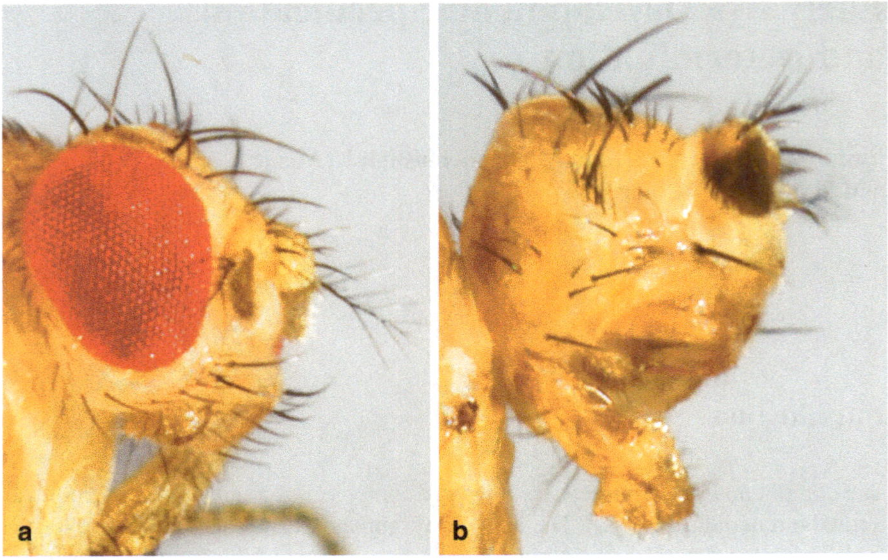

Fig. 1 **a)** The *Drosophila* compound eye. **b)** *eya* mutant flies

Walther and Gruss 1991; Hill et al. 1991; Ton et al. 1991). Support for the homologous function was further provided by cross-phylum genetic experiments in which the mouse *Pax6* gene was shown to be able to replace the *Drosophila ey* gene, since its targeted expression in *Drosophila* results in the formation of ectopic eyes (Halder et al. 1995). While the evolutionary origin of eyes was still widely subject of debates in the field, this discovery challenged much of previous beliefs in the independent convergent evolution of various different eye types across species (Gehring 2002). The finding that both *Drosophila* and vertebrate genes share the same function in governing the formation of eyes strongly indicated that these organs have evolved from a common ancestral prototypic eye and therefore supports the theory of a monophyletic origin of the eye (Gehring 2002; Halder et al. 1995).

Moreover, in addition to *ey,* other members of the RDN that specify the eye field in *Drosophila* are homologs to the corresponding genes in vertebrates (reviewed in Wawersik and Maas 2000). Thus, even though the camera-type eye of vertebrates and the compound eye of *Drosophila* are morphologically very distinct, the molecular mechanisms underlying the early specification of an eye field are surprisingly conserved. This discovery has turned the *Drosophila* eye-antennal disc into an excellent model system to analyze the formation of vertebrate eyes and to model human diseases. Also, studying the specification of an eye field can teach us how early determination genes integrate multiple signaling pathways during the course of development.

In this chapter we will focus on the RDN, the genetic network underlying eye field specification and determination. In particular each of the members of the RDN and their interactions with each other will be discussed in detail. Further attention is paid to the development of the eye precursor cells during embryogenesis, and the establishment of a distinct eye field in the developing eye-antennal disc during the first and second instar stages. Finally, we will briefly summarize the knowledge gained so far on how some of the members of the RDN contribute to the specification of the extraretinal photoreceptors in *Drosophila.*

The Retinal Determination Network

Development of the eye field requires three distinct phases: specification, determination, and differentiation. A highly interconnected regulatory network called the RDN is responsible for the initial specification and determination steps (Fig. 2). This network consists of selector genes that are regulated by extracellular signaling pathways and interact in highly complex regulatory and autoregulatory feedback loops. Thus, apart from the initiation of RDN gene expression, this network does not form a linear hierarchal gene regulatory system. Many of the RDN genes are not exclusively functioning in eye development, but are also involved in the formation of other tissues. Hence, it is the combinatorial effect and the interaction of these genes with each other that serve as a base to build an eye. Since most RDN genes also provide essential functions during embryogenesis, eye-specific mutant phenotypes are mostly the result of deletions in a regulatory region of an enhancer that is necessary to drive reporter expression only in the eye imaginal disc (Bui et al. 2000; Zimmerman et al. 2000).

Which genes are parts of the RDN and what are their characteristics? While there is no clear definition of a RDN gene, the network members have to fulfill certain criteria. Clearly, these genes have to be active at some point between the formation of the eye disc and beginning of the differentiation of the retinal cells. Most of the genes are able to induce ectopic eyes when overexpressed in other imaginal discs, though they do not all have the same potential in doing so. Conversely, mutational inactivation of RDN genes should interfere with the development of the eye (Fig. 1b). Most of the RDN members are also nuclear proteins directly controlling or affecting transcription. Obviously, there are more factors than just the RDN genes involved in eye development; however, they do not fill the criteria to be considered as selector genes.

To current knowledge and applying the definition above, the RDN consists of following factors: the homologs of the vertebrate *Pax6 eyeless* (*ey*) and *twin-of-eyeless* (*toy*), the six family members *sine oculis* (*so*) and *optix*, the tyrosine phosphatase Eyes absent (Eya), a winged helix containing transcription factor Dachshund (Dac), zinc-finger transcription factors Teashirt (Tsh) and Tiptop (Tio), the nuclear factors Eyegone (Eyg) and Twin-of-eyegone (Toe), homeobox containing transcription cofactor Homothorax (Hth), serine-threonine kinase Nemo (Nmo) and the Pipsqueak motif containing genes *distal antenna* (*dan*) and *distal antenna related* (*danr*). The RDN also requires five canonical extracellular signaling pathways: the Notch, epidermal growth factor receptor (EGFR), Hedgehog (hh), Transforming growth factor beta (TGFβ), and Wnt signaling pathways. The factors of these signaling pathways will be presented in context and will not be described in detail.

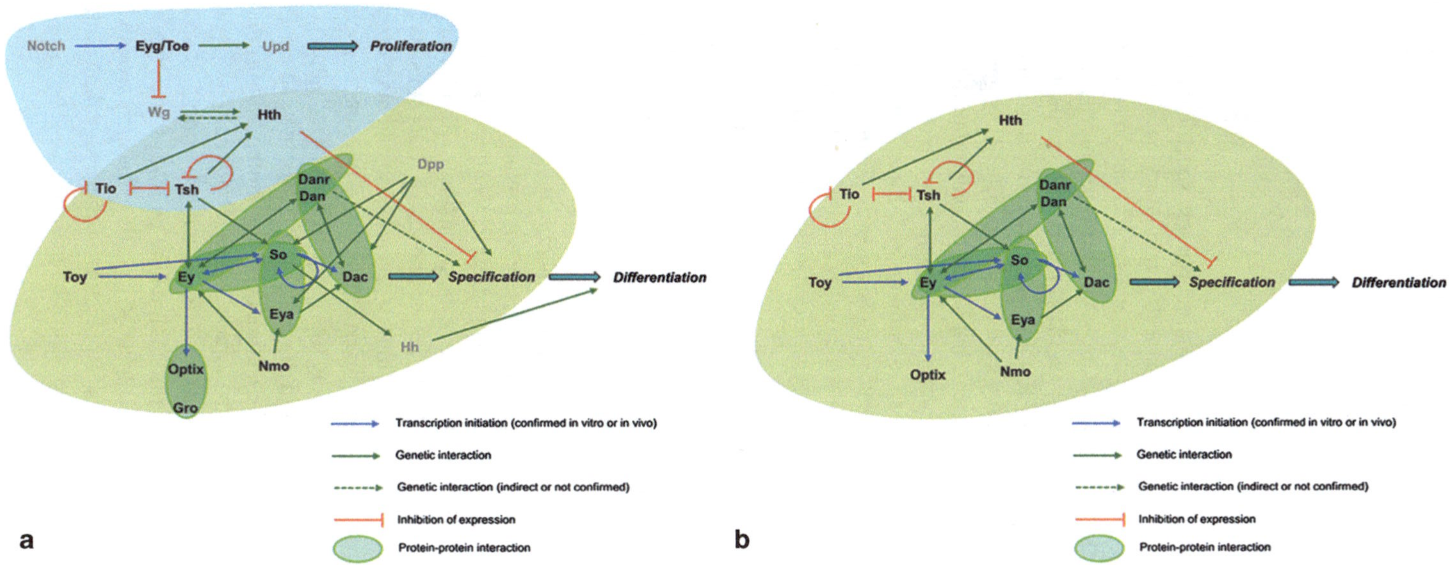

Fig. 2 The retinal determination network (RDN). Schematic representation of the key players of the RDN showing **a)** the RDN regulating both specification and differentiation with members of the extracellular signaling pathways and **b)** the core network involved in specification. Note that the genes are not placed in a chronological order. *Blue arrows* indicate a direct transcriptional activation and the *green arrows* indicate a genetic interaction that has been confirmed in vivo or in vitro. *Dashed green lines* indicate a putative direct or nondirect genetic interaction. Protein–protein complexes are marked with a circle. *Light green* area covers genes involved in specification; *light blue* area covers genes involved in proliferation; *grey factors* are members of different signaling pathways. (Figure adapted from Kumar 2009)

Players of the Retinal Determination Network

eyeless and *twin-of-eyeless*

Pax6 genes encode transcription factors containing two DNA binding domains; a 128 amino acid long paired domain and a homeodomain (Quiring et al. 1994; Treisman et al. 1991; Walther and Gruss 1991). Other Pax family members have a varying number of paired domains with homeodomains and all of them have important functions in nervous system development across different species (for references see Treisman et al. 1991). The paired domain can be divided in two subdomains, the N-terminal PAI and C-terminal RED. Both of them contain a helix-turn-helix (HTH) motif (for references see Jun et al. 1998).

Drosophila has two *Pax6* homologs; *ey* and *toy*. Ey and Toy share 95 % and 91% amino acid sequence identity, respectively, in the paired domain and 90% in their homeodomains with the murine Pax6 protein (Quiring et al. 1994; Czerny et al. 1999). Toy in comparison to Ey shares more sequence similarities to Pax6 outside these domains and has an additional transcription activation C-terminal domain that is present in Pax6 but missing in Ey. In addition, Ey has one amino acid substitution (Asn to Gly) in the paired domain, which affects its binding affinity to some *Pax6/toy* binding sites and thus explains the differential expression patterns and functions of *ey* and *toy* in the fly (Czerny et al. 1999).

Both *toy* and *ey* arose most likely due to a gene duplication event during late insect evolution, since two *Pax6*-like genes are only present in holometabolous insects (Czerny et al. 1999). Originally, these two genes were acting in parallel after the duplication event. *ey* eventually gained a mutation affecting the amino acid composition of the paired domain and radically changing the mode of action of the protein. The changes in the protein sequence abolished the capability of *ey* for autoregulation. *toy* and *Pax6* among other species still posses this character. (Plaza et al. 1993). Since *toy* shares more sequence similarities with the vertebrate *Pax6* gene and is essential for the head formation both in vertebrates and in the fly, it has been postulated that *toy* has been under stronger selective pressure than *ey* (Czerny et al. 1999).

Whole mount *in situ* hybridization experiments have shown that *toy* is the first RDN gene expressed in the embryo in areas giving rise to the eye precursors. The first transcripts of *toy* can be detected early in the posterior procephalic region of a stage 5 embryo (cellular blastoderm). After gastrulation, during stages 14–15, the expression pattern covers the optic lobe primordia, brain, parts of the ventral nerve cord (VNC), and eye-antennal imaginal disc anlagen. In the third instar larva *toy* expression is localized to both posterior and anterior of the MF and in the ocellar precursors (Czerny et al. 1999; Jacobsson et al. 2009).

The expression of *ey* first appears in the brain and VNC during germ band extension in the stage 9 embryo (Quiring et al. 1994; Czerny et al. 1999; Kammermeier et al. 2001). The *ey* transcripts can be localized earlier in the VNC, but later in the brain than those of *toy*. In addition, the *ey* expression pattern in the brain is more spatially restricted than that of *toy*. In the developing visual anlagen, *ey* expression is almost entirely overlapping with *toy* expression pattern (Czerny et al. 1999). Ey can also be detected in the mushroom bodies, optic lobes, and in the central complex of the adult brain, and different *ey* mutants show defects also in the development of these organs (Callaerts et al. 2001). These observations link Ey functionally closer to the vertebrate Pax6 which is crucial for the formation of certain brain areas (Callaerts et al. 2001 and references therein). The *ey* expression persists until the third instar larval stage and is detected similarly to *toy* in the anterior boundaries of the eye disc as well as in the regions anterior to the MF (Quiring et al. 1994; Czerny et al. 1999).

The *ey* loss-of-function mutants have different eye and head phenotypes depending on the severity of the mutation. In general, hypomorphic alleles of *ey* lead to eyeless flies, whereas in eye-specific null mutants the head is completely missing due to the lack of the whole eye-antennal disc (Halder et al. 1998; Kronhamn et al. 2002; Quiring et al. 1994). Eye-specific *toy* null and hypomorphic mutants (toy^{hdl} and $toy^{G7.39}$) are also mostly headless (Kronhamn et al. 2002; Punzo et al. 2002). Some escapers are able to form a head and even compound eyes, but the ocelli are always missing (Kronhamn et al. 2002). The phenotype of the toy^{hdl} is the result of a truncation in the homeodomain, yet the transcript pattern of toy^{hdl} corresponds to wild-type transcripts (Kronhamn et al. 2002).

Forced overexpression of *ey* using different enhancer lines can effectively induce ectopic eyes on the wings, legs, halters, and antenna (Halder et al. 1995), and none of the other RDN members have the ability to function as a potent ectopic eye inducer as *ey*. Targeted overexpression of *toy* can induce ectopic eyes in the wings, legs, and halteres (Czerny et al. 1999; Salzer and Kumar 2010).

sine oculis and *optix*

so and *optix* are both part of the RDN since mutants of these genes show severe defects in eye development, and their forced expression is able to induce ectopic eyes (Cheyette et al. 1994; Pignoni et al. 1997; Seimiya and Gehring 2000; Serikaku and O'Tousa 1994; Weasner et al. 2007). These two genes are a homeodomain and a SIX domain containing transcription factors and members of the SIX (Sine oculis box) gene family (Cheyette et al. 1994; Oliver et al. 1995b). Vertebrates have six members in this family called *Six1* to *Six6* (Kawakami et al. 1996a; Kawakami et al. 1996b; Oliver et al. 1995a; Oliver et al. 1995b; Toy et al. 1998). The *Drosophila so* gene is most closely related to the murine *Six1/Six2* genes. It was long believed to be the homolog of *Six3*, because it is the only gene of the SIX family involved in eye development in the mouse (Oliver et al. 1995a; Seimiya and Gehring 2000; Seo et al. 1999). However, later it became evident that *optix* instead is the homolog of *Six3* (Seo et al. 1999; Toy et al. 1998). In addition to *so* and *optix*, *Drosophila* has one more gene belonging to this group but representing a different subclass, *DSix4*, which has no shown functions in the formation of the retina neither in the fly nor in vertebrates (Seo et al. 1999).

so is required not only for the development of the adult compound eye but also for the formation of other structures of the visual system including the larval eye, termed Bolwig's organ, the ocelli and the optic ganglia in the brain (Cheyette et al. 1994). In the developing retina *so* is required for the initiation and propagation of the MF as well as for the formation of the PRs posterior to the furrow (Pignoni et al. 1997). The gene locus of *optix* is near the *so* locus, suggesting that these two genes went through a gene duplication event. *optix* is an orthologue of the mouse *Six3*, which also has specific functions during the eye development. Although these two genes share extensive sequence similarities even in the DNA binding homeodomain, they show distinct functions during the eye formation in *Drosophila* and are not redundant.

In the embryo, *so* is expressed in the optic lobe primordia anterior to the cephalic furrow. Later, *so* is detected also bilaterally at the segment boundaries. At stage 16, *so* expression is restricted to the anterior region of the head, including the Bolwig's organ (Serikaku and O'Tousa 1994). The eye-specific *so* expression in the eye discs is activated during the late second larval stage, right before the MF initiation takes place in the third instar stage. *so* is expressed in a gradient starting at the anterior and decreasing towards the posterior end of the disc, and persists until the end of the larval stage (Cheyette et al. 1994; Kenyon et al. 2003). The expression of *so* is not restricted to the anterior site of the MF but is also expressed within and posterior to it. In addition to the eye-antennal disc, *so* expression is also present in the optic lobes and the Bolwig's organ primordia of the embryo, and in the leg discs (Cheyette et al. 1994; Seimiya and Gehring 2000). The expression pattern of *so* during early larval stages, however, remains unclear. Depending on the *lacZ* reporter lines used, β-galactosidase expression is detected in the first instar (Wang and Sun 2012) or starting from mid-to-late second instar stage (Kenyon et al. 2003).

optix is expressed in a similar, but not identical pattern to *so*. *optix* transcripts can be detected early in the blastoderm embryo on its anterior end. During germ band extension, when *so* expression can already be detected in the optic lobe primordia, *optix* is still detectable only at the very anterior region of the embryo. During the second instar stage, *optix* is expressed in the whole eye disc, but in the third instar, contrasting *so*, it becomes restricted to an area anterior to the MF resembling *ey* and *toy* expression. *optix* is also present in the wing and haltere discs and in a part of the antennal disc (Seimiya and Gehring 2000).

so loss-of-function mutant animals show severe developmental defects in the retina, impairing cell proliferation and PR formation, leading to extensive cell death anterior to the MF (Cheyette et al. 1994; Pignoni et al. 1997; Serikaku and O'Tousa 1994). However, analysis of so^3 null mutant clones revealed that the cell death is a consequence and not a cause, since the clones initially show massive overproliferation (Pignoni et al. 1997). so^1 mutants are in most cases lacking compound eyes and ocelli, being the strongest non-lethal and eye-specific loss-of-function mutation. Other mutants cause less severe effects in the eye formation including a reduced eye size or rough eyes. Ocelli in these mutants are mostly depleted or reduced (Heitzler et al. 1993; Cheyette et al. 1994). so^D (*droplet*, *drl*) is a dominant negative allele of *so* that is able to recruit an *optix* co-factor Optix binding protein (Obp) (Kenyon et al. 2005). This mutation causes a phenotype similar to so^1 mutation with the exception that ocelli are not reduced and the mutation is homozygous lethal (Heitzler et al. 1993).

A loss-of-function mutation of *optix* has not been generated so far, leaving the phenotype of such a mutant under speculation.

The initial experiments to produce ectopic eyes by *so* overexpression failed, and it was assumed that So needs the co-activator Eya for its function in the eye. However, a screening combining upstream activation sequence (UAS)-*so* responder line with 219 distinct Gal4 driver lines revealed ectopic eye formation in four cases in the antennal part of the eye-antenna imaginal disc. This portion of the disc normally lacks Eya, suggesting that *so* is able to induce eye tissue without the presence of intrinsic Eya. Interestingly, So even induces *eya* expression in these ectopic eye tissues. It still remains to be cleared if this induction is direct or not (Weasner et al. 2007). Another Gal4 screen revealed *so*-induced ectopic eye tissue also in the head (Salzer and Kumar 2010). Nevertheless, compared to the ability of *ey* to induce eye tissue ectopically, *so* is a considerably less potent inducer.

Forced expression of *optix* in all imaginal discs leads to ectopic eye formation only in parts of the antennal disc, and to formation of extra ocelli (Seimiya and Gehring 2000). Using the same set of the 219 driver lines that were crossed with UAS-*so*, it was demonstrated that *optix* is also capable to induce ectopic eye tissue in the wing and haltere discs (Weasner et al. 2007). Interestingly, Optix does not require the presence of Ey to induce ectopic eyes (Seimiya and Gehring 2000), but *optix* is not able to induce ectopic eyes in *so* or *eya* mutant background (an observation of Seimiya and Gehring 2000).

Strikingly, only two of the 219 driver lines were able to induce ectopic eyes with both UAS-*so* and UAS-*optix*. Further analysis of the regulatory domains of *so* and *optix* indicate that although these genes do not considerably differ in their homeodomains and probably even bind the same target genes, they differ significantly in their SIX-domains and in the amino acid composition at the C-terminus. The SIX-domain is involved in the selection of a binding partner, thus with distinct partners So and Optix could also have different targets. Interestingly, the C-terminal end of Optix is involved in repression of retinal formation (Weasner et al. 2007) and one common binding partner shared by *so* and *optix* is the transcriptional co-repressor Groucho (Gro), suggesting that also *so* might function as a repressor of transcription (Silver et al. 2003; Kenyon et al. 2005).

Two genes interacting distinctively with *so* and *optix* are the *so binding protein* (*sbp*) and *obp*, respectively. The *sbp* expression is detected posterior to the MF where *so* but no *optix* is present. *obp* in contrast shows co-expression with both *so* and *optix*, but does not bind to *so* in a yeast-two-hybrid assay and has no effect on eye development when expressed in regions where only So is present. The exact functions of these two genes in combination with their binding partners are yet to be discovered (Kenyon et al. 2005). In addition, only *so* has been associated with *eya* (Seimiya and Gehring 2000).

Taken together, the existence of different factors that can bind *so* in distinct genetic environments suggests that *so* probably has a dynamic role in eye development, acting both as an activator and repressor of transcription. A similar kind of a mechanism has been demonstrated in vertebrates. Phosphatase activity of Eya changes the activity of Dach1, which is bound to *Six1*, to become a transcription co-activator (Li et al. 2003;

reviewed in Dominguez and Casares 2005). The *so* binding sites have also been located in *ey, hh,* and *lozenge* (*lz*) genes (Pauli et al. 2005; Niimi et al. 1999). The *lz* as a target is of particular interest, since it is expressed and functions posterior to the MF and is involved in the first differentiation steps of the PRs, thus linking cell fate specification with differentiation.

eyes absent

The *eya* gene belongs to the phosphatase subgroup of the haloacid dehalogenase (HAD) family of transcription co-activators including or having one member in *Drosophila* (Eya) and four in vertebrates (Eya1-4) (Bonini et al. 1993; Hanson 2001; Tootle et al. 2003). The phosphatase activity is located at the C-terminus of the gene, which is also involved in interacting with So and Dac (Pignoni et al. 1997; Chen et al. 1997). The transactivator domain lays at the N-terminus, containing a proline-serine-threonine-rich region (Silver et al. 2003). The vertebrate *Eya2* can rescue the eye-specific *eya* null mutant phenotype, causing eyeless flies and thus revealing a high degree of conservation in the *eya* gene sequence (Bonini et al. 1997). *eya* is involved in the inhibition of cell death and promotes cell specification in the eye imaginal disc anterior to the MF (Bonini et al. 1993). It was also demonstrated that *eya*, together with *so*, is required for the initiation and propagation of the MF, and for the development of the PRs posterior to the furrow (Pignoni et al. 1997). Since *eya* is able to produce ectopic eyes it belongs to the group of key factors in the RDN.

The *eya* transcription starts in the blastoderm embryo in the future head segments, but cannot be detected in the eye primordial cells at this stage. In the second instar stage *eya* shows expression in the posterior regions of the eye disc that decreases toward the anterior and middle parts of the disc. In the third instar stage, *eya* is expressed both in the anterior and posterior regions of the MF and in the region giving rise to the ocelli (Bonini et al. 1993).

The *eya* null mutants are embryonic lethal. Viable eye-specific eya^{1} and eya^{2} mutants display an eyeless phenotype, but develop all the other head structures (Bonini et al. 1993; Fig. 1b). The eye-specific mutant phenotype is the result of a deletion in a regulatory region of the *eya* enhancer (Bui et al. 2000; Zimmerman et al. 2000). Other less specific *eya* mutants (eya^{3cs} and eya^{4}) lack, in addition to the reduced eyes, also ocelli (Zimmerman et al. 2000). Strong loss-of-function eya^{clift1} mutant clones cause overgrowth of the disc followed by a massive cell death, as observed in so^{3} mutants (Pignoni et al. 1997).

The eye-specific enhancer deleted in eya^{1} and eya^{2} is sufficient to recapitulate the *eya* expression pattern in the eye disc. This enhancer is not expressed in ey^{2} mutant background, and is not affected in so^{1} and dac^{3} mutants (Bui et al. 2000). Ectopic expression of *dac* and *ey*, but not *eya* or *so*, is able to induce ectopic expression of this enhancer.

Overexpression of *eya* is able to cause ectopic eye formation in the antenna, and with more copies of UAS-*eya*, ectopic eyes form also in the legs and wings (Bonini et al. 1997). Using other imaginal disc-specific drivers *eya* can also initiate eye formation in the halters and in the head (Salzer and Kumar 2010).

dachshund

dac was discovered in an enhancer trap screen that was conducted to find novel regulators of PR specification. The gene identified in this screen was named *dachshund*, since the mutants show a recessive short-leg phenotype (Mardon et al. 1994). *dac* is required for the progression of the MF and in the formation of PRs. It belongs to a gene superfamily containing the vertebrate *dac* homologs *Dach1/2* and proto-oncogenes *Ski/Sno* of the Ski-family that are involved in transcriptional repression (Hammond et al. 1998). *Dach* genes in the vertebrates are also involved in eye and limb development (Hammond et al. 1998). The *Drosophila* Dac protein contains two conserved domains, a DNA binding, winged helix domain called DD1 and another domain called DD2. DD1 is essential for the function of Dac in the eye. DD2 is required to facilitate the function of DD1 and to interact with Eya, although neither of these functions is required for eye development (Pappu et al. 2005; Tavsanli et al. 2004).

dac is expressed in the third instar disc at the posterior edge anterior to the MF (Mardon et al. 1994). Posterior to the MF, *dac* is expressed in PRs R1, R6, and R7 and in cone cells. *dac* expression can be detected in other imaginal discs including leg, antenna, and wing as well as in the developing nervous system of the embryo and larval optic lobes.

dac mutants have severely reduced or absent eyes, they are not able to form PRs and the progression of the MF is impaired (Mardon et al. 1994). Overexpression of *dac* leads to eye tissue formation in the antennal and leg discs, confirming its place in the RDN (Shen and Mardon 1997).

The *dac* expression can be induced by ectopic expression of *ey*, *eya,* and *eya/so* complex, suggesting that these genes control *dac* activity also in the normal developmental conditions (Chen 1997). In addition, *dac* expression requires the presence of the extracellular signal Decaplentaplegic (Dpp) (Chen et al. 1999).

teashirt and *tiptop*

Drosophila Tsh and Tio are nuclear proteins containing three and four zinc finger motifs, respectively (Laugier et al. 2005). Forced expression of these genes can induce ectopic eyes, and this is why these genes were included in the RDN in the first place (Rubin 1998). The *tsh* expression was initially identified in the trunk region of an early embryo regulating patterning together with homeotic genes. The embryonic role of *tsh* is to repress head development, and certain loss-of-function mutations lead to trunk-to-head transformations (Fasano et al. 1991; Roder et al. 1992; de Zulueta et al. 1994). *tsh* promotes the growth and specification of the dorsal part of the eye disc mainly during second instar larval stage. On the ventral half *tsh* suppresses eye specification together with *hth* (Singh et al. 2002). *tio* is a paralog of *tsh*, and has analogous functions in the eye formation as *tsh*. These two genes most likely evolved after a duplication event that occurred during the evolution of *Drosophilidae* (Laugier et al. 2005; Bessa et al. 2009).

tsh expression begins in the trunk at an early stage 6 embryo proceeding *tio* expression, which commences at the posterior region at stage 10. In the early embryo their expression pattern is not overlapping. Co-expression of the genes begins at stage 12 embryo in certain regions of the CNS and epidermis, and increases during the development, but they also maintain distinct expression patterns (Laugier et al. 2005). In the eye disc *tsh* expression begins already in the first instar stage and occupies the whole disc, thus overlapping with *ey* and *hth* expression (Singh et al. 2002). During the second larval stage, *tsh* expression retracts towards the anterior part of the disc, and by the late third instar larval stage *tsh* expression is restricted in the eye disc anterior to the MF, symmetrically on the dorsal and ventral halves of the disc (Pan and Rubin 1998; Singh et al. 2002). *tio* is co-expressed with *tsh* in the third instar larval eye disc (Bessa and Casares 2005).

Weak loss-of-function mutations of *tsh* do not show any phenotype in the fly eye. The *tsh* null mutants are embryonic lethal. To study the role of *tsh* in eye development, *tsh* null mutant clones were generated using an x-ray-induced mitotic recombination (Pan and Rubin 1998). These mutant clones in the eye disc do not display any abnormal phenotypes, suggesting that *tsh* has a redundant function in the eye disc.

Overexpression of *tsh* induces ectopic eyes in head tissues and antenna, and these extra eyes have nearly normal ommatidia (Pan and Rubin 1998). Interestingly, also the opposite has been demonstrated, where suppression of *tsh* actually promotes ectopic eye formation and forced expression suppresses eye development. This, however, only occurs at the ventral margin of the eye disc. In the dorsal part of the disc *tsh* promotes eye development. Thus, *tsh* plays a dual role in regulating the growth of the eye disc. The level of *tsh* expression might play a significant role in the decision whether the eye development is suppressed or promoted, since overexpression of more than one copy of *tsh* enhances the phenotypes (Singh et al. 2002).

The *tio* null mutants are viable and show no aberrant phenotype in the adult (Laugier et al. 2005). Nevertheless, as in the case of *tsh*, forced overexpression of *tio* causes ectopic eye formation in the head region (Bessa et al. 2009). *tsh* is also able to induce *ey* expression in the antennal disc when ectopically expressed. Reciprocally, *ey* expression is necessary for *tsh* expression (Singh et al. 2002). *tsh* might also function upstream of *so*, *eya*, and *dac* since these genes are ectopically expressed upon forced *tsh* expression (Pan and Rubin 1998).

Tsh and Tio negatively regulate themselves (negative autoregulatory feedback loop) and each other (Bessa et al. 2009). They can also compensate for each other's function. For example, *tio* is able to rescue embryonic lethality of *tsh* mutants when expressed in the regions normally occupied by *tsh* (Bessa et al. 2009).

eyegone and *twin-of-eyegone*

eyg and *toe* encode for nuclear proteins showing similarities to a splice isoform of the vertebrate *Pax6* gene termed *Pax6(5a)*, containing only one DNA-binding domain, the RED domain (and no PAI domain) (Jun et al. 1998). These genes have a different function during the retinal specification compared to the *Pax6* homologs *ey* and *toy*. *eyg* is involved in the eye and salivary duct development. It positively regulates the growth of the eye disc, but also acts as a transcriptional repressor (Jang et al. 2003; Yao et al. 2008). *toe* acts as a repressor as well, but in somewhat different regions than *eyg*. The *eyg* possesses two, and *toe* only one repressor domain, which may account for their different functional outcomes (Yao et al. 2008).

eyg and *toe* are expressed in the embryo in the salivary gland placode (SGP), in the dorsal head, and in the segments of the trunk. Starting at embryonic stage 17, *eyg* and *toe* are expressed in the eye-antennal disc precursors in the same cells that are also expressing *ey* and *toy* (Czerny et al. 1999; Jones et al. 1998; Jun et al. 1998; Quiring et al. 1994; Yao et al. 2008). During the growth of the disc *eyg* and *toe* are expressed both in the antennal and eye part. In the eye disc they continue to be expressed anterior to the MF until the third instar stage. By then the expression is restricted to a small cluster of cells at the dorsal–ventral boundary (Yao et al. 2008). This expression pattern is different to that of *ey* and *toy*, which are expressed in a much larger area anterior to the furrow (Quiring et al. 1994; Czerny et al. 1999; Yao et al. 2008). *eyg* and *toe* are not expressed in equal levels; *eyg* contributes the clear majority to the total amount of transcript.

Weak loss-of-function mutants of *eyg* have reduced or missing eyes, whereas strong mutations fail to form a head and cannot hatch from their pupal cases (Jang et al. 2003). The eye discs of such mutants are highly reduced in size and show extensive cell death. However, these mutants cannot be rescued by overexpression of the apoptosis inhibitor p35, indicating that apoptosis is not the sole reason causing the *eyg* phenotype. *toe* alone does not seem to play an essential role in the eye since downregulation of *toe* transcripts by miRNA does not cause any obvious phenotype (Yao et al. 2008).

Forced expression of *eyg* is capable producing ectopic eyes on the ventral head region, deriving from the formation of extra eye fields on the eye disc (Jang et al. 2003). Forced *toe* expression in an *eyg* mutant background is also not able to rescue this phenotype. This suggests that these genes do not act redundantly in eye development and are differentially regulated, since *eyg* mutants do not affect *toe* transcription levels.

eyg is involved in a pathway independent of *ey*, since both genes are unable to launch each other's expression. Moreover, they are also functionally independent, since they do not require each other to induce ectopic eyes (Jang et al. 2003). This functional independency resembles that of *optix* (Seimiya and Gehring 2000). Nevertheless, *eyg* and *ey* can functionally substitute for each other, by partially rescuing each other's loss of function phenotype. Co-expression of these genes has synergistic effects in inducing ectopic eyes, and it has been postulated that they even form a heterodimer to commonly regulate gene expression (Jang et al. 2003) .

homothorax

hth encodes for the *Drosophila* homolog of the mouse *Meis1* proto-oncogene containing a TALE class homeodomain. Hth provides an important function during embryonic patterning as a cofactor for the homeotic genes (Rieckhof et al. 1997; Pai et al. 1998).

hth is expressed in several imaginal discs including the eye-antennal disc. In the second larval stage, *hth* is ubiquitously expressed in both the eye and antennal part (Jang et al. 2003; Pichaud and Casares 2000). Later, the expression is restricted to the anterior parts of the disc including the ocellar region and head capsule. Weak expression can be detected in the posterior boundaries of the disc and in already differentiated pigment cells (Pai et al. 1998; Pichaud and Casares 2000).

hth mutants show ectopic eye formation in the ventral side of the head, suggesting that *hth* is normally needed to suppress eye formation (Pai et al. 1998; Pichaud and Casares 2000). The ectopic eye tissue induced in *hth* mutant clones have nearly normal pigment cells, but the ommatidia are abnormally arranged, showing an orthogonal shape instead of a hexagonal one (Pichaud and Casares 2000). However, in the dorsal region of the disc *hth* is not able to do the same. Forced expression of *hth* within the eye disc causes suppression of eye formation. The function of *hth* is dependent on its interactions with several RDN genes that will be described later in this chapter.

nemo

nmo encodes a proline-directed serine/threonine kinase and is the founding member of the Nemo-like kinase (NLK) family of the MAPK superfamily. In eye development it is required for positive regulation of its downstream target *eya* in the RDN cascade.

nmo shows a dynamic expression pattern in the eye disc. At second instar stage, *nmo* is ubiquitously expressed showing co-expression with *ey* in the whole disc, and with *eya* in the posterior part of the disc. Later in the third instar stage the expression becomes restricted to the anterior edge of the MF, showing co-expression with *eya*. *nmo* can also be detected in ocellar precursor cells at the anterior-dorsal side of the disc. At the posterior boundary of the disc *nmo* is expressed together with *hth* (Braid and Verheyen 2008).

nmo mutants have slim eyes and show defects in the patterning of the ommatidia (Choi and Benzer 1994). Strong forced expression of *nmo* leads to expansion of the dorsal eye, to the induction of *dac* and *eya* expression and to *hth* repression in the antennal discs, leading to ectopic PR formation. Ectopic PRs were not detected in other imaginal discs (Braid and Verheyen 2008). However, *nmo* acts synergistically with *ey*, *eya*, *so*, and *dac* to promote eye specification, and enhances their ability to form ectopic eyes (Braid and Verheyen 2008). *nmo* does not seem to be a direct regulator of these genes but rather interacts with them at the protein level.

distal antenna and *distal antenna related*

Dan and Danr are transcription factors containing a DNA binding motif called Pipsqueak (Psq) that is found in proteins involved in chromatin modification (Baonza and Freeman 2002). Misexpression of *dan* and *danr* with an antenna-specific driver *Distal-less*(*Dll*)-Gal4 leads to ectopic eye formation in antennal precursors (Curtiss et al. 2007). During the eye specification, *dan* and *danr* are particularly involved in the onset of differentiation, positively regulating the expression of one of the first differentiation genes *atonal* (*ato*). In addition, they interact with EGFR signaling to ensure correct ommatidial patterning (Curtiss et al. 2007).

dan and *danr* are required already during the early embryogenesis for the correct development of the nervous system. During larval stages their expression is restricted to the eye-antennal disc (Suzanne 2004). The expression in the early third instar eye disc starts in cells surrounding the MF. As the furrow starts moving, both factors are expressed anteriorly to it and to a weaker extent in the differentiated cells posterior to the furrow, whereby *dan* shows a stronger expression than *danr* (Curtiss et al. 2007). The expression of *dan* and *danr* overlaps with *ey* and shows similarities to *eya*, *so*, and *dac* expression (Curtiss et al. 2007). As both genes are later expressed also in the antennal disc, the timing of expression is critical in determining when these genes promote eye formation and when antennal formation (Curtiss and Mlodzik 2000; Suzanne 2004; Suzanne et al. 2003).

dan and *danr* loss-of-function mutants display small and rough eyes, depending on the severity of the mutation. The rough eye phenotype is a consequence of defects in ommatidial spacing and photoreceptor formation, which is probably associated with the lack of *ato* expression (Curtiss et al. 2007). Misexpression of *dan* and *danr* in the eye disc causes a similar phenotype to the loss-of-function mutants. This demonstrates the importance of a controlled regulation of transcription levels in a system that consists of many interconnected members like the one the RDN represents (Curtiss et al. 2007).

dan and *danr* are able to induce ectopic eyes in the antennal disc (Curtiss et al. 2007). In addition, as they are also expressed in the antenna, ectopic expression in the leg discs can induce leg-to-antenna transformation and interestingly, loss-of-function mutations cause an antenna-to-leg transformation (Emerald et al. 2003; Suzanne et al. 2003).

Dan and Danr cross-regulate each other; Dan is required for *danr* expression on the anterior side of the MF whereas Danr represses *dan*. Both genes are able to induce or maintain *ey* expression in the ectopic eye tissue (Curtiss et al. 2007). The discrete expression patterns and differences in eye phenotypes induced by *dan* and *danr* mutations propose that these two genes do not act totally in parallel during the eye formation, and that they might have even antagonistic roles (Curtiss et al. 2007). These two genes are acting downstream of most of the RDN genes, since *so*, *eya*, and *dac* are required for their expression (Curtiss et al. 2007) .

Teamwork: Genetic Interactions Between the Retinal Determination Network Genes

The eye of *Drosophila* represents a masterpiece when it comes to finding an example of a regularly patterned organ. To generate such a structure, a robust network of genes has to work behind it. The RDN offers an interesting and challenging model to study the early steps of tissue specification. The RDN consists not only of a linear, hierarchical cascade of transcription activation, but also it is imprinted by feedback loops, protein complexes, and dynamic gene expression (Fig. 2). Correct protein levels seem to be essential, especially in the formation of multiple protein complexes between the members of the RDN. In this part we will discuss the interactions of the RDN genes with each other and their putative target genes outside of the RDN.

toy and *ey* Initiate the Retinal Determination Network Cascade

toy is the first RD gene to be expressed in the eye precursors. Toy binds to the *ey* enhancer region and activates its expression. Forced expression of *ey* does not cause ectopic *toy* expression, placing *toy* upstream of *ey* in the genetic cascade (Czerny et al. 1999). *ey* rather than *toy* is more essential for the initiation of the eye development in general, since *toy* is not able to produce ectopic eyes in an *ey* mutant background (Czerny et al. 1999).

ey has been shown to be able to initiate *so* and *eya* expression in the wing disc when expressed ectopically. These genes contain Ey-specific binding sites on their enhancers, suggesting that Ey is a direct regulator of their expression. However, *toy* is able to initiate *so* expression even in mutants where the Ey-specific enhancer region is deleted (so^1), indicating that Ey and Toy can regulate *so* distinctively. Ey and Toy both bind the eye-specific *so* enhancer (so^{10}) with their paired domains, albeit at different sites (Niimi et al. 1999; Punzo et al. 2002). Surprisingly, in the few regions of antenna where forced *so* can induce ectopic eye formation, ectopically expressed *ey* and *toy* are not able to do the same. Therefore, there must be additional factors that mediate the interaction between *ey/toy* and *so* at least in the formation of ectopic eye tissue (Weasner et al. 2007).

Ey is required for normal *eya* activity, since eye discs mutant for *ey* do not express *eya*. Moreover, ectopic expression of *ey* induces *eya* expression (Bonini et al. 1997; Halder et al. 1998) and *eya* is able to induce ectopic eyes only in an intact *ey* background (Bonini et al. 1997). It is assumed that Ey binds to regulatory regions in the *eya* gene that have been shown to be responsible for the eye-specific *eya* expression (Bui et al. 2000; Zimmerman et al. 2000). However, although it was demonstrated with an electrophoretic mobility shift assay (EMSA) that *eya* is a direct target of Ey, no *ey* binding site was located in the enhancer regions required for *eya* expression in the eye disc. Moreover, this *ey* binding locus of *eya* does not drive *eya* expression in a reporter assay. Nonetheless, it cannot be excluded that other regulatory regions left out from this assay are also needed for the eye-specific expression of *eya* (Ostrin et al. 2006). This is supported by the observation, that the *eya* enhancer is expressed also in *ey*, *so*, and *dac* mutant backgrounds in a pattern comparable to wild type (Bui 2000).

Ey can also bind directly to the eye-specific enhancer of *optix* and forced expression of *ey* induces ectopic expression of *optix* (Ostrin et al. 2006). However, in hypomorphic ey^2 mutants *optix* expression is not affected in the eye disc and ectopic eyes are produced regardless of *ey* expression (Seimiya and Gehring 2000). Thus, it is not clear what the function of this regulatory site in the enhancer of *optix* is.

dac was shown to be downstream of *ey*, since *ey* expression is not lost in a *dac* mutant background, but to a limited level Dac is also able to induce *ey* expression in the antennal disc, suggesting an existence of a feedback loop (Chen et al. 1997).

So and Eya Regulate *ey* Expression in a Positive Feedback Loop

Since *eya* and *so* require Ey for the induction of ectopic eyes in tissues where *ey* usually is not present (Pignoni et al. 1997), Eya and So must be able to bind to an eye-specific enhancer on the *ey* gene. Indeed, the *ey* enhancer contains a So binding site (Niimi et al. 1999; Punzo et al. 2002). Physical interaction between Ey and So has also been demonstrated and these two genes do form a protein complex together (Pauli et al. 2005; Niimi et al. 1999; Zhang et al. 2006). This complex has been associated with *ato* regulation, linking eye specification directly with differentiation (Zhang et al. 2006).

eya acts downstream of *ey*, since *ey* expression is not affected in *eya* mutants and forced *ey* expression causes ectopic *eya* expression. Yet reciprocally, *eya* is needed for the formation of ectopic eyes induced by forced *ey* expression, since eye formation cannot be launched in tissues mutant for *eya*. Forced expression of *eya* alone is able to induce ectopic eyes only in an intact *ey* background (Bonini et al. 1997). Hence, Eya and perhaps also So are able to induce *ey* expression only as a complex and not individually.

So and Eya Form a Protein Complex Acting as a Transcriptional Regulator

So forms a transcription activating protein complex with Eya. Together they control multiple steps during the development of the eye disc towards differentiation by regulating cell proliferation, MF formation and propagation, and later during neuronal development (Niimi et al. 1999; Pignoni et al. 1997; Serikaku and O'Tousa 1994). Forced expression of this complex is able to produce ectopic eyes in the antennal, wing, and leg discs in an *ey* dependent manner (Pignoni et al. 1997). The potency of *eya* and *so* to induce ectopic eyes is much higher when these are simultaneously rather than individually misexpressed. As Eya lacks a DNA binding domain, it is likely acting as a transcriptional co-activator in this complex (Pignoni et al. 1997). Another possibility is that Eya is involved in regulating the phosphorylation states of So, hence influencing the activity of So (Weasner et al. 2007).

Forced expression of *so* induces the expression of *eya* and *dac*. It is not yet clear if the interaction is direct or not, but this finding has inspired the following proposal: Ey and Toy initiate the expression of *so*, which in turn activates *eya* expression. Further,

So forms a complex with Eya regulating other downstream genes involved in the RDN (such as *dac*). It should be noted however, that So-induced *eya* expression is restricted to a certain subset of cells, suggesting the existence of additional factors in this pathway (Weasner et al. 2007). One of these factors could be Nmo (see the following section).

As previously mentioned, So alone physically binds the repressor protein Gro. The binding is, however, inhibited in the presence of Eya, although Eya and Gro are not competing for the same binding sites. This suggests that *so* has differential roles in (eye) development depending on the presence of its binding partner (Silver et al. 2003).

eya and *dac* Have a Synergistic Effect on Eye Formation

eya was placed upstream of *dac*, since *eya* mutant eye discs show a reduced *dac* expression, and *dac* mutation does not affect *eya* expression. *eya* is not able to form ectopic eyes in the absence of *dac* and forced coexpression of *eya* and *dac* can induce ectopic eyes in a higher rate than when these genes are independently misexpressed (Chen et al. 1997; Tavsanli et al. 2004). *eya* is expressed in a nearly identical pattern with *dac* (and *so*), which led to the logical assumption that their gene products are molecularly interacting. Indeed, a physical interaction between Eya and the DD2 domain of Dac was demonstrated by *in vitro* biochemistry and yeast-one-hybrid experiments (Chen et al. 1997; Tavsanli et al. 2004). Unexpectedly, this physical interaction of Eya with the DD2 domain is not essential for the synergic effect observed by *eya* and *dac* co-expression, suggesting that Eya-Dac complex might not have any function in eye development (Chen et al. 1997; Kumar 2009). However, a possible trimeric complex formation between So, Eya, and Dac should not be excluded (Kumar 2009).

Nmo Promotes the Activity of the So and Eya Complex

Loss of one copy of *eya* in a *nmo* mutant background shows defects in the formation of the ventral eye and leads to an eye-to-head transformation. Overexpression of *eya* driver causes ectopic eye formation in the head; however, this phenotype is significantly reduced in *nmo* mutant background. The same is true for *ey* and *dac*. Overexpression of these genes cannot significantly reduce the small eye phenotype manifested by *nmo* mutants. Loss of *nmo* in general restricts the ability of these genes to induce ectopic eyes. In addition, loss of *nmo* dramatically reduces viability on *dac* mutants, causing their death as early larvae.

Reciprocally, overexpression of *nmo* together with *eya* enhances the formation of ectopic eyes. The synergistic relationship between *eya* and *nmo* is depended on the *nmo* kinase domain. Nmo-mediated phosphorylation of two MAPK sites on Eya promotes the activation of the Eya–So complex. Nmo is perhaps an intrinsic component of the Eya–So complex allowing fast and dynamic modulation of transcriptional output and could regulate the overall activity of the complex, but does this only in specific cellular contexts (Braid and Verheyen 2008; Morillo et al. 2012).

It is yet to be defined if Nmo first binds to a target DNA (trough other factors) and then recruits the complex or if the Eya–So complex recruits Nmo (Morillo et al. 2012). Co-expression of *nmo* and *eya* induce an increased expression of *dac* and *lz*. However, this increase is restricted into small *dpp* expression domains in the antennal and wing discs; hence, the role of the extracellular pathways should not be diminished. The increase in *dac* expression is also *so*-dependent, since in *so* mutants such an increase does not occur (Morillo et al. 2012).

tsh, tio, and *hth* Act Together to Suppress Eye Formation and Promote Proliferation

Tsh induces *hth* expression together with *wingless* (*wg*) signaling, and Tsh is required for the maintenance of Hth. Hth is a known repressor of eye development, thus acting as a mediator that causes eye suppression launched by *tsh*. Indeed, *hth* expression correlates with the severity of the split eye phenotype observed in some cases of *tsh* overexpression (Singh et al. 2002). *hth* and *wg* together suppress ventral eye formation, but do not affect the dorsal half of the eye (Pichaud and Casares 2000; Singh et al. 2002). Also Tio is able to maintain Hth protein levels after *hth* transcription ceases (Bessa et al. 2009).

Hth acts together with Extradenticle (Exd) in the same pathway by enabling nuclear localization of Exd (Pai et al. 1998). It was postulated that either Hth suppresses Dpp signaling or activates *wg* to suppress the initiation of the MF, or it interacts with a nuclear protein of the RDN (Pai et al. 1998). *wg* expression is lost in *hth* mutants, but only in the ventral head regions (Pichaud and Casares 2000). *hth* expression mimics that of *wg*, and they indeed are involved in a positive regulatory feedback loop in the ventral head capsule (Pichaud and Casares 2000). *hth* suppresses MF movement downstream of *dpp* (Pichaud and Casares 2000).

A complex consisting of Ey, Hth and Tsh functions from the early eye disc to third instar larval stage to promote cell proliferation in the domain anterior to pre-proneural (PPN, region anterior to the MF) and thus suppress differentiation. This inhibition might also be a result of suppression of *eya* and *dac* in this region by this complex. Later in the PPN, where Hth is no longer present, it is likely that Ey possibly together with Tsh promotes eye formation by initiating *so* expression (Bessa et al. 2002).

Co-operation Between *nmo* and *ey*

It has been speculated that *nmo* contributes to the eye development by modulating the gene activity of other RDN genes, thus emphasizing the importance of the levels of gene transcription in the development of the eye. For example, loss-of-function mutants of *nmo* can compensate the severity of small-eyed *ey* loss-of-function mutant phenotype (ey^R), leading to bigger eyes, yet it acts together with *ey* to induce ectopic eye tissue. It has been postulated that Nmo co-operates with Ey in a context-dependent manner. In the first instar eye disc, they both promote eye specification by activating downstream genes, whereas in the third instar disc they would have antagonizing effects (Braid and Verheyen 2008).

Dan and Danr Act Physically With Ey and Dac Regulating the First Steps of Differentiation

After their expression has been initiated, Dan and Danr interact physically and genetically with their activators Ey and Dac. Dan and danr might function in a complex with Ey to regulate *ato* expression in a protein level-dependent manner, since mis-expression of *ey*, *dan*, or *danr* lead to deformation of the eye (Curtiss et al. 2007). It is not clear what the role of the Dan-Danr-Dac complex is, but it is associated with the chromatin modifying function of Dan and Danr making the chromosome around the eye-specific genes more accessible for transcription (Curtiss et al. 2007).

Identifying Targets of the Retinal Determination Network Genes

The compound eye of *Drosophila* is the result of the interactions of numerous genes, starting early in the embryo and lasting until the eclosion of the adult fly. The eye antennal imaginal disc expresses over 370 genes that are involved in the first steps of eye development (Michaut et al. 2003). What is the role of the RDN genes in the regulation of the rest? Are the RDN genes only controlling the early specification steps until the end of the larval stages or are they affecting the expression of the genes during the later steps as well, the ones that built the photoreceptors and confer them their function? Although a lot of progress has been made in the past years in understanding how the RDN functions, its targets remain still largely unknown. In the few studies conducted on RDN targets, microarray-based analysis of transcriptomes has been used as a tool. Two studies have searched for targets of *ey* by ectopically expressing *ey* in different imaginal discs and then comparing the gene expression patterns to wild-type discs (Michaut et al. 2003; Ostrin et al. 2006). The first study discovered 371 putative downstream targets of *ey*, without further analyzing them (Michaut et al. 2003). The second screen aimed to identify targets containing an Ey binding site, and found such sites, in *so*, *eya*, *optix*, and *shifted* (*shf*), which is involved in the extracellular transport of the signaling molecule Hh. *hh* itself is important during the first steps of differentiation (Ostrin et al. 2006).

Nevertheless, targets whose expression depends on mutual expression of more than one RDN gene or on interaction with extracellular signaling cannot be detected with this approach. Recently, this was acknowledged in a study where forced expression of *ey* was combined either with Hh, Dpp or Notch signaling (Nfonsam et al. 2012). Notably, a significant amount of putative targets would have remained undiscovered when the mutual overexpression would have not been applied. As it has become clear now that the RDN consists mainly of versatile feedback loops and interactions among its members, it is easy to appreciate the benefits this adapted microarray screen offers. Using this method, *CG4721*, a new target of *ey* with yet an unknown function was identified. This gene appears to play a role in the first steps of differentiation by regulating the expression of *ato* (Nfonsam et al. 2012).

A microarray-based screen was also conducted to search new targets for So–Eya complex. The screen revealed one new target, a cell cycle regulator String (Stg). *stg* was also upregulated in the first two *ey* target screens mentioned above (Michaut et al. 2003; Ostrin et al. 2006), probably because of Ey-mediated expression activation of *eya* and *so* and, thus indirectly, *stg*. Stg is a positive cell-cycle regulator (Jemc and Rebay 2007).

Similar screens would be necessary to reveal new targets of the other RDN members as well. Alternatively, sequence analysis has been relatively popular to compare known binding sites with putative target genes.

Eye Field Determination and the Primordial Cells of the Eye Imaginal Disc

While the adult imago emerges after metamorphosis of the larva, the primordia of adult structures are already set aside during embryogenesis. These primordial cells of the presumptive adult tissue are organized in imaginal discs. Imaginal disc cells are dedicated to form exclusively adult-specific structures, including the adult feeding organs, eyes, antennae, legs, halteres, wings, internal and external genitalia, as well as the epidermis. They develop stereotypically at defined positions in the larval body and can be identified by location, size, and developmental pattern in the larva. How and when are these cells in the embryo specified to form a particular structure?

Imaginal disc formation is initiated during cellular blastoderm stage (Simcox and Sang 1983). The discs originate from groups of founder cells at specific locations along the anterior–posterior body axes (Crick and Lawrence 1975). During early stages of embryogenesis, the eye field, giving rise to the eye-antennal disc, is formed as an elongated strip of cells in the developing dorsal pouch where the left and right primordium of the presumptive disc forms a V-shaped structure. The eye primordium can be identified by the expression of a Zinc-finger transcription factor Escargot (esg), which is required to maintain diploidy in imaginal disc cells (Hartenstein and Jan 1992; Hayashi et al. 1993). The presumptive eye-antennal imaginal disc develops from antennal, intercalary, and gnathal segments of the early embryo and also includes cells of the nonsegmental acron. Fatemap studies and lineage tracing have revealed that the eye-antennal imaginal disc originates from about 5–20 cells located at the anterior dorsolateral part of the early embryo. These cells form the presumptive eye field, which contains not only the eye-antennal disc but also most parts of the larval and adult visual system, such as the larval eye and the primordium of the adult optic lobes (Jurgens and Hartenstein 1993).

The disc is composed of two epithelial layers: the main epithelium (ME), a columnar layer also known as disc proper, and the peripodial epithelium (PE), a squamous epithelial layer (Haynie and Bryant 1986). The main epithelium gives rise to the eye, whereas the peripodial epithelium develops into the surrounding head capsule (Bessa and Casares 2005). The development of the eye-antennal disc proper takes place during late embryogenesis at stage 17, when the dorsal pouch shortens and the cells of the presumptive eye-antennal disc are compressed into a small cluster of cells. The inner layer of the dorsal pouch forms the medial wall of the eye disc, whereas the outer layer forms the PE. The PE participates in the fusion of the two bilateral symmetrical discs during metamorphosis (Fristrom et al. 1977; Pastor-Pareja et al. 2004). In freshly hatched larvae, the eye-antennal disc primordia are positioned in the third thoracic segment as a paired structure.

Genetic studies of several transcription factors and signaling pathways have identified the developmental program which is required for the proper specification and determination of the eye-antennal disc. *ey* and *toy* are expressed in the presumptive eye-antennal disc primordia and are both necessary and sufficient to turn the cell fate towards eye imaginal disc development. During embryogenesis, *ey* is expressed in a large domain covering the eye field giving rise to the eye imaginal discs and optic lobe primordia and in some distinct region of the brain and in VNC (Halder et al. 1995; Quiring et al. 1994). During early embryogenesis *toy* is expressed in a similar pattern as *ey*, however, *toy* expression precedes the expression of *ey* (Czerny et al. 1999). During the course of development, the *toy* expression domain gives rise to the brain and, if not all, to the most parts of the visual system including the optic lobe and eye imaginal disc primordia (Green et al. 1993; Younossi-Hartenstein et al. 1993). After germband retraction *toy* expression is restricted to the head region. At embryonic stage 13, *toy* expression marks the optic lobe primordia in a broader region that includes the brain and the presumptive eye imaginal discs. In contrast to *toy*, *ey* transcripts were first detected at stage 10 during germband extension in every segment of the developing VNC. However, during late embryogenesis at stage 16, both genes are expressed in different subset of cells in developing central nervous system. *ey* and *toy* seemed to be coexpressed in the optic lobe and the eye primordia of the late embryo (Czerny et al. 1999).

toy activation occurs very early during cellular blastoderm by the combined action of maternal active genes and gap genes. Based on the genetic analysis, a model has been proposed to demonstrate the onset of *toy* activation during embryogenesis. According to this model, after fertilization, maternally contributed *bicoid* (*bcd*) mRNA is translated generating a gradient of Bcd protein at the A/P axis with its highest peak at the anterior pole. This gradient activates transcription of *hunchback* (*hb*), which marks anterior part of the embryo and also restricts premature *toy* activation. During mid-cellular blastoderm stage, at the anterior pole, *toy* expression is suppressed due to the activation of *knirps* (*kni*) by the combined action of Bcd and Dorsal (Dl), which is a maternally contributed transcription factor and expressed as a gradient at the D/V axis. At the late blastoderm stage anterior part of the embryo is divided and the cephalic region is formed. *toy* is expressed in the cephalic region of the embryo

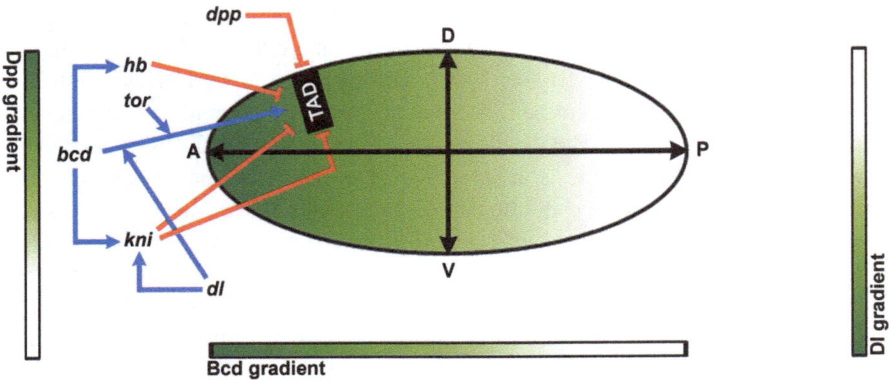

Fig. 3 A model explaining the onset of *toy* activation in the embryo. A gradient of Bcd protein forms at the anterior-posterior axis having its maximum peak at the anterior pole. Shortly after *hb* becomes transcriptionally active, it restricts the premature activation of *toy*. Bcd and Dl activate the transcription of *kni* at the anterior pole. Kni suppresses *toy* expression during mid-cellular blastoderm stage. The cephalic region including the Toy activation domain (TAD) forms at the anterior pole during late blastoderm stage. The translation of *toy* is initiated by the combined action of Bcd, Tor and Dl. Blue arrows indicate activation and red bars indicate repression of transcription. (Modified from Blanco and Gehring 2008)

by the combined action of Bcd, Torso (a maternally expressed Tyrosine kinase receptor) and Dl (Fig. 3). Thus, the genetic analysis shows that *toy* is activated at the cellular blastoderm stage as a result of a combined action of maternally contributed morphogens and zygotically expressed transcription factors (Blanco and Gehring 2008).

The Development of the Eye-Antennal Disc During First and Second Larval Stage: Determination of the Eye Primordia and Segregation of Eye and Antenna Fates

Developmental Plasticity Within the Eye-antennal Disc Is Maintained Until Second Larval Stage

While the disc giving rise to antenna and eyes originate from a single cluster of cells during embryonic development, within the eye-antennal disc, developmental plasticity is maintained until second instar stage. Clones induced up to second larval stage in the eye antennal disc could be found in any of the different structures the disc develops into (Morata and Lawrence 1979). Thus, at the time of clonal induction the cells marked by mitotic recombination do not segregate to the particular domain giving rise to the antenna or the eye. This is unusual compared to other discs, where the different domains are defined as early as blastoderm stage (Lawrence and Morata 1977). However, the exact timing of eye and antenna fate segregation has been subject to debate and has been placed either during late or early second larval stage (Kumar and Moses 2001; Kenyon et al. 2003).

The Onset of Gene Expression Patterns Correlates With the Late Eye-antennal Fate Segregation

The fact that the eye-antenna disc originally develops as a uniform field and is only late specified into different eye and antennal parts is reflected in its gene expression patterns of identified cell fate determinants (Fig. 3). Before fate segregation occurs, the early eye-antennal disc expresses the transcription factors that are necessary for both eye and antennal development. Only later, those genes become expressed in their respective domains. *ey* and *toy* are uniformly expressed in the early eye-antennal primordia during embryogenesis (Kammermeier et al. 2001). During second larval stage, the time of fate segregation between eye and antenna part, *ey* and *toy* are only expressed in the posterior domain of the disc, which marks the future eye part (Kenyon et al. 2003; Kumar and Moses 2001). Likewise, *hth* is expressed in the first larval stage in the whole disc but its expression becomes withdrawn from the posterior part during second larval stage (Bessa et al. 2002). Interestingly, *hth* remains expressed in the region of the eye disc that will develop into the head cuticle (Pai et al. 1998). Therefore, *hth* provides a later function during development by acting in the formation of sub-compartments within the eye disc.

The homeobox gene *cut*, essential for antennal development, is known to be the first marker of the antennal part of the eye-antennal disc (Bodmer et al. 1987). In contrast to *ey* and *toy*, it is not expressed in the eye-antennal disc before fate segregation but starts to be expressed exclusively in the anterior antennal domain by mid-second larval stage (Kenyon et al. 2003). *cut* expression is followed by the expression of *Distal-less* (*Dll*) (Kenyon et al. 2003). Both genes mark the future antennal part and are required for antennal development (Bodmer et al. 1987; Dong et al. 2000).

The Delayed Co-expression of Early Retinal Genes Locks in Eye Fate

A second group of genes starts to be co-expressed in the posterior eye field at the time of eye-antennal segregation during second larval stage (Kumar and Moses 2001). These genes, often referred to as early retinal genes, include the nuclear factors Eya, So, and Dac who work in a tight network specifying the formation of the eye (Desplan 1997; Kenyon et al. 2003; Kumar and Moses 2001). *eya, so*, and *dac* have been suggested to finally restrict the eye primordia's competence for retinal differentiation upon the interaction with extracellular signaling pathways (Kumar and Moses 2001; Baker and Firth 2011).

Even though the eye selector genes *ey* and *toy* are expressed in the early disc, regional eye identity is only established during second larval stage with the co-expression of *eya*, *so*, and *dac*. How is the delay of regional eye fate achieved and how is the late onset of co-expression of the early retinal genes *eya* and *so and dac* established?

The Role of Extracellular Signalling Pathways in Separating Eye and Antennal Domains

Extracellular signaling pathways are repeatedly used in a spatial and temporal manner during *Drosophila* development (reviewed in Pires-daSilva and Sommer 2003). They include factors that define boundaries and axes or control cell proliferation. It is known that extracellular signals induce different developmental outcomes depending on the type of cells they act on (Baker and Firth 2011). Somehow these cells must combine the information that they receive and the developmental program that they are undergoing at a specific point in time. The *Drosophila* eye antennal disc serves as a model system to understand how the same extracellular signaling pathways acting in all imaginal discs results in different developmental outcomes. In case of the eye, how do the members of the RDN, specifying the early eye field, interact with these cell–cell signals to regulate *Drosophila* eye development?

While the distinct roles of signaling pathways in retinal differentiation and photoreceptor specification during third instar stage have been and still are extensively studied, their function and mechanism of action in the specification of the eye primordia during first and second larval stage has been less well characterized (Dominguez and Casares 2005).

Ectopically expressing genes of the RDN either alone or in combination can induce the formation of ectopic eyes. However, this ability is spatially restricted, which strongly indicates that additional factors are needed for proper eye formation (Kango-Singh et al. 2003). Most of the ectopic eyes are formed by overexpressing *ey* and are found in domains expressing *dpp* (Salzer and Kumar 2010). In addition, ectopically expressing *ey* with *dpp* and or *hh* (a potential activator of *dpp*) increases the range of cell populations that can transform into ectopic eyes (Kango-Singh et al. 2003).

dpp is a *Drosophila* homolog of the transforming growth factor β family (TGF β), which encode for secreted molecules acting in wide range of developmental processes (reviewed in Massague and Wotton 2000). There is evidence suggesting that *dpp* is required for the initiation of the early retinal genes during second larval stage. The expression of the early retinal genes *eya*, *so*, and *dac* is greatly reduced in second and third larval instar of *dpp* mutant eye discs, as has been shown by measuring their respective mRNA levels (Chen et al. 1999). The same is true for homozygous clone mutant for *Mad*, an effector of Dpp signaling pathway (Curtiss and Mlodzik 2000). Even though Dpp is necessary for activating *so*, *eya*, and *dac* expression, it is not required to maintain their expression (Curtiss and Mlodzik 2000). Originally it has been shown that ectopic eye formation by *ey* is observed in *dpp* expression domains (Chen et al. 1999). However, it was found recently that ectopic eyes can also form in cells outside of the *dpp* expression domain (Salzer and Kumar 2010). Even though this finding does not exclude the fact that *dpp* is required for normal eye development, it appears that the ability to form ectopic eyes does not require Dpp signaling.

The signaling molecule Wg is known to antagonize Dpp and to act as a suppressor of eye development (Hazelett et al. 1998). Ectopic expression of *wg* results in the loss of early retinal gene expression (Baonza and Freeman 2002). Moreover, loss of Wg signaling by blocking its receptors, results in ectopic expression of *eya, so,* and *dac* (Baonza and Freeman 2002). In line with the antagonistic actions of *dpp* and *wg* in controlling the expression of the early retinal genes, both genes are expressed in opposite sides of the eye-antennal disc in the early second instar larvae (Cho et al. 2000), *wg* along the anterior dorsal end and *dpp* along the posterior dorsal end (which will give rise to the future eye disc).

The Notch signaling pathway provides another essential function for proper eye development (Kenyon et al. 2003). The Notch receptor is activated by the ligands Delta and Serrate in the dorsoventral part of the eye disc, also referred to as the signaling centre of eye disc growth (Cho and Choi 1998). Notch activates the expression of *eyg* (Dominguez et al. 2004) which in turn activates the expression of *unpaired* (*upd*), a ligand of the Jak/Stat signaling pathway. As Upd is secreted, it may act over long distances to promote growth in the entire eye disc (Chao et al. 2004).

How do Notch, dpp, and wg signaling contribute to the late onset of early retinal genes and thus the final segregation of eye and antenna fates? It has been suggested that Notch signaling, contrary to previous findings, is not involved in directing the formation of eye and antenna fields by acting genetically upstream of *ey* (Kumar and Moses 2001). Notch signaling rather indirectly contributes to the regional specification of eye and antenna through its control of cell proliferation and disc size (Kenyon et al. 2003; Fig. 4). According to this model, cellular proliferation and thus, the increase of disc size initiated by Notch causes the opposing *dpp* and *wg* expression domains to be set apart. Cells that thereby no longer receive Wg signaling but Dpp in the posterior domain initiate the expression of *eya*. Confirming this model, reducing Wg signaling in small eye discs (initiated through antagonizing Notch) is sufficient to restore *eya* expression (Kenyon et al. 2003). However, this model does not explain the finding that *eya* and *so* are still expressed in *eyg* mutant eye discs that are severely reduced in size due to the prevention of eye growth (Dominguez and Casares 2005; Dominguez et al. 2004). Moreover, Eyg is negatively regulating *wg* expression, hence supporting *eya* and *so* expression, so there must be additional factors that are involved in the pathway suppressing *wg*.

Interestingly, similarly to the model described above, the proximal-distal segregation during vertebrate limb development has been proposed to be induced by cell proliferation (Tabin and Wolpert 2007). Through the proliferation of the limb morphogenetic field, the presumptive distal domains are moved away out of the range of the proximal signal retinoic acid. Thereby the future distal cells are now able to respond to the distal signal fibroblast growth factor. Taken together, these models

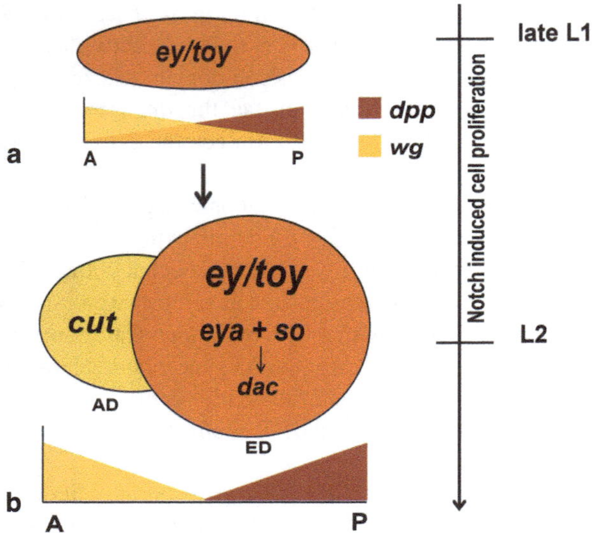

Fig. 4 Schematic representation of the *Notch*-induced proliferation model explaining the activation of the So, Eya and Dac complex. **a)** In the late L1 disc *ey* and *toy* are expressed uniformly in the eye-antennal disc. The underlying graph shows the graded expression of *wg* and *dpp* initiated at the anterior and posterior domain, respectively. At this time point all cells lie in both the *wg* and *dpp* expression domains. **b)** L2 eye antennal disc with the antennal domain (AD), expressing *cut*, and the eye domain (ED), expressing *ey* and *toy*. Note that for simplicity events from early to late L2 are represented, since the exact timing of the individual events remains unclear. Notch-induced proliferation causes the disc to increase in size during L1 to L2 transformation resulting in the *wg* and *dpp* expression domains to be torn apart. The cells that no longer receive *wg* signaling initiate the expression of *eya*. This event causes the So, Eya and Dac complex to be activated and prepares the eye disc for final differentiation in L3. *L1* first instar, *L2* second instar, *L3* third instar

shed a new light on how the morphogenetic field size can indirectly control the establishment of separate domains within an originally uniform field. In fact, it has been suggested that more complex body parts can only be formed during development by allowing proliferation and specification to interconnect (Amore and Casares 2010). Proliferation would enable cells to exit the regulatory state imposed by the original field and change their developmental fate. Therefore, studying the early developmental events in the formation of the *Drosophila* eye helps to understand more general schemes in developmental biology that appear to be common to many species.

Yet, it still remains to be investigated how *ey* and *toy* expression become restricted to the posterior part of the eye antennal disc as this occurs prior to the onset of *dpp* expression (Doroquez and Rebay 2006). The EGFR pathway has been suggested to suppress *ey* in the future antennal domain (Kumar and Moses 2001), which could explain the posterior restriction of *ey* and *toy*. In addition, EGFR has been shown to be required for the regulation of *eya* (Salzer et al. 2010). Thus, EGFR signaling has multiple roles during development depending on temporal and spatial clues. This makes the pathway difficult to investigate in early development by the simple study of loss of function phenotypes (Shilo 2003).

Similarly, in segregating eye and antennal fates, the eye only develops from one of the two epithelia in the disc, the main epithelium (ME), even though, *ey* and *toy* are expressed also in the second layer, in the peripodial epithelium (PE). Tsh is a candidate transcription factor to restrict eye development to the ME as it is expressed only in this epithelium (Bessa and Casares 2005). When expressed ectopically in the PE, it initiates the expression of *eya* and *dac*, but is no longer able to do so when Dpp signaling is blocked. It appears that *tsh* enables the ME to mature into an eye by allowing it to respond to Dpp signaling (Bessa and Casares 2005).

Eye and Antenna Fates are Maintained by Mutual Repression

Once different eye and antennal primordial have been established, their fates have to be maintained. A way to achieve stable fate maintenance is to antagonize each other, for instance by the repression of transcription factors conferring the opposing fate. Indeed, it has been shown that the division of antennal and eye fate is preserved by the reciprocal repression between the eye and antenna determination genes (Wang and Sun 2012). Cut and Hth repress *ey* transcription in the antennal domain by directly binding to its promoter. Similarly, So represses *cut* and *hth* in the eye domain of the disc.

While Ey and Toy are the earliest expressed transcription factors specifying the future eye disc during embryogenesis, the eye-antennal disc remains a uniformly specified field until second larval stage. During this stage the eye primordium becomes determined posteriorly by the expression of *eya, so,* and *dac*. Cell proliferation participates in the formation of an eye primordium that is competent to undergo retinal differentiation to form a fully differentiated compound eye.

The Role of the Retinal Determination Network in the Development of the Extraretinal Photoreceptors of *Drosophila*

The Retinal Determination Network in Larval Eye Formation

The *Drosophila* larvae sense light through the Bolwig's organ, which is composed of 12 photoreceptor cells (Sprecher et al. 2007). The developmental mechanisms controlling the larval and adult visual systems have some overlapping features. They both originate from the same ectodermal invagination during embryogenesis (Green et al. 1993) and they have the same pattern of photoreceptor axon projection through the optic stalk in a stereotypical pathway (Zipursky et al. 1984; Schmucker et al. 1997).

When comparing the genes involved in adult and larval visual system development, we find a number of RDN genes "shared" by these systems. The development of the larval eye begins with the invagination of optic lobe placode during stage 12 embryo (Green et al. 1993). *toy* is expressed in the entire developing eye fields including the presumptive larval eye and adult eye primordia during embryogenesis (Czerny et al. 1999), but *toy* is not required for the formation of the larval eye (Suzuki and Saigo 2000). Conversely, *so* has been shown to be required for both larval and adult visual system development (Serikaku and O'Tousa 1994). *so* is expressed in the larval eye precursors at stage 10 during embryogenesis and mutations in *so* lead to the absence of the larval eye (Cheyette et al. 1994). *eya* has been shown to be co-expressed with *so* at stage 10, and similar to *so*, no Bolwig's organ is formed in *eya* mutants. Eya and So act together in a complex that activates *ato* expression in the larval eye precursors. *ato* is a proneural gene that is required for the development of the larval eye (Suzuki and Saigo 2000).

In summary, some of the retinal determining genes act not only during adult visual system development, but also seem to be important for the development of non-retinal visual organs like the larval eye.

The Retinal Determination Network in the Development of Ocelli

In addition to the compound eyes, the adult *Drosophila* has three simple light sensing organs called ocelli located on the dorsal head. These extra eyes are able to sense ultraviolet (UV) and blue light and serve as navigational help for the fly mainly during flying (Pollock and Benzer 1988; Yoon et al. 1996). The development of these visual organs begins in the third instar larvae from a dorsal anterior margin of the eye imaginal disc (Garcia-Alonso et al. 1996; Royet and Finkelstein 1996). This region not only gives rise to the ocelli, but to the whole vertex region including mechanosensory chetae and bristles (reviewed in Friedrich 2006).

In the developing eye imaginal disc, one gene, which is not part of the RDN, is exclusively expressed in the prospective ocelli forming region and is essential for the formation of the ocelli. The gene is called *orthodenticle* (*otd*) and belongs to the conserved *otd/Otx* gene family that is involved in the head formation both in insects and in vertebrates (Simeone et al. 1993). Otd is a transcription factor containing a homeodomain of the paired class (Finkelstein 1990). Viable *otd* mutants are ocelliless (Finkelstein et al. 1990). *otd* expression in the eye disc begins in the second instar larva and is restricted to the ocellar primordial cells by the third instar stage. This restriction is initiated by Wg and Hh signaling (Blanco et al. 2009). Subsequently, *otd* maintains its own expression via an autoregulatory feedback loop (Blanco et al. 2009).

The development of the ocelli is not as well understood as the development of the compound eye. Nevertheless, many of the RDN genes essential for the eye formation are also required in the specification of the ocelli. *toy* is one of the first genes expressed in the ocellar primordial cells (Brockmann et al. 2011) and *toy* mutants that are able to develop head and eyes are always lacking the ocelli (Jacobsson et al. 2009). *ey* transcripts are not present in this region (Brockmann et al. 2011). The regulation of the transcription initiation of *toy* in the embryonic eye primordium is still largely unknown, but *otd* seems to be involved in this process. Null mutants of *otd* lack *toy* expression in the eye precursor cells of the embryo, but the formation of the primordium itself is not affected, as one of the early determination genes, *eyg,* is expressed normally. *otd* expression is not abolished in *toy* mutants (Blanco et al. 2010).

otd and *toy* are not the only essential genes for the formation of the ocelli. Two other RD genes, *so* and *eya,* are also expressed in the precursors of the ocelli and mutations in these genes also lead to an ocelliless phenotype (Cheyette et al. 1994; Zimmerman et al. 2000). The expression of these genes is initiated during the third instar larval stage. The regulation of these genes in the ocellar region differs from their regulation during the retinal development. Toy was initially thought to be responsible only for *so* activation because of the presence of Toy binding sites in the *so* enhancer (Punzo et al. 2002). However, So is not totally lost in a *toy* mutant background. In addition, forced *toy* expression activates not only *so* but also *eya* expression. Toy is not able to initiate *so* expression in an *eya* mutant background, suggesting that *eya* is involved in *so* activation, as is the case in the compound eye formation (Blanco et al. 2010).

Although Wg is initially required for *otd* expression, Otd itself represses *wg*. *hh* expression on the contrary is positively regulated by Otd, enabling *eya* activation. *eya* expression is then initiated by the activator form of Cubitus interruptus (Ci155) which is a transcription factor activated by the *hh* signaling pathway. Interestingly, Ci is not needed for the expression of *eya* in the compound eye (Blanco et al. 2009).

The current model suggests that Toy and Otd are involved in the initiation of *eya* expression and then together with Eya, perhaps by forming a protein complex, initiate the transcription of *so* (Blanco et al. 2010). Once initiated, *so* can maintain its own expression via an *eya* dependent autoregulatory loop (Pauli et al. 2005). Also, *eya* expression maintenance is *so* dependent (Brockmann et al. 2011).

It is intriguing how the RDN genes can act almost in parallel in two different genetic networks, having distinct responses depending on the genetic environment and cause different structural outcomes (retina and ocelli). This again enhances the view that the regulation of the RDN is highly dynamic and adaptable to the changing needs of the visual system of the fly during the course of evolution.

Final Remarks

The formation of the *Drosophila* compound eye is a highly dynamic and stereotypically orchestrated developmental process. Using *Drosophila* as a model system, we can not only learn more about the similarities or differences in eye development between vertebrates and invertebrates, but the *Drosophila* eye serves as model to discover new principles of early developmental biology. Yet, many open questions considering the early eye development remain and certainly not all possibilities to study it have been explored. Although many signaling pathways and cell fate determinants acting during eye development have been identified and characterized, it is still surprising that a network of evolutionarily conserved transcription factors act as initial step to make an eye. While the genetic and molecular interactions between the individual players of the RDN are being studied in detail, the basic question of how the connections between the RDN members are orchestrated remains still largely unanswered. How are the members of this network able to induce ectopic eyes? This feature cannot be simply explained by the fact that the RDN genes induce the expression of genes involved in differentiation. If it was so, then why are the genes downstream of the RDN genes not able to do this? Despite the intensive research conducted in this field during the past two decades, we are still learning the rules of this game. The coming years will reveal how long it takes us to learn to play.

References

Amore G, Casares F (2010) Size matters: the contribution of cell proliferation to the progression of the specification Drosophila eye gene regulatory network. Dev Biol 344(2):569–577. doi:10.1016/j.ydbio.2010.06.015

Baker NE, Firth LC (2011) Retinal determination genes function along with cell-cell signals to regulate Drosophila eye development: examples of multi-layered regulation by master regulators. Bioessays 33(7):538–546. doi:10.1002/bies.201000131

Baonza A, Freeman M (2002) Control of Drosophila eye specification by wingless signaling. Development 129(23):5313–5322

Bessa J, Carmona L, Casares F (2009) Zinc-finger paralogues tsh and tio are functionally equivalent during imaginal development in Drosophila and maintain their expression levels through auto- and cross-negative feedback loops. Dev Dyn 238(1):19–28. doi:10.1002/dvdy.21808

Bessa J, Casares F (2005) Restricted teashirt expression confers eye-specific responsiveness to Dpp and Wg signals during eye specification in Drosophila. Development 132(22):5011–5020. doi:10.1242/dev.02082

Bessa J, Gebelein B, Pichaud F, Casares F, Mann RS (2002) Combinatorial control of Drosophila eye development by eyeless, homothorax, and teashirt. Genes Dev 16(18):2415–2427. doi:10.1101/gad.1009002

Blanco J, Gehring WJ (2008) Analysis of twin of eyeless regulation during early embryogenesis in Drosophila melanogaster. Gene Expr Patterns 8(7–8):523–527. doi:10.1016/j.gep.2008.06.002

Blanco J, Pauli T, Seimiya M, Udolph G, Gehring WJ (2010) Genetic interactions of eyes absent, twin of eyeless and orthodenticle regulate sine oculis expression during ocellar development in Drosophila. Dev Biol 344(2):1088–1099. doi:10.1016/j.ydbio.2010.05.494

Blanco J, Seimiya M, Pauli T, Reichert H, Gehring WJ (2009) Wingless and hedgehog signaling pathways regulate orthodenticle and eyes absent during ocelli development in Drosophila. Dev Biol 329(1):104–115. doi:10.1016/j.ydbio.2009.02.027

Bodmer R, Barbel S, Sheperd S, Jack JW, Jan LY, Jan YN (1987) Transformation of sensory organs by mutations of the cut locus of D. melanogaster. Cell 51(2):293–307

Bonini NM, Bui QT, Gray-Board GL, Warrick JM (1997) The Drosophila eyes absent gene directs ectopic eye formation in a pathway conserved between flies and vertebrates. Development 124(23):4819–4826

Bonini NM, Leiserson WM, Benzer S (1993) The eyes absent gene: genetic control of cell survival and differentiation in the developing Drosophila eye. Cell 72(3):379–395

Braid LR, Verheyen EM (2008) Drosophila nemo promotes eye specification directed by the retinal determination gene network. Genetics 180(1):283–299. doi:10.1534/genetics.108.092155

Brockmann A, Dominguez-Cejudo MA, Amore G, Casares F (2011) Regulation of ocellar specification and size by twin of eyeless and homothorax. Dev Dyn 240(1):75–85. doi:10.1002/dvdy.22494
Bui QT, Zimmerman JE, Liu H, Gray-Board GL, Bonini NM (2000) Functional analysis of an eye enhancer of the Drosophila eyes absent gene: differential regulation by eye specification genes. Dev Biol 221(2):355–364. doi:10.1006/dbio.2000.9688
Callaerts P, Leng S, Clements J, Benassayag C, Cribbs D, Kang YY, Walldorf U, Fischbach KF, Strauss R (2001) Drosophila Pax-6/eyeless is essential for normal adult brain structure and function. J Neurobiol 46(2):73–88. doi:10.1002/1097-4695(20010205)46:2<73::AID-NEU10>3.0.CO;2-N
Chao JL, Tsai YC, Chiu SJ, Sun YH (2004) Localized Notch signal acts through eyg and upd to promote global growth in Drosophila eye. Development 131(16):3839–3847. doi:10.1242/dev.01258
Chen R, Amoui M, Zhang Z, Mardon G (1997) Dachshund and eyes absent proteins form a complex and function synergistically to induce ectopic eye development in Drosophila. Cell 91(7):893–903
Chen R, Halder G, Zhang Z, Mardon G (1999) Signaling by the TGF-beta homolog decapentaplegic functions reiteratively within the network of genes controlling retinal cell fate determination in Drosophila. Development 126(5):935–943
Cheyette BN, Green PJ, Martin K, Garren H, Hartenstein V, Zipursky SL (1994) The Drosophila sine oculis locus encodes a homeodomain-containing protein required for the development of the entire visual system. Neuron 12(5):977–996
Cho KO, Chern J, Izaddoost S, Choi KW (2000) Novel signaling from the peripodial membrane is essential for eye disc patterning in Drosophila. Cell 103(2):331–342
Cho KO, Choi KW (1998) Fringe is essential for mirror symmetry and morphogenesis in the Drosophila eye. Nature 396(6708):272–276. doi:10.1038/24394
Choi KW, Benzer S (1994) Rotation of photoreceptor clusters in the developing Drosophila eye requires the nemo gene. Cell 78(1):125–136
Crick FH, Lawrence PA (1975) Compartments and polyclones in insect development. Science 189(4200):340–347
Curtiss J, Burnett M, Mlodzik M (2007) Distal antenna and distal antenna-related function in the retinal determination network during eye development in Drosophila. Dev Biol 306(2):685–702. doi:10.1016/j.ydbio.2007.04.006
Curtiss J, Mlodzik M (2000) Morphogenetic furrow initiation and progression during eye development in Drosophila: the roles of decapentaplegic, hedgehog and eyes absent. Development 127(6):1325–1336
Czerny T, Halder G, Kloter U, Souabni A, Gehring WJ, Busslinger M (1999) twin of eyeless, a second Pax-6 gene of Drosophila, acts upstream of eyeless in the control of eye development. Mol Cell 3(3):297–307
de Zulueta P, Alexandre E, Jacq B, Kerridge S (1994) Homeotic complex and teashirt genes co-operate to establish trunk segmental identities in Drosophila. Development 120(8):2287–2296
Desplan C (1997) Eye development: governed by a dictator or a junta? Cell 91(7):861–864
Dominguez M, Casares F (2005) Organ specification-growth control connection: new in-sights from the Drosophila eye-antennal disc. Dev Dyn 232(3):673–684. doi:10.1002/dvdy.20311
Dominguez M, Ferres-Marco D, Gutierrez-Avino FJ, Speicher SA, Beneyto M (2004) Growth and specification of the eye are controlled independently by eyegone and eyeless in Drosophila melanogaster. Nat Genet 36(1):31–39. doi:10.1038/ng1281
Dong PD, Chu J, Panganiban G (2000) Coexpression of the homeobox genes Distal-less and homothorax determines Drosophila antennal identity. Development 127(2):209–216
Doroquez DB, Rebay I (2006) Signal integration during development: mechanisms of EGFR and Notch pathway function and cross-talk. Crit Rev Biochem Mol Biol 41(6):339–385. doi:10.1080/10409230600914344

Emerald BS, Curtiss J, Mlodzik M, Cohen SM (2003) Distal antenna and distal antenna related encode nuclear proteins containing pipsqueak motifs involved in antenna development in Drosophila. Development 130(6):1171–1180

Fasano L, Roder L, Core N, Alexandre E, Vola C, Jacq B, Kerridge S (1991) The gene teashirt is required for the development of Drosophila embryonic trunk segments and encodes a protein with widely spaced zinc finger motifs. Cell 64(1):63–79

Finkelstein R, Smouse D, Capaci TM, Spradling AC, Perrimon N (1990) The orthodenticle gene encodes a novel homeo domain protein involved in the development of the Drosophila nervous system and ocellar visual structures. Genes Dev 4(9):1516–1527

Friedrich M (2006) Ancient mechanisms of visual sense organ development based on comparison of the gene networks controlling larval eye, ocellus, and compound eye specification in Drosophila. Arthropod Struc Dev 35(4):357–378. doi:10.1016/j.asd.2006.08.010

Fristrom JW, Fristrom DK, Fekete E, Kuniyuki AH (1977) The mechanism of evagination of imaginal discs of Drosophila melanogaster. Am Zool 17(3):671–684. doi:10.1093/icb/17.3.671

Garcia-Alonso L, Fetter RD, Goodman CS (1996) Genetic analysis of laminin A in Drosophila: extracellular matrix containing laminin A is required for ocellar axon pathfinding. Development 122(9):2611–2621

Gehring WJ (2002) The genetic control of eye development and its implications for the evolution of the various eye-types. Int J Dev Biol 46(1):65–73

Green P, Hartenstein AY, Hartenstein V (1993) The embryonic development of the Drosophila visual system. Cell Tissue Res 273(3):583–598

Halder G, Callaerts P, Flister S, Walldorf U, Kloter U, Gehring WJ (1998) Eyeless initiates the expression of both sine oculis and eyes absent during Drosophila compound eye development. Development 125(12):2181–2191

Halder G, Callaerts P, Gehring WJ (1995) Induction of ectopic eyes by targeted expression of the eyeless gene in Drosophila. Science 267(5205):1788–1792

Hammond KL, Hanson IM, Brown AG, Lettice LA, Hill RE (1998) Mammalian and Drosophila dachshund genes are related to the ski proto-oncogene and are expressed in eye and limb. Mech Dev 74(1–2):121–131

Hanson IM (2001) Mammalian homologues of the Drosophila eye specification genes. Semin Cell Dev Biol 12(6):475–484. doi:10.1006/scdb.2001.0271

Hartenstein V, Jan Y (1992) Studying Drosophila embryogenesis with P-lacZ enhancer trap lines. Roux's Arch Dev Biol 201(4):194–220. doi:10.1007/BF00188752

Hayashi S, Hirose S, Metcalfe T, Shirras AD (1993) Control of imaginal cell development by the escargot gene of Drosophila. Development 118(1):105–115

Haynie JL, Bryant PJ (1986) Development of the eye-antenna imaginal disc and morphogenesis of the adult head in Drosophila melanogaster. J Exp Zool 237(3):293–308. doi:10.1002/jez.1402370302

Hazelett DJ, Bourouis M, Walldorf U, Treisman JE (1998) Decapentaplegic and wingless are regulated by eyes absent and eyegone and interact to direct the pattern of retinal differentiation in the eye disc. Development 125(18):3741–3751

Heitzler P, Coulson D, Saenz-Robles MT, Ashburner M, Roote J, Simpson P, Gubb D (1993) Genetic and cytogenetic analysis of the 43A-E region containing the segment polarity gene costa and the cellular polarity genes prickle and spiny-legs in Drosophila melanogaster. Genetics 135(1):105–115

Hill RE, Favor J, Hogan BL, Ton CC, Saunders GF, Hanson IM, Prosser J, Jordan T, Hastie ND, van Heyningen V (1991) Mouse small eye results from mutations in a paired-like homeobox-containing gene. Nature 354(6354):522–525. doi:10.1038/354522a0

Hoge MA (1915) Another gene in the fourth chromosome of Drosophila. Am Nat 49:47–49

Jacobsson L, Kronhamn J, Rasmuson-Lestander A (2009) The Drosophila Pax6 paralogs have different functions in head development but can partially substitute for each other. Mol Gen Genomics 282(3):217–231. doi:10.1007/s00438-009-0458-2

Jang CC, Chao JL, Jones N, Yao LC, Bessarab DA, Kuo YM, Jun S, Desplan C, Beckendorf SK, Sun YH (2003) Two Pax genes, eye gone and eyeless, act cooperatively in promoting Drosophila eye development. Development 130(13):2939–2951

Jemc J, Rebay I (2007) Identification of transcriptional targets of the dual-function transcription factor/phosphatase eyes absent. Dev Biol 310(2):416–429. doi:10.1016/j.ydbio.2007.07.024

Jones NA, Kuo YM, Sun YH, Beckendorf SK (1998) The Drosophila Pax gene eye gone is required for embryonic salivary duct development. Development 125(21):4163–4174

Jun S, Wallen RV, Goriely A, Kalionis B, Desplan C (1998) Lune/eye gone, a Pax-like protein, uses a partial paired domain and a homeodomain for DNA recognition. Proc Natl Acad Sci U S A 95(23):13720–13725

Jurgens G, Hartenstein V (1993) The terminal regions of the body pattern. In: Bate M, Martinez AA (eds) The development of Dorsophilia melanogaster, Cold Spring Harbor Laboratory Press, Cold Spring Harbor, pp. 687–746

Kammermeier L, Leemans R, Hirth F, Flister S, Wenger U, Walldorf U, Gehring WJ, Reichert H (2001) Differential expression and function of the Drosophila Pax6 genes eyeless and twin of eyeless in embryonic central nervous system development. Mech Dev 103(1–2):71–78

Kango-Singh M, Singh A, Henry Sun Y (2003) Eyeless collaborates with hedgehog and decapentaplegic signaling in Drosophila eye induction. Dev Biol 256(1):49–60

Kawakami K, Ohto H, Ikeda K, Roeder RG (1996a) Structure, function and expression of a murine homeobox protein AREC3, a homologue of Drosophila sine oculis gene product, and implication in development. Nucleic Acids Res 24(2):303–310

Kawakami K, Ohto H, Takizawa T, Saito T (1996b) Identification and expression of six family genes in mouse retina. FEBS Lett 393(2–3):259–263

Kenyon KL, Ranade SS, Curtiss J, Mlodzik M, Pignoni F (2003) Coordinating proliferation and tissue specification to promote regional identity in the Drosophila head. Dev Cell 5(3):403–414

Kenyon KL, Yang-Zhou D, Cai CQ, Tran S, Clouser C, Decene G, Ranade S, Pignoni F (2005) Partner specificity is essential for proper function of the SIX-type homeodomain proteins Sine oculis and Optix during fly eye development. Dev Biol 286(1):158–168. doi:10.1016/j.ydbio.2005.07.017

Kronhamn J, Frei E, Daube M, Jiao R, Shi Y, Noll M, Rasmuson-Lestander A (2002) Headless flies produced by mutations in the paralogous Pax6 genes eyeless and twin of eyeless. Development 129(4):1015–1026

Kumar JP (2009) The molecular circuitry governing retinal determination. Biochim Biophys Acta 1789(4):306–314. doi:10.1016/j.bbagrm.2008.10.001

Kumar JP, Moses K (2001) EGF receptor and Notch signaling act upstream of eyeless/Pax6 to control eye specification. Cell 104(5):687–697

Laugier E, Yang Z, Fasano L, Kerridge S, Vola C (2005) A critical role of teashirt for patterning the ventral epidermis is masked by ectopic expression of tiptop, a paralog of teashirt in Drosophila. Dev Biol 283(2):446–458. doi:10.1016/j.ydbio.2005.05.005

Lawrence PA, Morata G (1977) The early development of mesothoracic compartments in Drosophila. An analysis of cell lineage and fate mapping and an assessment of methods. Dev Biol 56(1):40–51

Li X, Oghi KA, Zhang J, Krones A, Bush KT, Glass CK, Nigam SK, Aggarwal AK, Maas R, Rose DW, Rosenfeld MG (2003) Eya protein phosphatase activity regulates Six1-Dach-Eya transcriptional effects in mammalian organogenesis. Nature 426(6964):247–254. doi:10.1038/nature02083

Mardon G, Solomon NM, Rubin GM (1994) dachshund encodes a nuclear protein required for normal eye and leg development in Drosophila. Development 120(12):3473–3486

Massague J, Wotton D (2000) Transcriptional control by the TGF-beta/Smad signaling system. EMBO J 19(8):1745–1754. doi:10.1093/emboj/19.8.1745

Michaut L, Flister S, Neeb M, White KP, Certa U, Gehring WJ (2003) Analysis of the eye developmental pathway in Drosophila using DNA microarrays. Proc Natl Acad Sci U S A 100(7):4024–4029. doi:10.1073/pnas.0630561100

Morata G, Lawrence PA (1979) Development of the eye-antenna imaginal disc of Drosophila. Dev Biol 70(2):355–371

Morillo SA, Braid LR, Verheyen EM, Rebay I (2012) Nemo phosphorylates eyes absent and enhances output from the Eya-Sine oculis transcriptional complex during Drosophila retinal determination. Dev Biol 365(1):267–276. doi:10.1016/j.ydbio.2012.02.030

Nfonsam LE, Cano C, Mudge J, Schilkey FD, Curtiss J (2012) Analysis of the transcriptomes downstream of eyeless and the hedgehog, Decapentaplegic and Notch signaling pathways in Drosophila melanogaster. PLoS ONE 7(8):e44583. doi:10.1371/journal.pone.0044583

Niimi T, Seimiya M, Kloter U, Flister S, Gehring WJ (1999) Direct regulatory interaction of the eyeless protein with an eye-specific enhancer in the sine oculis gene during eye induction in Drosophila. Development 126(10):2253–2260

Oliver G, Mailhos A, Wehr R, Copeland NG, Jenkins NA, Gruss P (1995a) Six3, a murine homologue of the sine oculis gene, demarcates the most anterior border of the developing neural plate and is expressed during eye development. Development 121(12):4045–4055

Oliver G, Wehr R, Jenkins NA, Copeland NG, Cheyette BN, Hartenstein V, Zipursky SL, Gruss P (1995b) Homeobox genes and connective tissue patterning. Development 121(3):693–705

Ostrin EJ, Li Y, Hoffman K, Liu J, Wang K, Zhang L, Mardon G, Chen R (2006) Genome-wide identification of direct targets of the Drosophila retinal determination protein eyeless. Genome Res 16(4):466–476. doi:10.1101/gr.4673006

Pai CY, Kuo TS, Jaw TJ, Kurant E, Chen CT, Bessarab DA, Salzberg A, Sun YH (1998) The Homothorax homeoprotein activates the nuclear localization of another homeoprotein, extradenticle, and suppresses eye development in Drosophila. Genes Dev 12(3):435–446

Pan D, Rubin GM (1998) Targeted expression of teashirt induces ectopic eyes in Drosophila. Proc Natl Acad Sci U S A 95(26):15508–15512

Pappu KS, Ostrin EJ, Middlebrooks BW, Sili BT, Chen R, Atkins MR, Gibbs R, Mardon G (2005) Dual regulation and redundant function of two eye-specific enhancers of the Drosophila retinal determination gene dachshund. Development 132(12):2895–2905. doi:10.1242/dev.01869

Pastor-Pareja JC, Grawe F, Martin-Blanco E, Garcia-Bellido A (2004) Invasive cell behavior during Drosophila imaginal disc eversion is mediated by the JNK signaling cascade. Dev Cell 7(3):387–399. doi:10.1016/j.devcel.2004.07.022

Pauli T, Seimiya M, Blanco J, Gehring WJ (2005) Identification of functional sine oculis motifs in the autoregulatory element of its own gene, in the eyeless enhancer and in the signaling gene hedgehog. Development 132(12):2771–2782. doi:10.1242/dev.01841

Pichaud F, Casares F (2000) homothorax and iroquois-C genes are required for the establishment of territories within the developing eye disc. Mech Dev 96(1):15–25

Pignoni F, Hu B, Zavitz KH, Xiao J, Garrity PA, Zipursky SL (1997) The eye-specification proteins So and Eya form a complex and regulate multiple steps in Drosophila eye development. Cell 91(7):881–891

Pires-daSilva A, Sommer RJ (2003) The evolution of signaling pathways in animal development. Nat Rev Genet 4(1):39–49. doi:10.1038/nrg977

Plaza S, Dozier C, Saule S (1993) Quail Pax-6 (Pax-QNR) encodes a transcription factor able to bind and trans-activate its own promoter. Cell Growth Differ 4(12):1041–1050

Pollock JA, Benzer S (1988) Transcript localization of four opsin genes in the three visual organs of Drosophila; RH2 is ocellus specific. Nature 333(6175):779–782. doi:10.1038/333779a0

Punzo C, Seimiya M, Flister S, Gehring WJ, Plaza S (2002) Differential interactions of eyeless and twin of eyeless with the sine oculis enhancer. Development 129(3):625–634

Quiring R, Walldorf U, Kloter U, Gehring WJ (1994) Homology of the eyeless gene of Drosophila to the small eye gene in mice and Aniridia in humans. Science 265(5173):785–789

Rieckhof GE, Casares F, Ryoo HD, Abu-Shaar M, Mann RS (1997) Nuclear translocation of extradenticle requires homothorax, which encodes an extradenticle-related homeodomain protein. Cell 91(2):171–183

Roder L, Vola C, Kerridge S (1992) The role of the teashirt gene in trunk segmental identity in Drosophila. Development 115(4):1017–1033

Royet J, Finkelstein R (1996) Hedgehog, wingless and orthodenticle specify adult head development in Drosophila. Development 122(6):1849–1858

Salzer CL, Kumar JP (2010) Identification of retinal transformation hot spots in developing Drosophila epithelia. PLoS ONE 5(1):e8510. doi:10.1371/journal.pone.0008510

Schmucker D, Jackle H, Gaul U (1997) Genetic analysis of the larval optic nerve projection in Drosophila. Development 124(5):937–948

Seimiya M, Gehring WJ (2000) The Drosophila homeobox gene optix is capable of inducing ectopic eyes by an eyeless-independent mechanism. Development 127(9):1879–1886

Seo HC, Curtiss J, Mlodzik M, Fjose A (1999) Six class homeobox genes in drosophila belong to three distinct families and are involved in head development. Mech Dev 83(1–2):127–139

Serikaku MA, O'Tousa JE (1994) Sine oculis is a homeobox gene required for Drosophila visual system development. Genetics 138(4):1137–1150

Shen W, Mardon G (1997) Ectopic eye development in Drosophila induced by directed dachshund expression. Development 124(1):45–52

Shilo BZ (2003) Signaling by the Drosophila epidermal growth factor receptor pathway during development. Exp Cell Res 284(1):140–149

Silver SJ, Davies EL, Doyon L, Rebay I (2003) Functional dissection of eyes absent reveals new modes of regulation within the retinal determination gene network. Mol Cell Biol 23(17):5989–5999

Simcox AA, Sang JH (1983) When does determination occur in Drosophila embryos? Dev Biol 97(1):212–221

Simeone A, Acampora D, Mallamaci A, Stornaiuolo A, D'Apice MR, Nigro V, Boncinelli E (1993) A vertebrate gene related to orthodenticle contains a homeodomain of the bicoid class and demarcates anterior neuroectoderm in the gastrulating mouse embryo. EMBO J 12(7):2735–2747

Singh A, Kango-Singh M, Sun YH (2002) Eye suppression, a novel function of teashirt, requires wingless signaling. Development 129(18):4271–4280

Sprecher SG, Pichaud F, Desplan C (2007) Adult and larval photoreceptors use different mechanisms to specify the same Rhodopsin fates. Genes Dev 21(17):2182–2195. doi:10.1101/gad.1565407

Suzanne M (2004) Expression analysis of the Drosophila pipsqueak family members fernandez/distal antenna and hernandez/distal antenna related. Dev Dyn 230(2):361–365. doi:10.1002/dvdy.20046

Suzanne M, Estella C, Calleja M, Sanchez-Herrero E (2003) The Hernandez and Fernandez genes of Drosophila specify eye and antenna. Dev Biol 260(2):465–483

Suzuki T, Saigo K (2000) Transcriptional regulation of atonal required for Drosophila larval eye development by concerted action of eyes absent, sine oculis and hedgehog signaling independent of fused kinase and cubitus interruptus. Development 127(7):1531–1540

Tabin C, Wolpert L (2007) Rethinking the proximodistal axis of the vertebrate limb in the molecular era. Genes Dev 21(12):1433–1442. doi:10.1101/gad.1547407

Tavsanli BC, Ostrin EJ, Burgess HK, Middlebrooks BW, Pham TA, Mardon G (2004) Structure-function analysis of the Drosophila retinal determination protein Dachshund. Dev Biol 272(1):231–247. doi:10.1016/j.ydbio.2004.05.005

Ton CC, Hirvonen H, Miwa H, Weil MM, Monaghan P, Jordan T, van Heyningen V, Hastie ND, Meijers-Heijboer H, Drechsler M et al. (1991) Positional cloning and characterization of a paired box- and homeobox-containing gene from the aniridia region. Cell 67(6):1059–1074

Tootle TL, Silver SJ, Davies EL, Newman V, Latek RR, Mills IA, Selengut JD, Parlikar BE, Rebay I (2003) The transcription factor Eyes absent is a protein tyrosine phosphatase. Nature 426(6964):299–302. doi:10.1038/nature02097

Toy J, Yang JM, Leppert GS, Sundin OH (1998) The optx2 homeobox gene is expressed in early precursors of the eye and activates retina-specific genes. Proc Natl Acad Sci U S A 95(18):10643–10648

Treisman J, Harris E, Desplan C (1991) The paired box encodes a second DNA-binding domain in the paired homeo domain protein. Genes Dev 5(4):594–604

Walther C, Gruss P (1991) Pax-6, a murine paired box gene, is expressed in the developing CNS. Development 113(4):1435–1449

Wang CW, Sun YH (2012) Segregation of eye and antenna fates maintained by mutual antagonism in Drosophila. Development 139(18):3413–3421. doi:10.1242/dev.078857

Wawersik S, Maas RL (2000) Vertebrate eye development as modeled in Drosophila. Hum Mol Genet 9(6):917–925

Weasner B, Salzer C, Kumar JP (2007) Sine oculis, a member of the SIX family of transcription factors, directs eye formation. Dev Biol 303(2):756–771. doi:10.1016/j.ydbio.2006.10.040

Wolff T, Ready DF (1993) Pattern formation in the Drosophila retina. In: Bate M, Martinez-Arias A (eds) The development of Drosophila melanogaster, vol 2. Cold Spring Harbor Laboratory Press, Cold Spring Harbor, pp 1277–1325

Yao JG, Weasner BM, Wang LH, Jang CC, Weasner B, Tang CY, Salzer CL, Chen CH, Hay B, Sun YH, Kumar JP (2008) Differential requirements for the Pax6(5a) genes eyegone and twin of eyegone during eye development in Drosophila. Dev Biol 315(2):535–551. doi:10.1016/j.ydbio.2007.12.037

Yoon CS, Hirosawa K, Suzuki E (1996) Studies on the structure of ocellar photoreceptor cells of Drosophila melanogaster with special reference to subrhabdomeric cisternae. Cell Tissue Res 284(1):77–85

Younossi-Hartenstein A, Tepass U, Hartenstein V (1993) Embryonic origin of the imaginal discs of the head of Drosophila melanogaster. Roux's Arch Dev Biol 203(1–2):60–73. doi:10.1007/BF00539891

Zhang T, Ranade S, Cai CQ, Clouser C, Pignoni F (2006) Direct control of neurogenesis by selector factors in the fly eye: regulation of atonal by Ey and So. Development 133(24):4881–4889. doi:10.1242/dev.02669

Zimmerman JE, Bui QT, Liu H, Bonini NM (2000) Molecular genetic analysis of Drosophila eyes absent mutants reveals an eye enhancer element. Genetics 154(1):237–246

Zipursky SL, Venkatesh TR, Teplow DB, Benzer S (1984) Neuronal development in the Drosophila retina: monoclonal antibodies as molecular probes. Cell 36(1):15–26

Molecular Genetic Mechanisms of Axial Patterning: Mechanistic Insights into Generation of Axes in the Developing Eye

Meghana Tare, Oorvashi Roy Puli and Amit Singh

The hallmark of development of all multicellular organisms is transition of the organ primordium cells into a three-dimensional adult organ. Most tissues are derived from epithelial cell sheets that form highly organized structures. These structures exhibit polarization of apical–baso–lateral axes along with planar polarity. During organogenesis, many genetically programmed events that are sensitive to environmental cues play major roles. Various models like yeast (*Saccharomyces cerevisiae*), worm (*Caenorhabditis elegans*), fruit fly (*Drosophila melanogaster)*, newts (*Notophthalmus viridescens*), mouse (*Mus musculus*), rabbit (*Oryctolagus cuniculus*), guinea pig (*Cavia porcellus*), etc. are being used to understand the genetic basis of organogenesis. Studies in different model systems have revealed that the process of organogenesis involves important events of specification, determination, and differentiation. Any deviation in these events can impair the processes of axes specification, cell proliferation, cell death, and cell differentiation. These cell biological processes work in tandem like part of a genetic orchestra, which results in final sculpting of the organ. Any perturbation in these processes leads to growth and patterning defects. During organogenesis, the determination of antero-posterior (AP), dorso-ventral (DV), and proximo-distal (PD) axes is referred to as axial patterning. We will focus on contributions from the *Drosophila* eye model to understand these important questions of developmental biology.

A. Singh (✉) · M. Tare · O. R. Puli
Department of Biology, University of Dayton, Dayton, OH, 45469, USA
e-mail: asingh1@udayton.edu

A. Singh
Premedical Program, University of Dayton, Dayton, OH, 45469, USA

Center for Tissue Regeneration and Engineering at Dayton (TREND),
University of Dayton, Dayton, OH, 45469, USA

A. Singh, M. Kango-Singh (eds.), *Molecular Genetics of Axial Patterning, Growth and Disease in the Drosophila Eye,* DOI 10.1007/978-1-4614-8232-1_2,

Axial Patterning: Means to Generate Third Dimension to an Organ Primordium

Axial patterning marks a lineage restriction event that results in the generation of AP, DV, and PD domains in the developing organ primordium (Cohen et al. 1993; Cohen 1993). These domains are an outcome of progressive restriction of cell fates due to subdivision of the developing field into smaller fields with a more or less rigid developmental potential. These smaller fields within a larger developing field are referred to as compartments (Blair 2001; Curtiss et al. 2002; Dahmann et al. 2011; Held 2002b; Singh et al. 2012). Thus, compartments are unique territories within a bigger developing field (Blair 2001; Curtiss et al. 2002). The cells within a compartment commonly follow looser edicts such as "*You may make any portion of region 'R' but nothing outside it*" (Held 2002a; Wilkins 1993). The compartment boundaries are defined by the spatiotemporal expression or function of cell fate selector genes (Blair 2001; Curtiss et al. 2002; Dahmann et al. 2011). Therefore, "Selector" genes are responsible for attributing a unique property to the cells within their expression/functional domains. In the wing imaginal disc, *engrailed* (*en*) expressed in the posterior compartment and *apterous* (*ap*) expressed in the dorsal compartment (Brower 1986; Cohen et al. 1992; Held 2002b; Hidalgo 1998) serve as the selector for the posterior and dorsal fate, respectively (Table 1). The boundary between the cell populations of two compartments is the site for initiation of the signaling center that regulates patterning, growth, and differentiation of the developing field (Blair 2001; Meinhardt 1983). Activation of the signaling centers at these developmental boundaries are responsible in maintaining the downstream patterning events (Blair 2001; Curtiss et al. 2002; Dahmann et al. 2011; Singh et al. 2012). This leads to an important question: How are these boundaries generated and maintained during the development of a patterning field or an organ? In this chapter, we will provide an overview of recent advances on the genetic circuitry involved in generation of the boundary between the dorsal and ventral compartments, and its significance on the development of an organ using the *Drosophila* eye model. In this chapter, we will focus on the role of axial patterning genes in *Drosophila* eye development.

Drosophila Eye Model to Study Axial Patterning

The power of *Drosophila* as a model organism for patterning and disease lies in its large repertoire of genetic tools, making it a highly tractable model organism (Bier 2005; Singh and Irvine 2012). The *Drosophila* eye has been extensively used (a) to investigate tissue patterning, growth, cell–cell communication, cell survival, and cell death mechanisms during organogenesis and (b) to understand the genetic mechanism responsible for positional fate restrictions within a developing field that leads to formation of compartments (Dominguez and Casares 2005; Singh and Irvine 2012; Singh et al. 2005b; Singh et al. 2012). Interestingly, the eye as an organ has evolved independently as many as 40 different times (Land and Fernald 1992).

Table 1 Gene involved in axial patterning in developing imaginal discs of *Drosophila melanogaster*

Imaginal disc	Axis	Time	Selector genes	Reference
Wing	AP	L1	Anterior: *cubitus interruptus* Posterior: *engrailed, invected*	Lawrence and Morata 1976; Morata and Lawrence 1975; Sanicola et al. 1995
	DV	L2	Dorsal: *apterous, Capricious, tartan, fringe, Serrate* Ventral: *Delta, wingless*	Blair et al. 1994; Cohen et al. 1992; Cohen et al. 1993; Cohen 1993; Diaz-Benjumea and Cohen 1993
	PD	L3	Proximal: *homothorax, teashirt* Distal: *nubbin, elbow, no ocelli*	Blair et al. 1994; Cohen et al. 1992; Cohen et al. 1993; Diaz-Benjumea and Cohen 1993; Zirin and Mann 2007
Leg	AP	L1	Anterior: *cubitus interruptus* Posterior: *engrailed, invected*	Dominguez et al. 1996; Eaton and Kornberg 1990; Kornberg et al. 1985; Masucci et al. 1990; Raftery et al. 1991; Zecca et al. 1995
	DV	L2	Dorsal: *decapentaplegic* Ventral: *wingless*	Baker, 1988a, b; Couso et al. 1993; Irvine and Vogt 1997; Zirin and Mann 2007
	PD	L3	Proximal: *teashirt, homothorax* Distal: *Distalless*	Diaz-Benjumea et al. 1994; Irvine and Vogt 1997; Lecuit et al. 1996
Eye	DV	L2	Ventral: *Lobe, Serrate* Dorsal: *pannier, Iroquois-Complex (araucan, caupolican and mirror), wingless*	Maurel-Zaffran and Treisman 2000; Oros et al. 2010; Singh and Choi 2003
	AP	L3	Anterior: *eyeless* Posterior: *hedgehog*	Dominguez and Casares 2005; Halder et al. 1995; Lee and Treisman 2001
	PD	L3	Proximodistal: Not fully understood	

AP anteroposterior, *DV* dorsoventral, *PD* proximodistal

Despite the differences in the structure of the *Drosophila's* compound eye and a vertebrate eye of a single lens and a retina with multiple layers of neurons, there is similarity in the underlying genetic pathways controlling eye fate specification and differentiation. Thus, the genetic machinery involved in eye development is highly conserved and exhibits structural and functional similarity between insects and humans (Erclik et al. 2009; Gehring 2005; Hartenstein and Reh 2002; Kumar 2009; Wawersik and Maas 2000). This suggests that information generated in the fly eye can be extrapolated to higher organisms. Therefore, *Drosophila* has proved to be an excellent model system for identifying new genes that are conserved in vertebrate retinal development (Singh et al. 2012).

Embryonic Eye Primordium Develops Into the Larval Eye Disc in *Drosophila*

Drosophila, a dipteran, is a holometabolous insect (Anderson 1972b; Miall and Hammond 1892) where the primordia for all adult structures are first specified during embryonic development. The embryonic precursors grow asynchronously from the rest of the developing embryo (Anderson 1972a, b; Cohen et al. 1993; Cohen 1993; Crick and Lawrence 1975; Held 2002b; Kumar 2011; Singh et al. 2012). These embryonic primordia grow inside the larva as epidermal invaginations called imaginal discs (Atkins and Mardon 2009; Bodentstein 1950; Ferris 1950; Held 2002b). The *Drosophila* embryonic eye primordium originates from five embryonic head segments and the acron (Jurgens and Hartenstein 1993; Younossi-Hartenstein and Hartenstein 1993), and is specified by expression of *twin of eyeless (toy)* and *eyeless (ey)*, a *Drosophila* homolog of human PAX6 (Quiring et al. 1994). The embryonic eye primordium begins as an antero-dorsal sac comprising of approximately 20 cells that are set aside during mid-embryogenesis (Garcia-Bellido and Merriam 1969; Held 2002b; Poulson 1950; Tsachaki and Sprecher 2012; Yamamoto 1996).

During larval development, the embryonic eye primordium develops into a monolayer epithelium called the eye–antennal imaginal disc (Fig. 1a). The monolayer epithelium does not accurately reflect the sac-like anatomy of the imaginal discs (Gibson and Schubiger 2001). *Drosophila* imaginal discs are a contiguous cell sheet of flattened epithelial cells with two opposing surfaces comprising a columnar epithelium called the disc proper (DP) and a squamous epithelium called the peripodial membrane (PM) (Atkins and Mardon 2009; McClure and Schubiger 2005). Fate map studies have revealed that the DP of the eye–antennal imaginal disc gives rise to the retina, whereas the PM forms the adult head structures (Atkins and Mardon 2009; Haynie and Bryant 1986; Milner et al. 1983; Singh et al. 2012). Earlier, it was postulated that the PM is required during metamorphosis events of eversion and fusion. However, recent findings suggest that the PM is involved in sending signals to the DP and is required for cell survival and proliferation in the DP (Atkins and Mardon 2009). The eye–antennal imaginal disc upon differentiation gives rise to the adult eye, antenna, head cuticle, and other head structures (Cohen 1993; Held 2002b). In the second instar larva, the division of the complex eye–antennal disc into the eye and antennal field occurs due to restriction of developmental potentials. This division occurs due to activation of the genetic circuitry required to initiate specification followed by differentiation of the eye and antenna (Atkins and Mardon 2009; Dominguez and Casares 2005; Kenyon et al. 2003; Kumar and Moses 2001). The developing eye field gives rise to the eye proper, head cuticle, and the ocelli, whereas the antennal field develops into the antenna and head cuticle (Haynie and Bryant 1986).

Drosophila, like other dipteran insects, has compound eyes for vision (Fig. 1d). The compound eye of the adult fly develops from the larval eye imaginal disc (Garcia-Bellido and Merriam 1969; Haynie and Bryant 1986). The growth spurt occurs during early larval (first and second instar) eye development. During this stage, the undifferentiated cells of the eye–antennal imaginal disc cells divide and undergo rapid proliferation. During late second or early third larval instar stage, a synchronous

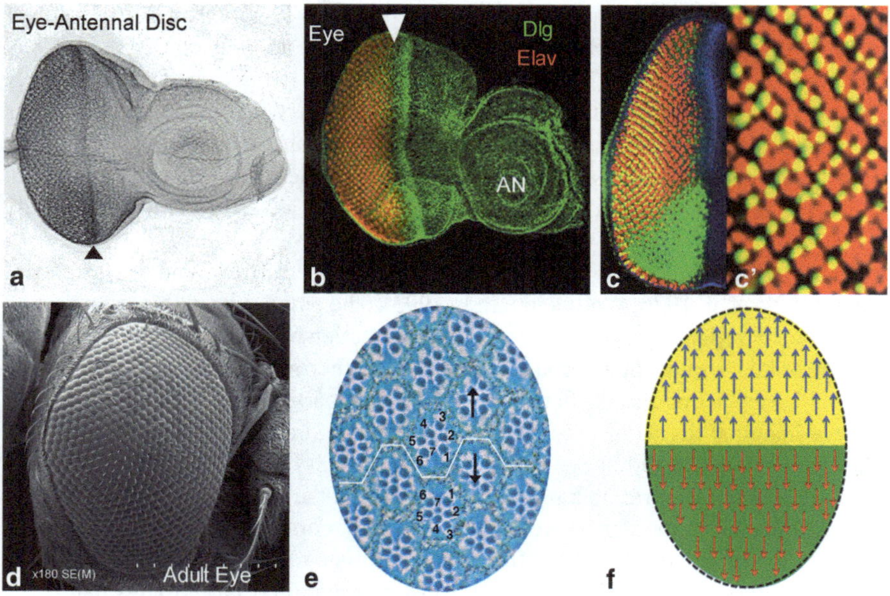

Fig. 1 Dorsoventral (DV) patterning of the *Drosophila* compound eye begins in larval eye imaginal disc. **a–c** Eye–antennal imaginal disc of a third instar larva. **a** Eye imaginal disc. **b** Eye–antennal imaginal disc stained for membrane-specific marker disc large (*Dlg*, *green*), and pan-neural marker Elav (*red*). Elav marks the photoreceptor neurons in the eye. *Arrowhead* in **a** and **b** marks the position of the morphogenetic furrow. **c, c'** Photoreceptor neurons exhibit DV polarity in the eye imaginal disc indicated by staining of Bar (B) antibody (*green*). **d** Scanning electron micrograph (SEM) of a wild-type adult *Drosophila* eye. The adult compound eye of *Drosophila* is made up of 750–800 unit eyes, each referred to as an ommatidium (Ready et al. 1976). All ommatidia are arranged in mirror image symmetry along the DV axis. **e** Each ommatidium consists of eight rhabdomeres, which are organized as an asymmetric hexagonal structure. The DV polarity is determined by two different orientations of the ommatidia based on the orientation of R3 rhabdomere. R3 pointing upward represents a dorsal ommatidium, whereas R3 pointing downward represents a ventral ommatidium. *White line* marks the equator. **f** Cartoon representing the mirror image symmetry of the ommatidia along the DV margins in the adult eye. *Blue arrows* in the yellow background mark the dorsal ommatidia, whereas *red arrows* in the green background mark the ventral ommatidia. Orientation of all images is dorsal up, ventral down, anterior right, and posterior left. AN: antenna

wave of retinal differentiation is initiated in the eye imaginal disc. This progressive pattern of differentiation results in the transition of an undifferentiated epithelium of retinal precursor cells to differentiated cell types comprising regularly spaced photoreceptor clusters (Ready et al. 1976; Wolff and Ready 1993). The differentiating cells undergo an apical constriction and apico-basal contraction, which results in an indentation in the eye imaginal disc. This indentation corresponds to the wave of retinal differentiation, which initiates on the posterior margin of the eye disc and moves anteriorly and is referred to as the morphogenetic furrow (MF, Fig. 1a, b, arrowhead). The photoreceptor clusters are generated posterior to the furrow by a sequence of events including the selection of the R8 founder neuron and recruitment of additional photoreceptor precursors in the order of R2/5, R3/4, and R1/6/7 (Kumar 2011; Wolff and Ready 1993). The compound eyes in the adult fly consist of 750–800 unit eyes called ommatidia (Fig. 1d). Each ommatidium is made up of approximately 20 cells. Of these, eight are distinct photoreceptor neurons (Fig. 1e) that project axons to the optic lobe of the brain. The remaining non-neuronal cells in an ommatidium are pigment cells, cone cells, and mechanosensory bristles (Fig. 1d, e) (Held 2002b; Kumar 2011; Roignant and Treisman 2009; Singh et al. 2012; Wolff and Ready 1993).

The photoreceptor neurons are arranged in an asymmetric trapezoid or a hexagonal facet (Ready et al. 1976; Wolff and Ready 1993). The ommatidia within a compound eye are polarized in opposite directions. The orientation of one group of ommatidia is the mirror image of the other group (Fig. 1e, f). Furthermore, their orientation serves as a marker to distinguish the dorsal and ventral compartment-specific fate in the larval eye imaginal disc (Fig. 1c, c') as well as the adult compound eye (Fig. 1d–f). The ommatidia in the adult eye possess mirror image symmetry along the DV axis. These dorsal and ventral domains are referred to as the dorsal (D) and ventral (V) compartments. The border between these D and V compartments is referred to as an equator. Since photoreceptor differentiation initiates at the intersection of the DV midline of the eye imaginal disc from the posterior margin (Lee and Treisman 2002; Moses 2002), the delineation of DV midline or equator is crucial for differentiation of photoreceptors. Thus, DV patterning is an important facet of axial patterning during organogenesis. The DV polarity has been attributed to play a role in targeting of the retinal axons to the brain. This ommatidial configuration along with the targeting of the axons from the retina to the brain is a masterpiece of microptics and microcircuitry, and enhances visual acuity and thereby forms the equator a sensitive "fovea" (Held 2002b). Interestingly, the eye imaginal disc is largely undifferentiated until second instar of development. It raises an interesting issue of how and when the compartments are established in the *Drosophila* eye imaginal disc.

Sequence of Events During Axis Determination

The sequence of events during axial patterning of the wing and the leg imaginal discs first involves the division of a field into anterior and posterior compartments of independent cell lineages, which is defined by selector genes (Table 1). Generation of AP lineage is followed by subdivision of the wing and leg imaginal discs into dorsal and ventral compartments (Blair 2001; Singh et al. 2012). However, this sequence of division is not followed in the eye imaginal disc. The AP axis is generated with the onset of differentiation marked by the MF in early third instar of larval development. The MF sweeps across the eye imaginal disc from the posterior margin toward anterior, resulting in the formation of posterior fate behind the furrow. The entire early eye primordium is ventral in fate and on which the dorsal fate is established in early second instar of larval eye development (Singh and Choi 2003). Therefore, Dorso-Ventral (DV) patterning, which is established during early second instar of eye development, is the first lineage restriction in the eye imaginal disc

(Singh and Choi 2003; Singh et al. 2005b; Singh et al. 2012). Although there are differences in the sequence of events, evidence suggests that some aspects of the DV patterning mechanism are highly conserved in the developing eye and the wing. One of the common features among all these organ primordia is the generation of the DV boundary, which serves as the site for activation of the signaling pathways to trigger growth and patterning of the imaginal disc. Here our emphasis will be on the genetic mechanism of the generation of DV domains and how it regulates growth and patterning in the developing eye.

Generation of Dorsal and Ventral Compartments in the Developing Eye Disc

The *Drosophila* eye is a polarized tissue. The polarity in *Drosophila* eye is reflected by mirror image arrangement of ommatidia across the DV midline or equator (Fig. 1f). The relation between the equator and DV compartmental boundary has been a matter of debate for a long time. The equator was first reported by Wilhelm Dietrich (Dietrich 1909). In many insect eyes, the equator has been described as the boundary between the photoreceptor neurons of the dorsal and ventral compartments (Dietrich 1909). The equator is generated upon specification of dorsal and ventral compartments and serves as the signaling centre, which is crucial for cell proliferation and differentiation of the eye as an organ. The *Drosophila* eye model has been extensively used to unravel the molecular genetic mechanisms underlying this crucial process of generation of DV compartments in the eye (Singh et al. 2005b; Singh et al. 2012). Since the developmental mechanisms underlying the DV pattern are not fully understood, it raises an interesting question of how the dorsal and ventral pattern is established in the developing eye.

Earlier studies employed the genetic mosaic approach to study the generation of the DV pattern in the developing eye. Hans Becker reported that clones respect the equator and do not cross the DV lineage boundary (Becker 1966; Held 2002b). The pioneering studies authored by Donald Ready, Thomas Hansen, and Seymour Benzer (1976) entitled "Development of the *Drosophila* retina, a neurocrystalline lattice," provided insights into patterning of the *Drosophila* eye (Ready et al. 1976). They rejected the clonal analysis model of ommatidial lineage (Kankel et al. 1980). They employed a genetic mosaic approach to generate mitotic recombination between the $white^{+}$ (w^{+}) wild type and w^{-}mutant chromosomes. Their aim was to generate two new cell populations $w^{-}/w^{-}$and w^{+}/w^{+} clones in a w^{+}/w^{-} paternal heterozygous background. The w^{+} gene is essential for red eye pigment uptake in the cells and serves as an excellent cell autonomous marker for photoreceptors and pigment cells (Lawrence and Green 1979; Ready et al. 1976). They found that in genetic mosaic, w^{-} clones generated in the dorsal half of the eye can cross a few cells into the ventral half and vice versa. The results from these studies in the *Drosophila* eye suggested that the equator is not determined as the boundary between the D and V cell lineages (Ready et al. 1976). Although the result from this study does not exclude the possibility that the dorsal and the ventral domains of the eye derive from two independent cell lineages, the lineage boundary may not precisely correspond to the equator (Netter et al. 1998).

In a series of elegant genetic analysis experiments involving a large number of mosaic clones in the adult eye, Baker (1978) demonstrated that clones strictly follow the DV boundary and do not intermingle near the DV border (Held 2002b; Singh et al. 2012). These studies validated the hypothesis that the *Drosophila* eye is derived from D and V compartments. To analyze whether the eye and the head are also subdivided into different domains by sequential compartmentalization, a mosaic analysis was carried out. Nearly all clones (96 %) respected the DV boundary (did not cross the boundary) and were restricted to either dorsal or ventral domain of the eye. A few clones (4 %) did cross the DV border, which is probably because of the fact that such clones may have been induced prior to formation of dorsal and ventral compartment boundary. Alternatively, two independent dorsal and ventral clones may have juxtaposed at the equator region, thereby giving a false notion of a single clone not respecting the DV boundary (Baker 1978; Singh et al. 2005b; Singh et al. 2012). The DV lineage restriction observed in the adult eye was also confirmed in the developing eye imaginal disc where large clones did not cross the DV midline in the larval eye imaginal disc. These clones showed a sharp outline along the DV midline, and the clones located within the dorsal or ventral domain had wiggly borders (Dominguez and de Celis 1998). Later, it was established that DV lineage specification is the first event that occurs during organogenesis of the eye (Singh et al. 2012). Therefore, identification of the major developmental landmarks along the temporal axis is important to understand patterning and growth of this organ.

Genesis of the Eye

Activation of Notch (N) signaling at equator, the boundary between dorsal and ventral compartments, has been shown to promote growth, in establishing planar polarity, in spacing of ommatidial clusters, and in cell fate specification and differentiation (Baonza and Garcia-Bellido 2000; Cagan and Ready 1989; de Celis et al. 1996; Go et al. 1998; Singh et al. 2012). However, this argument of DV patterning being crucial for growth, does not fit the timeline of developmental events (Singh et al. 2012). If ommatidial orientation corresponds to the generation of the DV axis, then on the basis of the time point when ommatidial rotation occurs, the majority of the growth and cell proliferation of the developing eye field is already accomplished. The ommatidial orientation of the photoreceptors occurs in the pupal retina, and growth spurt occurs during early larval instars of eye imaginal disc development. Based on the earlier notion, if DV patterning occurs in the pupal retina, then its role in growth and differentiation cannot be explained as majority of both growth and differentiation occurs prior to it during imaginal disc development, and not in the adult eye. Thus, efforts were channeled toward investigating the timeline and the genetic control that initiates DV patterning during eye development. Therefore, efforts were directed to (a) understand the time point of generation of DV axis in the developing eye or (b) identify the developmental event that corresponds to the onset of N signaling in the developing eye (Singh et al. 2012).

Three different groups provided evidences in their independent publications that DV lineage restriction takes place earlier in the larval eye imaginal disc because of domain-specific expression of the genes. These genes are referred to as the DV patterning genes (Cho and Choi 1998; Dominguez and de Celis 1998; Papayannopoulos et al. 1998). These genes may be involved in assigning, generating, and maintaining the DV lineage in the developing eye imaginal disc. A new timeline assigned the time window of initiation of DV patterning to early larval development. This hypothesis also fits with the logic of a growth spurt. They identified the domain-specific expression of these genes whose function also follows the DV domain constraint that is established during early larval stages of development (Cho et al. 2000; Cho and Choi 1998; Dominguez and de Celis 1998; Papayannopoulos et al. 1998; Singh et al. 2012).

These studies raised a new question: if DV patterning occurs so early in the developing eye disc, then what is the default state of the early eye primordium? During embryonic development, the eye primordium begins as a homogenous group of cells that continue to grow during first larval instar to form the eye imaginal disc. Several studies have reported the genes that are expressed in the early larval eye primordium. It is known that the generation of MF marks the formation of AP axis in early third instar of larval eye imaginal disc development (Ready et al. 1976; Wolff and Ready 1993). However, the DV axis is determined as early as late first instar of larval development by domain-specific expression of genes along the DV axis (Cho and Choi 1998; Dominguez and de Celis 1998; Papayannopoulos et al. 1998; Singh and Choi 2003; Singh et al. 2012). Another interesting outcome from the Singh and Choi (2003) studies was that early eye primordium begins from a default ventral state (Fig. 2), which depends on the function of ventral genes like *Lobe* (*L*) and its downstream target *Serrate (Ser)* (Kumar 2011; Singh et al. 2005a; Singh and Choi 2003; Singh et al. 2005b). It has been shown that loss of function of *L/Ser* results in preferential loss of ventral eye (Figs. 2, 3b, c). *L* is expressed uniformly in the entire eye imaginal disc (Figs. 2, 3a). The loss-of-function studies suggested that the requirement of *L* function evolves along the temporal axis (Singh and Choi 2003; Singh et al. 2005b; Singh et al. 2012). During early eye development, the loss-of-function of *L* results in the complete loss of the eye field (Figs. 2, 3b). However, loss of the *L* gene function later during eye development causes selective loss of the ventral half of the eye (Fig. 2) (Singh et al. 2012). Loss of function of *Ser* also results in the similar loss of ventral eye phenotype (Kumar and Moses 2001; Singh and Choi 2003; Singh et al. 2005b; Singh et al. 2012). Interestingly, the timing of restriction of the *L/Ser* functional domain from the entire developing eye field (Fig. 3e, f) to only the ventral half of eye (Fig. 3c, d) corresponds to the onset of *pannier* (*pnr*) gene expression along the dorsal margin of the eye (Table 2, Fig. 2). During late first instar larval eye development, the entire homogenous population of the ventral cells of the eye primordium transitions into two distinct dorsal and ventral lineages with the onset of *pnr* expression on the dorsal eye margin (Singh and Choi 2003; Singh et al. 2012). This suggests that the ventral fate is the ground state of the larval

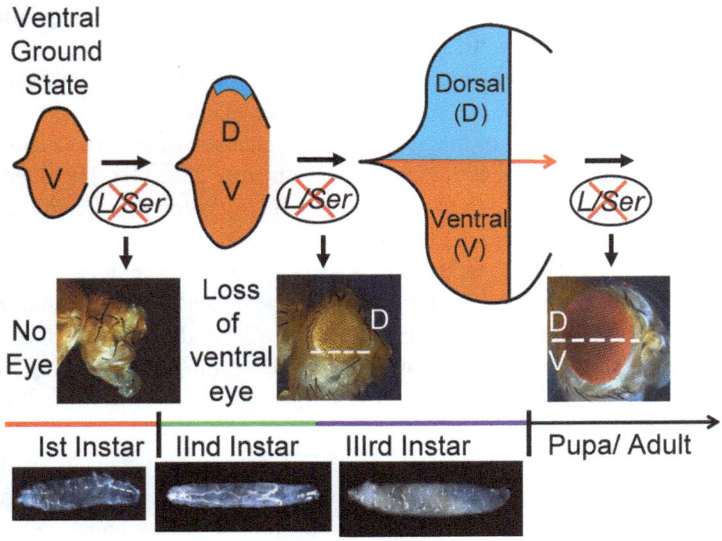

Fig. 2 Ventral is the default state of the developing *Drosophila* eye. Larval eye primordium begins with a default ventral state where all the cells of the eye primordium require ventral genes *L/Ser* function for growth and proliferation (Singh and Choi 2003; Singh et al. 2012). Loss-of-function phenotype of *Lobe/Serrate* (*L/Ser*) in the developing eye imaginal disc evolves progressively along the temporal scale. During early first instar of larval development, loss-of-function of *L/Ser* results in complete loss of the eye field. During early second instar of larval eye development, a few cells start expressing *pannier* (*pnr*) and the dorsal fate is specified. By the end of the second instar stage, DV lineage is established, and at this stage, loss of *L/Ser* results in loss of only the ventral half of the eye. In the late third instar stage of development, when retinal differentiation is almost complete, loss of *L/Ser* does not have significant effect on the overall adult eye morphology. These results clearly indicate that the entire early eye primordium, prior to onset of *pnr* expression, is ventral in fate (Singh and Choi 2003). DV dorsoventral

eye imaginal disc, and *L* and *Ser* are essential for survival and/or maintenance of this ventral state (Singh and Choi 2003; Singh et al. 2005b; Singh et al. 2006). In the subsequent parts of this chapter, we will focus on specific functions of DV patterning genes responsible for patterning of the in the developing eye.

DV Patterning During Imaginal Disc Development

The DV axis is determined by domain-specific expression or function of DV patterning genes. However, their localization may not be identical in all the imaginal discs. Unlike the wing imaginal disc where *Ser* and *Delta (Dl)* are preferentially expressed in the dorsal and ventral domains, respectively, their expression domains are reversed in the eye imaginal disc (Table 1). In the wing imaginal disc, the LIM homeodomain protein Apterous (Ap) acts as a dorsal fate selector (Table 1) (Blair et al. 1994; Cohen et al. 1992). It is known that Ap can induce Fringe (Fng) and Ser in the dorsal compartment of wing imaginal disc (Bachmann and Knust 1998; Cohen

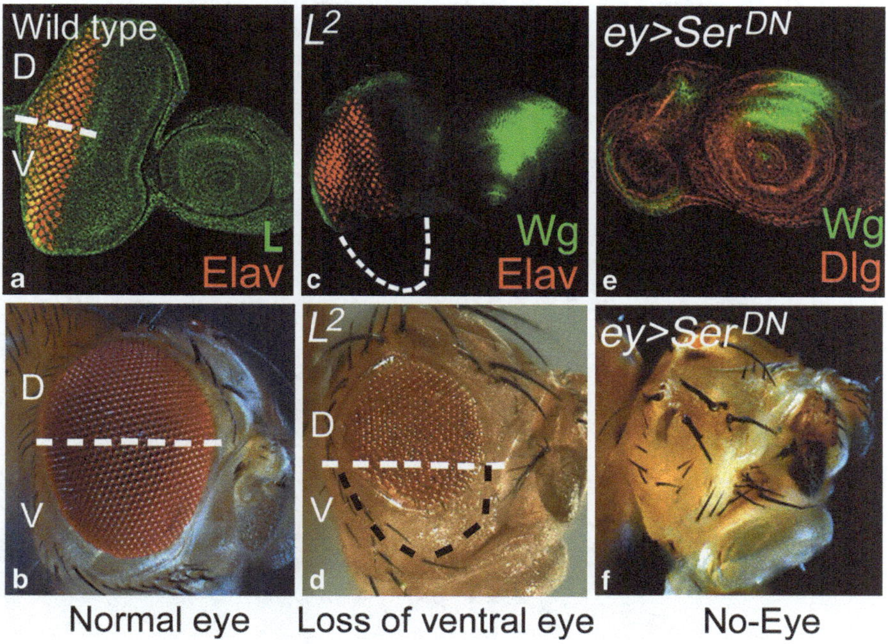

Fig. 3 *Lobe* (*L*) and *Serrate* (*Ser*) are required to promote cell survival in the developing eye imaginal disc (Singh et al. 2006). **a** In the wild-type eye imaginal disc, L (*green*) is expressed ubiquitously throughout the eye disc and Elav (*red*) marks the photoreceptor neurons. **b** Wild-type adult eye. *White dotted line* separates dorsal (*D*) half of the eye from ventral (*V*). **c**, **d** Loss of *L* results in the preferential loss of ventral half of the **c** developing eye imaginal disc and **d** the adult eye. **c** Eye imaginal discs stained for Wg (*green*) to identify dorsal versus ventral eye imaginal disc compartments. The boundary of the eye field is as outlined in **c** (*white*) and **d** (*black*) showing preferential loss of ventral eye. **e**, **f** Early loss-of-function of *Ser* by misexpressing dominant negative form of Ser in the entire eye imaginal disc using an *ey*-Gal4 driver results in complete loss of eye field both in **e** the early eye imaginal disc and **f** the adult. (Adapted from Singh et al., 2012)

et al. 1992; de Celis et al. 1996; Diaz-Benjumea and Cohen 1995). The distribution of N ligands in the eye disc is reversed when compared with the wing imaginal disc, as Ser expression is restricted to the dorsal compartment of the wing, whereas Dl expression is observed in the ventral compartment of the wing imaginal disc. In the eye imaginal disc, Dl and Ser are preferentially expressed in the dorsal and ventral domains, respectively (Cho et al. 2000; Cho and Choi 1998; Struhl 1981). Therefore, in the wing imaginal disc, Ser functions as an N ligand in the dorsal cells, whereas Dl is the N ligand in the ventral cells. Furthermore, *fng* is ventral-specific in the eye imaginal disc (Cho and Choi 1998) but dorsal-specific in the wing imaginal disc (Singh et al. 2012). This mirror image reversal in the distribution of the dorsal and ventral genes in the eye versus wing imaginal disc is probably due to the fact that the eye disc rotates 180° during embryogenesis, and as a result, the DV axis is inverted in the eye with respect to the wing disc (Struhl 1981). Despite the differences in distribution, similar to the developing eye imaginal disc, Fng is still required for N

Table 2 Genes involved in dorsoventral (DV) patterning and domain-specific expression and growth

Drosophila	Vertebrate homolog	Nature	Function in eye	References
Ventral Genes				
Serrate (*Ser*)	Jagged-1	N ligand in the ventral eye	Growth and development of ventral eye	Cho et al. 2000;Cho and Choi 1998; Dominguez and de Celis 1998; Papayannopoulos et al. 1998; Speicher et al. 1994
Lobe (*L*)		Proline-rich Akt substrate	Ventral eye growth and survival; has no effect on dorsal eye growth	Chern and Choi 2002; Singh et al. 2005a; Singh and Choi 2003; Singh et al. 2005b; Singh et al. 2006; Wang and Huang 2009
fringe (*fng*)	Lunatic fringe	Glycosyl transferase	Secreted signaling protein, DV boundary formation	Cho and Choi 1998; Dominguez and de Celis 1998; Irvine and Wieschaus 1994; Papayannopoulos et al. 1998
Chip	Nli/Ldb1/Clim-2	Ubiquitin ligase, transcription cofactor	Define ventral eye boundary	Roignant et al. 2010
sloppy paired 2 (*slp2*)	BF-1 (not complete homology)	Forkhead transcription factor	Ventral eye growth	Sato and Tomlinson 2007
decapentaplegic (*dpp*)	BMP	TGF-β	Ventral growth	Chanut and Heberlein 1997a; Singh et al. 2005b
Dorsal Genes				
pannier (pnr)	GATA	Zinc finger, GATA family	Dorsal eye fate selector	Gomez-Skarmeta and Modolell 2002; Maurel-Zaffran and Treisman 2000; Oros et al. 2010; Ramain et al. 1993; Singh et al. 2005b
araucan (ara)	Irx 1, 3	Homeodomain	Dorsal eye fate selector	Cavodeassi et al. 1999; Gomez-Skarmeta and Modolell 1996, 2002; Pichaud and Casares 2000

Table 2 (continued)

Drosophila	Vertebrate homolog	Nature	Function in eye	References
caupolican (caup)	Irx2, 5	Homeodomain	Dorsal eye fate selector	Cavodeassi et al. 1999; Gomez-Skarmeta and Modolell 1996, 2002; Pichaud and Casares 2000
Delta (Dl)	Delta like 3 (DLL3)	Transmembrane Notch ligand	Dorsal Notch (N) ligand	Cho et al. 2000; Cho and Choi 1998; Dominguez and Casares 2005; Dominguez and de Celis 1998; Papayannopoulos et al. 1998; Singh et al. 2005b
Asymmetrically expressed genes regulating domain-specific growth				
(1) Functional domain is restricted only to the ventral half of the eye				
homothorax (*hth*)	Meis	Homeodomain	Negative regulator of eye, ventral eye suppression	Bessa et al. 2002; Bessa et al. 2008; Pai et al. 1998; Pichaud and Casares 2000; Singh et al. 2005b; Singh et al. 2011; Singh et al. 2012
(2) Functions differently in the dorsal and ventral half of the eye				
teashirt (*tsh*)	TSH1, TSH2, TSH3	C2H2 zinc finger transcription factor	DV asymmetric function, promote dorsal eye growth, ventral eye suppression	Bessa and Casares 2005; Bessa et al. 2002; Datta et al. 2009; Pan and Rubin 1998; Singh et al. 2004; Singh et al. 2002; Singh et al. 2005b; Singh et al. 2012
(3.1) Marginally expressed genes				
optomotor blind (*omb*)	Tbx5	Transcription factor	Cell proliferation	Calleja et al. 1996; Singh et al. 2004; Tare et al. 2013
wingless (*wg*)	Wnt	Signaling/secreted morphogen	Eye growth, negative regulator of eye differentiation, promote head fate	Heslip et al. 1997; Legent and Treisman 2008; Ma and Moses 1995; Treisman and Rubin 1995

Table 2 (continued)

Drosophila	Vertebrate homolog	Nature	Function in eye	References
Dacshsous (*Ds*)	DCHS1	Cell adhesion and molecule binding	Ds transcription is regulated by Wg that negatively regulates Fj gradient. Regulation of planar cell polarity	Simon 2004; Singh and Mlodzik 2012; Willecke et al. 2008; Zeidler et al. 1999a
(3.2) Equatorially expressed genes				
four-jointed (*fj*)	FJX1	Partially secreted transmembrane (TM) type II glyco-protein/ golgi kinase	Proliferation, planar cell polarity, regulate its own expression	Bosveld et al. 2012; Brodsky and Steller 1996; Zeidler et al. 1999a
unpaired (*upd*) also known as *outstretched* (*os*)	Leptin family of pathway ligands	Secreted glycosylated protein	Posterior midline, acts downstream of *four-jointed*, regulates eye size through the JAK/STAT signaling, regulate cell cycle and cell proliferation	Bach et al. 2007; Langer et al. 2004; Tsai and Sun 2004

activation at the DV border in the wing imaginal disc. Fng, a glycosyltransferase, elongates O-linked fucose residues to EGF domains of N to promote N–Dl interaction and thereby modulate N signaling (Okajima and Irvine 2002). Contrary to it, Fng inhibits Ser-N interaction (Ju et al. 2000; Moloney et al. 2000). The genes evolved in DV patterning can be categorized in three broad categories.

Genes Regulating Ventral Eye Growth

The DV patterning genes have been classified into dorsal or ventral genes based on their domain-specific expression, function, or both (Table 2). The ventral eye genes include several genes like *fng*, *L*, *Ser*, *Chip*, and *sloppy-paired (slp)* (Table 2). Among these genes *L* was first reported in 1925, as a gene required for eye growth (Morgan et al. 1925). Based on the mutant phenotypes, *L* was shown to be required for growth and differentiation of only the ventral half of the eye (Chern and Choi 2002; Singh et al. 2005b; Singh et al. 2012). Genetic analysis demonstrated that ventral eye-specific function of *L* was downstream to N signaling (Chern and Choi 2002). These results further validated the hypothesis that early eye disc growth is regulated by asymmetric function of DV patterning genes.

The genetic epistasis analysis revealed that *L* acts upstream of *Ser*, an N ligand in the ventral eye (Chern and Choi 2002; Cho and Choi 1998; Dominguez and Casares 2005; Dominguez and de Celis 1998; Papayannopoulos et al. 1998; Speicher et al. 1994). Furthermore, *Ser* transcription (based on *Ser*-lacZ reporter expression) is repressed in early eye discs from L^{si} homozygous larvae (Chern and Choi 2002). Loss-of-function clones of *L* in the eye imaginal disc resulted in strong downregulation of *Ser* in the ventral eye, whereas increased levels of *L* using the random "flp-out" approach induced *Ser* expression even in the dorsal domain of eye imaginal disc (Chern and Choi 2002). These studies assigned *L* to a genetic hierarchy of ventral eye genes (Table 2) where *L* acts downstream of N and upstream of *Ser* in the developing eye imaginal disc (Chern and Choi 2002). The reduced eye size seen in the hypomorphic alleles of *Ser* further validated the role of *Ser* in early eye development. Surprisingly, loss-of-function clones of *Ser* in the eye did not result in a reduced eye phenotype (Chern and Choi 2002; Papayannopoulos et al. 1998; Sun and Artavanis-Tsakonas 1996). However, misexpression of dominant negative form of *Ser* (Ser^{DN}) (Fleming et al. 1997) in the entire early eye imaginal disc using *ey*-Gal4 driver (Hazelett et al. 1998) results in either preferential loss of ventral eye or loss of the entire eye (Kumar and Moses 2001; Singh and Choi 2003; Singh et al. 2005b; Singh et al. 2012). Random gain-of-function clones of Ser^{DN} generated by the "flp-out" method (Pignoni and Zipursky 1997) also resulted in suppression of eye fate in the ventral eye. Lack of phenotype in *Ser* mutant clones can be attributed to compensation of *Ser* function by another factor. Alternatively, cell culture experiments suggested that Ser may be secreted or transendocytosed into neighboring cells (Klueg and Muskavitch 1999; Kumar and Moses 2001; Singh et al. 2005b; Singh et al. 2012). Similar phenotypes of Ser^{DN} misexpression and *L* mutants in the eye disc further validate that *L* and *Ser* work in the same pathway to regulate the growth of ventral eye domain (Singh et al. 2012).

Fng is known to bind N to promote N–Dl interaction and is required to restrict N activation at the DV border (Fleming et al. 1997; Irvine and Wieschaus 1994; Kim et al. 1995). Contrary to the positive function of Fng in N–Dl interaction, Fng inhibits Ser–N interaction when it is bound to N protein (Ju et al. 2000; Moloney et al. 2000; Singh et al. 2005b; Singh et al. 2012). As a result, the N activation by Dl is enhanced only at the DV border. The expression pattern of DV patterning genes changes dynamically in the developing eye imaginal disc. As a result, striking differences exists in the expression patterns before and after the initiation of retinal differentiation. For example, in the developing eye imaginal disc, *fng* is expressed in the ventral domain, but as the eye imaginal disc undergoes retinal differentiation and the morphogenetic furrow proceeds anteriorly, *fng* exhibits preferential localization anterior to the furrow in both the dorsal and ventral eye domain (Cho and Choi 1998). These results validate the conclusion of genetic mosaic studies, which suggested that DV pattern is established during early eye development prior to retinal differentiation. The loss of function clone of *fng* further emphasized the important role of Fng in DV patterning of eye. Loss-of-function clones of *fng* in the ventral eye exhibit reorganization of DV polarity near the ectopic fng^{+}/fng^{-} border that results in nonautonomous polarity reversals. It results in the generation of de novo equators and ectopic localized activation of N at the fng^{+}/fng^{-} boundary (Baonza and Garcia-Bellido 2000; Cho and Choi 1998; de Celis et al. 1996; Go et al. 1998). Taken together, these observations suggest that (a) Fng has an essential role in DV patterning and (b) the DV pattern is established prior to retinal differentiation during the early eye development.

Other candidate genes involved in ventral eye development are *Chip* and *slp1* or *slp2* (Table 2). *Chip* is an ubiquitin ligase that acts as a ubiquitous transcriptional cofactor. *Chip* interacts with classes of transcription factor during neural development. *Chip* has been reported to establish the ventral boundary of the eye and the head tissue (Roignant et al. 2010). Loss-of-function of *Chip* has been shown to induce ectopic retinal differentiation in the ventral eye. Therefore, possible function of *Chip* is to prevent ectopic retinal differentiation at the ventral eye–antennal disc boundary and thereby promote the head-specific fate. The eye repression function of *Chip* is mediated via interactions with LIM homeodomain proteins: Arrowhead (Awh) and Lim 1. *Chip* and *Lim1* repress the selector gene *eyeless* (*ey*) to prevent ectopic differentiation (Roignant et al. 2010). Thus, LIM-HD/Chip complex is required for defining the boundary between eye and head field. However, the eye suppression activity of *Chip* is independent of Meis class protein Hth or its cofactor Exd. However, it is predicted that Chip and Hth act independently, but parallel to each other in order to suppress eye fate on the ventral eye margin (Roignant et al. 2010). Other ventral eye genes *slp1* and *slp2* encode homologous Forkhead transcription factors that are known to have redundant roles in embryonic patterning (Grossniklaus et al. 1992). In the developing eye, Slp proteins are expressed in a ventral-specific manner and are required to repress Iro-C proteins in the dorsal compartment. During early stages of development, Slp and Iro-C abut the DV border. N signaling activation at the equator results in downregulation of *slp* and a gap is generated between the expression domains of Slp and Iro-C, which is necessary for induction of N ligands Dl and Ser in the dorsal and ventral cells, respectively. Thus, repressive interaction between *slp* and *N* promote the emergence of Ser and Dl expressions in the eye (Sato and Tomlinson 2007).

A member of the TGF-β family, *decapentaplegic* (*dpp*), is another possible ventral eye gene. It exhibits preferential expression in the ventral eye domain of the early eye imaginal disc (Cho et al. 2000). Dpp acts as a long-range secreted morphogen (Chanut and Heberlein 1997b; Nellen et al. 1996). It is known to form a morphogen gradient in the early eye anlage (anterior brain and eye field) (Chang et al. 2001). Mutants of *dpp* exhibit similar pattern defects in the ventral eye disc as seen in *L* mutants. This *dpp* mutant phenotype may be an outcome of ectopic induction of dorsal eye genes *pnr*, *iro-C members*,or *wingless* (*wg*) expression in the ventral domain as observed in *L* mutants (Singh et al. 2005a). Dpp, Hedgehog (Hh), and Wg signaling from the PM is required to trigger N activation in the DP of early eye imaginal disc. During eye imaginal disc development Dpp antagonizes Wg. This developmental interaction between Wg and Dpp in the eye is similar to that observed during limb development (Brook and Cohen 1996; Penton and Hoffmann 1996; Theisen et al. 1996). This antagonistic interaction occurs in the PM across the DV border (Cho et al. 2000). Thus, Dpp signaling plays a role in inducing DV polarity from PM.

Dorsal Fate Selector Genes

The compartment boundaries are defined by the spatio-temporal expression or function of fate selector genes. Loss-of-function of these selector genes results in the loss/elimination of that particular fate in the developing field (Blair 2001; Curtiss et al. 2002; Dahmann et al. 2011). In the *Drosophila* eye, these selector genes were identified in the earlier enhancer trap screens (Bhojwani et al. 1995; Bier et al. 1989; Singh et al. 2012; Sun et al. 1995). These enhancer trap lines had *mini-white* (*w*) and *lacZ* reporter gene (*P-lacW*) (Bellen et al. 1989; Bhojwani et al. 1995; Bier et al. 1989; Sun et al. 1995; Wilson et al. 1989), which exhibited domain-specific expression in the developing as well as the adult eye. These enhancer trap lines have made significant contributions toward understanding the DV patterning in the eye (Choi et al. 1996; Kehl et al. 1998; McNeill et al. 1997; Morrison and Halder 2010; Sun and Artavanis-Tsakonas 1996). Some of the lines identified had w^{+} expression restricted only to the dorsal half of the adult eye. Most of these dorsal-specific P element insertion lines were mapped to the chromosomal region 69CD. The molecular characterization of this 69CD chromosomal region, which was identified as a hot spot for P-lacW insertions that show dorsal eye-specific expression, revealed the existence of a cluster of homeobox genes, *araucan* (*ara*), *caupolican* (*caup*), and *mirror* (*mirr*) (Table 2) (Gomez-Skarmeta and Modolell 1996; Grillenzoni et al. 1998; Heberlein et al. 1998; Kehl et al. 1998; McNeill et al. 1997; Singh et al. 2005b). This cluster of the homeobox gene is located within an approximately 140 Kb region (Netter et al. 1998), are expressed in the dorsal half of the eye (Fig. 7b, b′).

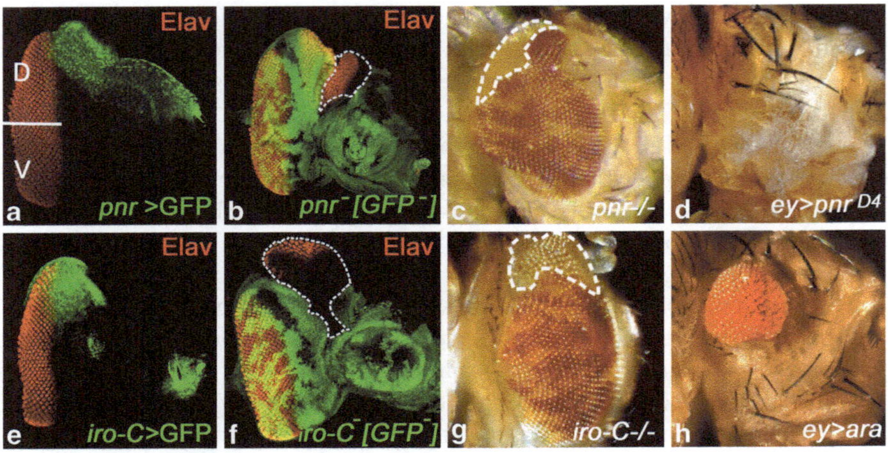

Fig. 4 Pnr and Iro-C members function as dorsal eye fate selectors. **a** Pnr expression (*green*) is restricted to the dorsal eye margin of the developing eye imaginal disc. Elav (*red*) marks the photoreceptor neurons. **b**, **c** Loss-of-function clones of *pnr*, marked by absence of green fluorescent protein (GFP) reporter (*green*), results in the enlargement of existing dorsal eye field (e.g., in the clone outlined in **b**) in the eye imaginal disc (**b**) and adult eye (**c**). **b** Note that there is a nonautonomous eye enlargement in the anterior region of the eye imaginal disc, which is attributed by generation of de novo equator in the dorsal compartment of eye imaginal disc. **d** Misexpression of *pnr* ($ey > pnr^{D4}$) in the eye imaginal discsuppresses the eye fate, validating a late function of *pnr* in defining the eye field boundary (Maurel-Zaffran and Treisman 2000; Oros et al. 2010). **e** The expression domain of the members of Iroquois complex (Iro-C > GFP, *green*) spans the dorsal region of the eye imaginal disc (**g**, **h**). Loss-of-function of *Iro-C* causes dorsal eye enlargements in the **g** eye imaginal disc and in **h** adult eye. These phenotypes are similar to the **b**, **c** *pnr* loss-of-function phenotypes. **h** Misexpression of *ara*, a member of Iro-C, in the eye imaginal disc (*ey* > *ara*) results in small eye. *D* dorsal, *V* ventral. (Adapted from Singh et al., 2012)

They are referred to as Iroquois-Complex (Iro-C), as the mutation in these genes lack lateral thoracic bristles and resemble the hair style of the Indian tribe, the Iroquois (a native tribe which shaved all but a medial stripe of hair on the head and are also called Mohawks) (Gomez-Skarmeta and Modolell 1996; Leyns et al. 1996). They named the genes Araucan and Caupolican in honor of an Amerindian tribes, Aracaunians, and one of their heroes, Caupolican. The third member of this complex was named *mirror* (*mirr*).

The members of Iro-C are highly conserved essential genes and exhibit significant differences in their expression pattern (Gomez-Skarmeta and Modolell 2002). However, there is functional redundancy in terms of Iro-C members between flies and higher vertebrates. Mirr is strongly and dynamically expressed in the central nervous system (Netter et al. 1998; Urbach and Technau 2003), and is essential for follicle cell patterning (Jordan et al. 2000). The other members Ara and Caup are preferentially expressed in mesodermal tissues in the embryos (Netter et al. 1998). The expression of all three Iro-C members is restricted to the dorsal half of the eye imaginal discs (Figs. 4e, 7b, b′), raising a possibility that they may be functionally redundant. Loss-of-function of *mirr* using $mirr^{e48}$ allele showed weak but significant defects of nonautonomous DV polarity reversals in comparison with *mirr*$^{+}$ ommatidia in the dorsal half of the eye (McNeill et al. 1997). The cells from two different compartments are of different cell lineages and do not intermingle because of differences in cell identities and affinities (Dahmann et al. 2011; Garcia-Bellido et al. 1973; Irvine 1999). Loss-of-function clones of *mirr* in the dorsal half of eye exhibit smooth clone borders, whereas those in the ventral half of the eye show wiggly clone borders (Yang et al. 1999). It suggests that dorsal eye cells lacking *mirr* avoid mixing with the neighboring *mirr* expressing cells. Furthermore, the dorsal clones exhibit dorsal eye enlargements and the polarity of the ommatidia in *mirr* loss-of-function clones is reversed. This analysis suggests that *mirr* functions as a dorsal fate selector. The phenotype of *mirr* clones was not strong enough. It raised the possibility that *ara* and *caup*, the other two members of Iro-C, can partly compensate for the loss of *mirr* function in the eye. The issue of functional redundancy was resolved when a deficiency *iro*^{DMF3}, which uncovers all three *Iro-C* genes by the deletion of *ara* and *caup* as well as a 5′-region of *mirr*, was employed for clonal analysis (Diez del Corral et al. 1999; Gomez-Skarmeta et al. 1996; Gomez-Skarmeta and Modolell 1996). Loss-of-function clones of *iro*^{DMF3} in the eye showed repolarization of the ommatidial polarity in the dorsal clones along with dorsal eye enlargement or formation of an ectopic eye field on the dorsal margin (Fig. 4f, g). There was no phenotype in the ventral half of the eye. Gain of function of Iro-C members in the eye result in reduced eye phenotype (Fig. 4h). These results further highlighted the importance of the boundary between the dorsal and ventral cell types. These results strongly support that the three members of *Iro-C* are partially redundant and the *Iro-C* as a whole is required for organizing the DV polarity pattern and growth of the eye.

Loss-of-function of *iro*^{DMF3} also suggested that *Iro-C* genes function as dorsal selectors for head structures as well, as mutant clones in the dorsal region induce the formation of ventral head structures (Cavodeassi et al. 2000). Ectopic ventral head tissues resulting from loss of *Iro-C* genes are cell-autonomous and therefore accompanied by loss of corresponding dorsal structures. In contrast, ectopic ventral eyes are generated non-cell autonomously, as reversals of DV ommatidial polarity are detected in the *Iro-C*$^{+}$ wild-type region adjacent to the mutant clones. This also supports the idea that the DV boundary is an organizing center for DV pattern and growth in the eye imaginal disc. Furthermore, DV patterning of the eye occurs in earlier larval stages than the head patterning. In the *Drosophila* eye, *pnr is* another dorsal gene, expressed in the dorsal eye margin (Figs. 7a; 7a, 4), which exhibits similar loss-of-function (Table 2; Fig. 4b, c) and gain-of-function (Fig. 4d) phenotypes as observed with Iro-C in the eye and the head (Maurel-Zaffran and Treisman 2000; Oros et al. 2010; Pichaud and Casares 2000; Singh et al. 2005b). Pnr, a GATA-1 transcription factor, plays an important role in the dorsal eye development and acts as a selector for the dorsal eye fate (Dominguez and Casares 2005; Maurel-Zaffran and Treisman 2000; Oros et al. 2010; Pichaud and Casares 2000; Ramain et al. 1993; Singh et al. 2005b). In the hierarchy of dorsal genes, *pnr* is the top-most gene and induces Wg, which in turn induces the expression of downstream target genes *mirr* in the dorsal half of the eye (Dominguez and Casares 2005; Maurel-Zaffran and Treisman 2000; Singh et al. 2005b). During later stages of development, which corresponds to the retinal differentiation stage in late second instar and third instar of larval eye development, *pnr* is involved in defining the dorsal eye margin by regulating the retinal determination (RD) genes (Oros et al. 2010).

Wg, a secretory protein and a morphogen, is expressed along the anterolateral margins of the third instar eye imaginal disc (Fig. 7h) (Baker 1988a). Wg plays multiple roles during eye development. One of these roles of Wg is to promote growth of early eye imaginal disc. During early eye development, Wg expression is restricted to the dorsal eye domain (Chang et al. 2001; Cho et al. 2000; Maurel-Zaffran and Treisman 2000). During the retinal differentiation stage, Wg is known to prevent ectopic induction of retinal differentiation from the lateral eye imaginal disc margin (Ma and Moses 1995; Treisman and Rubin 1995). Thus, Wg, which acts as a negative regulator of eye during retinal differentiation, also functions as a dorsal eye fate gene. In the dorsal eye imaginal disc, an N ligand, Dl, has been assigned to the dorsal gene category (Table 2). Dl is preferentially expressed in the dorsal domain of eye imaginal discs during first and second instar stages (Cho and Choi 1998). Apart from genes with domain-specific expression, there are genes that although expressed in broader domains, exhibit DV domain-specific functions (Table 2).

Asymmetrically Expressed Genes Regulating Domain-Specific Growth

A group of DV patterning genes exhibits differential functions in the dorsal–ventral compartments even though they are not expressed in a DV-specific pattern. The members of this group are further classified into three categories:(1) Genes expressed uniformly in the eye imaginal disc but their functional domain is restricted only to the ventral half of the eye, for example, *L* and *homothorax* (*hth*) (Fig. 7e, e′; Table 2). (2) Genes that are expressed uniformly in the early eye imaginal disc and function differently in the dorsal and ventral half of the eye, for example, *teashirt* (*tsh*) (Fig. 7f, f′; Table 2; (Singh et al. 2012). (3) Class of genes expressed in a domain-specific manner are involved in generating morphogen gradient across the developing eye imaginal disc. They are (a) marginally expressed genes like *optomotor blind* (*omb*) (Fig. 7g, g′; Table 2) and Wg (Fig. 7h, h′; Table 2) and (b) equatorially expressed genes like *four jointed* (*fj*) (Fig. 7i, i′; Table 2) and *unpaired* (*upd*) (Fig. 7j, j′; Table 2).

1. Hth is a vertebrate homolog of murine proto-oncogene MEIS1 (Moskow et al. 1995). It encodes a homeodomain transcription factor of the three-amino-acid extension loop (TALE) subfamily (Rieckhof et al. 1997). The expression of *hth* is present in the entire early eye primordium (Bessa et al. 2002; Singh et al. 2002, 2012), which is similar to *L* expression in the early eye (Singh et al. 2012). However, unlike *L*, which is uniformly expressed in the entire eye imaginal disc during all stages of eye development (Singh and Choi 2003), *hth* expression evolves with the onset of differentiation in the eye. Hth expression gets restricted to the cells anterior to the MF (morphogenetic furrow; Bessa et al. 2002; Pai et al.

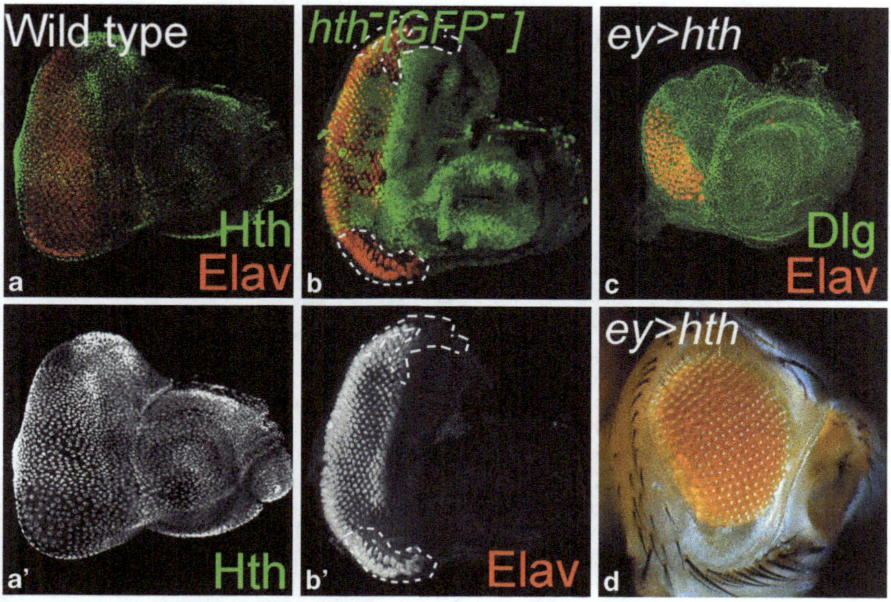

Fig. 5 A domain-specific function of *homothorax (hth)* in the ventral eye margin. **a, a'** Hth (*green*) is expressed anterior to the furrow both in the dorsal as well as ventral domain of the eye imaginal disc (Bessa et al. 2002; Pai et al. 1998; Pichaud and Casares 2000; Rieckhof et al. 1997; Singh et al. 2002). Elav (*red*), a pan neural marker, marks the photoreceptor neurons in the eye imaginal disc. **a'** Note that Hth is expressed in the peripodial membrane (PM). **b, b'** Loss-of-function clones of *hth* marked by the absence of the GFP reporter (*green*, clonal boundary marked by *white dotted line*) in the ventral eye result in eye enlargements whereas in the dorsal eye these clones do not have any effect. **c, d** Misexpression of *hth* in the eye using *ey*-Gal4 driver (*ey* > *hth*) results in a reduced eye field as seen in the **c** eye imaginal disc and the **d** adult eye (Pai et al. 1998). (Adapted from Singh et al., 2012)

1998; Pichaud and Casares 2000; Singh et al. 2002). Although *hth* is expressed anterior to the furrow, its expression is uniform both in the dorsal and ventral half of the eye imaginal disc (Figs. 5a, a′; 7e, e′; Table 2). *hth* is expressed uniformly in the PM of the eye imaginal disc (Fig. 5a). Surprisingly, the loss-of-function phenotypes of *hth* are restricted only to the ventral eye margins (Pai et al. 1998). Loss-of-function clones of *hth* causes eye enlargement only in the ventral eye margin (Fig. 5b, b′) whereas the loss-of-function clones of *hth* in the dorsal compartment do not show any phenotype in the eye imaginal disc (Pai et al. 1998; Pichaud and Casares 2000; Singh et al. 2011, 2012). Furthermore, *hth* mutant cells do not survive in the anterior eye (Bessa et al. 2002, 2008; Pichaud and Casares 2000). Therefore, despite the uniform expression of *hth* in developing eye imaginal disc the loss-of-function phenotype exhibits DV constraint. Misexpression of *hth* in the eye imaginal disc suppresses the eye fate (Pai et al. 1998). Furthermore, eye suppression function of Hth is independent of any domain constraint (Singh et al. 2011, 2012). During development, *hth* is involved in multiple

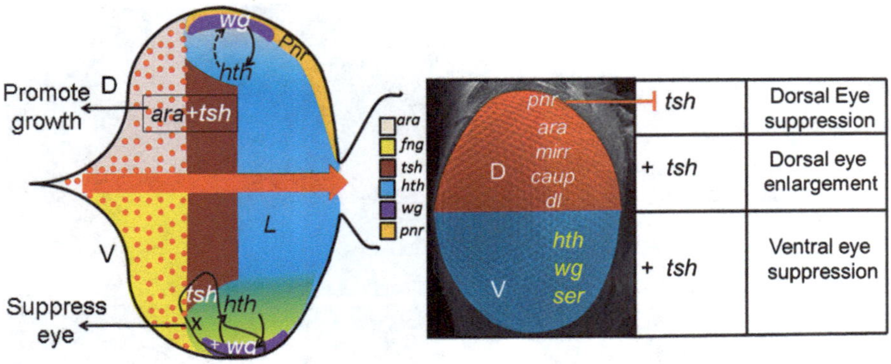

Fig. 6 Dorsoventral asymmetric function of homeotic gene *tsh* depends on its partners (genes) in D and V compartment of the developing eye imaginal disc. Gain of function of *tsh* suppresses the eye fate in the ventral eye (Singh et al. 2002). Tsh collaborates with Wg and Ser and is required for suppression of the ventral eye by ectopic induction of hth. However, this eye suppression function of *tsh* in ventral eye is independent of genes *L* and *fng*. Dorsal eye enlargement function of *tsh* depends on collaboration of members of Iro-C family and the N ligand (Singh et al. 2004). In the dorsal eye, *pnr* is required to suppress *tsh* in order to suppress the dorsal eye fate (Oros et al. 2010)

functions and is required for nuclear localization of a homeo-protein Extradenticle (Exd). Hth encodes a protein with nuclear localization signal (NLS) and two conserved domains: the N terminal evolutionarily conserved MH domain (for Meis and Hth) and a C-terminal region including the homeodomain (HD; Kurant et al. 1998; Noro et al. 2006; Pai et al. 1998; Rieckhof et al. 1997). Alternative splicing is known to provide additional complexity to the genes encoding the Hth transcription factors (Glazov et al. 2005; Noro et al. 2006). Hth forms a heterodimer with Exd through its MH domain and translocates into the nucleus to regulate transcription (Jaw et al. 2000; Ryoo et al. 1999; Stevens and Mann 2007). Since Exd is expressed uniformly in the eye, the ventral-specific function of *hth* has been proposed through its interaction with Wg and Tsh (Fig. 6). Together they are involved in suppression of eye fate on the ventral margin. Furthermore, *hth* plays an important role in delineating the boundary between the eye and the head cuticle on the ventral eye margin (Singh et al. 2011, 2012). We have found that *hth* antagonizes another ventral gene *L* function in the ventral eye margins to define the developing eye field boundary (Singh et al. 2011, 2012). However, this antagonizing activity is independent of Exd (Singh et al. 2011).

2. The homeotic gene *tsh* belongs to the second category. It is expressed uniformly in the early eye imaginal disc but its function exhibits DV domain constraint. Tsh encodes a C_2H_2 zinc-finger transcription factor with three widely spaced Zinc finger domains (Fasano et al. 1991). Tsh plays an important role during *Drosophila* eye development (Bessa et al. 2002; Datta et al. 2009; Kumar 2009, 2011; Pan and Rubin 1998; Singh et al. 2002, 2012). *tsh* is expressed anterior to the furrow both in dorsal and ventral eye (Fig. 7f, F′) and exhibits a DV constraint

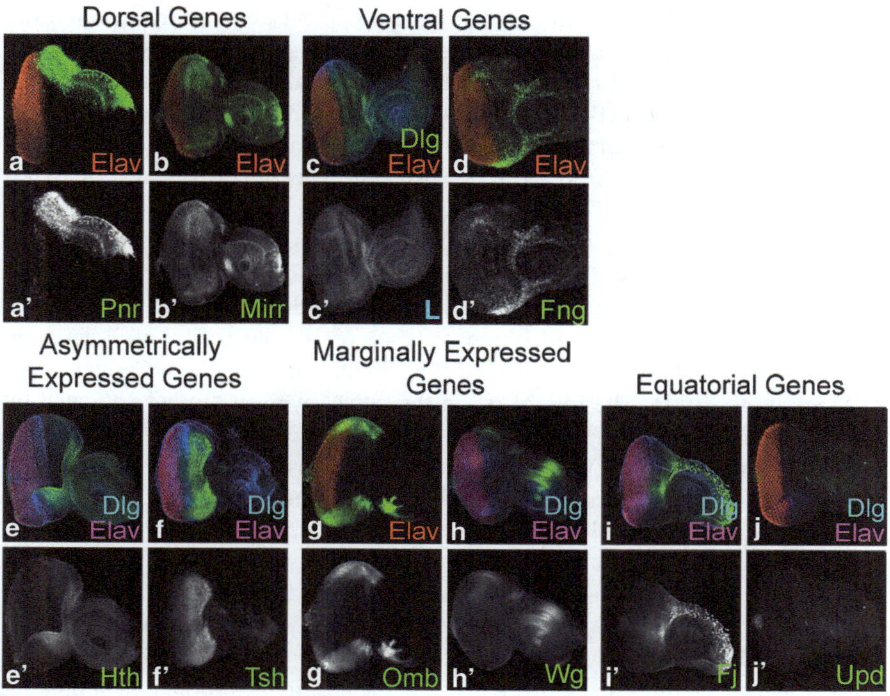

Fig. 7 Expression profiles of different genes required for axis determination during eye organogenesis. **a, a'** Pnr and **b, b'** Mirr (*green*) are expressed in the dorsal domain of the eye. **c, c'** L is expressed ubiquitously in the entire eye imaginal disc (*blue*), whereas Fng (**d, d'**; *green*) is expressed only in the ventral domain of the eye. In the images **a–d**, Elav (*red*), the pan-neuronal marker, is used for marking the photoreceptor neurons. Among asymmetrically expressed genes, Hth and Tsh (*green*, **e** and **f**), **e, e'** Hth is expressed in an asymmetric fashion anterior to the morphogenetic furrow in the head and antenna but not in the eye region. **f, f'** DV asymmetric gene Tsh is also expressed anterior to the furrow both in dorsal and ventral domains of the eye and is regulated differentially in these regions. The eye imaginal discs in **e** and **f** have been stained for membrane marker Dlg (*blue*) and pan-neuronal marker Elav (*magenta*). **g, g'** Among marginally expressed genes, Omb and Wg (*green*; **g, h'**) are expressed exclusively on dorsal and the ventral margins. **h, h'** Wg, a secreted morphogen, is expressed along dorsolateral margins and in the antenna (*green*). Equatorial genes are expressed on equator or the borderline of dorsal and ventral compartments. **i, i'** Fj (*green*) forms a gradient which is more concentrated on the equator and closer to the antennal region. **j, j'** Upd, (*green*) the JAK STAT ligand, is expressed on posterior boundary as a dot on the equator of the eye

in its function [Fig. 6; Table 2; (Singh et al. 2002, 2004)]. In the dorsal eye, *tsh* promotes eye development, whereas in the ventral eye it acts as a repressor of eye fate (Singh et al. 2002, 2004, 2005b, 2012). Interestingly, the DV constraint in *tsh* function in the eye depends on the partners with which it collaborate in the dorsal or the ventral eye disc (Singh et al. 2004). Tsh cooperates with Iro-C members and N ligand Dl in the dorsal eye for its growth promotion function (Singh et al. 2004).The ventral eye-specific function of *tsh* is dependent on Hth and Ser. The expression of *tsh* overlaps with *hth* in the eye imaginal disc, and like *hth*, the *tsh* expression also evolves during larval eye development. Initially, in the first instar eye imaginal disc, *tsh* is expressed in the entire eye imaginal disc but its expression retracts anteriorly to nearly three quarters of the eye imaginal disc when the retinal differentiation begins (Bessa et al. 2002; Singh et al. 2002). Furthermore, Tsh and Hth physically interact with each other (along with Pax-6 homolog, Eyeless (Ey)) to repress the expression of downstream target genes (Bessa et al. 2002; Dominguez and Casares 2005). Further insights into the potential mechanism of *tsh* and *hth* in regulating growth and differentiation in the eye came from analysis of expression patterns of the retinal determination (RD) gene network members (Bessa et al. 2002). It has been proposed that Tsh, Hth, and Ey coexpress in the proliferating cells anterior to furrow to block precocious retinal differentiation and promote cell proliferation (Bessa et al. 2002; Dominguez and Casares 2005; Singh et al. 2002). The role of *tsh* in the dorsal eye was further validated by studies on interaction of *tsh* with the dorsal fate selector *pnr* (Fig. 6 (Oros et al. 2010)). It was shown that *pnr* suppresses the eye fate on the dorsal eye margin by downregulating *tsh* expression in the dorsal eye (Oros et al. 2010). Tsh is known to act upstream of retinal differentiation genes eyes *absent (eya), sine oculis (so), and dacshund (dac)* (Pan and Rubin 1998). Thus, *pnr*, which is expressed in the dorsal PM (Fig. 7a, a′) suppresses *tsh* in the dorsal eye. It results in the suppression of eye fate on the dorsal margin of the eye field (Oros et al. 2010; Singh et al. 2012).

3. During the patterning and growth of a field, the positional information or polarizing signals are provided in a concentration-dependent manner. These signals are determined by concentration of signaling molecules or morphogens. Genetic studies of the polarity genes in *Drosophila* suggested that planar polarity in the dorsal and ventral eye fields is dependent on gradients of the polarizing signals (Wehrli and Tomlinson 1998; Zeidler et al. 1999b). It has been shown that noncanonical Wg/Wnt pathway is important for determining planar polarity (Boutros et al. 1998, 2000; Mlodzik 1999; Reifegerste and Moses 1999; Singh et al. 2005b). In the developing *Drosophila* eye, Wg is responsible for the pole to equator gradient (Legent and Treisman 2008; Zecca et al. 1996). In the third instar eye imaginal disc, Wg is strongly expressed on the anterolateral margins (Fig. 7h, h′; Table 2). It results in a diffusible pole to equator gradient that originates from the dorsal and ventral margins of the eye disc epithelium (Fig. 7h, h′). It suggests that Wg can function as a primary polarizing signal. Since Wg is involved in other events during eye development and loss of *wg* causes defects in the eye (Wehrli and Tomlinson 1998), it may not suffice to state that Wg is strictly required only for planar polarity. It is possible that DV patterning in the eye also requires the Wg gradient to interpret patterning cues in the developing eye imaginal disc. The local cues within the cell are crucial for the Wg gradient but the secondary cell interactions shape the morphogen gradient by interpreting the information and setup differential expression.

Another candidate which may be participating in pole to equator gradient can be the T-box transcription factor Omb, also known as bifid (bi). Omb is a target of Wg signaling in the wing (Zecca et al. 1996). Its expression is regulated in the wing by Dpp and Wg. In the eye imaginal disc epithelium, *omb* is expressed in an pole to equator gradient where it has highest level on the dorsal and ventral margins and its levels decrease toward the equator (Fig. 7g, g′; Table 2; (Tare et al. 2013)). It is also expressed in some glial cells (Poeck et al. 1993). Gain of function of *omb* results in the reduction of eye size and loss of function of *omb* exhibits enhanced proliferation in the ventral eye disc (Porsch et al. 2005). Omb functions to delimit the extent of the DV eye (Poeck et al. 1993). The insertion of a P-element carrying a *white*$^+$ gene in the *omb* locus results in pigmentation on the dorsal and ventral eye margins in the adult eye. An *omb*-Gal4 line that was later isolated by Calleja and colleagues and Tang and Sun has been used vastly as an important tool to drive expression of genes on the dorsal and ventral borders margins of the eye imaginal disc (Calleja et al. 1996; Tang and Sun 2002; Tare et al. 2013).

In addition to pole to equator gradient of Wg, Fj, a Golgi associated protein, and a member of the planar cell polarity pathway (PCP pathway) has been proposed to be involved in equator to pole signaling (Zeidler et al. 1999a). It is expressed in a broad equatorial domain (Fig. 7i, i′; Brodsky and Steller 1996; Heberlein et al. 1998). The Fj expression gradient provides directional cues in ommatidial polarity (Zeidler et al. 1999a). Its gradient of expression is highest at the equator and decreases toward the margins. This graded expression of Fj is opposite to that of the pole to equator gradient of Wg, Omb and Dacshous (Ds), which are highest at the dorsal and ventral margins (poles) of the eye imaginal epithelium and decrease toward the equator. Fj acts upstream to Ds, therefore modulating and restricting its gradient expression.

Upd, a ligand of the JAK/STAT pathway (Harrison et al. 1998) is also required in the developing eye (Zeidler et al. 1999b). In the *Drosophila* eye, Upd is expressed at the posterior margin of the eye disc (Fig. 7j, j′) and is required to repress Wg and to promote differentiation in the eye (Table 2, (Singh et al. 2012; Tsai and Sun 2004; Tsai et al. 2007). Therefore, Wg downregulates Fj, and N and Upd act as positive regulators of Fj (Reynolds-Kenneally and Mlodzik 2005; Zeidler et al. 1999a). Fj and Upd, together or in parallel, are candidates for the secondary signal. These studies suggest that DV patterning genes not only contribute toward the growth of the eye field but also in delineation of boundary between the eye and head field.

Boundary Formation During Organogenesis

One of the important questions is how DV patterning genes regulate size and growth of the eye as an organ. The dorsal selector genes like *pnr*, Iro-C members, which are expressed in the dorsal eye margin (Fig. 7a, b), assign a dorsal fate in a group of early eye primordial cells that are basically ventral in fate. These dorsal fate selectors generate a group of dorsal cells with unique properties. The boundary between the dorsal and ventral cells (equator) is maintained by the antagonistic interactions between the dorsal and ventral eye genes (Singh et al. 2005a). It has been shown that *L/Ser* are essential for growth of the ventral eye tissue but is dispensable

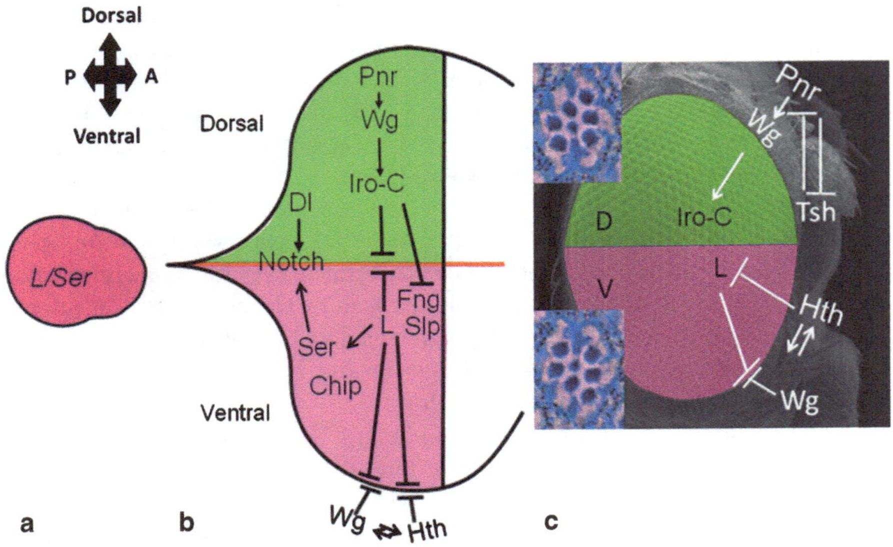

Fig. 8 Genetic basis of DV asymmetry in the developing *Drosophila* eye. **a** During early eye development (first instar larva), the entire eye primordium belongs to a homogenous ventral state assigned by functions of *L/Ser.* **b** In later stages (second instar and early third instar stages), dorsal lineage is specified upon onset of *pnr* expression. *pnr* acts upstream of Wg and this interaction is required for triggering expression of downstream genes like members of Iro-C complex (Maurel-Zaffran and Treisman 2000). Dl, a ligand of N pathway, is also required for the development of the dorsal eye. The default state of ventral eye is maintained by expression of *L/Ser*. *L* has been shown to antagonize functions of genes in the dorsal eye to define the eye boundary between dorsal and ventral compartment of the eye. Furthermore, functions of *L/Ser* are also required for antagonizing Hth and Wg and define the ventral eye margin to prevent cell death, respectively (Singh et al. 2011). However, there is a positive feedback loop between Hth and Wg in the ventral eye margin. The other genes important for ventral eye development are *chip*, *fng*, and *slp* (Cho and Choi 1998; Dominguez and de Celis 1998; Legent and Treisman 2008; Roignant et al. 2010; Sato and Tomlinson 2007). The interactions of dorsal and ventral genes are responsible for structural and functional organization of the adult compound eye of the *Drosophila*. **c** The ommatidia in an adult compound eye are organized into mirror image symmetry which are polarized into opposite directions of dorsal and ventral half

in the dorsal region specified by *pnr* function (Singh and Choi 2003). In addition to a boundary between the dorsal and ventral compartment within the eye, a boundary is defined between the developing eye field and the surrounding head cuticle on the dorsal and ventral margins (Fig. 8). Since the adult eye, head cuticle, and other mouthparts are generated from the eye-antennal imaginal disc, there is a sequential fate restriction between the developing eye and head cuticle. These DV patterning genes play an important role in defining the boundary of the eye field on the dorsal and the ventral margins (Oros et al. 2010).

The boundary between the eye field and the head cuticle on the dorsal margin is regulated by *pnr* (Fig. 8). It has been suggested that *pnr* is required for two different functions during eye development : (a) DV axis determination during early eye development (Maurel-Zaffran and Treisman 2000; Singh and Choi 2003) and (b) suppression of retinal determination to define the dorsal eye field margin (Oros et al. 2010; Singh and Choi 2003; Singh et al. 2005b Singh et al. 2012). During early second instar of larval development, *pnr* is required for defining the dorsal lineage, before the onset of retinal differentiation by inducing Wg and members of the Iro-C complex (Maurel-Zaffran and Treisman 2000; Oros et al. 2010; Singh and Choi 2003; Singh et al. 2005b). However, during the late second instar stage of eye development, *pnr* suppresses the photoreceptor differentiation at the dorsal eye margin (Oros et al. 2010; Singh et al. 2012). The expression of *pnr* is restricted to the peripodial membrane of the dorsal eye margin, which gives rise to the adult head cuticle. Loss-of-function clones of *pnr* exhibit dorsal eye enlargement. It suggests that absence of *pnr* function promotes ectopic eye formation in the dorsal eye margin. Therefore, *pnr* defines the boundary between the head cuticle and the dorsal margin of the developing eye field (Oros et al. 2010; Singh et al. 2012). There is a different mechanism for delineating the boundary between the eye and head cuticle on the ventral eye margin as *pnr* is not expressed in the ventral eye (Singh et al. 2011). The boundary of eye field on the ventral eye margin is defined by the antagonistic interaction of *L* with *hth* (Singh et al. 2011; Singh et al. 2012). Antagonistic interaction between L and Hth is not the exclusive mechanism to define the ventral eye margin. In the ventral eye, transcriptional cofactor Chip interacts with the LIM-homeodomain proteins to define the boundary of the eye field (Roignant et al. 2010). Interestingly, Chip mediated regulation of the ventral eye boundary is independent of *hth* (Roignant et al. 2010). Thus, the genetic cascade regulating the boundary of eye field on the dorsal and the ventral margin of the eye is different.

Similarities With Vertebrate Eye

There are remarkable similarities in general developmental design based on functional and structural homologies between the *Drosophila* eye genes and the vertebrate eye field transcription factors (EFTFs; Wawersik and Maas 2000). Furthermore, the basic sensory epithelium design of the vertebrate and most invertebrate eyes, including the *Drosophila* eye is similar (Charlton-Perkins and Cook 2010; Sanes and Zipursky 2010; Singh et al. 2012). The morphogenetic furrow (MF) in the fly eye is analogous to the wave of neurogenesis in the vertebrate retina (Hartenstein and Reh 2002; Neumann and Nuesslein-Volhard 2000). However, the MF in the *Drosophila* eye initiates from the posterior margin and moves toward the anterior part of the developing eye imaginal disc whereas differentiation in vertebrate retina initiates from centre and moves radially out (Hartenstein and Reh 2002). Several genes that are expressed in a DV domain-specific manner in the retina have been identified in the vertebrate visual system. In the dorsal half of the eye, BMP4, a TGF-β closely related to Dpp, has been implicated in development of progenitor cells. It has also been shown to function in establishment of the DV axis of the Xenopus retina (Papalopulu and Kintner 1996). In the vertebrate eye, the dorsal selectors BMP-4 and TbX5 restrict the expression of Vax2 and Pax2 to the ventral domain of the eye (Koshiba-Takeuchi et al. 2000; Mui et al. 2002; Peters 2002; Peters and Cepko 2002). These DV expression domains correspond to the developmental compartments (Peters 2002; Peters and Cepko 2002). The DV patterning plays an important role in the retinotectal projection pattern (Koshiba-Takeuchi et al. 2000; McLaughlin et al. 2003). The R-cell projections, form a precise topographic connection with the optic lobe, and are referred to as retinotopy, which is common to both the vertebrate and the insect visual system (Gaul 2002). Jagged-1 (Jag1), a vertebrate homolog of the *Drosophila* ventral eye gene Ser, exhibits a DV asymmetric expression pattern in the retina. In addition, loss-of-function of Jag1 results in Alagille's syndrome, which also affects the eye (Kim and Fulton 2007; Oda et al. 1997; Xue et al. 1999). It has also been suggested that mouse retina also begins with a default ventral like state (Murali et al. 2005). Therefore, the DV boundary may play conserved roles in organizing the growth and pattern of visual system in higher animals, and studies in *Drosophila* will upgrade our knowledge in the area of animal development mechanisms and help to unravel the genetic underpinnings of developmental defects caused by mutations in human homologs of *Drosophila* DV patterning genes.

Summary

In this chapter, we have focused on the key developmental events and genes that are involved in DV patterning of the *Drosophila* eye. It has been established that formation of the DV compartment formation is a key event in initiating patterning and growth of the early eye imaginal disc. This may also hold true in primordia of other adult appendages. It is clear that DV patterning is required to initiate the DV asymmetry within a homogenous default ventral fate of early eye primordial cells. Even though our understanding of the DV patterning in the eye has dramatically increased in recent years, our understanding of the axial patterning of the *Drosophila* eye is far from complete, and we are still not aware of all the members of genetic circuitry and the molecular interactions between them which are important for the regulation of DV patterning. There is a room for identification of many more novel genes that are involved in DV patterning. The future studies using novel genetic and bioinformatics approaches should help in defining the full complement of genes involved in this intricate process. These studies will help in addressing the age old question of how a small number of cells in the disc primordium grow to form a precise pattern of mirror symmetry in the compound eye. In addition, the possibility of crosstalk of the DV patterning pathway with other signaling pathways to regulate growth during early phase of eye development cannot be refuted. All this information will lay a foundation about understanding the process of organogenesis as loss-of-function of the genes involved in DV patterning results in the loss of the eye field or a part of the eye field. The complexity and precision of the neural connectivity in the adult visual system has fascinated researchers for a long time. The DV polarity of the retina is responsible for controlling the targeting of the retinal axon projections to the brain in humans and other higher vertebrates. Thus, DV patterning genes also contribute toward the wiring of the brain to the retina. How all these different facets work together to define the final form of this complex structure eye is an open question and is of fundamental importance.

Acknowledgements The authors thank Erika Wittkorn, Timothy Cutler, and Aditi Singh for their help and comments on the manuscript. We apologize to all authors whose work could not be cited due to space limitations. AS is supported by a NIH grant (1R15 HD064557–01), start-up support from the University of Dayton.

References

Anderson DT (1972a) The development of hemimetabolous insects. In: Counce S, Waddington CH (eds) Developmental systems: insects. Academic Press, New York, pp 165–242

Anderson DT (1972b) The development of hemimetabolous insects. In: Counce S, Waddington CH (eds) Developmental systems: insects. Academic Press, New York, pp 96–163

Atkins M, Mardon G (2009) Signaling in the third dimension: the peripodial epithelium in eye disc development. Dev Dyn 238:2139–2148

Bach EA, Ekas LA, Ayala-Camargo A, Flaherty MS, Lee H, Perrimon N, Baeg GH (2007) GFP reporters detect the activation of the *Drosophila* JAK/STAT pathway in vivo. Gene Expr Patterns 7:323–331

Bachmann A, Knust E (1998) Positive and negative control of Serrate expression during early development of the *Drosophila* wing. Mech Dev 76:67–78

Baker NE (1988a) Embryonic and imaginal requirements for wingless, a segment-polarity gene in *Drosophila*. Dev Biol 125:96–108

Baker NE (1988b) Transcription of the segment-polarity gene wingless in the imaginal discs of *Drosophila*, and the phenotype of a pupal-lethal wg mutation. Development 102:489–497

Baker WK (1978) A clonal analysis reveals early developmental restrictions in the *Drosophila* head. Dev Biol 62:447–463

Baonza A, Garcia-Bellido A (2000) Notch signaling directly controls cell proliferation in the *Drosophila* wing disc. Proc Natl Acad Sci U S A 97:2609–2614

Becker HJ (1966) Genetic and variegation mosaics in the eye of *Drosophila*. Curr Top Dev Biol 1:155–171

Bellen HJ, O'Kane CJ, Wilson C, Grossniklaus U, Pearson RK, Gehring WJ (1989) P-element-mediated enhancer detection: a versatile method to study development in *Drosophila*. Genes Dev 3:1288–1300

Bessa J, Casares F (2005) Restricted teashirt expression confers eye-specific responsiveness to Dpp and Wg signals during eye specification in *Drosophila*. Development 132:5011–5020

Bessa J, Gebelein B, Pichaud F, Casares F, Mann RS (2002) Combinatorial control of *Drosophila* eye development by eyeless, homothorax, and teashirt. Genes Dev 16:2415–2427

Bessa J, Tavares MJ, Santos J, Kikuta H, Laplante M, Becker TS, Gomez-Skarmeta JL, Casares F (2008) meis1 regulates cyclin D1 and c-myc expression, and controls the proliferation of the multipotent cells in the early developing zebrafish eye. Development 135:799–803

Bhojwani J, Singh A, Misquitta L, Mishra A, Sinha P (1995) Search for the *Drosophila* genes based on patterned expression of mini-*white* reporter gene of a P lacW vector in adult eyes. Roux's Arch Dev Biol 205:114–121

Bier E (2005) *Drosophila*, the golden bug, emerges as a tool for human genetics. Nat Rev Genet 6:9–23

Bier E, Vaessin H, Shepherd S, Lee K, McCall K, Barbel S, Ackerman L, Carretto R, Uemura T, Grell E et al (1989) Searching for pattern and mutation in the *Drosophila* genome with a P-lacZ vector. Genes Dev 3:1273–1287

Blair SS (2001) Cell lineage: compartments and capricious. Curr Biol 11:R1017–1021
Blair SS, Brower DL, Thomas JB, Zavortink M (1994) The role of apterous in the control of dorsoventral compartmentalization and PS integrin gene expression in the developing wing of *Drosophila*. Development 120:1805–1815
Bodentstein D (1950) The postembryonic development of *Drosophila melanogaster*. In: Demerec M (ed) Biology of *Drosophila*. Wiley, New York, pp 275–367
Bosveld F, Bonnet I, Guirao B, Tlili S, Wang Z, Petitalot A, Marchand R, Bardet PL, Marcq P, Graner F et al (2012) Mechanical control of morphogenesis by Fat/Dachsous/Four-jointed planar cell polarity pathway. Science 336:724–727
Boutros M, Paricio N, Strutt DI, Mlodzik M (1998) Dishevelled activates JNK and discriminates between JNK pathways in planar polarity and wingless signaling. Cell 94:109–118
Boutros M, Mihaly J, Bouwmeester T, Mlodzik M (2000) Signaling specificity by frizzled receptors in *Drosophila*. Science 288:1825–1828
Brodsky MH, Steller H (1996) Positional information along the dorsal-ventral axis of the *Drosophila* eye: graded expression of the four-jointed gene. Dev Biol 173:428–446
Brook WJ, Cohen SM (1996) Antagonistic interactions between wingless and decapentaplegic responsible for dorsal-ventral pattern in the *Drosophila* leg. Science 273:1373–1377
Brower DL (1986) Engrailed gene expression in *Drosophila* imaginal discs. EMBO J 5:2649–2656
Cagan RL, Ready DF (1989) Notch is required for successive cell decisions in the developing *Drosophila* retina. Genes Dev 3:1099–1112
Calleja M, Moreno E, Pelaz S, Morata G (1996) Visualization of gene expression in living adult *Drosophila*. Science 274:252–255
Cavodeassi F, Diez Del Corral R, Campuzano S, Dominguez M (1999) Compartments and organising boundaries in the *Drosophila* eye: the role of the homeodomain Iroquois proteins. Development 126:4933–4942
Cavodeassi F, Modolell J, Campuzano S (2000) The Iroquois homeobox genes function as dorsal selectors in the *Drosophila* head. Development 127:1921–1929
Chang T, Mazotta J, Dumstrei K, Dumitrescu A, Hartenstein V (2001) Dpp and Hh signaling in the *Drosophila* embryonic eye field. Development 128:4691–4704
Chanut F, Heberlein U (1997a) Retinal morphogenesis in *Drosophila*: hints from an eye-specific decapentaplegic allele. Dev Genet 20:197–207
Chanut F, Heberlein U (1997b) Role of decapentaplegic in initiation and progression of the morphogenetic furrow in the developing *Drosophila* retina. Development 124:559–567
Charlton-Perkins M, Cook TA (2010) Building a fly eye: terminal differentiation events of the retina, corneal lens, and pigmented epithelia. Curr Top Dev Biol 93:129–173
Chern JJ, Choi KW (2002) Lobe mediates Notch signaling to control domain-specific growth in the *Drosophila* eye disc. Development 129:4005–4013
Cho KO, Choi KW (1998) Fringe is essential for mirror symmetry and morphogenesis in the *Drosophila* eye. Nature 396:272–276
Choi KW, Mozer B, Benzer S (1996) Independent determination of symmetry and polarity in the *Drosophila* eye. Proc Natl Acad Sci U S A 93:5737–5741
Cho KO, Chern J, Izaddoost S, Choi KW (2000) Novel signaling from the peripodial membrane is essential for eye disc patterning in *Drosophila*. Cell 103:331–342
Cohen SM (1993) Imaginal disc development. In: Bate M, Martinez-Arias A (eds) The development of *Drosophila* melanogaster. Cold Spring Harbor Laboratory Press, New York
Cohen B, McGuffin ME, Pfeifle C, Segal D, Cohen SM (1992) Apterous, a gene required for imaginal disc development in *Drosophila* encodes a member of the LIM family of developmental regulatory proteins. Genes Dev 6:715–729
Cohen B, Simcox AA, Cohen SM (1993) Allocation of the thoracic imaginal primordia in the *Drosophila* embryo. Development 117:597–608
Couso JP, Bate M, Martinez-Arias A (1993) A wingless-dependent polar coordinate system in *Drosophila* imaginal discs. Science 259:484–489

Crick FH, Lawrence PA (1975) Compartments and polyclones in insect development. Science 189:340–347
Curtiss J, Halder G, Mlodzik M (2002) Selector and signalling molecules cooperate in organ patterning. Nat Cell Biol 4:E48–E51
Dahmann C, Oates AC, Brand M (2011) Boundary formation and maintenance in tissue development. Nat Rev Genet 12:43–55
Datta RR, Lurye JM, Kumar JP (2009) Restriction of ectopic eye formation by *Drosophila* teashirt and tiptop to the developing antenna. Dev Dyn 238:2202–2210
de Celis JF, Garcia-Bellido A, Bray SJ (1996) Activation and function of Notch at the dorsal-ventral boundary of the wing imaginal disc. Development 122:359–369
Diaz-Benjumea FJ, Cohen SM (1993) Interaction between dorsal and ventral cells in the imaginal disc directs wing development in *Drosophila*. Cell 75:741–752
Diaz-Benjumea FJ, Cohen SM (1995) Serrate signals through Notch to establish a Wingless-dependent organizer at the dorsal/ventral compartment boundary of the *Drosophila* wing. Development 121:4215–4225
Diaz-Benjumea FJ, Cohen B, Cohen SM (1994) Cell interaction between compartments establishes the proximal-distal axis of *Drosophila* legs. Nature 372:175–179
Dietrich W (1909) Die Facettenaugen der Dipteran. Z Wiss Zool 92:465–539
Diez del Corral R, Aroca P, JL G.m.-S., Cavodeassi F, Modolell J (1999) The Iroquois homeodomain proteins are required to specify body wall identity in *Drosophila*. Genes Dev 13:1754–1761
Dominguez M, Casares F (2005) Organ specification-growth control connection: new in-sights from the *Drosophila* eye-antennal disc. Dev Dyn 232:673–684
Dominguez M, de Celis JF (1998) A dorsal/ventral boundary established by Notch controls growth and polarity in the *Drosophila* eye. Nature 396:276–278
Dominguez M, Brunner M, Hafen E, Basler K (1996) Sending and receiving the hedgehog signal: control by the *Drosophila* Gli protein Cubitus interruptus. Science 272:1621–1625
Eaton S, Kornberg TB (1990) Repression of ci-D in posterior compartments of *Drosophila* by engrailed. Genes Dev 4:1068–1077
Erclik T, Hartenstein V, McInnes RR, Lipshitz HD (2009) Eye evolution at high resolution: the neuron as a unit of homology. Dev Biol 332:70–79
Fasano L, Roder L, Core N, Alexandre E, Vola C, Jacq B, Kerridge S (1991) The gene teashirt is required for the development of *Drosophila* embryonic trunk segments and encodes a protein with widely spaced zinc finger motifs. Cell 64:63–79
Ferris GF (1950) External morphology of the adult. In: Demerec M (ed) Biology of *Drosophila*. Wiley, New York, pp 368–419
Fleming RJ, Gu Y, Hukriede NA (1997) Serrate-mediated activation of Notch is specifically blocked by the product of the gene fringe in the dorsal compartment of the *Drosophila* wing imaginal disc. Development 124:2973–2981
Garcia-Bellido A, Merriam JR (1969) Cell lineage of the imaginal discs in *Drosophila* gynandromorphs. J Exp Zool 170:61–75
Garcia-Bellido A, Ripoll P, Morata G (1973) Developmental compartmentalisation of the wing disk of *Drosophila*. Nat New Biol 245:251–253
Gaul U (2002) The establishment of retinal connectivity. In: Moses K (ed) *Drosophila* eye development. Springer, Berlin, pp. 205–216
Gehring WJ (2005) New perspectives on eye development and the evolution of eyes and photoreceptors. J Hered 96:171–184
Gibson MC, Schubiger G (2001) *Drosophila* peripodial cells, more than meets the eye? Bioessays 23:691–697
Glazov EA, Pheasant M, McGraw EA, Bejerano G, Mattick JS (2005) Ultraconserved elements in insect genomes: a highly conserved intronic sequence implicated in the control of homothorax mRNA splicing. Genome Res 15:800–808
Go MJ, Eastman DS, Artavanis-Tsakonas S (1998) Cell proliferation control by Notch signaling in *Drosophila* development. Development 125:2031–2040

Gomez-Skarmeta JL, Modolell J (1996) araucan and caupolican provide a link between compartment subdivisions and patterning of sensory organs and veins in the *Drosophila* wing. Genes Dev 10:2935–2945

Gomez-Skarmeta JL, Modolell J (2002) Iroquois genes: genomic organization and function in vertebrate neural development. Curr Opin Genet Dev 12:403–408

Gomez-Skarmeta JL, Diez del Corral R, de la Calle-Mustienes E, Ferre-Marco D, Modolell J (1996) Araucan and caupolican, two members of the novel iroquois complex, encode homeoproteins that control proneural and vein-forming genes. Cell 85:95–105

Grillenzoni N, van Helden J, Dambly-Chaudiere C, Ghysen A (1998) The iroquois complex controls the somatotopy of *Drosophila* notum mechanosensory projections. Development 125:3563–3569

Grossniklaus U, Pearson RK, Gehring WJ (1992) The *Drosophila* sloppy paired locus encodes two proteins involved in segmentation that show homology to mammalian transcription factors. Genes Dev 6:1030–1051

Halder G, Callaerts P, Gehring WJ (1995) Induction of ectopic eyes by targeted expression of the eyeless gene in *Drosophila*. Science 267:1788–1792

Harrison DA, McCoon PE, Binari R, Gilman M, Perrimon N (1998) *Drosophila* unpaired encodes a secreted protein that activates the JAK signaling pathway. Genes Dev 12:3252–3263

Hartenstein V, Reh TA (2002) Homologies between vertebrate and invertebrate eyes. In: Moses K (ed) *Drosophila* eye development. Springer, Heidelberg, pp 219–251

Haynie JL, Bryant PJ (1986) Development of the eye-antenna imaginal disc and morphogenesis of the adult head in *Drosophila melanogaster*. J Exp Zool 237:293–308

Hazelett DJ, Bourouis M, Walldorf U, Treisman JE (1998) decapentaplegic and wingless are regulated by eyes absent and eyegone and interact to direct the pattern of retinal differentiation in the eye disc. Development 125:3741–3751

Heberlein U, Borod ER, Chanut FA (1998) Dorsoventral patterning in the *Drosophila* retina by wingless. Development 125:567–577

Held LIJ (2002a) Cell lineage vs. intercellular signaling. In: Held LI (ed) Imaginal disc. Cambridge University Press, Cambridge, pp 1–4

Held LIJ (2002b) The eye disc. In: Held LI (ed) Imaginal disc. Cambridge University Press, Cambridge, pp 197–236

Heslip TR, Theisen H, Walker H, Marsh JL (1997) Shaggy and dishevelled exert opposite effects on Wingless and Decapentaplegic expression and on positional identity in imaginal discs. Development 124:1069–1078

Hidalgo A (1998) Growth and patterning from the engrailed interface. Int J Dev Biol 42:317–324

Irvine KD (1999) Fringe, Notch, and making developmental boundaries. Curr Opin Genet Dev 9:434–441

Irvine KD, Vogt TF (1997) Dorsal-ventral signaling in limb development. Curr Opin Cell Biol 9:867–876

Irvine KD, Wieschaus E (1994) Fringe, a Boundary-specific signaling molecule, mediates interactions between dorsal and ventral cells during *Drosophila* wing development. Cell 79:595–606

Jaw TJ, You LR, Knoepfler PS, Yao LC, Pai CY, Tang CY, Chang LP, Berthelsen J, Blasi F, Kamps MP et al (2000) Direct interaction of two homeoproteins, homothorax and extradenticle, is essential for EXD nuclear localization and function. Mech Dev 91:279–291

Jordan KC, Clegg NJ, Blasi JA, Morimoto AM, Sen J, Stein D, McNeill H, Deng WM, Tworoger M, Ruohola-Baker H (2000) The homeobox gene mirror links EGF signalling to embryonic dorso-ventral axis formation through notch activation. Nat Genet 24:429–433

Ju BG, Jeong S, Bae E, Hyun S, Carroll SB, Yim J, Kim J (2000) Fringe forms a complex with Notch. Nature 405:191–195

Jurgens J, Hartenstein V (1993) The terminal regions of body pattern. In: Martinez-Arias MBaA (ed) The development of *Drosophila melanogaster*. Cold Spring Harbor Laboratory Press, Cold-Spring Harbor, pp 687–746

Kankel DR, Ferrus A, Garen SH, Harte PJ, Lewis PE (1980) The structure and development of the nervous system. In: Ashburner M, Wright TRF (ed) The genetics and biology of *Drosophila*. Academic Press, New York, pp 295–368

Kehl BT, Cho KO, Choi KW (1998) Mirror, a *Drosophila* homeobox gene in the Iroquois complex, is required for sensory organ and alula formation. Development 125:1217–1227

Kenyon KL, Ranade SS, Curtiss J, Mlodzik M, Pignoni F (2003) Coordinating proliferation and tissue specification to promote regional identity in the *Drosophila* head. Dev Cell 5:403–414

Kim BJ, Fulton AB (2007) The genetics and ocular findings of Alagille syndrome. Semin Ophthalmol 22:205–210

Kim J, Irvine KD, Carroll SB (1995) Cell recognition, signal induction, and symmetrical gene activation at the dorsal-ventral boundary of the developing *Drosophila* wing. Cell 82:795–802

Klueg KM, Muskavitch MA (1999) Ligand-receptor interactions and trans-endocytosis of Delta, Serrate and Notch: members of the Notch signalling pathway in *Drosophila*. J Cell Sci 112(Pt 19):3289–3297

Kornberg T, Siden I, O'Farrell P, Simon M (1985) The engrailed locus of *Drosophila*: in situ localization of transcripts reveals compartment-specific expression. Cell 40:45–53

Koshiba-Takeuchi K, Takeuchi JK, Matsumoto K, Momose T, Uno K, Hoepker V, Ogura K, Takahashi N, Nakamura H, Yasuda K et al (2000) Tbx5 and the retinotectum projection. Science 287:134–137

Kumar JP (2009) The molecular circuitry governing retinal determination. Biochim Biophys Acta 1789:306–314

Kumar JP (2011) My what big eyes you have: how the *Drosophila* retina grows. Dev Neurobiol 71:1133–1152

Kumar JP, Moses K (2001) EGF receptor and Notch signaling act upstream of Eyeless/Pax6 to control eye specification. Cell 104:687–697

Kurant E, Pai CY, Sharf R, Halachmi N, Sun YH, Salzberg A (1998) Dorsotonals/homothorax, the *Drosophila* homologue of meis1, interacts with extradenticle in patterning of the embryonic PNS. Development 125:1037–1048

Land MF, Fernald RD (1992) The evolution of eyes. Annu Rev Neurosci 15:1–29

Langer JA, Cutrone EC, Kotenko S (2004) The Class II cytokine receptor (CRF2) family: overview and patterns of receptor-ligand interactions. Cytokine Growth Factor Rev 15:33–48

Lawrence PA, Green SM (1979) Cell lineage in the developing retina of *Drosophila*. Dev Biol 71:142–152

Lawrence PA, Morata G (1976) Compartments in the wing of *Drosophila*: a study of the engrailed gene. Dev Biol 50:321–337

Lecuit T, Brook WJ, Ng M, Calleja M, Sun H, Cohen SM (1996) Two distinct mechanisms for long-range patterning by Decapentaplegic in the *Drosophila* wing. Nature 381:387–393

Lee JD, Treisman JE (2001) The role of Wingless signaling in establishing the anteroposterior and dorsoventral axes of the eye disc. Development 128:1519–1529

Lee JD, Treisman JE (2002) Regulators of the morphogenetic furrow. In: Moses K (ed) *Drosophila* eye development. Springer, Berlin, pp 21–29

Legent K, Treisman JE (2008) Wingless signaling in *Drosophila* eye development. Methods Mol Biol 469:141–161

Leyns L, Gomez-Skarmeta JL, Dambly-Chaudiere C (1996) Iroquois: a prepattern gene that controls the formation of bristles on the thorax of *Drosophila*. Mech Dev 59:63–72

Ma C, Moses K (1995) Wingless and patched are negative regulators of the morphogenetic furrow and can affect tissue polarity in the developing *Drosophila* compound eye. Development 121:2279–2289

Masucci JD, Miltenberger RJ, Hoffmann FM (1990) Pattern-specific expression of the *Drosophila* decapentaplegic gene in imaginal disks is regulated by 3' cis-regulatory elements. Genes Dev 4:2011–2023

Maurel-Zaffran C, Treisman JE (2000) Pannier acts upstream of wingless to direct dorsal eye disc development in *Drosophila*. Development 127:1007–1016

McClure KD, Schubiger G (2005) Developmental analysis and squamous morphogenesis of the peripodial epithelium in *Drosophila* imaginal discs. Development 132:5033–5042

McLaughlin T, Hindges R, O'Leary DD (2003) Regulation of axial patterning of the retina and its topographic mapping in the brain. Curr Opin Neurobiol 13:57–69

McNeill H, Yang CH, Brodsky M, Ungos J, Simon MA (1997) Mirror encodes a novel PBX-class homeoprotein that functions in the definition of the dorsal-ventral border in the *Drosophila* eye. Genes Dev 11:1073–1082

Meinhardt H (1983) Cell determination boundaries as organizing regions for secondary embryonic fields. Dev Biol 96:375–385

Miall LC, Hammond AR (1892) The development of the head of the imago of chironomus. Trans Linn Soc Lond 5:265–279

Milner M, Bleasby A, Pyott A (1983) The role of the peripodial membrane in the morphogenesis of the eye antennal disc of *Drosophila melanogaster*. Roux's Arch Dev Biol 192:164–170

Mlodzik M (1999) Planar polarity in the *Drosophila* eye: a multifaceted view of signaling specificity and cross-talk. EMBO J 18:6873–6879

Moloney DJ, Panin VM, Johnston SH, Chen J, Shao L, Wilson R, Wang Y, Stanley P, Irvine KD, Haltiwanger RS et al (2000) Fringe is a glycosyltransferase that modifies Notch. Nature 406:369–375

Morata G, Lawrence PA (1975) Control of compartment development by the engrailed gene in *Drosophila*. Nature 255:614–617

Morgan TH, Bridges CB, Strutevant AH (1925) The genetics of *Drosophila*. Bibliog Genet 2:1–262

Morrison CM, Halder G (2010) Characterization of a dorsal-eye Gal4 Line in *Drosophila*. Genesis 48:3–7

Moses K (2002) *Drosophila* eye development. Springer, Berlin

Moskow JJ, Bullrich F, Huebner K, Daar IO, Buchberg AM (1995) Meis1, a PBX1-related homeobox gene involved in myeloid leukemia in BXH-2 mice. Mol Cell Biol 15:5434–5443

Mui SH, Hindges R, O'Leary DD, Lemke G, Bertuzzi S (2002) The homeodomain protein Vax2 patterns the dorsoventral and nasotemporal axes of the eye. Development 129:797–804

Murali D, Yoshikawa S, Corrigan RR, Plas DJ, Crair MC, Oliver G, Lyons KM, Mishina Y, Furuta Y (2005) Distinct developmental programs require different levels of Bmp signaling during mouse retinal development. Development 132:913–923

Nellen D, Burke R, Struhl G, Basler K (1996) Direct and long-range action of a DPP morphogen gradient. Cell 85:357–368

Netter S, Fauvarque MO, Diez del Corral R, Dura JM, Coen D (1998) White + transgene insertions presenting a dorsal/ventral pattern define a single cluster of homeobox genes that is silenced by the polycomb-group proteins in *Drosophila melanogaster*. Genetics 149:257–275

Neumann CJ, Nuesslein-Volhard C (2000) Patterning of the zebrafish retina by a wave of sonic hedgehog activity. Science 289:2137–2139

Noro B, Culi J, McKay DJ, Zhang W, Mann RS (2006) Distinct functions of homeodomain-containing and homeodomain-less isoforms encoded by homothorax. Genes Dev 20:1636–1650

Oda T, Elkahloun AG, Pike BL, Okajima K, Krantz ID, Genin A, Piccoli DA, Meltzer PS, Spinner NB, Collins FS et al (1997) Mutations in the human Jagged1 gene are responsible for Alagille syndrome. Nat Genet 16:235–242

Okajima T, Irvine KD (2002) Regulation of notch signaling by o-linked fucose. Cell 111:893–904

Oros SM, Tare M, Kango-Singh M, Singh A (2010) Dorsal eye selector pannier (pnr) suppresses the eye fate to define dorsal margin of the *Drosophila* eye. Dev Biol 346:258–271

Pai CY, Kuo TS, Jaw TJ, Kurant E, Chen CT, Bessarab DA, Salzberg A, Sun YH (1998) The Homothorax homeoprotein activates the nuclear localization of another homeoprotein, extradenticle, and suppresses eye development in *Drosophila*. Genes Dev 12:435–446

Pan D, Rubin GM (1998) Targeted expression of teashirt induces ectopic eyes in *Drosophila*. Proc Natl Acad Sci U S A 95:15508–15512

Papalopulu N, Kintner C (1996) A Xenopus gene, Xbr-1, defines a novel class of homeobox genes and is expressed in the dorsal ciliary margin of the eye. Dev Biol 174:104–114

Papayannopoulos V, Tomlinson A, Panin VM, Rauskolb C, Irvine KD (1998) Dorsal-ventral signaling in the *Drosophila* eye. Science 281:2031–2034

Penton A, Hoffmann FM (1996) Decapentaplegic restricts the domain of wingless during *Drosophila* limb patterning. Nature 382:162–164

Peters MA (2002) Patterning the neural retina. Curr Opin Neurobiol 12:43–48

Peters MA, Cepko CL (2002) The dorsal-ventral axis of the neural retina is divided into multiple domains of restricted gene expression which exhibit features of lineage compartments. Dev Biol 251:59–73

Pichaud F, Casares F (2000) Homothorax and iroquois-C genes are required for the establishment of territories within the developing eye disc. Mech Dev 96:15–25

Pignoni F, Zipursky SL (1997) Induction of *Drosophila* eye development by decapentaplegic. Development 124:271–278

Poeck B, Hofbauer A, Pflugfelder GO (1993) Expression of the *Drosophila* optomotor-blind gene transcript in neuronal and glial cells of the developing nervous system. Development 117:1017–1029

Porsch M, Sauer M, Schulze S, Bahlo A, Roth M, Pflugfelder GO (2005) The relative role of the T-domain and flanking sequences for developmental control and transcriptional regulation in protein chimeras of *Drosophila* OMB and ORG-1. Mech Dev 122:81–96

Poulson DF (1950). Histogenesis, oogenesis, and differentiation in the embryo of *Drosophila melanogaster* meigen. In: Demerec M (ed) Biology of *Drosophila*. Wiley, New York, pp 168–274

Quiring R, Walldorf U, Kloter U, Gehring WJ (1994) Homology of the eyeless gene of *Drosophila* to the Small eye gene in mice and Aniridia in humans. Science 265:785–789

Raftery LA, Sanicola M, Blackman RK, Gelbart WM (1991) The relationship of decapentaplegic and engrailed expression in *Drosophila* imaginal disks: do these genes mark the anterior-posterior compartment boundary? Development 113:27–33

Ramain P, Heitzler P, Haenlin M, Simpson P (1993) Pannier, a negative regulator of achaete and scute in *Drosophila*, encodes a zinc finger protein with homology to the vertebrate transcription factor GATA-1. Development 119:1277–1291

Ready DF, Hanson TE, Benzer S (1976) Development of the *Drosophila* retina, a neurocrystalline lattice. Dev Biol 53:217–240

Reifegerste R, Moses K (1999) Genetics of epithelial polarity and pattern in the *Drosophila* retina. Bioessays 21:275–285

Reynolds-Kenneally J, Mlodzik M (2005) Notch signaling controls proliferation through cell-autonomous and non-autonomous mechanisms in the *Drosophila* eye. Dev Biol 285:38–48

Rieckhof GE, Casares F, Ryoo HD, Abu-Shaar M, Mann RS (1997) Nuclear translocation of extradenticle requires homothorax, which encodes an extradenticle-related homeodomain protein. Cell 91:171–183

Roignant JY, Treisman JE (2009) Pattern formation in the *Drosophila* eye disc. Int J Dev Biol 53:795–804

Roignant JY, Legent K, Janody F, Treisman JE (2010) The transcriptional co-factor Chip acts with LIM-homeodomain proteins to set the boundary of the eye field in *Drosophila*. Development 137:273–281

Ryoo HD, Marty T, Casares F, Affolter M, Mann RS (1999) Regulation of Hox target genes by a DNA bound Homothorax/Hox/Extradenticle complex. Development 126:5137–5148

Sanes JR, Zipursky SL (2010) Design principles of insect and vertebrate visual systems. Neuron 66:15–36

Sanicola M, Sekelsky J, Elson S, Gelbart WM (1995) Drawing a stripe in *Drosophila* imaginal disks: negative regulation of decapentaplegic and patched expression by engrailed. Genetics 139:745–756

Sato A, Tomlinson A (2007) Dorsal-ventral midline signaling in the developing *Drosophila* eye. Development 134:659–667

Simon MA (2004) Planar cell polarity in the *Drosophila* eye is directed by graded Four-jointed and Dachsous expression. Development 131:6175–6184

Singh A, Choi KW (2003) Initial state of the *Drosophila* eye before dorsoventral specification is equivalent to ventral. Development 130:6351–6360

Singh A, Irvine KD (2012) *Drosophila* as a model for understanding development and disease. Dev Dyn 241:1–2

Singh J, Mlodzik M (2012) Planar cell polarity signaling: coordination of cellular orientation across tissues. Wiley Interdiscip Rev Dev Biol 1:479–499

Singh A, Kango-Singh M, Sun YH (2002) Eye suppression, a novel function of teashirt, requires wingless signaling. Development 129:4271–4280

Singh A, Kango-Singh M, Choi KW, Sun YH (2004) Dorso-ventral asymmetric functions of teashirt in *Drosophila* eye development depend on spatial cues provided by early DV patterning genes. Mech Dev 121:365–370

Singh A, Chan J, Chern JJ, Choi KW (2005a) Genetic interaction of Lobe with its modifiers in dorsoventral patterning and growth of the *Drosophila* eye. Genetics 171:169–183

Singh A, Lim J, Choi K-W (2005b). Dorso-ventral boundary is required for organizing growth and planar polarity in the *Drosophila* eye. In: Mlodzik M (ed) Planar cell polarization during development: advances in developmental biology and biochemistry. Elsevier Science & Technology Books, San Diego, pp 59–91

Singh A, Shi X, Choi KW (2006) Lobe and Serrate are required for cell survival during early eye development in *Drosophila*. Development 133:4771–4781

Singh A, Tare M, Kango-Singh M, Son WS, Cho KO, Choi KW (2011) Opposing interactions between homothorax and Lobe define the ventral eye margin of *Drosophila* eye. Dev Biol 359:199–208

Singh A, Tare M, Puli OR, Kango-Singh M (2012) A glimpse into dorso-ventral patterning of the *Drosophila* eye. Dev Dyn 241:69–84

Speicher SA, Thomas U, Hinz U, Knust E (1994) The Serrate locus of *Drosophila* and its role in morphogenesis of the wing imaginal discs: control of cell proliferation. Development 120:535–544

Stevens KE, Mann RS (2007) A balance between two nuclear localization sequences and a nuclear export sequence governs extradenticle subcellular localization. Genetics 175:1625–1636

Struhl G (1981) A blastoderm fate map of compartments and segments of the *Drosophila* head. Dev Biol 84:386–396

Sun X, Artavanis-Tsakonas S (1996) The intracellular deletions of Delta and Serrate define dominant negative forms of the *Drosophila* Notch ligands. Development 122:2465–2474

Sun YH, Tsai CJ, Green MM, Chao JL, Yu CT, Jaw TJ, Yeh JY, Bolshakov VN (1995) White as a reporter gene to detect transcriptional silencers specifying position-specific gene expression during *Drosophila melanogaster* eye development. Genetics 141:1075–1086

Tang CY, Sun YH (2002) Use of mini-white as a reporter gene to screen for GAL4 insertions with spatially restricted expression pattern in the developing eye in drosophila. Genesis 34:39–45

Tare M, Puli OR, Moran MT, Kango-Singh M, Singh A (2013) Domain specific genetic mosaic system in the *Drosophila* eye. Genesis 51:68–74

Theisen H, Haerry TE, O'Connor MB, Marsh JL (1996) Developmental territories created by mutual antagonism between Wingless and Decapentaplegic. Development 122:3939–3948

Treisman JE, Rubin GM (1995) wingless inhibits morphogenetic furrow movement in the *Drosophila* eye disc. Development 121:3519–3527

Tsachaki M, Sprecher SG (2012) Genetic and developmental mechanisms underlying the formation of the *Drosophila* compound eye. Dev Dyn 241:40–56

Tsai YC, Sun YH (2004) Long-range effect of upd, a ligand for Jak/STAT pathway, on cell cycle in *Drosophila* eye development. Genesis 39:141–153

Tsai YC, Yao JG, Chen PH, Posakony JW, Barolo S, Kim J, Sun YH (2007) Upd/Jak/STAT signaling represses wg transcription to allow initiation of morphogenetic furrow in *Drosophila* eye development. Dev Biol 306:760–771

Urbach R, Technau GM (2003) Segment polarity and DV patterning gene expression reveals segmental organization of the *Drosophila* brain. Development 130:3607–3620

Wang YH, Huang ML (2009) Reduction of Lobe leads to TORC1 hypoactivation that induces ectopic Jak/STAT signaling to impair *Drosophila* eye development. Mech Dev 126:781–790

Wawersik S, Maas RL (2000) Vertebrate eye development as modeled in *Drosophila*. Hum Mol Genet 9:917–925

Wehrli M, Tomlinson A (1998) Independent regulation of anterior/posterior and equatorial/polar polarity in the *Drosophila* eye; evidence for the involvement of Wnt signaling in the equatorial/polar axis. Development 125:1421–1432

Wilkins AS (1993) Genetic analysis of animal development, 2nd edn. Wiley-Liss., New York

Willecke M, Hamaratoglu F, Sansores-Garcia L, Tao C, Halder G (2008) Boundaries of Dachsous Cadherin activity modulate the Hippo signaling pathway to induce cell proliferation. Proc Natl Acad Sci U S A 105:14897–14902

Wilson C, Pearson RK, Bellen HJ, O'Kane CJ, Grossniklaus U, Gehring WJ (1989) P-element-mediated enhancer detection: an efficient method for isolating and characterizing developmentally regulated genes in *Drosophila*. Genes Dev 3:1301–1313

Wolff T, Ready DF (1993) Pattern formation in the *Drosophila* retina. In: Bate M, Arias AM (eds) The development of *Drosophila melanogaster*. Cold Spring Harbor Laboratory Press, New York, pp 1277–1325

Xue Y, Gao X, Lindsell CE, Norton CR, Chang B, Hicks C, Gendron-Maguire M, Rand EB, Weinmaster G, Gridley T (1999) Embryonic lethality and vascular defects in mice lacking the Notch ligand Jagged1. Hum Mol Genet 8:723–730

Yamamoto D (1996). Architecture of the adult compound eye and the developing eye disks. In: Yamamoto D (ed) Molecular dynamics in the developing drosophila eye. Chapman and Hall, Austin, p 169

Yang CH, Simon MA, McNeill H (1999) mirror controls planar polarity and equator formation through repression of fringe expression and through control of cell affinities. Development 126:5857–5866

Younossi-Hartenstein A, Hartenstein V (1993) The role of the tracheae and musculature during pathfinding of *Drosophila* embryonic sensory axons. Dev Biol 158:430–447

Zecca M, Basler K, Struhl G (1995) Sequential organizing activities of engrailed, hedgehog and decapentaplegic in the *Drosophila* wing. Development 121:2265–2278

Zecca M, Basler K, Struhl G (1996) Direct and long-range action of a wingless morphogen gradient. Cell 87:833–844

Zeidler MP, Perrimon N, Strutt DI (1999a) The four-jointed gene is required in the *Drosophila* eye for ommatidial polarity specification. Curr Biol 9:1363–1372

Zeidler MP, Perrimon N, Strutt DI (1999b) Polarity determination in the *Drosophila* eye: a novel role for unpaired and JAK/STAT signaling. Genes Dev 13:1342–1353

Zirin JD, Mann RS (2007) Nubbin and Teashirt mark barriers to clonal growth along the proximal-distal axis of the *Drosophila* wing. Dev Biol 304:745–758

Catching the Next Wave: Patterning of the *Drosophila* Eye by the Morphogenetic Furrow

Justin P. Kumar

Introduction

In 1864, August Weismann published the first drawing of the insect eye–antennal disc complex. In this image he drew a line within the eye and described it as the border between the eye and antennal portions of the disc complex. One hundred and twelve years later, Donald Ready, Thomas Hanson, and Seymour Benzer demonstrated that this line, which they called the morphogenetic furrow, is actually the leading edge of a differentiating wave that traverses the eye disc of the fruit fly, *Drosophila melanogaster*, and transforms a field of undifferentiated and nonpatterned cells into an ordered array of periodically spaced ommatidia or unit eyes. In the 36 years since this seminal discovery, dozens of papers have focused on elucidating the molecular mechanisms that underlie the initiation and progression of the furrow as well as the many cellular changes that cells undergo as they enter, temporarily reside, and then exit the furrow. This review will summarize what is currently known about the cellular architecture of the furrow and the mechanisms that control its birth and propagation across the eye primordium. This chapter will also discuss the means by which the initiation of the furrow is restricted to a single point along the posterior margin.

The Adult Eye: A Product of Pattern Formation

The compound eye of *Drosophila* is a simple nervous system of such extraordinary precision that it has been described as a "neurocrystalline lattice" (Fig. 1; Ready et al. 1976). The adult retina consists of approximately 800 unit eyes or ommatidia that are organized into nearly three dozen vertical columns. Each unit eye is constructed as a hexagon, thus adjoining columns of ommatidia appear to be interlocked with each other. The number of unit eyes per column is variable with columns in the center of the eye containing the largest number of ommatidia while those that lie at increasing

J. P. Kumar (✉)
Department of Biology, Indiana University, Bloomington, IN 47405, USA
e-mail: jkumar@indiana.edu

A. Singh, M. Kango-Singh (eds.), *Molecular Genetics of Axial Patterning, Growth and Disease in the Drosophila Eye,* DOI 10.1007/978-1-4614-8232-1_3,

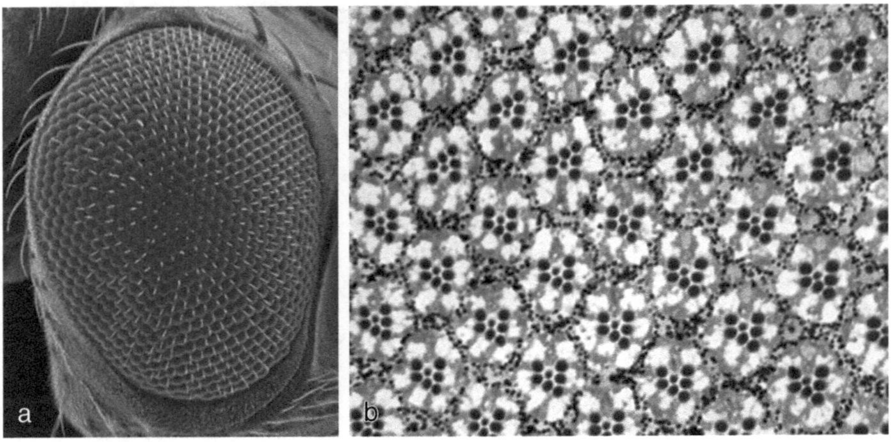

Fig. 1 Structure of the adult compound eye of *Drosophila*: external view and retinal section. **a** Scanning electron micrograph of the adult compound eye. Note the approximately three dozen columns of unit eyes or ommatidia. **b** Light microscope section of the adult retina. The photoreceptors are organized into an asymmetrical trapezoid. A line of mirror symmetry (the equator) divides the eye into dorsal and ventral compartments. The ommatidia on either side of the equator exist in two chiral forms. Anterior is to the *right* in all images

distances from the center have fewer and fewer unit eyes. This arrangement gives the compound eye an overall egg or oval shape. During larval development, the first column, which is set down at the posterior margin of the retina, serves as a template upon which the next ommatidial column is added. Subsequent columns are similarly added, with each preceding column serving as a template for the next, until the approximately three dozen columns of unit eyes are set within the eye primordium. It is the responsibility of the morphogenetic furrow to inlay each column of ommatidia onto the epithelium (Fig. 2b; Ready et al. 1976; Lebovitz and Ready 1986; Wolff and Ready 1991).

Each unit eye consists of eight photoreceptor neurons, four lens secreting cone cells, and a cadre of optically insulating pigment cells (Fig. 3c; Dietrich 1909; Waddington and Perry 1960; Ready et al. 1976; Tomlinson and Ready 1987; Cagan and Ready 1989a). The photoreceptors lie at the core of the ommatidium and are arranged as an asymmetric trapezoid. Ommatidia exist in two chiral forms with the equator serving as the transition point between the two forms. Ommatidia in the dorsal half of the retina point "north" while those in the ventral half point "south" (Fig. 3a). The outer photoreceptors, R1–6, occupy the entire length of the ommatidium while the inner neurons R7/8 reside with the distal and proximal sections, respectively (Fig. 3b). The cone cells lie atop the photoreceptor cluster and secrete the overlying lens while the pigment cells surround and optically insulate the photoreceptors. Their physical arrangement gives the ommatidium its hexagonal shape. The furrow not only organizes the eye into columns of unit eyes but it also contributes to the earliest step in ommatidial assembly—the specification and recruitment of the R8 photoreceptor (Figs. 8 and 11; see further discussion). The specification of the R8 then begins the recruitment of the remaining photoreceptors, cone cells, and pigment cells (Fig. 8; reviewed in Kumar 2012).

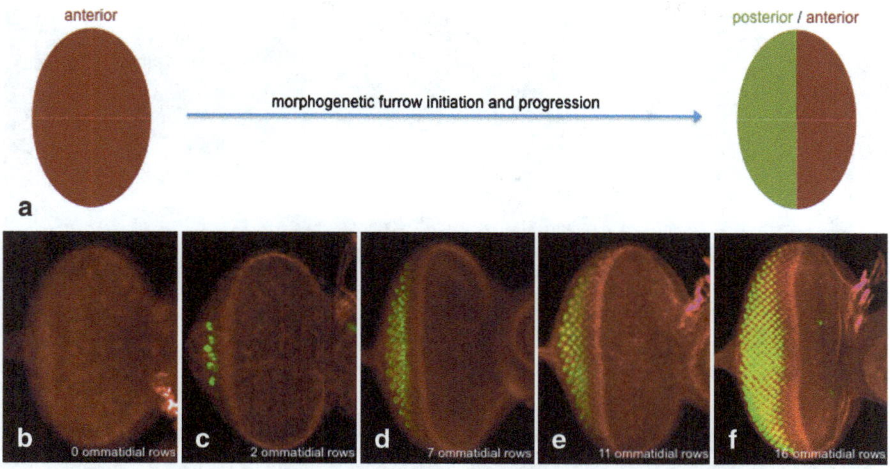

Fig. 2 Progression of the furrow transforms a field of undifferentiated cells into an organized array of unit eyes. **a** Schematic depiction of the anterior compartment boundaries in the developing eye. During early development, the entire eye is made up of anterior fated tissue (*red*). After the furrow initiates and progresses across the eye field, the eye is slowly converted into all posterior tissue (*green*). The schematic shows an eye in which the furrow has migrated across half of the epithelium. **b–f** Confocal images of third instar eye discs in which the furrow has progressed to various points. The number of ommatidial rows is shown in each panel. *Red* = F-actin and *green* = Elav. Anterior is to the *right* in all panels

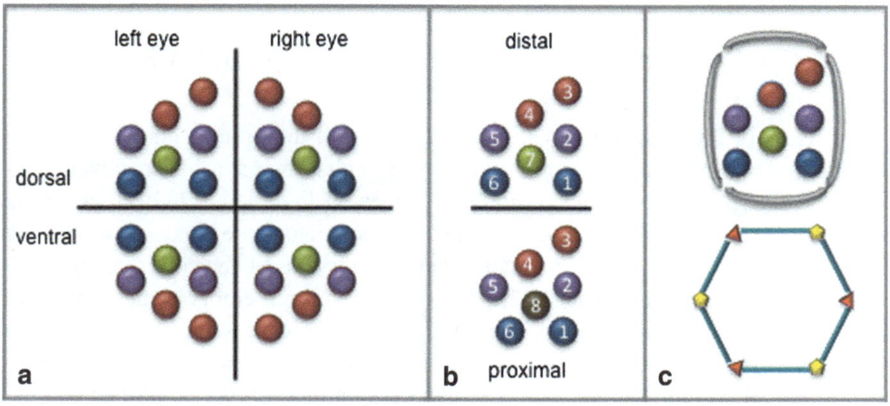

Fig. 3 Schematic of cell types and orientation within the adult retina. **a** Schematic depicting the different chiral forms that are found in the four quadrants of the two adult compound eyes. **b** Schematic depicting the photoreceptors that are found in the distal and proximal layers of the retina. The identity of each photoreceptor is listed within the figure. **c** Schematic depicting the different cell types that are present within each ommatidium. In the *top* portion of the panel, the *circles* and *gray brackets* represent the photoreceptor neurons and cone cells, respectively. In the *lower* portion of the panel, the *blue bars* are the secondary pigment cells, the *red triangles* are the tertiary pigment cells, and the *yellow hexagons* are the mechanosensory bristles. Anterior is to the *right* in all panels

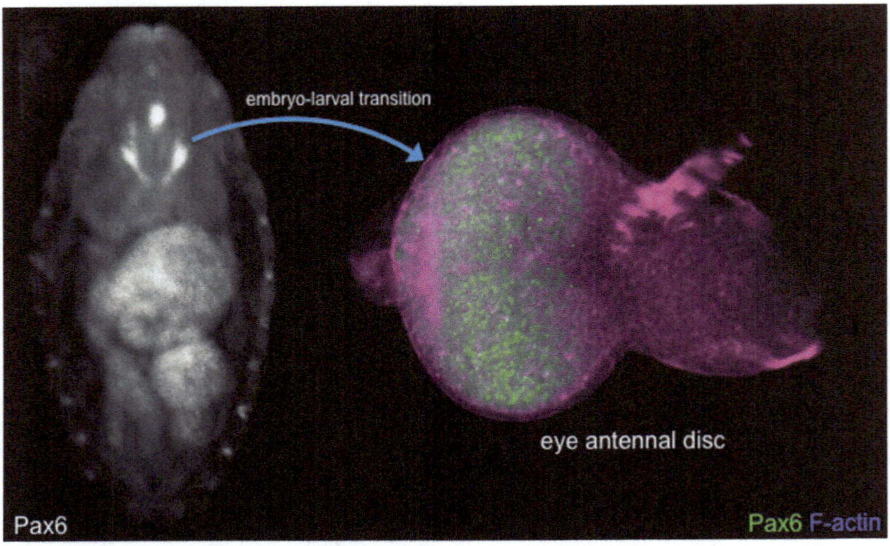

Fig. 4 Birth of the eye–antennal disc. (*Right* portion) Confocal image of a stage 16 embryo in which the eye–antennal disc is visualized by the presence of the Ey/Pax6 protein. Note that Ey is distributed throughout the entire eye disc. Anterior is to the *top* in this panel. (*Left* portion) Confocal image of a mid/late second instar eye–antennal disc. Note that by this point, Ey protein is segregated to just the eye portion of the epithelium. Visualized molecules are listed within the images. Anterior is to the *right* in this image

Early Eye Development: A Prologue to Furrow Initiation

The origins of the adult retina can be traced back to the optic primordium of the embryonic blastoderm when approximately 20 cells are set aside to develop independently from the rest of the embryo (reviewed in Cohen 1993; Held 2002). While these cells are morphologically indistinct and fail to express any tissue-specific genes at these early times, their existence and ancestry have been confirmed through fate-mapping experiments (Struhl 1981). Midway through embryogenesis, the eye anlage is fused to the antennal primordium to form a single monolayer epithelium called the eye–antennal disc (reviewed in Jurgens and Hartenstein 1993). Shortly after this fusion, the eye–antennal disc can be identified by the combined expression of several markers (Fig. 4) including *escargot* (*esg*), which marks all imaginal discs, and several Pax6 genes such as *eyeless* (*ey*; Quiring et al. 1994), *twin of eyeless* (*toy*; Czerny et al. 1999), *eyegone* (*eyg*; Jones et al. 1998; Jun et al. 1998), and *twin of eyegone* (*toe*; Yao et al. 2008). By the first larval instar stage, the eye portion of the disc is attached to the brain via the Bolwig's nerve while the antennal segments are directly attached to the dorsal pouch (reviewed in Jurgens and Hartenstein 1993).

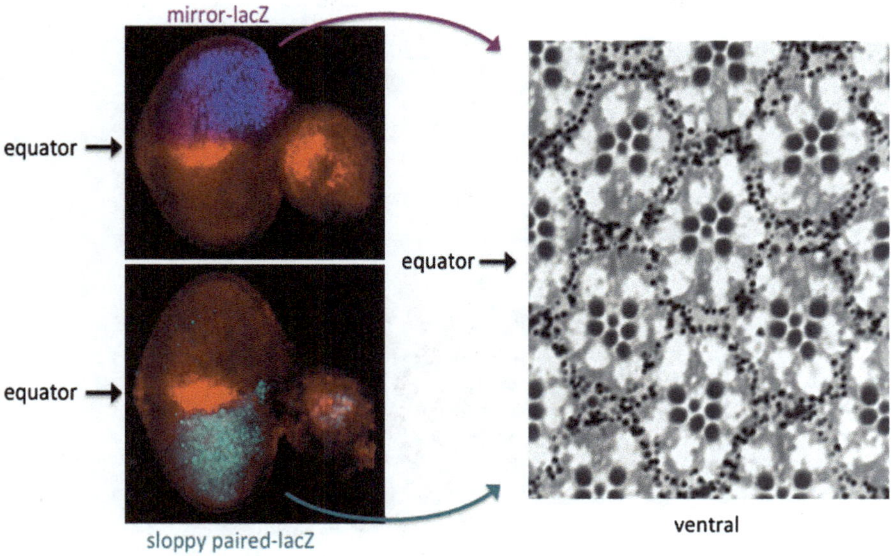

Fig. 5 The dorsal and ventral compartments are set prior to furrow initiation. Confocal images of mid/late second instar larval eye–antennal discs depicting expression of the mirror-lacZ (*purple*) and sloppy paired-lacZ (*aqua blue*) transcriptional reporters. The midline is marked by the expression of an emc-GFP (*red*) transcriptional reporter. The division of the eye into dorsal and ventral compartments early in development results in the adult eye containing different chiral forms of ommatidia. In the *right* portion of the figure a light microscope section of the adult retina is shown. In this image the different chiral forms are seen in the dorsal and ventral halves of the eye. Anterior is to the *right* in all images

During the first and second larval instars, the eye field is primarily concerned with growth (reviewed in Kumar 2011). At the end of the second instar, the eye disc contains approximately 2,000 cells, a 100-fold increase in size from its embryonic origins (Becker 1957). Since pattern formation via the morphogenetic furrow has thus far not initiated, all cells within the eye disc are still undifferentiated and not patterned at this point in development. Yet, some basic features of tissue organization can be discerned. For instance, the field itself has already taken on its characteristic oval shape and can be easily distinguished from all other imaginal discs. Additionally, by this stage the eye primordium, which was born with only ventral identity, is subdivided into dorsal and ventral compartments (Fig. 5; Cho and Choi 1998; Dominguez and de Celis 1998; Papayannopoulos et al. 1998; Singh and Choi 2003). And finally, while cells within the disc have not taken on specific cellular identities the tissue itself is committed to adopting an eye fate as members of the retinal determination network, which are first expressed broadly throughout the entire eye–antennal disc are segregated to just the eye field (Fig. 4; Kumar and Moses 2001a; Kenyon et al. 2003; Kumar 2010).

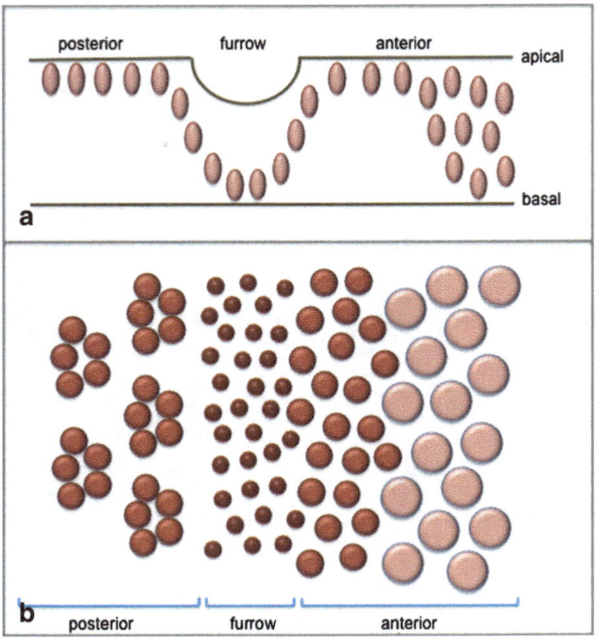

Fig. 6 Nuclear migration and apical constriction in the furrow. **a** A schematic depiction of the eye disc shown in cross section. Ahead of the furrow, nuclei are randomly distributed. As cells approach the furrow, their nuclei migrate to the apical surface. As cells enter the furrow, the nuclei plunge basally but ascend again as they exit the furrow and begin differentiation. **b** Schematic of the eye disc showing the apical profiles of cells ahead within and behind the furrow. As cells enter the furrow, their apical profiles are constricted. Behind the furrow, cells are organized into periodically spaced clusters—their apical profiles expand. Anterior is to the *right* in all images

Cellular Architecture of the Furrow

Apical Constrictions and Tissue Ingression

At the transition to the third instar larval stage, overt patterning of the retina begins when the furrow initiates at the posterior margin of the eye primordium and proceeds toward the anterior edge where the eye and antennal fields meet (Fig. 2). As the name implies, the furrow is an actual physical groove in the epithelium. When the developing retina is viewed in cross section, cells within the furrow have a bottle-like appearance and undergo a slight ingression (Fig. 6; Ready et al. 1976). Cell shape changes and invaginations of this kind are seen in a variety of tissues, and it is thought that both are caused by the constriction of apical cell surfaces (Kimberly and Hardin 1998). Similar cellular events appear to be in play within the eye disc as cells that are approaching and entering the furrow undergo dramatic constriction of their apical profiles (Fig. 6b; Ready et al. 1976). Mechanistically, Hedgehog (Hh) signaling from developing photoreceptors induces apical surface constriction by triggering a reorganization of apically positioned microtubules and actin filaments (Heberlein et al. 1993; Benlali et al. 2000; Corrigall et al. 2007). This event is essential for the ordered development of the eye as mutations that interfere with apical cell constriction cause precocious neuronal development (Benlali et al. 2000).

How do changes in cell shape affect cell fate decisions? The answer to this question comes from an analysis of the nature of the Hh signaling itself. The Hh ligand is a signaling molecule that can function at both short and long ranges (Lee et al. 1992; Basler and Struhl 1994; Heemskerk and DiNardo 1994; Tabata and Kornberg 1994). In the retina, it is expressed in and secreted from developing photoreceptor cells and influences patterning and cell shape changes over only a short range: just in a small stripe of approximately 10-cell diameters within and just ahead of the furrow (Heberlein et al. 1993; Ma et al. 1993). Restricting its range for patterning is conveniently accomplished by the constriction of apical profiles in cells just ahead and within the furrow. It is thought that this constriction results in a concentration of apical membrane, the site of the Patched (Ptc) receptor. This allows for the efficient capture of Hh molecules and prevents its signal from traveling too far ahead of the furrow. Thus, cells lying just ahead and within the furrow receive the Hh signal, constrict their apical profiles, increase the capture of the ligand and thereby prevent the signal from travelling further. This model is supported by the observation that the loss of *act up* (*acu*), which encodes the fly homolog of cyclase-associated protein (CAP), results in the retention of large apical profiles and an accumulation of the active form of Cubitus Interruptus (Ci^{ACT}), the zinc finger transcription factor that transduces the Hh signal, in cells that lie in more anterior regions of the disc than found in normal retinas (Benlali et al. 2000).

Nuclear Migration

The cell shape changes that are seen within the furrow are also influenced by the position of the nucleus. In anterior regions of the disc, nuclei are randomly positioned within the apical–basal plane of the epithelium. Then, much like cars on a roller coaster, nuclei will first rise to the apical surface, rapidly plunge to the basement of the epithelium and then ascend again. These nuclear movements are choreographed with the approach, entrance and exit of cells from the furrow (Fig. 6a; Ready et al. 1976; Tomlinson 1985). As in other developmental contexts, nuclear migration in the eye is dependent upon microtubules and the activity of the cytoplasmic motor protein Dynein (Fan and Ready 1997; Swan et al. 1999; Patterson et al. 2004; Houalla et al. 2005). The current model is that the KASH domain-containing protein Klarsicht interacts with both nuclear Lamin and the microtubule-organizing center (MTOC) to physically link nuclei to microtubules (Patterson et al. 2004). Dynein is then thought to aid in the movement of the nucleus during its migration. It is not clear if the other major microtubule motor protein, Kinesin, is also involved in nuclear migrations within the eye.

The basal migration of nuclei within the furrow contributes to the bottle-like appearance of cells within this zone while the subsequent rise of nuclei as cells exit the furrow is coordinated with the adoption of individual cellular fates. While pattern formation appears to be coordinated with changes in nuclear position, is the former dependent upon the latter? The answer to this question is mixed. Cells that comprise the precluster (R8, R2/5, R3/4) appear to differentiate normally despite disruptions in nuclear migration and positioning. The only visible defects within these neurons are malformations of the rhabdomere, the light capturing organelle of the insect photoreceptor (Fischer-Vize and Mosley 1994; Mosley-Bishop et al. 1999). However, when nuclear positioning is disrupted within the second mitotic wave, differentiation of R1/6 and R7 fails to occur correctly (Fan and Ready 1997). Neither it is clear why such a difference exists between the two classes of photoreceptor neurons nor it is completely settled that the loss of the final three photoreceptors is due to nuclear positioning and not another, yet to be described, defect in microtubule dynamics.

Cell Cycle Synchronization

Another feature is the cell cycle synchronization of cells within and just ahead of the furrow. In the most anterior regions of the eye disc, cells proliferate asynchronously and express markers for all four phases (G1, S, G2, and M) of the cell cycle (Fig. 7; Ready et al. 1976; Thomas et al. 1994; Richardson et al. 1995). However, just ahead of the furrow the cell cycle profiles begin to synchronize. Approximately 10-cell diameters anterior to the furrow, cells cease to express *cyclin E* (*cycE*) and fail to incorporate the thymidine analog 5-bromo-2-deoxyuridine (BrdU) thereby suggesting that all cells within this zone have progressed through and past S phase (Thomas et al. 1994; 1997). This is closely followed by the termination of *cyclin A* (*cycA*) and *cyclin B* (*cycB*) transcription, which signals the transition through G2. And finally, levels of *string* (*stg*), the fly homolog of yeast cdc25, are elevated in cells that are just about to enter the furrow (Edgar and O'Farrell 1989; Thomas et al. 1994). This last step signals cells to undergo a final mitosis before entering the morphogenetic furrow (Penton et al. 1997; Horsfield et al. 1998). Within the furrow, all cells are arrested in G1 and do not express appreciable levels of the above discussed cyclins (Fig. 7).

Synchronization at G1 within the furrow is an important step as key decisions regarding cell cycle reentry and exit are made shortly after cells leave the furrow. A subset of cells will exit the cell cycle and form periodically spaced clusters that contain five photoreceptors (Fig. 8; R8, R2/5, and R3/4). Any cell that does not exit and differentiate will reenter the cell cycle, undergo one final round of division, and then give rise to the final three photoreceptors (R1/6 and R7), the lens secreting cone cells, and the optically insulating pigment cells (Fig. 8). As this is happening across the entire dorsal–ventral axis of the eye disc, it is important for all cells to have been arrested in G1 within the furrow so that decisions regarding cell cycle exit and differentiation as well as cell cycle reentry can be synchronized along the length of the furrow.

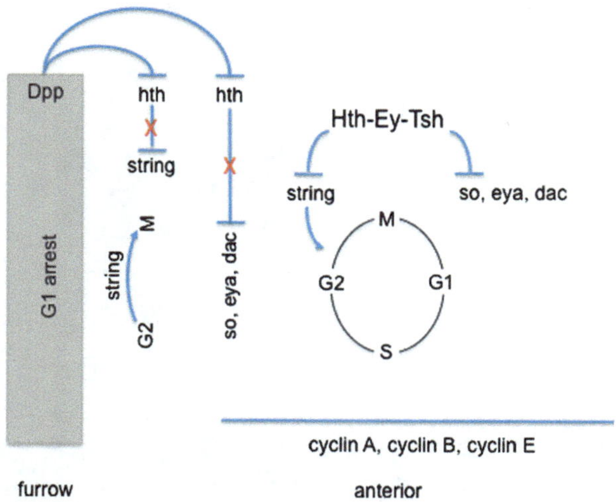

Fig. 7 Regulation of the cell cycle ahead and within the furrow. Schematic depicting the role that decapentaplegic (Dpp) signaling plays in arresting cells in the G1 phase of the cell cycle. Far ahead of the furrow, cells divide asynchronously. As cells approach the furrow, they simultaneously express *string* while ceasing to express *cyclin A*, *B*, and *E* and are thus arrested in G1 within the furrow. Dpp signaling counteracts the activity of homothorax (Hth), which normally represses string transcription. Anterior is to the *right*

Given that pattern formation and cell cycle synchronization are coordinated within the furrow, a prime candidate for regulating G1 arrest is *decapentaplegic* (*dpp*), which encodes a Transforming Growth Factor-β (TGF-β) family member (Padgett et al. 1987). The TGF-β superfamily has been implicated in the regulation of the cell cycle in a number of tissues and organisms (Massague and Polyak 1995). Prior to the initiation of pattern formation, *dpp* is expressed along the posterior margins of the eye disc and is essential for proper initiation of the furrow (Blackman et al. 1991; Heberlein et al. 1993; Chanut and Heberlein 1997a, b). In later stages, *dpp* expression is found exclusively within the furrow and is primarily tasked with repressing the expression of *wingless* (*wg*), a negative regulator of the furrow (Burke and Basler 1996; Dominguez and Hafen 1997; Greenwood and Struhl 1999; Curtiss and Mlodzik 2000). Mutations in the *thickvein* (*tkv*) or *saxophone* (*sax*) genes, which both encode cell surface receptors, render cells unresponsive to Dpp signaling (Massague 1996). Cells within clones that span the furrow and lack either of these receptors maintain high CycA, CycB, and CycE protein levels and also show inappropriate entrance into S phase (Penton et al. 1997; Horsfield et al. 1998). In contrast, ectopic expression of *dpp* in cells ahead of the furrow leads to a transient and reversible reduction in the number of cells entering S phase (Horsfield et al. 1998). These results implicate Dpp in the G1 arrest of cells within the furrow. Paradoxically, *dpp* is also required for growth of the early eye disc, a task that would seemingly be at odds with its role in preventing cells from making the G1/S transition. This apparent contradiction was resolved by the demonstration that Dpp forms a gradient in the eye and that a threshold level of Dpp protein is required to induce cell cycle arrest with the furrow (Firth et al. 2010).

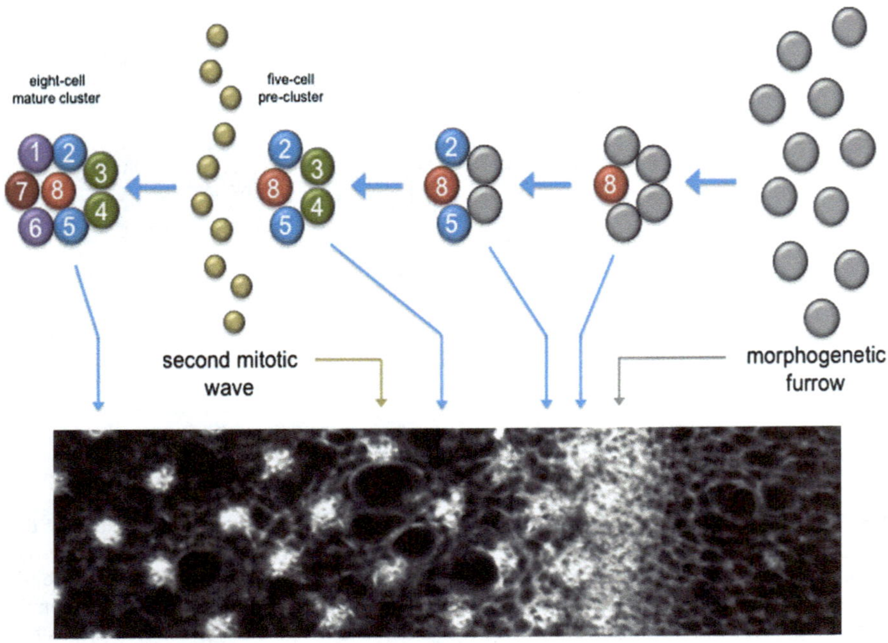

Fig. 8 The morphogenetic furrow, the second mitotic wave, and ommatidial assembly. As cells exit the furrow, a subset will exit the cell cycle and will adopt the fates of the first five photoreceptor clusters. All remaining cells will undergo a single round of mitosis and then adopt the fates of the last three photoreceptors as well as the cone and pigment cells. The schematic drawing marks the position of various events with the confocal image of the developing eye disc. Anterior is to the *right* in all images

Synchronization of the cell cycle within the furrow also requires *homothorax* (*hth*), which encodes a homeodomain-containing transcription factor. Hth, a member of the retinal determination network, is expressed in the most anterior regions of the eye disc and is part of a biochemical complex that also contains the zinc finger transcription factor Teashirt (Tsh) and Ey (Bessa et al. 2002). The Ey-Tsh-Hth complex is required to repress the transcription of several other retinal determination genes such as *sine oculis* (*so*), *eyes absent* (*eya*), and *dachshund* (*dac*) thereby allowing cells in this zone to rapidly proliferate (Fig. 7). As cells begin to synchronize their cell cycles, *hth* expression is eliminated and *stg* expression is elevated. These mutually exclusive expression patterns hinted at potential regulation of *stg* by *hth*. Indeed, ectopic expression of *hth* within the eye leads to repression of *stg* transcription, maintenance of CycB levels, and a release from G1 arrest (Lopes and Casares 2009; Peng et al. 2009). These results suggest that, in order for cells to properly complete their last mitosis and then arrest in G1 within the furrow, *hth* expression must be repressed within the *stg* expression domain. How is the repression of *hth* transcription ahead of the furrow achieved? A prime candidate is Dpp signaling since the phenotypes associated with the loss of either *tkv* or *sax* mimic those that result from ectopic *hth* expression. A direct support to this model comes from the observation that removal of either receptor (and thus a reduction in Dpp signaling) results in the maintenance of *hth* expression in cells within the furrow. Thus, the current model is that long-range signaling by Dpp represses *hth* which in turn leads to the activation of *so*, *eya*, *dac*, and *stg* as well as the termination of *cycE*, *cycA*, and *cycB* (Fig. 7). These combined effects lead to cell cycle synchronization within the furrow.

Furrow Initiation and Progression

Birth of the Furrow

The eye disc is unusual in that, unlike all other imaginal discs, it is born without an established anterior–posterior (A/P) compartment boundary. Instead, the early eye disc contains just one of the two compartments; it is solely comprised of anterior tissue. Surprisingly, by the time that patterning of the eye is finished the entire field has undergone a complete change in compartment identity. What used to be the anterior compartment is now the posterior compartment. This transition is mediated by the passage of the furrow across the eye disc and thus it represents a mobile compartment boundary (Fig. 2a). While differences in the use of mobile and stationary A/P boundaries exist, all imaginal discs use a common molecular mechanism to signal across and maintain compartment identities. Patterning of the embryo as well as the imaginal discs is dependent upon the activities of the *hh*, *dpp*, and *wg* genes. All encode secreted proteins with varying signaling ranges (Lee et al. 1992; Tabata and Kornberg 1994; Panganiban et al. 1990a, b; van den Heuvel et al. 1989; Gonzalez et al. 1991; Peiffer and Vincent 1999). *hh* is expressed in the posterior compartment and activates both *dpp* and *wg* expression in adjacent cells along the A/P compartment boundary (Basler and Struhl 1994; Capdevila et al. 1994). All three signaling pathways are present in the eye and play roles in both furrow initiation and progression (Fig. 9).

During the third and final instar, the morphogenetic furrow initiates at the posterior margin of the eye disc and begins its long journey across the eye primordium (Ready et al. 1976). Its initiation is restricted to a single point: the intersection of the posterior margin and the midline, which is called the posterior center (Tsai et al. 2007). Although the early eye lacks a posterior compartment and an A/P boundary, both *hh* and *dpp* are expressed along the posterior margin of the eye field prior to the initiation of the furrow. Just prior to furrow initiation, *hh* expression overlaps with the posterior center. *dpp*, on the other hand, while present along most of the posterior-lateral margins, is distinctly absent from the posterior center (Fig. 9; Masucci et al. 1990; Blackman et al. 1991; Dominguez and Hafen 1997; Borod and Heberlein 1998). Loss of either gene inhibits initiation of the endogenous furrow while ectopic expression induces formation of ectopic furrows and neuronal differentiation within anterior quadrants of the eye field (Heberlein et al. 1995; Ma and Moses 1995; Pan and Rubin 1995; Strutt et al. 1995; Wehrli and Tomlinson 1995; Wiersdorff et al. 1996; Dominguez and Hafen 1997; Pignoni and Zipursky 1997; Borod and

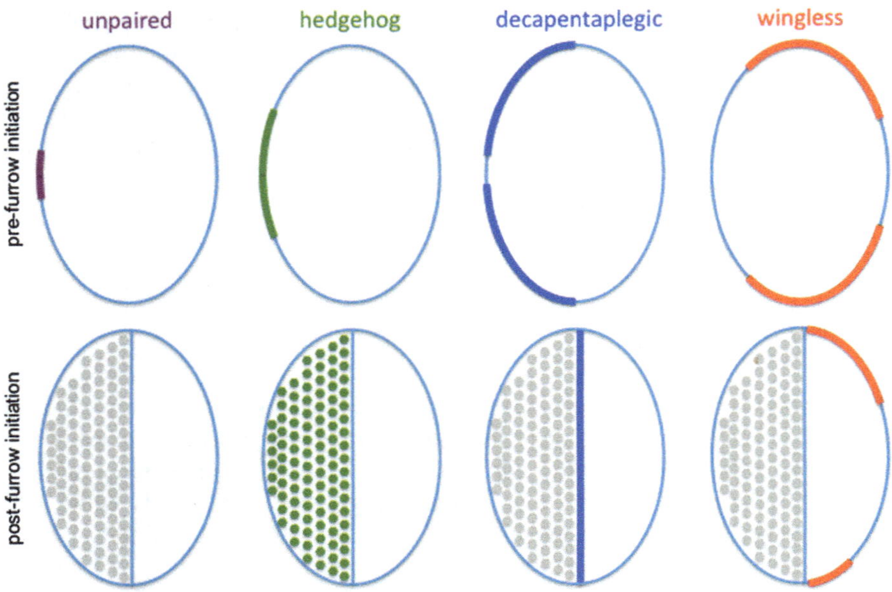

Fig. 9 Expression patterns of signaling molecules that regulate furrow and progression. The JAK/STAT, Hh, Dpp, and Wg signaling pathways are critical for regulating furrow initiation and progression. These schematics depict the expression patterns of the ligands for these pathways in early pre-furrow discs (*upper row*) and late post-furrow discs (*bottom row*). Anterior is to the *right* in all images

Heberlein 1998). Despite the apparent requirement for both genes, Hh signaling alone is necessary and sufficient to initiate the furrow. This conclusion is based on the ability of Hh to induce neuronal differentiation in clones lacking *dpp* (Dominguez and Hafen 1997). What, if any, role does *dpp* then play in furrow initiation and why is it expressed at the posterior margin in the early eye? Interestingly, it may play a novel role in the maintenance of *hh* expression. Unlike the developing embryo and both wing and leg imaginal discs where Hh in the posterior compartment signals forward and activates *dpp* expression, in the eye there is a feedback loop between *hh* and *dpp*. Ectopic expression of *dpp* activates *hh* transcription, which is in turn required for *dpp*-induced neuronal differentiation. Thus, the ability of *dpp* to initiate ectopic furrows and neural development is actually due to the initiation of *hh* transcription (Fig. 10; Pignoni and Zipursky 1997; Borod and Heberlein 1998).

The restriction of furrow initiation to the posterior center ensures that the retina is correctly patterned. Situations in which additional furrows are initiated at the anterior and/or lateral margins leave the eye disorganized and significantly smaller in size (Ma and Moses 1995; Treisman and Rubin 1995; Pignoni and Zipursky 1997). The Wg signaling pathway is tasked with limiting furrow initiation to the posterior center. *wg* is transcribed along the lateral margins of the disc and its loss results in the initiation of ectopic furrows (Figs. 9 and 10; Ma and Moses 1995). Wg signaling appears to be sufficient to block the furrow as ectopic expression of *wg* within the eye field inhibits progression of the endogenous furrow (Treisman and Rubin 1995). In

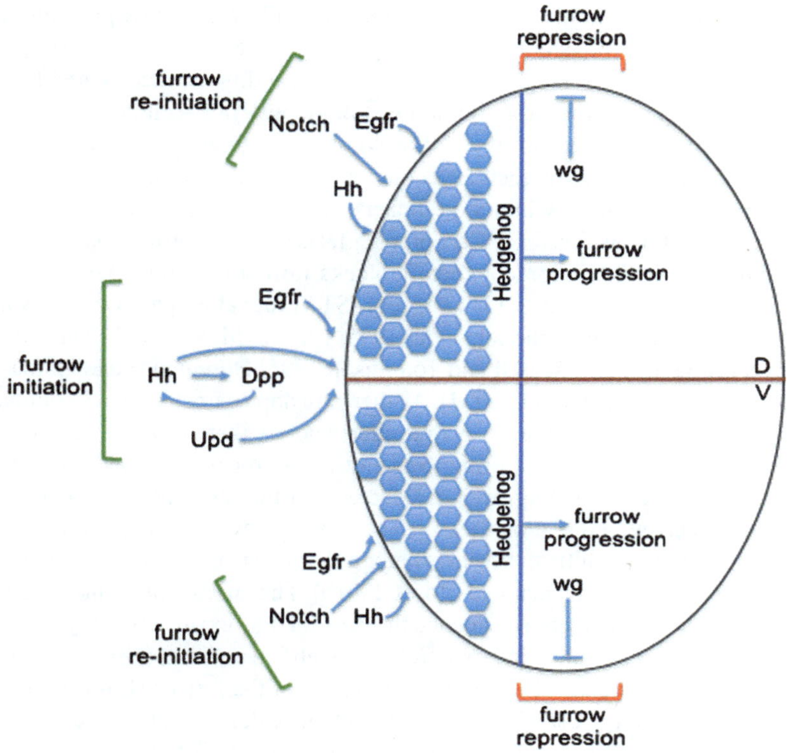

Fig. 10 Signaling pathways involved in birth, reinitiation, and progression. Schematic summarizing the position that the Hh, Dpp, Wg, Egfr, Notch, and JAK/STAT pathways occupy during the birth, reinitiation, and progression of the furrow across the eye disc. Note that Wg signaling is used to repress ectopic furrow initiation while all other pathways play roles in promoting the movement of the furrow. Anterior is to the *right*

addition to its role in furrow initiation, *wg* is also essential for delimiting the border between the compound eye and the surrounding head capsule (Royet and Finkelstein 1996; 1997).

In order for the furrow to initiate, *wg* expression must be repressed at the posterior center. A candidate for repressing *wg* transcription is the *unpaired* (*upd*) gene, which encodes a ligand for the JAK/STAT signaling pathway. In the developing eye, JAK/STAT signaling plays a major role in promoting cell proliferation (Bach et al. 2003; Chao et al. 2004; Tsai and Sun 2004; Ekas et al. 2006; Gutierrez-Avino et al. 2009). Prior to the initiation of the furrow, it is expressed exclusively at the posterior center (Fig. 9; Sun et al. 1995; Pignoni and Zipursky 1997; Tsai and Sun 2004). Reductions in *upd* expression lead to derepression of *wg* along the posterior margin and a block in furrow initiation. Conversely, overexpression of *upd* along the lateral margins downregulates *wg* transcription and induces ectopic furrow initiation (Ekas et al. 2006; Tsai et al. 2007). Thus, the combined activity of JAK/STAT and Hh signaling leads to the initiation of the furrow at the posterior center (with the Dpp pathway functioning to maintain *hh* expression), while Wg activity prevents ectopic furrow initiation at the lateral margins (Fig. 10). Interestingly, as *wg* expression is relegated to the anterior-lateral margins of the disc, additional factors are likely used to prevent ectopic furrows from initiating at more posterior sections of the margins.

The initial birth of the furrow is followed by its continuous reinitiation along the posterior-lateral margins as each new column of ommatidia is added to the growing eye. An analysis of furrow initiation indicates that a critical control point precedes the birth of the furrow. The EGF Receptor (Egfr) is required during this developmental window, as its inhibition completely blocks furrow initiation (Fig. 10; Kumar and Moses 2001b). It joins the Hh and JAK/STAT signaling pathways as being required for furrow birth (Heberlein et al. 1995; Ma and Moses 1995; Pan and Rubin 1995; Strutt et al. 1995; Wehrli and Tomlinson 1995; Dominguez and Hafen 1997; Ekas et al. 2006; Tsai and Sun 2004). As there are approximately 32–34 ommatidial columns in a typical eye, the furrow reinitiates nearly three dozen times during the course of eye development. During the reinitiation process a second control point, also requiring Egfr signaling, was discovered. Further evidence implicated Notch signaling in furrow reinitiation as well. These two pathways, along with Hh signaling, are required to reinitiate the furrow along the posterior-lateral margins (Fig. 10; Wiersdorff et al. 1996; Kumar and Moses 2001b). The number of reinitiation control points is unknown but there is evidence that several may exist. The study that uncovered a role for Egfr signaling in furrow rebirth identified a control point for reinitiation as existing approximately 12 h after the initiation of the furrow (Kumar and Moses 2001b). In a mutant allele of *hh*, one that contains a deletion of an eye-specific enhancer lying within the first intron (hh^{bar3}), the furrow fails to progress beyond the first 8–10 rows of ommatidia (Ives 1950; Mohler 1988; Heberlein et al. 1993; Lee et al. 1992; Ma et al. 1996; Pauli et al. 2005; Rogers et al. 2005). The resulting small eye is certainly due in large part to the reductions in *hh* levels within developing photoreceptor clusters. However, reductions in Hh signaling at the margins and a disruption in a reinitiation control point cannot be ruled out. Additional checkpoints may exist as the furrow stops short in several mutants such as $Drop^1$, $Wedge^1$, and ro^{Dom} (Heberlein et al. 1991; 1993; Tearle et al. 1994; Mozer 2001). As with hh^{bar3} mutants, it has been shown that a block in furrow progression is main underlying cause for the small eye phenotype of these mutants. Still, whether defects in furrow rebirth also contribute to the furrow stop phenotype remains to be determined.

Progression of the Furrow

Once the furrow has initiated and started to progress across the epithelium, the eye contains both anterior and posterior compartments as well as an A/P boundary. Like other tissues, *hh* is transcribed in the posterior compartment, which in the eye lies behind the morphogenetic furrow and is comprised of developing photoreceptor clusters (Fig. 9; Heberlein et al. 1993; Ma et al. 1993). Hh signaling, emanating from the photoreceptor neurons, is required for progression of the furrow as its loss leads to a furrow stop phenotype and a small eye (Fig. 10; Ives 1950; Mohler 1988; Heberlein et al. 1993; Ma et al. 1993). How is its expression in developing photoreceptors regulated? An analysis of the eye-specific enhancer that is deleted in the hh^{bar3} and hh^{fse} mutants is particularly informative. The retinal determination protein Sine Oculis (So) and the Ets transcription factor, which mediates Egfr signaling, both bind to this enhancer and are required for the activation of *hh* transcription (Pauli et al. 2005; Rogers et al. 2005).

Similar regulatory mechanisms that exist between *hh* and *dpp* in other developmental contexts are in place during furrow progression. Hh signaling, originating from the photoreceptors, activates transcription of *dpp* (Masucci et al. 1990; Blackman et al. 1991; Heberlein et al. 1993; Borod and Heberlein 1998; Greenwald and Struhl 1999). However, mutations that interfere with Dpp signaling result in only a small retardation in furrow progression, thus the major role of *dpp* in the furrow is to coordinate the synchronization of the cell cycle of cells anterior to the furrow (Burke and Basler 1996; Wiersdorff et al. 1996; Horsfield et al. 1998; Greenwald and Struhl 1999; Curtiss and Mlodzik 2000; Firth et al. 2010). The atypical relationship that leads to Dpp activation of *hh* at the margin during furrow activation does not appear to exist in the eye field proper during furrow progression.

Over the years, several studies have provided differing accounts of how quickly the furrow traverses the eye disc. One study has documented the furrow laying down a new column of ommatidia approximately every 2 h (Campos-Ortega and Hofbauer 1977). Another report has clocked the furrow building an ommatidial column every 70 min in the posterior half of the eye but then slowing down to 100 min in the anterior half (Basler and Hafen 1989). There are also suggestions that the furrow moves much more dynamically, alternating between periods of accelerations and decelerations (Spratford and Kumar, unpublished data). It will be important to definitively determine which rate is correct in order for an accurate understanding of how pattern formation and cell proliferation are coordinated in the eye. If the former outpaces the latter, as it happens when Wg signaling is blocked at the margins, the resulting eye will be small and disorganized (Ma and Moses 1995; Treisman and Rubin 1995). Thus, it is important for the furrow to travel across the eye disc at a rate that allows for the eye to generate the requisite number of cells needed to generate approximately 800 ommatidia. This rate is likely to be influenced by several factors including the overall developmental timing and the physical dimensions of the eye disc.

Putting a Brake on Furrow Progression

As we have seen, mutations in several signaling pathways can either block the furrow from initiating or from progressing across the eye field. There are also many instances in which the furrow will slow without stopping (Strutt and Mlodzik 1997; Zelhof et al. 1997; Brennan et al. 1998; Dominguez 1999; Greenwood and Struhl 1999). However, there are very few mutations that result in the opposite phenotype, namely the acceleration of the furrow. Two genes that do appear to be involved in slowing the rate of furrow progression are *hairy* (*h*) and *extramacrochaetae* (*emc*). *h* encodes a basic helix–loop–helix (bHLH) DNA-binding protein while *emc* encodes a helix–loop–helix (HLH) transcription factor that regulates transcription, not by binding to DNA, but by interacting with other bHLH proteins and sequestering them away from target enhancer/promoter sequences (Ellis et al. 1990; Garrell and Modolell 1990; Van Doren et al. 1991). Both proteins are enriched within a stripe of cells ahead of the morphogenetic furrow. Individual loss of either gene has no affect on furrow progression (Brown et al. 1991, 1995; Bhattacharya and Baker 2009). But surprisingly, the combined reduction in both proteins leads to an advancement in the furrow through mutant tissue (Brown et al. 1995). This led to the conclusion that both genes were simultaneously required to slow the furrow. However, in this experiment, levels of *emc* were just reduced and not eliminated. In contrast, when *emc* is completely eliminated, the furrow will accelerate without the need for alterations in *h* expression (Bhattacharya and Baker 2009). It appears that the Emc controls the rate at which the furrow progresses by regulating the levels of Ci^{ACT} (Spratford and Kumar, unpublished data).

The Furrow and Ommatidial Assembly

The first cell to be specified within each developing ommatidial cluster is the R8 photoreceptor (Ready et al. 1976; Tomlinson and Ready 1987; reviewed in Frankfort and Mardon 2002). This neuron is considered the founder cell and its initial determination (but not complete differentiation) is required for the correct recruitment and specification of subsequent photoreceptors (Frankfort et al. 2001). The decision as to which cell within each cluster will become the R8 neuron begins deep within the furrow and depends upon *atonal* (*ato*), which encodes a bHLH transcription factor (Fig. 11; Jarman et al. 1994, 1995; Dokucu et al. 1996). Ato protein is found within a stripe of cells at the entire edge of the furrow. Well within the furrow, *ato* expression is retained in periodically spaced clusters of approximately 10–12 cells that are called intermediate groups. Bridges consisting of 3–4 *ato* positive cells connect intermediate groups to each other. As cells exit the furrow, the number of *ato* positive cells per cluster is reduced to 2–3 and is now referred to as the R8 equivalence group. From this cluster, a single cell retains *ato* expression and is selected to adopt the R8 cell fate. The pruning of *ato* expression is dependent upon a number of inputs including Notch signaling and the transcription factors Rough and Senseless (Fig. 11; Cagan and Ready 1989b; Baker et al. 1996; Dokucu et al. 1996; Chanut et al. 2000; Frankfort et al. 2001; Pepple et al. 2008). The R8 will then activate the Egfr signaling pathway in two neighboring cells inducing them to adopt the R2/5 cell fate (Freeman 1994; Tio et al. 1994; Freeman 1996; Tio and Moses 1997; Kumar et al. 1998). Egfr signaling is used reiteratively to recruit the remaining cell types, thus from this point onward ommatidial assembly becomes a self-sustaining process and the furrow no longer plays a role in cell fate specification (Freeman 1996, 1997).

The loss of *ato* expression in the eye disc results in a no-eye phenotype that is characterized by the complete elimination of photoreceptor, cone, and pigment cell development (Jarman et al. 1994). However, despite the lack of photoreceptor

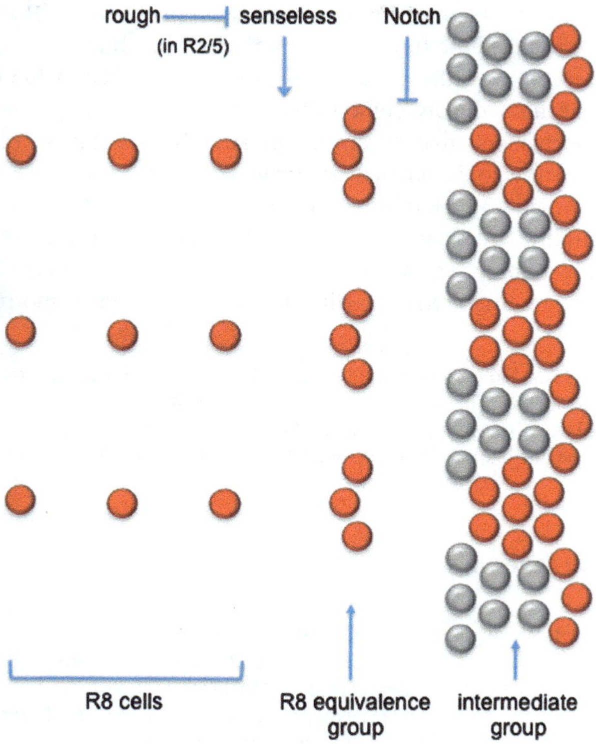

Fig. 11 R8 specification begins in the morphogenetic furrow. The Atonal bHLH transcription factor is required for the specification of the R8 founder cell. Its expression goes through successive waves of refinement until it is found within a single cell in each ommatidium. These pruning steps are mediated by the Notch pathway as well as two transcription factors Rough and Senseless. Loss of *ato* leads to the complete disruption in ommatidial assembly while overexpression of *ato* leads to ommatidia containing multiple R8 cells. Anterior is to the *right*

development, the morphogenetic furrow still initiates and progresses a considerable distance across the eye field (Jarman et al. 1995). It is not entirely clear how this occurs but presumably the levels of *hh* transcription at the margins are sufficient to initiate and propel the furrow. Interestingly, mutations in several retinal determination genes (*ey*, *so*, *eya*, *dac*) are associated with no-eye phenotypes, but in contrast to *ato* mutants the furrow fails to initiate in these instances (Bonini et al. 1993; Cheyette et al. 1994; Quiring et al. 1994; Serikaku and O'Tousa 1994). These differences are likely to be attributed to the fact that at least three retinal determination genes regulate the expression of either *hh* and/or *dpp* (Hazelett et al. 1998; Pauli et al. 2005). The loss of either signaling pathway is amplified since both genes are in turn required for the proper functioning of the retinal determination network itself (Chen et al. 1999; Kango-Singh et al. 2003).

Concluding Remarks

Patterning of the *Drosophila* compound eye by the morphogenetic furrow has fascinated biologists for decades. This review is an attempt to briefly summarize our current knowledge of the mechanisms that underlie its movement across the eye field and some of its cellular characteristics. Despite the intense scrutiny that has surrounded the furrow, a number of outstanding questions continue to exist and are certainly worthy of future investigations. Some of the issues (just to name a few) that immediately come to mind include (1) discovering the identity of the timing mechanisms that govern the initiation of the furrow, (2) elucidating the means by which ectopic furrows are prevented from initiating outside of the *wg* expression domain, (3) determining the link between the rates of pattern formation and cellular proliferation, and (4) establishing the position of the control points that regulate furrow reinitiation and progression. These are just some of the questions that hopefully will be answered by the time the next review on the morphogenetic furrow is written.

Acknowledgments I would like to thank everyone who has contributed to our understanding of the mechanisms that regulate the initiation and progression of the morphogenetic furrow. I would like to also apologize to anyone whose work is not cited here. This work was supported by a grant from the National Eye Institute (R01 EY4863) to Justin P. Kumar.

References

Bach EA, Vincent S, Zeidler MP, Perrimon N (2003) A sensitized genetic screen to identify novel regulators and components of the *Drosophila* janus kinase/signal transducer and activator of transcription pathway. Genetics 165:1149–1166

Baker NE, Yu S, Han D (1996) Evolution of proneural atonal expression during distinct regulatory phases in the developing *Drosophila* eye. Curr Biol 6:1290–1301

Basler K, Hafen E (1989) Dynamics of *Drosophila* eye development and temporal requirements of sevenless expression. Development 107:723–731

Basler K, Struhl G (1994) Compartment boundaries and the control of *Drosophila* limb pattern by hedgehog protein. Nature 368:208–214

Becker HJ (1957) Uber Rontgenmossaikflecken und Defektmutationen am Auge von *Drosophila* und die Entwicklungsphysiologie des Auges. Z Induk Abst Vererb Lehre 88:333–373

Benlali A, Draskovic I, Hazelett DJ, Treisman JE (2000) act up controls actin polymerization to alter cell shape and restrict Hedgehog signaling in the *Drosophila* eye disc. Cell 101:271–281

Bessa J, Gebelein B, Pichaud F, Casares F, Mann RS (2002) Combinatorial control of *Drosophila* eye development by eyeless, homothorax, and teashirt. Genes Dev 16:2415–2427

Bhattacharya A, Baker NE (2009) The HLH protein Extramacrochaetae is required for R7 cell and cone cell fates in the *Drosophila* eye. Dev Biol 327:288–300

Blackman RK, Sanicola M, Raftery LA, Gillevet T, Gelbart WM (1991) An extensive 3′ cis-regulatory region directs the imaginal disk expression of decapentaplegic, a member of the TGF-b family in *Drosophila*. Development 111:657–665

Bonini NM, Leiserson WM, Benzer S (1993) The eyes absent gene: genetic control of cell survival and differentiation in the developing *Drosophila* eye. Cell 72:379–395

Borod ER, Heberlein U (1998) Mutual regulation of decapentaplegic and hedgehog during the initiation of differentiation in the *Drosophila* retina. Dev Biol 197:187–197

Brennan CA, Ashburner M, Moses K (1998) Ecdysone pathway is required for furrow progression in the developing *Drosophila* eye. Development 125:2653–2664

Brown NL, Sattler CA, Markey DR, Carroll SB (1991) hairy gene function in the *Drosophila* eye: normal expression is dispensable but ectopic expression alters cell fates. Development 113:1245–1256

Brown NL, Sattler CA, Paddock SW, Carroll SB (1995) Hairy and emc negatively regulate morphogenetic furrow progression in the *Drosophila* eye. Cell 80:879–887
Burke R, Basler K (1996) Hedgehog-dependent patterning in the *Drosophila* eye can occur in the absence of Dpp signaling. Dev Biol 179:360–368
Cagan RL, Ready DF (1989a) The emergence of order in the *Drosophila* pupal retina. Dev Biol 136:346–362
Cagan RL, Ready DF (1989b) Notch is required for successive cell decisions in the developing *Drosophila* retina. Genes Dev 3:1099–1112
Campos-Ortega JA, Hofbauer A (1977) Cell clones and pattern formation on the lineage of photoreceptor cells in the compound eye of *Drosophila*. Roux Arch Dev Biol 181:227–245
Capdevila J, Estrada MP, Sanchez-Herrero E, Guerrero I (1994) The Drosophila segment polarity gene patched interacts with decapentaplegic in wing development. EMBO 13:71–82
Chanut F, Heberlein U (1997a) Retinal morphogenesis in *Drosophila*: hints from an eye-specific decapentaplegic allele. Dev Genet 20:197–207
Chanut F, Heberlein U (1997b) Role of decapentaplegic in initiation and progression of the morphogenetic furrow in the developing *Drosophila* retina. Development 124:559–567
Chanut F, Luk A, Heberlein U (2000) A screen for dominant modifiers of ro(Dom), a mutation that disrupts morphogenetic furrow progression in *Drosophila*, identifies groucho and hairless as regulators of atonal expression. Genetics 156:1203–1217
Chao JL, Tsai YC, Chiu SJ, Sun YH (2004) Localized Notch signal acts through eyg and upd to promote global growth in *Drosophila* eye. Development 131:3839–3847
Chen R, Halder G, Zhang Z, Mardon G (1999) Signaling by the TGF-b homolog decapentaplegic functions reiteratively within the network of genes controlling retinal cell fate determination in *Drosophila*. Development 126:935–943
Cheyette BN, Green PJ, Martin K, Garren H, Hartenstein V, Zipursky SL (1994) The *Drosophila* sine oculis locus encodes a homeodomain-containing protein required for the development of the entire visual system. Neuron 12:977–996
Cho KO, Choi KW (1998) Fringe is essential for mirror symmetry and morphogenesis in the *Drosophila* eye. Nature 396:272–276
Cohen SM (1993) Imaginal disc development. In: Bate M, Martinez Arias A (eds) The development of *Drosophila melanogaster*, Cold Spring Harbor Laboratory Press, Cold Spring Harbor, NY, pp 747–841
Corrigall D, Walther RF, Rodriguez L, Fichelson P, Pichaud F (2007) Hedgehog signaling is a principal inducer of Myosin-II-driven cell ingression in *Drosophila* epithelia. Dev Cell 13:730–742
Curtiss J, Mlodzik M (2000) Morphogenetic furrow initiation and progression during eye development in *Drosophila*: the roles of decapentaplegic, hedgehog and eyes absent. Development 127:1325–1336
Czerny T, Halder G, Kloter U, Souabni A, Gehring WJ, Busslinger M (1999) twin of eyeless, a second Pax-6 gene of *Drosophila*, acts upstream of eyeless in the control of eye development. Mol Cell 3:297–307
Dietrich W (1909) Die Fazettenaugen der Dipteren. Z Wiss Zool 92:465–539
Dokucu ME, Zipursky SL, Cagan RL (1996) Atonal, rough and the resolution of proneural clusters in the developing *Drosophila* retina. Development 122:4139–4147
Dominguez M (1999) Dual role for hedgehog in the regulation of the proneural gene atonal during ommatidia development. Development 126:2345–2353
Dominguez M, Celis JF de (1998) A dorsal/ventral boundary established by Notch controls growth and polarity in the *Drosophila* eye. Nature 396:276–278
Dominguez M, Hafen E (1997) Hedgehog directly controls initiation and propagation of retinal differentiation in the *Drosophila* eye. Genes Dev 11:3254–3264
Edgar BA, O'Farrell PH (1989) Genetic control of cell division patterns in the *Drosophila* embryo. Cell 57:177–187

Ekas LA, Baeg GH, Flaherty MS, Ayala-Camargo A, Bach EA (2006) JAK/STAT signaling promotes regional specification by negatively regulating wingless expression in *Drosophila*. Development 133:4721–4729

Ellis HM, Spann DR, Posakony JW (1990) extramacrochaetae, a negative regulator of sensory organ development in *Drosophila*, defines a new class of helix-loop-helix proteins. Cell 61:27–38

Fan SS, Ready DF (1997) Glued participates in distinct microtubule-based activities in *Drosophila* eye development. Development 124:1497–1507

Firth LC, Bhattacharya A, Baker NE (2010). Cell cycle arrest by a gradient of Dpp signaling during *Drosophila* eye development. BMC Dev Biol 10:28

Fischer-Vize JA, Mosley KL (1994) Marbles mutants: uncoupling cell determination and nuclear migration in the developing *Drosophila* eye. Development 120:2609–2618

Frankfort BJ, Mardon G (2002) R8 development in the *Drosophila* eye: a paradigm for neural selection and differentiation. Development 129:1295–1306

Frankfort BJ, Nolo R, Zhang Z, Bellen H, Mardon G (2001) senseless repression of rough is required for R8 photoreceptor differentiation in the developing *Drosophila* eye. Neuron 32:403–414

Freeman M (1994) The spitz gene is required for photoreceptor determination in the *Drosophila* eye where it interacts with the EGF receptor. Mech Dev 48:25–33

Freeman M (1996) Reiterative use of the EGF Receptor triggers differentiation of all cell types in the *Drosophila* eye. Cell 87:651–660

Freeman M (1997) Cell determination strategies in the *Drosophila* eye. Development 124:261–270

Garrell J, Modolell J (1990) The *Drosophila* extramacrochaetae locus, an antagonist of proneural genes that, like these genes, encodes a helix-loop-helix protein. Cell 61:39–48

Gonzalez F, Swales L, Bejsovec A, Skaer H, Martinez Arias A (1991) Secretion and movement of wingless protein in the epidermis of the *Drosophila* embryo. Mech Dev 35:43–54

Greenwood S, Struhl G (1999) Progression of the morphogenetic furrow in the *Drosophila* eye: the roles of Hedgehog, Decapentaplegic and the Raf pathway. Development 126:5795–5808

Gutierrez-Avino FJ, Ferres-Marco D, Dominguez M (2009) The position and function of the Notch-mediated eye growth organizer: the roles of JAK/STAT and four-jointed. EMBO Rep 10:1051–1058

Hazelett DJ, Bourouis M, Walldorf U, Treisman JE (1998) decapentaplegic and wingless are regulated by eyes absent and eyegone and interact to direct the pattern of retinal differentiation in the eye disc. Development 125:3741–3751

Heberlein U, Mlodzik M, Rubin GM (1991) Cell-fate determination in the developing *Drosophila* eye: role of the rough gene. Development 112:703–712

Heberlein U, Singh CM, Luk AY, Donohoe TJ (1995) Growth and differentiation in the *Drosophila* eye coordinated by hedgehog. Nature 373:709–711

Heberlein U, Wolff T, Rubin GM (1993) The TGF beta homolog dpp and the segment polarity gene hedgehog are required for propagation of a morphogenetic wave in the *Drosophila* retina. Cell 75:913–926

Heemskerk J, DiNardo S (1994) *Drosophila* hedgehog acts as a morphogen in cellular patterning. Cell 76:449–460

Held LI (2002) Imaginal discs: the genetic and cellular logic of pattern formation. Developmental and cell biology series, vol 39. Cambridge Press, Cambridge, p 460

Horsfield J, Penton A, Secombe J, Hoffman FM, Richardson H (1998) decapentaplegic is required for arrest in G1 phase during *Drosophila* eye development. Development 125:5069–5078

Houalla T, Hien Vuong D, Ruan W, Suter B, Rao Y (2005) The Ste20-like kinase misshapen functions together with Bicaudal-D and dynein in driving nuclear migration in the developing drosophila eye. Mech Dev 122:97–108

Ives P (1950) New mutant report: bar-3. Dros Info Serv 24:58

Jarman AP, Grell EH, Ackerman L, Jan LY, Jan YN (1994) atonal is the proneural gene for *Drosophila* photoreceptors. Nature 369:398–400

Jarman AP, Sun Y, Jan LY, Jan YN (1995) Role of the proneural gene, atonal, in formation of *Drosophila* chordotonal organs and photoreceptors. Development 121:2019–2030

Jones NA, Kuo YM, Sun YH, Beckendorf SK (1998) The *Drosophila* Pax gene eye gone is required for embryonic salivary duct development. Development 125:4163–4174
Jun S, Wallen RV, Goriely A, Kalionis B, Desplan C (1998) Lune/eye gone, a Pax-like protein, uses a partial paired domain and a homeodomain for DNA recognition. Proc Natl Acad Sci U S A 95:13720–13725
Jurgens G, Hartenstein V (1993) The terminal regions of the body plan. In: Bate M, Martinez Arias A (eds) The development of *Drosophila melanogaster*, Cold Spring Harbor Laboratory Press, Cold Spring Harbor, NY, pp 687–746
Kango-Singh M, Singh A, Henry Sun Y (2003) Eyeless collaborates with Hedgehog and Decapentaplegic signaling in *Drosophila* eye induction. Dev Biol 256:49–60
Kenyon KL, Ranade SS, Curtiss J, Mlodzik M, Pignoni F (2003) Coordinating proliferation and tissue specification to promote regional identity in the *Drosophila* head. Dev Cell 5:403–414
Kimberly EL, Hardin J (1998) Bottle cells are required for the initiation of primary invagination in the sea urchin embryo. Dev Biol 204:235–250
Kumar JP (2010) Retinal determination the beginning of eye development. Curr Top Dev Biol 93:1–28
Kumar JP (2011) My what big eyes you have: how the *Drosophila* retina grows. Dev Neurobiol 71:1133–1152
Kumar JP (2012) Building an ommatidium one cell at a time. Dev Dyn 241:136–149
Kumar JP, Moses K (2001a) EGF receptor and Notch signaling act upstream of Eyeless/Pax6 to control eye specification. Cell 104:687–697
Kumar JP, Moses K (2001b) The EGF receptor and notch signaling pathways control the initiation of the morphogenetic furrow during *Drosophila* eye development. Development 128:2689–2697
Kumar JP, Tio M, Hsiung F, Akopyan S, Gabay L, Seger R, Shilo BZ, Moses K (1998) Dissecting the roles of the *Drosophila* EGF receptor in eye development and MAP kinase activation. Development 125:3875–3885
Lebovitz RM, Ready DF (1986) Ommatidial development in *Drosophila* eye disc fragments. Dev Biol 117:663–671
Lee JJ, Kessler DP von, Parks S, Beachy PA (1992) Secretion and localized transcription suggest a role in positional signaling for products of the segmentation gene hedgehog. Cell 71:33–50
Lopes CS, Casares F (2009) hth maintains the pool of eye progenitors and its downregulation by Dpp and Hh couples retinal fate acquisition with cell cycle exit. Dev Biol 339:78–88
Ma C, Liu H, Zhou Y, Moses K (1996) Identification and characterization of autosomal genes that interact with glass in the developing *Drosophila* eye. Genetics 142:1199–1213
Ma C, Moses K (1995) wingless and patched are negative regulators of the morphogenetic furrow and can affect tissue polarity in the developing *Drosophila* compound eye. Development 121:2279–2289
Ma C, Zhou Y, Beachy PA, Moses K (1993) The segment polarity gene hedgehog is required for progression of the morphogenetic furrow in the developing *Drosophila* eye. Cell 75:927–938
Massague J (1996) TGFbeta signaling: receptors, transducers, and Mad proteins. Cell 85:947–950
Massague J, Polyak K (1995) Mammalian antiproliferative signals and their targets. Curr Opin Genet Dev 5:91–96
Masucci JD, Miltenberger RJ, Hoffmann FM (1990) Pattern-specific expression of the *Drosophila* decapentaplegic gene in imaginal disks is regulated by 3' cis-regulatory elements. Genes Dev 4:2011–2023
Mohler J (1988) Requirements for hedgehog, a segmental polarity gene, in patterning larval and adult cuticle of *Drosophila*. Genetics 120:1061–1072
Mosley-Bishop KL, Li Q, Patterson L, Fischer JA (1999) Molecular analysis of the klarsicht gene and its role in nuclear migration within differentiating cells of the *Drosophila* eye. Curr Biol 9:1211–1220
Mozer BA (2001) Dominant Drop mutants are gain-of-function alleles of the muscle segment homeobox gene (msh) whose overexpression leads to the arrest of eye development. Dev Biol 233:380–393

Padgett RW, St Johnston RD, Gelbart WM (1987) A transcript from a *Drosophila* pattern gene predicts a protein homologous to the transforming growth factor-beta family. Nature 325:81–84
Pan D, Rubin GM (1995) cAMP-dependent protein kinase and hedgehog act antagonistically in regulating decapentaplegic transcription in *Drosophila* imaginal discs. Cell 80:543–552
Panganiban GE, Rashka KE, Neitzel MD, Hoffmann FM (1990a) Biochemical characterization of the *Drosophila* dpp protein, a member of the transforming growth factor beta family of growth factors. Mol Cell Biol 10:2669–2677
Panganiban GE, Reuter R, Scott MP, Hoffmann FM (1990b) A *Drosophila* growth factor homolog, decapentaplegic, regulates homeotic gene expression within and across germ layers during midgut morphogenesis. Development 110:1041–1050
Papayannopoulos V, Tomlinson A, Panin VM, Rauskolb C, Irvine KD (1998) Dorsal-ventral signaling in the *Drosophila* eye. Science 281:2031–2034
Patterson K, Molofsky AB, Robinson C, Acosta S, Cater C, Fischer JA (2004) The functions of Klarsicht and nuclear lamin in developmentally regulated nuclear migrations of photoreceptor cells in the *Drosophila* eye. Mol Biol Cell 15(2):600–610
Pauli T, Seimiya M, Blanco J, Gehring WJ (2005) Identification of functional sine oculis motifs in the autoregulatory element of its own gene, in the eyeless enhancer and in the signalling gene hedgehog. Development 132:2771–2782
Peng HW, Slattery M, Mann RS (2009) Transcription factor choice in the Hippo signaling pathway: homothorax and yorkie regulation of the microRNA bantam in the progenitor domain of the *Drosophila* eye imaginal disc. Genes Dev 23:2307–2319
Penton A, Selleck SB, Hoffmann FM (1997) Regulation of cell cycle synchronization by decapentaplegic during *Drosophila* eye development. Science 275:203–206
Pepple KL, Atkins M, Venken K, Wellnitz K, Harding M, Frankfort B, Mardon G (2008) Two-step selection of a single R8 photoreceptor: a bistable loop between senseless and rough locks in R8 fate. Development 135:4071–4078
Pfeiffer S, Vincent JP (1999) Signalling at a distance: transport of Wingless in the embryonic epidermis of *Drosophila*. Semin Cell Dev Biol 10:303–309
Pignoni F, Zipursky SL (1997) Induction of *Drosophila* eye development by Decapentaplegic. Development 124:271–278
Quiring R, Walldorf U, Kloter U, Gehring WJ (1994) Homology of the eyeless gene of *Drosophila* to the Small eye gene in mice and Aniridia in humans (see comments). Science 265(5173):785–789
Ready DF, Hanson TE, Benzer S (1976) Development of the *Drosophila* retina, a neurocrystalline lattice. Dev Biol 53:217–240
Richardson H, O'Keefe LV, Marty T, Saint R (1995) Ectopic cyclin E expression induces premature entry into S phase and disrupts pattern formation in the *Drosophila* eye imaginal disc. Development 121:3371–3379
Rogers EM, Brennan CA, Mortimer NT, Cook S, Morris AR, Moses K (2005) Pointed regulates an eye-specific transcriptional enhancer in the *Drosophila* hedgehog gene, which is required for the movement of the morphogenetic furrow. Development 132:4833–4843
Royet J, Finkelstein R (1996) hedgehog, wingless and othrodenticle specify adult head development in *Drosophila*. Development 122:1849–1858
Royet J, Finkelstein R (1997) Establishing primordia in the *Drosophila* eye-antennal imaginal disc: the role of decapentaplegic, wingless and hedgehog. Development 124:4793–4800
Serikaku MA, O'Tousa JE (1994) sine oculis is a homeobox gene required for *Drosophila* visual system development. Genetics 138:1137–1150
Singh A, Choi KW (2003) Initial state of the *Drosophila* eye before dorsoventral specification is equivalent to ventral. Development 130:6351–6360
Struhl G (1981) A blastoderm fate map of compartments and segments of the *Drosophila* head. Dev Biol 84:386–396
Strutt DI, Mlodzik M (1997) Hedgehog is an indirect regulator of morphogenetic furrow progression in the *Drosophila* eye disc. Development 124:3233–3240

Strutt DI, Wiersdorff V, Mlodzik M (1995) Regulation of furrow progression in the *Drosophila* eye by cAMP-dependent protein kinase A. Nature 373:705–709

Sun YH, Tsai CJ, Green MM, Chao JL, Yu CT, Jaw TJ, Yeh JY, Bolshakov VN (1995) White as a reporter gene to detect transcriptional silencers specifying position-specific gene expression during *Drosophila melanogaster* eye development. Genetics 141:1075–1086

Swan A, Nguyen T, Suter B (1999) *Drosophila* Lissencephaly-1 functions with Bic-D and dynein in oocyte determination and nuclear positioning. Nat Cell Biol 1:444–449

Tabata T, Kornberg TB (1994) Hedgehog is a signaling protein with a key role in patterning *Drosophila* imaginal discs. Cell 76:89–102

Tearle R, Tomlinson A, Saint R (1994) The dominant Drop eye mutations of *Drosophila melanogaster* define two loci implicated in normal eye development. Mol Gen Genet 244:426–434

Thomas BJ, Gunning DA, Cho J, Zipursky L (1994) Cell cycle progression in the developing *Drosophila* eye: roughex encodes a novel protein required for the establishment of G1. Cell 77:1003–1014

Thomas BJ, Zavitz KH, Dong X, Lane ME, Weigmann K, Finley RL Jr, Brent R, Lehner CF, Zipursky SL (1997) Roughex down-regulates G2 cyclins in G1. Genes Dev 11:1289–1298

Tio M, Ma C, Moses K (1994) Spitz, a *Drosophila* homolog of transforming growth factor-alpha, is required in the founding photoreceptor cells of the compound eye facets. Mech Dev 48:13–23

Tio M, Moses K (1997) The *Drosophila* TGF alpha homolog Spitz acts in photoreceptor recruitment in the developing retina. Development 124:343–351

Tomlinson A (1985). The cellular dynamics of pattern formation in the eye of *Drosophila*. J Embryol Exp Morphol 89:313–331

Tomlinson A, Ready DF (1987) Neuronal differentiation in the *Drosophila* ommatidium. Dev Biol 120:366–376

Treisman JE, Rubin GM (1995) Wingless inhibits morphogenetic furrow movement in the *Drosophila* eye disc. Development 121:3519–3527

Tsai YC, Sun YH (2004) Long-range effect of upd, a ligand for Jak/STAT pathway, on cell cycle in *Drosophila* eye development. Genesis 39:141–153

Tsai YC, Yao JG, Chen PH, Posakony JW, Barolo S, Kim J, Sun YH (2007) Upd/Jak/STAT signaling represses wg transcription to allow initiation of morphogenetic furrow in *Drosophila* eye development. Dev Biol 306:760–771

den Heuvel M van, Nusse R, Johnston P, Lawrence PA (1989) Distribution of the wingless gene product in *Drosophila* embryos: a protein involved in cell-cell communication. Cell 59:739–749

Van Doren M, Ellis HM, Posakony JW (1991) The *Drosophila* extramacrochaetae protein antagonizes sequence-specific DNA binding by daughterless/achaete-scute protein complexes. Development 113:245–255

Waddington CH, Perry MM (1960) The ultrastructure of the developing eye of *Drosophila*. Proc Roy Soc Biol Sci 153:155–178

Wehrli M, Tomlinson A (1995) Epithelial planar polarity in the developing *Drosophila* eye. Development 121:2451–2459

Wiersdorff V, Lecuit T, Cohen SM, Mlodzik M (1996) Mad acts downstream of Dpp receptors, revealing a differential requirement for dpp signaling in initiation and propagation of morphogenesis in the *Drosophila* eye. Development 122:2153–2162

Wolff T, Ready DF (1991) The beginning of pattern formation in the *Drosophila* compound eye: the morphogenetic furrow and the second mitotic wave. Development 113:841–850

Yao JG, Weasner BM, Wang LH, Jang CC, Weasner B, Tang CY, Salzer CL, Chen CH, Hay B, Sun YH et al (2008) Differential requirements for the Pax6(5a) genes eyegone and twin of eyegone during eye development in *Drosophila*. Dev Biol 315(2):535–551

Zelhof AC, Ghbeish N, Tsai C, Evans RM, McKeown M (1997) A role for ultraspiracle, the *Drosophila* RXR, in morphogenetic furrow movement and photoreceptor cluster formation. Development 124:2499–2506

Cell Morphogenesis: Tracing the Paths of Induction During *Drosophila* Ommatidial Development

Jennifer Curtiss

Introduction

The study of the *Drosophila melanogaster* eye began in earnest in the late 1980s and continues to this day. Hundreds of fine papers have been published along the way. Thus, some may ask why it is important to continue to study the *Drosophila* eye. After all, what can possibly be left to discover that would lead to substantial advances in the field of Developmental Biology? One answer, as has been pointed out numerous times before, is that the *Drosophila* eye is a remarkable system for studying the role of induction in cell fate specification. And, in spite of the attention shown to it in the past, there are substantial gaps in our knowledge of how even this relatively simple system develops. Filling these gaps promises to unlock mysteries that no one yet suspects exist, which will lead to discoveries no one can yet fathom.

A Brief Summary of *Drosophila* Eye Structure and Development

Structure

The adult *Drosophila melanogaster* eye is composed of ~750 ommatidia (Fig. 1), each of which contains eight photoreceptors (R1–R8) plus a number of accessory cells (cone cells, pigment cells, and the group of four cells that form the bristle complex). A number of photoreceptor types are evident, based on their arrangement and type of rhodopsin expressed. The R1–R6 cells are called the "outer" photoreceptors. They extend the entire apical-basal length of the retina, express the broad-spectrum Rh1 rhodopsin, and develop larger rhabdomeres that form a trapezoidal arrangement

J. Curtiss (✉)
Department of Biology, New Mexico State University, MSC 3AF, PO Box 30001, Las Cruces, NM 88003, USA
e-mail: curtij01@nmsu.edu

A. Singh, M. Kango-Singh (eds.), *Molecular Genetics of Axial Patterning, Growth and Disease in the Drosophila Eye*, DOI 10.1007/978-1-4614-8232-1_4,

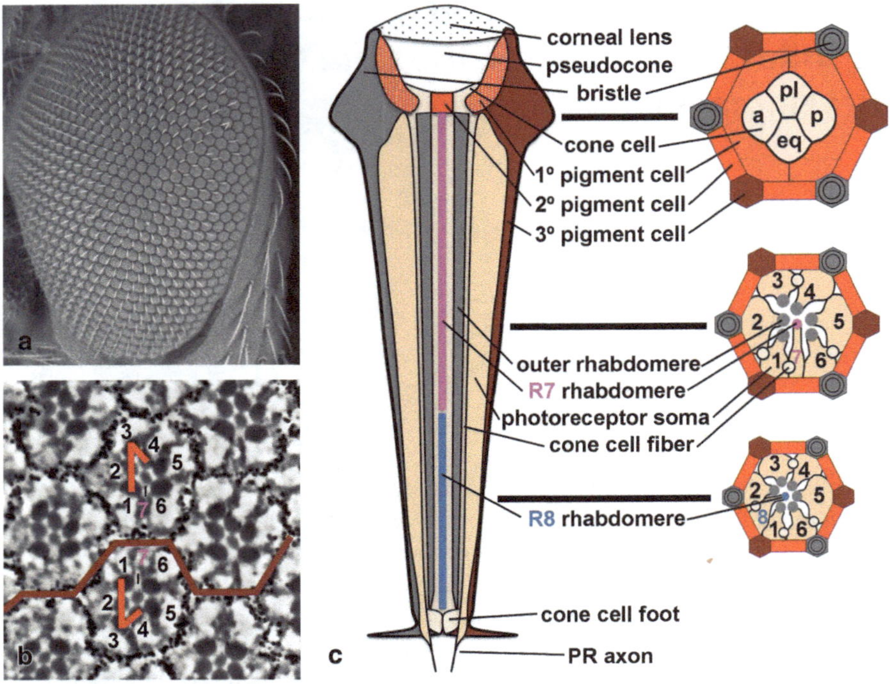

Fig. 1 Structure of the adult *Drosophila* eye. **a** Scanning electron micrograph of a wild-type *Drosophila melanogaster* eye containing approximately 750 facets, or ommatidia. Interommatidial bristles emerge from every other vertex of the hexagonally shaped ommatidia. In this and all other figures, anterior is towards the left and dorsal towards the top. **b** Tangential section of an eye taken at the R7 level. A *maroon-colored line* marks the equator-/dorsal-ventral axis. *Numbers* denote photoreceptor cell bodies for both a dorsal and ventral ommatidium. Dark-stained rhabdomeres project towards the center of each ommatidium. The large rhabdomeres of the outer photoreceptors (R1–R6) are arranged to form a trapezoidal shape. The small R7 rhabdomere is located in the center of the trapezoid. *Red arrows* mark dorsal and ventral chiral forms on opposite sides of the equator. **c** Diagrams of longitudinal (*left*) and tangential (*right*) sections of a dorsal-type ommatidium at the indicated levels ("*a*," "*p*," "*pl*," and "*eq*" indicate anterior, posterior, polar, and equatorial cone cells, respectively; after Wolff and Ready 1993; Charlton-Perkins and Cook 2010)

around the two inner photoreceptors (R7 and R8) that are responsible for motion detection and low light vision (Ready et al. 1976; Wolff and Ready 1993; Charlton-Perkins and Cook 2010).

R7 and R8 are the "inner" photoreceptors, with R7 lying in the center of each ommatidium in the distal part of the eye, and R8 lying beneath R7. It is the inner photoreceptors that allow for color vision in *Drosophila*: approximately 30 % of ommatidia contain an R7 that expresses Rh3 and an R8 that expresses Rh5; these are termed "pale ommatidia." The other 70 % or so contain an R7 that expresses Rh4 and an R8 that expresses Rh6 and are termed "yellow ommatidia." "Pale" and "yellow" ommatidia are arranged randomly within the eye. The tale of how rhodopsin expression is coordinately regulated in R7 and R8 cells is a fascinating one (Morante et al. 2007; Charlton-Perkins and Cook 2010), but I will not discuss it further here.

The two rhabdomeres of the R2/R5 and R1/R6 pairs of photoreceptors are arranged in a bilaterally symmetric fashion within the ommatidium, one on either side of the inner photoreceptors. In contrast, R3 and R4 are asymmetrically positioned, with R3 closer to the polar side of each ommatidium and R4 closer to the equatorial side (Fig. 1b). Ommatidia in the dorsal and ventral parts of the eye (i.e., on opposite sides of the equator) are mirror images of one another and have opposite chiral forms (Wolff and Ready 1993). As described in more detail below this asymmetry results from planar polarity signaling.

The accessory cells are non-neuronal cells with a number of support functions in the eye. Cone cells, primary (1°) pigment cells and secondary (2°) pigment cells are responsible for secreting the corneal lens associated with each ommatidium as well as the underlying pseudocone. Together, these structures focus light onto the photoreceptors. The 2° pigment cells and tertiary (3°) pigment cells lie in between and are shared among adjacent ommatidia. The 2° and 3° pigment cells produce pigment granules that limit light scattering between ommatidia and protect photoreceptors from light-induced damage. In addition, they produce critical components of the phototransduction pathway (Wolff and Ready 1993; Charlton-Perkins and Cook 2010). Finally, mechanosensory bristles are present at alternate ommatidial vertices (Fig. 1a). Their development occurs independently of that of other ommatidial cells (Ready et al. 1976; Cagan and Ready 1989b; Wolff and Ready 1993; Charlton-Perkins and Cook 2010), and I will not discuss them further here.

Development

Specification and differentiation of distinct types of ommatidial cells begin during larval stages, when an indentation called the morphogenetic furrow (MF) propagates from posterior to anterior across the epithelium that comprises the field of eye precursors (Fig. 2). Cells in the furrow undergo an apical constriction and begin to organize into clusters that will eventually develop into ommatidia. As the furrow moves from posterior to anterior, a new column of ommatidial precursors emerges from the furrow approximately every 2 h.

The first recognizable cluster of cells within the furrow is called a rosette (not shown). As the furrow continues to move forward, the posterior row of cells in each rosette forms an arc of ∼9 cells, which is located at the posterior edge of the furrow (Fig. 2a, b column 1). The arcs zipper shut to form a precluster of 6–7 cells, which include the future R8, R2, R5, R3, and R4 photoreceptors, along with two "mystery cells" (Fig. 2a, b column 2). R2 and R5 cells contact each other first on either side of R8. Subsequently, R3 and R4 precursors contact each other, followed by the mystery cells. Shortly thereafter, the "mystery cells" are ejected and likely rejoin the pool of cells surrounding each nascent ommatidium, leading to formation of the 5-cell cluster (Fig. 2a, b column 4; Ready et al. 1976; Tomlinson and Ready 1987b; Wolff and Ready 1991b, 1993).

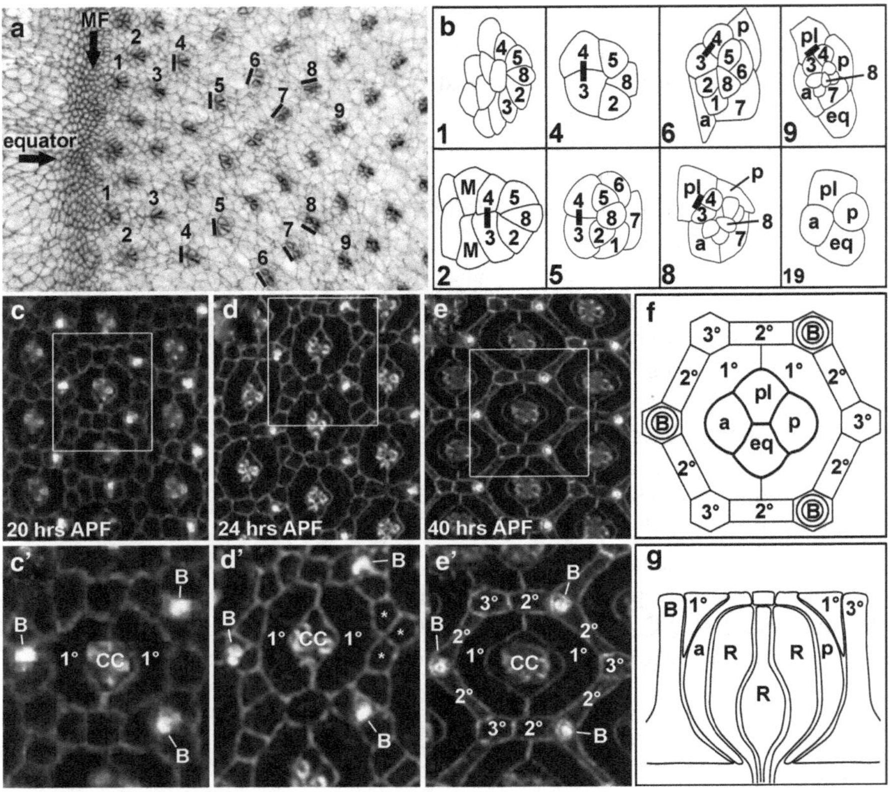

Fig. 2 Development of the *Drosophila* eye. **a** Larval eye precursor tissue stained with an antibody against *Drosophila* E-Cadherin to mark cellular outlines. *Arrows* mark the positions of the morphogenetic furrow and the equator. *Numbers* denote columns posterior to the furrow. *Bars* indicate rotation of ommatidial clusters. **b** Diagrams of typical ommatidial clusters from the indicated columns. All diagrams depict clusters from the region dorsal to the equator. *Numbers* indicate photoreceptor types. "*a*," "*p*," "*pl*," and "*eq*" indicate anterior, posterior, polar, and equatorial cone cells, respectively. *Bars* indicate rotation of ommatidial clusters. **c–e** Pupal eye precursor tissue at the indicated times after the start of the pupal stage (APF) stained with an antibody against *Drosophila* E-Cadherin to mark cellular outlines. **c**′, **d**′, and **e**′ are close-up views of the *boxed areas* in (**c**), (**d**), and (**e**). **f**, **g** Diagrams of tangential (**f**) and longitudinal (**g**) sections of pupal eye precursor tissue at 40 h APF. *M* mystery cell, *R* photoreceptor, *B* bristle, *a* anterior cone cells, *p* posterior cone cells, *pl* polar cone cells, *eq* equatorial cone cells, *CC* cone cells, *1°* primary pigment cells, *2°* secondary pigment cells, *3°* tertiary pigment cells. (After Wolff and Ready 1993; Bao and Cagan 2005; Tepass and Harris 2007)

Cells that are not part of the 5-cell cluster undergo a single additional round of mitosis (referred to as the second mitotic wave), and it is from this pool of cells that the R1, R6, and R7 photoreceptors, the cone cells, pigment cells, and bristle cells are eventually recruited (Ready et al. 1976; Tomlinson 1985; Wolff and Ready 1991b, 1993; Kumar 2011). R1, R6, and R7 precursors are physically added to the cluster at approximately the same time (Fig. 2a, b column 5; Ready et al. 1976; Tomlinson

1985; Wolff and Ready 1993; Mavromatakis and Tomlinson 2012a). At this stage, the R8 precursor is in the center of the cluster and is in contact with each of the other seven photoreceptor precursors, which surround it. Later, R4 will lose contact with R8, breaking the symmetry of the ommatidium (Tomlinson 1985) and leading to the asymmetrical chiral ommatidial form present in the adult.

Based on marker expression, the photoreceptor precursors acquire neuronal fate in a defined sequence that mostly mirrors recruitment of cells to the cluster, although based on the markers used, the onset of neuronal fate is somewhat delayed with respect to the morphogenetic changes that signal recruitment. R8 is first to acquire neuronal fate (column 2), followed by R2 and R5 (column 5), R3 and R4 (column 8), R1 and R6 (column 9–10), and finally R7 (column 13; Ready et al. 1976; Tomlinson 1985; Tomlinson and Ready 1987b; Wolff and Ready 1991b, 1993). As neuronal fate is acquired, each photoreceptor nucleus moves towards the apical side of the cell and subsequently returns to a more basal localization (Tomlinson 1985; Tomlinson and Ready 1987b; Wolff and Ready 1993).

Following addition of the final R7 photoreceptor, four non-neuronal cone cells are added to the cluster. The anterior and posterior cone cells are recruited first (Fig. 2a, b column 6), followed by the polar cone cell (Fig. 2a, b column 8), and eventually the equatorial cone cell (Fig. 2a, b column 9). As the photoreceptor cells sink within the epithelium, the cone cells move over them such that they eventually lie over the photoreceptors (Tomlinson 1985; Tomlinson and Ready 1987b; Wolff and Ready 1993).

These recruitment events continue through the rest of the third larval instar, as the furrow moves ever closer to the anterior edge of the eye precursor field, and are completed by approximately 10 h of pupal development. In addition, as a result of planar polarity signaling, ommatidial clusters rotate 90° during the period between the establishment of the 5-cell cluster through the end of cone cell recruitment (Ready et al. 1976; Tomlinson and Ready 1987b; Wolff and Ready 1993; Doroquez and Rebay 2006; Jenny 2010).

Between approximately 10 and 40 h of pupal development, the remainder of ommatidial cells—the pigment cells and bristle precursors—are recruited. The 1° pigment cells are recruited first: they form two half-moon shapes that surround the cone cells and separate them from nearby cells (Fig. 2c). Next, three cells positioned at every other ommatidial vertex compete to become 3° pigment cells (Fig. 3e, asterisks). The cell that manages to contact 1° pigment cells in three adjacent ommatidia becomes a 3° pigment cell. Finally, a single 2° pigment cell is established between adjacent bristle precursors and 3° pigment cells, forming the edge of each facet (Fig. 2e). Remaining cells that do not become 2° and 3° pigment cells are eliminated by apoptosis (Cagan and Ready 1989b; Wolff and Ready 1991a, 1993; Bao and Cagan 2005; Carthew 2007). Final differentiation of ommatidial cells occurs following establishment of the various ommatidial cell types (Charlton-Perkins and Cook 2010).

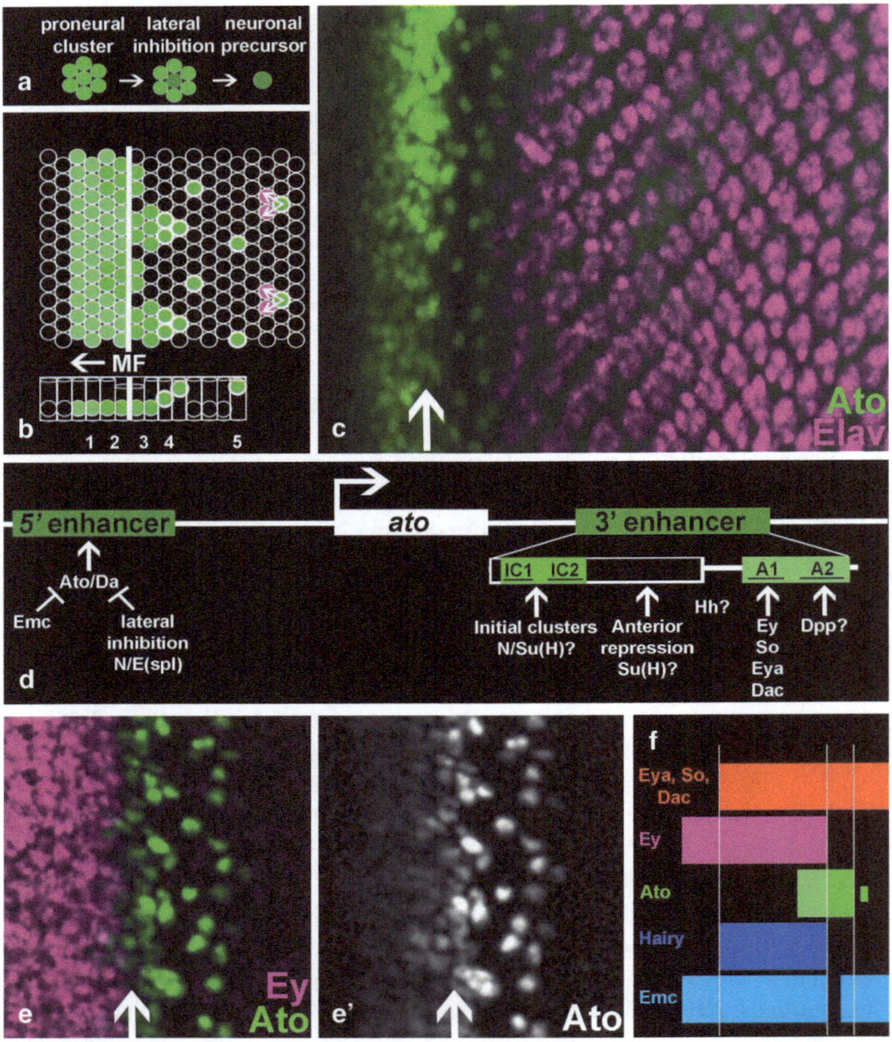

Fig. 3 Factors that regulate Ato expression. **a** Diagram of the role of lateral inhibition in the selection of neuronal precursors (see text for more details). **b** Diagram of the different phases of Ato expression (*green*). Photoreceptors other than R8 are colored *magenta*. "*MF*" indicates the position of the morphogenetic furrow (MF). **c** Image of a wild-type eye precursor field stained for Ato and Elav. The *white arrow* indicates the approximate position of the MF. **d** Diagram of the *ato* gene, showing the approximate positions of the 5′ and 3′ enhancers and the regions affected by factors that regulate *ato* expression. **e** Image of a wild-type eye precursor field stained for Ato and Ey. The *white arrow* indicates the approximate position of the MF. **e′** shows a black and white image of Ato, only. **f** Diagram of the expression patterns of factors that regulate *ato* expression with respect to the Ato expression pattern and the MF

RTK/Ras/MAPK and Notch Signaling in Ommatidial Cell Specification

Two signaling pathways are particularly important for specification of ommatidial cell types. One is the RTK/Ras/MAPK pathway, in which a receptor tyrosine kinase (RTK) is activated upon binding to its ligand, leading to activation of the Ras small GTPase, which in turn leads to activation of a MAP kinase (MAPK) cascade that culminates in regulation of transcriptional targets by Ets family transcription factors (Shilo 2005; Doroquez and Rebay 2006).

The other pathway is the Notch pathway (Doroquez and Rebay 2006; Fortini 2009; Kopan and Ilagan 2009; Artavanis-Tsakonas and Muskavitch 2010). Transmembrane DSL ligands Delta (Dl) and/or Serrate (known as Jagged in mammals) bind to Notch receptors on adjacent cells, leading to cleavage and release of a Notch intracellular fragment (N^{intra}), which enters the nucleus and binds to a CSL DNA-binding transcription factor [Su(H) in *Drosophila*]. In the absence of N^{intra}, Su(H) associates with corepressors and functions as a repressor. However, Su(H) in complex with N^{intra} and coactivators activates transcription of Notch targets. One important Notch pathway target is the E(spl) complex of genes, and in most contexts, E(spl) proteins mediate the effects of Notch signaling.

The important interplay that occurs between RTK/Ras/MAPK and Notch signaling pathways during specification of *Drosophila* ommatidial cell types has been noted elsewhere (Voas and Rebay 2004; Shilo 2005; Sundaram 2005; Doroquez and Rebay 2006), and is described in more detail below. In addition, the *split ends* gene has recently been shown to participate via an unknown mechanism in cross-talk between these two pathways (Doroquez et al. 2007).

Specification of R8: The Founding Cell of Each Ommatidium

The R8 photoreceptor is the founding cell of each ommatidium: it is the first cell to be specified, and as described below, it is required to initiate recruitment of other ommatidial cells. Thus, selection and establishment of R8 photoreceptors is a particularly critical step in eye development. As with other types of neuronal precursors, selection of R8s occurs via interactions between proneural proteins and Notch-mediated lateral inhibition (Baker 2002; Bertrand et al. 2002; Frankfort and Mardon 2002; Quan and Hassan 2005; Bray 2006; Doroquez and Rebay 2006; Powell and Jarman 2008; García-Bellido and de Celis 2009; Weinmaster and Fischer 2011).

Briefly, neuronal development begins with expression of proneural proteins, class II bHLH transcription factors required for neuronal fate (Massari and Murre 2000). Initially, particular proneural proteins are expressed at low levels in discrete clusters of cells that are positioned by the activities of signaling pathways and transcription factors to form a prepattern. All cells in each proneural cluster have the potential to develop as a neuronal precursor. However, through lateral inhibition, proneural expression is eventually switched off in all but one or a few cells in the cluster.

The cell(s) that ultimately expresses the proneural protein develops as a neuronal precursor(s) (Fig. 3a).

The precise mechanisms of lateral inhibition are still not completely clear (Baker 2000; Bray 2006; García-Bellido and de Celis 2009). To summarize, in each proneural cluster, the cell with the highest level of proneural protein becomes a Dl-signaling cell and activates Notch signaling in surrounding cells. As a result of Notch activity, proneural protein levels drop in Dl-receiving cells, but high levels of proneural protein are maintained in the Dl-signaling cell, which develops as a neuronal precursor. In the absence of components of the Notch signaling pathway, all cells in proneural clusters maintain high levels of proneural proteins and become neuronal precursors, resulting in a so-called neurogenic phenotype.

The class II bHLH transcription factor Atonal (Ato) serves as the proneural protein for R8 selection. Ato is required for establishment of the R8 photoreceptor: complete loss-of-function mutations in the *ato* gene result in failure of R8 development. Since R8 is required for recruitment of other ommatidial cells, loss of *ato* also results in complete loss of all other ommatidial cell types except for a few pigment cells and interommatidial bristles (Jarman et al. 1994).

The expression pattern of the Ato protein is very dynamic, reflecting the process by which R8s are selected from among surrounding cells (Jarman et al. 1994; Jarman et al. 1995; Baker et al. 1996; Dokucu et al. 1996; Baker and Yu 1997; Sun et al. 1998; Domínguez 1999; Baker 2002; Frankfort and Mardon 2002; Hsiung and Moses 2002; Brown et al. 2006; Zhang et al. 2006; Tanaka-Matakatsu and Du 2008). I will summarize the basic steps here and provide more details further.

1. Ato is first expressed in all cells in a striped pattern that spans $\sim$4 cell diameters ahead of the furrow. Within this stripe, all of the Ato-expressing nuclei are localized basally.
2. Ato is upregulated in initial clusters in a process termed proneural enhancement.
3. Within the furrow, Ato expression becomes limited to groups of $\sim$15 cells (equivalent to proneural clusters) referred to as intermediate groups.
4. A group of 2–3 Ato-expressing nuclei located posteriorly within the intermediate group migrate apically to form the R8 equivalence group. Ato expression is lost in the other, still basally localized nuclei in the intermediate group.
5. By the arc stage of ommatidial development, Ato expression becomes restricted to just the R8 precursor (Fig. 3b, c).

Two regions flanking the *ato* gene contain enhancer elements that regulate *ato* transcription during eye development (Fig. 3d; Sun et al. 1998; Zhang et al. 2006; Tanaka-Matakatsu and Du 2008). A 3′ enhancer controls the first two steps listed above: initiation of Ato in a stripe anterior to the furrow as well as upregulation and limitation of Ato to the intermediate groups. Both these steps depend on regulatory elements that respond to signals and transcription factors other than Ato itself. In contrast, a 5′ enhancer controls steps 3 and 4 and is subject to Ato autoregulation.

A number of reviews have discussed the mechanisms that regulate aspects of *ato* expression (Frankfort and Mardon 2002; Hsiung and Moses 2002; Doroquez and Rebay 2006; Bao 2010; Kumar 2012). What follows focuses on what is known about direct regulation of *ato* transcription.

Phase I, Ato Initiation: Activating and Repressive Elements in the 3′ Enhancer

Based on work from two groups (Zhang et al. 2006; Tanaka-Matakatsu and Du 2008), there are two *cis*-regulatory elements in the 3′ *ato* enhancer that drive initiation of *ato* transcription in a stripe of cells anterior to the furrow during furrow progression (Fig. 3d): (1) Two conserved elements necessary to activate *ato* expression (A1 and A2) are located approximately 5 kb downstream of the 3′ end of the *ato* coding region. (2) A repressor element is present approximately 4.5 kb downstream of *ato* that, when deleted, allows *ato* to be transcribed further anteriorly compared to wild type (Zhang et al. 2006; Tanaka-Matakatsu and Du 2008).

The Hedgehog (Hh) and Decapentaplegic (Dpp) signaling pathways, which regulate furrow progression, have both been implicated in initial *ato* activation (Heberlein et al. 1995; Baker et al. 1996; Baker and Yu 1997; Domínguez and Hafen 1997; Borod and Heberlein 1998; Domínguez 1999; Greenwood and Struhl 1999; Curtiss and Mlodzik 2000; Baonza and Freeman 2001; Fu and Baker 2003; Doroquez and Rebay 2006; Roignant and Treisman 2009; Bao 2010). Transcription factors known to affect *ato* initiation include the Retinal Determination (RD) factors Eyeless (Ey), Eyes absent (Eya), Sine oculis (So), and Dachshund (Dac; Suzuki and Saigo 2000; Silver and Rebay 2005; Zhang et al. 2006; Tanaka-Matakatsu and Du 2008; Kumar 2009, 2010). These transcription factors are expressed in overlapping zones that parallel the MF and propagate with it under the control of Hh, Dpp, and Notch signaling (Fig. 3e, f; Bonini et al. 1993; Cheyette et al. 1994; Halder et al. 1998; Curtiss and Mlodzik 2000; Bessa et al. 2002; Fu and Baker 2003; Firth and Baker 2009).

Both A1 and A2 are required to activate *ato* expression during initiation: neither sequence is sufficient to drive expression by itself. Zhang et al. 2006 identified a conserved binding site in A2 for the Dpp pathway transcription factor Mad, however whether this site functions during *ato* initiation is currently unknown. In contrast, Ey and So have been shown to bind directly to A1 in vitro. In addition, Ey, Eya+So, and Eya+So+Dac are all capable of ectopically activating *ato* expression via this element. Since the Ey, Eya, So, and Dac expression patterns overlap with initial low-level Ato expression (Fig. 3e, f), it is likely that the four RD factors directly participate in *ato* activation via the A1 element (Zhang et al. 2006; Tanaka-Matakatsu and Du 2008), perhaps with direct input from the Dpp pathway via the A2 element.

Clones of loss- or gain-of-function mutations in components of the Hh signaling pathway autonomously affect the initial low-level expression of Ato, suggesting that Hh signaling, in addition to the RD factors and Dpp signaling, is also directly involved in initiating *ato* transcription (Domínguez and Hafen 1997; Domínguez 1999; Greenwood and Struhl 1999). However, the mechanism involved (e.g., through binding of the A1/A2 element by the pathway-specific transcription factor Ci155) is currently unknown.

The identities of the factors that mediate repression via the 3′ enhancer have been elusive. Two candidates are the bHLH repressor protein Hairy and the class V HLH protein Extramacrochaete (Emc). Hairy and Emc repress proneural protein expression during development of the mechanosensory bristles (Ellis et al. 1990; Garrell and Modolell 1990; Van Doren et al. 1991, 1994; Ohsako et al. 1994; Campuzano 2001; Usui et al. 2008). Accordingly, in the eye, it initially appeared that Hairy/Emc and Ato are expressed in mutually exclusive domains, with Hairy/Emc expressed in an anterior zone that ends abruptly where Ato expression initiates (Brown et al. 1995; Greenwood and Struhl 1999). Using a hypomorphic allele of *emc*, Brown et al. 1995 found that whereas neither *hairy* nor *emc* single mutant clones affected Ato expression, double mutant clones resulted in premature Ato expression and faster furrow progression. These results suggested that Hairy and Emc are both involved in negatively regulating *ato* transcription.

Hairy in particular was thought to help define the "pre-proneural" (PPN) zone, which is marked on the anterior by the onset of expression of Hairy and the RD proteins Eya, So, and Dac, and on the posterior by Ato initiation. Cells in the anterior region of the PPN zone do not express Ato, in spite of the fact that Ey, Eya, So, Dac, and presumably Dpp are all present there, suggesting the presence of a repressor that is later switched off to allow Ato expression (Greenwood and Struhl 1999; Baonza and Freeman 2001; Bessa et al. 2002). Hairy appeared to be a good candidate for this repressor because it appeared to be switched off at the point at which Ato switched on. Furthermore, in various mutant backgrounds affecting furrow progression, loss of Ato or of Ato targets correlates with posterior maintenance of Hairy expression (Greenwood and Struhl 1999; Baonza and Freeman 2001; Fu and Baker 2003), suggesting that when furrow progression is compromised, Hairy is maintained and is capable of repressing *ato* transcription. One additional piece of evidence, consistent with the idea that Hairy could be the repressive factor preventing premature *ato* activation, is the fact that in the absence of the 3′ repressive elements *ato* expression expands all the way to the anterior edge of the PPN zone (Zhang et al. 2006; Tanaka-Matakatsu and Du 2008) where Hairy expression initiates.

Other results, however, call this hypothesis into question. First, Zhang et al. 2006 were unable to find a consensus Hairy binding site in the repressive element of the 3′ *ato* enhancer. Second, although loss of Ato and gain of Hairy correlate when furrow progression fails, it has never been directly shown that ectopic Hairy expression is capable of negatively regulating *ato* expression. Thus, the effects on Ato and Hairy of mutants that affect furrow progression could be independent. Third, rather than having mutually exclusive expression patterns as initially described, Hairy and Emc in fact overlap with initial low-level Ato expression, switching off only at the point at which Ato proneural enhancement occurs (Baonza and Freeman 2001; Li and Baker 2001; Bhattacharya and Baker 2012; Fig. 3f). Fourth, a recent report using a null allele of *emc* has shown that faster furrow progression can be explained solely by loss of *emc*, independently of *hairy* (Bhattacharya and Baker 2012). Thus, a role for Hairy in furrow progression and *ato* initiation appears to be limited at best. See below for more information about Emc's role in regulating *ato* expression.

Another possibility for control of the 3′ *ato* repressive element is Su(H), which is a transcriptional effector of the Notch signaling pathway. As mentioned above, in the absence of Notch signaling, Su(H) acts as a repressor of Notch targets, and is only converted to an activator in the presence of Notch signaling. Unlike Hh and Dpp signaling, Notch signaling is not required for initial *ato* activation (Baker and

Yu 1997; Baonza and Freeman 2001; Li and Baker 2001). In contrast, Li and Baker 2001 have shown that initial low-level Ato expression is found in more anterior regions in Su(H) clones. However, whether Ato repression by Su(H) occurs by direct binding to the 3′ *ato* repressive element or via some other unknown mechanism, and how repression by Su(H) is relieved to allow *ato* initiation are currently unanswered questions.

Phase 2, Proneural Enhancement

A few cell diameters after its expression initiates, Ato is upregulated and becomes restricted to initial clusters via proneural enhancement. Conserved regulatory elements located within the 3′ *ato* enhancer approximately 4 kb downstream of *ato* (IC1 and IC2) have been shown to promote its expression in intermediate groups (Sun et al. 1998; Zhang et al. 2006). This region of the 3′ *ato* enhancer contains a conserved binding site for Su(H), the Notch pathway transcription factor (Fig. 3d). Accordingly, Notch signaling activated by Dl is required for Ato proneural enhancement during this stage (Baker and Yu 1997; Baonza and Freeman 2001; Li and Baker 2001).

As discussed earlier, in most cases of Notch signaling, Su(H) is converted to an activator when bound to the Notch intracellular domain fragment released following ligand binding, leading to activation of E(spl) target genes, which mediate downstream effects. However, Su(H) activator function is not required for upregulation of Ato protein (Ligoxygakis et al. 1998). This has led Li and Baker 2001 to postulate that Notch signaling simply relieves repression of *ato* by Su(H), rather than through activation of E(spl)-C expression.

Ensuring High Levels of Ato: bHLH Proteins, Autoregulation, and the 5′ ato Enhancer

Recent findings suggest that Emc participates in a transcriptional network with another bHLH protein, Daughterless (Da), that regulates the onset of high levels of Ato expression (Bhattacharya and Baker 2011). Da is a broadly expressed class I bHLH protein that heterodimerizes with spatially regulated class II bHLH proteins including Ato, to regulate transcription of target genes (Jarman et al. 1993; Rosay et al. 1995; Massari and Murre 2000). One of the targets of the Ato/Da heterodimer is Ato itself (Jarman et al. 1995; Brown et al. 1996; Sun et al. 1998; Melicharek et al. 2008). Accordingly, Da is expressed in all eye precursor cells but at higher levels close to the furrow, and it is required for normal progression of Ato expression with the furrow (Brown et al. 1995, 1996; Bhattacharya and Baker 2011).

As a class V HLH protein (Massari and Murre 2000), Emc is capable of dimerizing with other HLH proteins, including Ato and Da, but lacks the basic domain that would allow it to interact with E-box sites in DNA. Emc has been shown in other contexts to prevent proneural gene autoactivation by binding directly to the proneural protein (Ellis et al. 1990; Garrell and Modolell 1990; Van Doren et al. 1991; Cubas and Modolell

1992; Campuzano 2001). As mentioned above, Emc is known to prevent premature Ato expression and speeding up of the furrow (Brown et al. 1995; Bhattacharya and Baker 2012). Conversely, ectopic Emc expression does not prevent Ato initiation but does prevent Ato proneural enhancement (Bhattacharya and Baker 2011).

Bhattacharya and Baker (2011) found that Emc and Da are both Da targets, and that Emc inhibits Da expression, presumably by binding to Da and preventing Da/Da homodimers from binding and activating the *da cis*-regulatory element. This feedback loop keeps Da levels relatively low anterior to the furrow. However, Emc is also a target of negative regulation by the Hh and Dpp signaling pathways. Thus, as the furrow approaches, Emc levels plummet, allowing Da levels to rise at the same time as Ato levels are rising in response to the same signals (Lim et al. 2008). This would be predicted to favor the formation of Ato/Da heterodimers, leading to regulation of Ato targets, including *ato*, and the onset of neuronal differentiation. It seems likely that these events regulate activity at the 5′ *ato* enhancer, which is known to mediate *ato* autoregulation (Sun et al. 1998; Fig. 3d).

Phase 3, Formation of Intermediate Groups

Earlier in the chapter, I have discussed the mechanisms by which *ato* expression is increased in preparation for lateral inhibition. Additional information is required to limit Ato to evenly spaced intermediate groups. What the precise mechanisms regulating this process are has been controversial and is another interesting story (Doroquez and Rebay 2006; Roignant and Treisman 2009), but is outside the scope of this review.

Phases 4, 5, Ato Becomes Restricted to Individual R8 Precursors via Lateral Inhibition

Historically, there have been two proposed mechanisms for how single R8s are selected from intermediate groups. One is lateral inhibition, the other is the R8 equivalence model. Consistent with a role for lateral inhibition in R8 selection, loss-of-function mutations in components of the Notch pathway (e.g., using temperature-sensitive alleles to block Notch signaling after proneural enhancement has already occurred) results in all intermediate group cells maintaining Ato expression and differentiating into R8 precursors—a neurogenic phenotype (Cagan and Ready 1989a; Baker and Zitron 1995; Baker et al. 1996; Lee et al. 1996; Baker and Yu 1997, 1998; Ligoxygakis et al. 1998; Li and Baker 2001; Frankfort and Mardon 2002; Hsiung and Moses 2002; Doroquez and Rebay 2006).

The R8 equivalence group model was introduced to explain observations that did not seem to fit with the lateral inhibition model. For instance, loss-of-function mutations in the *rough* (*ro*) gene result in formation of three R8s per intermediate group (Kimmel et al. 1990; Heberlein et al. 1991; Dokucu et al. 1996). In addition, Ro misexpression in R8 precursors prevents them from differentiating properly. Based on these data, the R8 equivalence group model postulates that three cells in each intermediate group have the potential to become R8, but Ro expression in two of them prevents them from developing as R8 cells, leaving a single R8 precursor (Kimmel et al. 1990; Dokucu et al. 1996).

Another piece of evidence that initially seemed to support the equivalence group model comes from studies of *senseless* (*sens*), a direct Ato target that encodes a bHLH transcription factor required for maintenance of the R8 fate following R8 selection (Nolo et al. 2000; Frankfort et al. 2001; Pepple et al. 2008). Sens protein is initially observed in the three-cell R8 equivalence group and subsequently becomes limited to just the R8 precursor. In *sens*$^-$ mutants, Ro is expressed ectopically in all three R8 equivalence group cells, and all three cells, but no more as might be expected from a failure of lateral inhibition, develop as R2/R5-type photoreceptors.

A recent report (Pepple et al. 2008) has made progress towards distinguishing between the lateral inhibition and R8 equivalence models for R8 selection. These authors demonstrate that single R8 cells *are* initially generated in *ro* mutants. The two extra R8s observed in *ro* mutants (Dokucu et al. 1996) are only added later. Thus, formation of R8s in the eye appears to occur via lateral inhibition, similar to the development of other neuronal precursors.

The reason that multiple R8s form in *ro*$^-$ mutants appears to be that, after the furrow moves on and the effects of lateral inhibition fade, Ro is necessary to repress Sens expression in the other two cells of the R8 equivalence group, allowing them to develop as R2 and R5. Accordingly, misexpressing Sens in all eye precursors leads to formation of multiple cells with R8 properties, and *sens*$^-$, *ro*$^-$ double mutant eye precursors are capable of developing as photoreceptors that resemble R8s in a number of respects (Frankfort et al. 2001). Ro binds directly to *sens* regulatory elements to prevent Sens expression in the R2 and R5 precursors (Pepple et al. 2008).

One candidate target of Notch signaling [via E(spl)] during lateral inhibition is likely to be Da. In spite of its probable earlier role in promoting high levels of *ato* expression (see previous discussion), Da is also required for *ato* repression during lateral inhibition, and high levels of ectopic Da are capable of repressing Ato expression (Lim et al. 2008; Melicharek et al. 2008).

Although lateral inhibition is likely to mediate R8 selection, there is no question that an R8 equivalence group exists. It may even have a counterpart during bristle development (Cubas et al. 1991; Culí and Modolell 1998): prior to selection of the neuronal precursor that gives rise to bristle components, the Achaete/Scute proneural proteins are expressed at higher levels in a few cells in each proneural cluster, and these cells also express Sens. As with the R8 equivalence group during R8 development, the position of the proneural field and subsequently the neuronal precursor within the proneural cluster in the context of bristle development is remarkably consistent. This suggests that unknown factors are involved in ensuring highest levels of proneural factor expression at a particular position.

Thus, because of the parallels with neuronal precursor selection in other contexts in both flies and in vertebrates, further study of R8 selection, of the regulation of Ato expression, and of lateral inhibition during eye development is relevant for neuronal precursor selection in other contexts. Other potential pieces of the puzzle have recently been identified that have not yet been well incorporated into current thinking (Lim et al. 2008; Melicharek et al. 2008; Steele et al. 2009; Distefano et al. 2012), and further study of their roles is likely to prove illuminating.

Switching Ato off

Ato is expressed in R8 precursors for 3–4 columns before it is switched off. The BarH1 and BarH2 homeodomain transcription factors are expressed in a complementary pattern to Ato, and are required for *ato* repression posterior to the furrow. This is important because Ato maintenance leads to formation of ectopic R8s that are capable of recruiting additional ommatidial cells and disrupt the precise ommatidial pattern (White and Jarman 2000; Lim and Choi 2003, 2004). Unlike Ato itself, Ato's target Sens is maintained in R8 cells throughout larval and into pupal development (Frankfort et al. 2001; Domingos et al. 2004a; Frankfort et al. 2004). Sens has roles in regulating a number of aspects of R8 differentiation (Domingos et al. 2004a; Xie et al. 2007; Morey et al. 2008). Sens also has roles in regulating recruitment of additional ommatidial cells (see further).

Egfr Signaling and Recruitment of Ommatidial Cells Other Than R8

The recruitment of additional cells to join R8 in forming an ommatidium involves intricate and reiterative control of the Epidermal growth factor receptor (Egfr)-signaling pathway (Freeman 1996, 1997; Kumar et al. 1998). The *Drosophila melanogaster* genome contains a single Egfr, which has greatly facilitated analysis of its roles during development. The canonical Ras-Raf-MAPK cascade, culminating in transcriptional regulation via Ets domain transcription factors, appears to be the predominant mechanism for Egfr signal transduction in *Drosophila* (Shilo 2005; Doroquez and Rebay 2006). There are several activating Egfr ligands in *Drosophila* (Shilo 2005; Doroquez and Rebay 2006), including Spitz (Spi) and Keren, which have redundant roles during eye development, with Spi being the major and Keren the minor ligand (Brown et al. 2007). There is also an inhibitory ligand, Argos (Aos: Schweitzer et al. 1995a), which is also a target of the Egfr signaling pathway (Golembo et al. 1996b).

The idea that Egfr signaling has a role in ommatidial cell recruitment first came from data obtained from eye precursor tissues containing clones homozygous for loss-of-function mutants for Egfr pathway components. In these clones, R8 photoreceptors are specified, but no other ommatidial cells develop (Domínguez et al. 1998; Kumar et al. 1998; Spencer et al. 1998; Lesokhin et al. 1999; Wasserman et al. 2000; Baonza et al. 2001; Yang and Baker 2001). Similar, though weaker, phenotypes are obtained in eye tissue homozygous for loss-of-function mutations in *spi* (Freeman 1994b; Tio et al. 1994; Tio and Moses 1997), with stronger effects when *spi* and *keren* are simultaneously removed (Brown et al. 2007). In contrast, loss-of-function mutations in *aos*, which encodes an inhibitory ligand, result in over-recruitment of photoreceptors, cone cells, and pigment cells (Freeman 1994a).

Use of a dominant negative Egfr by Freeman 1996 enabled a more controlled analysis of which ommatidial cells require Egfr activity for recruitment, and confirmed that, in addition to photoreceptors other than R8, Egfr signaling is required for recruitment of the cone and pigment cells. Conversely, Freeman 1996 showed that overactivation of the Egfr pathway causes recruitment of extra ommatidial cells, including photoreceptors, cone cells, and pigment cells. These data strongly show that the Egfr pathway is required to recruit ommatidial cells other than R8.

As mentioned above, the presence of an R8 precursor is necessary for recruitment of other ommatidial cells, and R8 development depends on Ato. At some level, then, Ato must control activation of Egfr signaling in cells surrounding R8. Accordingly, artificial maintenance of Ato in R8 past the point at which Ato is normally switched off results in excessive photoreceptor recruitment, and mutations in Egfr pathway components can suppress this effect (White and Jarman 2000).

One way that Ato could activate Egfr activity in surrounding cells is by activating expression of the ligands, Spi and Keren in R8. However, this does not appear to be likely because both appear to be expressed fairly broadly (Tio et al. 1994; Tio and Moses 1997; Reich and Shilo 2002; Urban et al. 2002). However, whereas the *spi* gene is broadly transcribed, its function is more limited: it is almost absolutely required in R8, strongly required in R2 and R5, and more weakly required in R3 and R4 (Freeman 1994b; Tio et al. 1994). Spi needs to undergo processing and cleavage via the chaperone Star and a Rhomboid family serine protease in order to be secreted and active (Schweitzer et al. 1995b; Golembo et al. 1996a; Guichard et al. 1999; Pickup and Banerjee 1999; Bang and Kintner 2000; Wasserman et al. 2000; Lee et al. 2001; Urban et al. 2001; Freeman 2002). In contrast to Spi/Keren, *Star* and *rhomboid-1* (*rho-1*) are specifically expressed in R8, R2, and R5 (Heberlein et al. 1991; Freeman et al. 1992a; Heberlein et al. 1993; Kolodkin et al. 1994; Spencer et al. 1998). Thus, it is possible that Ato regulates expression of *Star* and/or *rho-1* to control where Spi/Keren are secreted.

In *Drosophila* chordotonal (stretch) sensory organs, which develop in a similar fashion to ommatidia with initial selection of a subset of Ato-expressing precursors that subsequently use Egfr signaling to recruit additional precursors, *ato* is in fact required for *rho-1* expression, though it is not clear if this is direct or indirect (Lage et al. 1997; Okabe and Okano 1997). In addition, Ato can activate ectopic *rho-1* expression when ectopically expressed during eye development (Baonza et al. 2001).

However, in contrast to most other contexts including chordotonal organs, in the *Drosophila* eye *rho-1* functions redundantly with *rho-3* (aka *roughoid*). Mutations in *rho-1* have only subtle effects on photoreceptor development, and *rho-3* appears to be broadly expressed rather than being expressed specifically in R8, R2, and R5 (Freeman et al. 1992a; Wasserman et al. 2000). These two facts make it seem less likely that Ato controls Spi secretion solely by regulating expression of the Rhomboid proteases, though future analyses of the *rho-3* expression pattern may suggest otherwise. It is also possible that specific activation of Rho-1 is sufficient, and/or Ato regulates expression of the chaperone *Star* or some other unknown factor that regulates Spi/Keren secretion.

Double but not single mutants for the *Drosophila* Retinoblastoma family transcription factor gene, *rbf*, and the *rhinoceros* (*rno*) gene, which encodes a nuclear protein with a PHD zinc-finger domain, show no delay in R8 development but do show a delay in subsequent recruitment of other ommatidial cells (Voas and Rebay 2003; Sukhanova et al. 2011). This suggests that *rbf* and *rno* have synergistic roles in promoting ommatidial cell recruitment by R8. Accordingly, loss of *rbf* or of both *rbf* and *rno* results in loss of Rhomboid and pERK. Furthermore, whereas loss of *rbf* results in partial loss of the Pointed-P1 (pntP1) gene, which is an activating Ets family transcription factor in the Egfr pathway and is itself an Egfr signaling target (Gabay et al. 1996), loss of *rno* or of both *rno* and *rbf* results in a stronger loss of PntP1. Loss-of-function of *Drosophila* E2F1, which is bound and inhibited by RBF, can suppress the differentiation delay of *rbf*, *rno* double mutants, suggesting that deregulated E2F1 is one of the factors responsible for the delay (Sukhanova et al. 2011).

Limiting Egfr Activity to Allow for Sequential Photoreceptor Recruitment

Spitz/Keren are secreted ligands, and it is not immediately obvious why they do not recruit all cells surrounding R8 as photoreceptors at the same time, but instead in the precise sequence laid out by Tomlinson and Ready 1987b. One possible answer to this question was first clearly articulated by Matthew Freeman, in his now classic model for how sequential recruitment might occur (Freeman 1997). At the time, there was mounting evidence suggesting broad expression of Spi protein, but limited secretion and function only in the first few photoreceptors to become specified, as described previously. Additional experiments had shown *aos*, which as mentioned before encodes an inhibitory ligand, to be a direct target of Egfr signaling (Golembo et al. 1996b), and had demonstrated that Aos protein has a greater range of activity compared to Spi: Spi can only activate Egfr within 3–4 cell diameters, whereas the range of Aos is 10–12 cell diameters (Freeman et al. 1992b; Freeman 1994b).

Based on these data, Freeman 1997 proposed that Spi is initially released from the R8 precursor at the arc stage of ommatidial development when R8 lies at the apex of the arc, flanked by the future R2 and R5. Activation of Egfr in R2 and R5 leads these cells to secrete Spi as well, recruiting R3 and R4 precursors, which lie further out along the arc (Fig. 2b). Meanwhile, Aos is also expressed and secreted from each recruited ommatidial cell in response to Egfr pathway activity, and represses Egfr activity in cells farther away from the source of the Spi ligand, preventing premature recruitment of cells. This process then continues, leading to recruitment of R1 and R6, R7, the cone cells, and finally the pigment cells. Thus, in the Freeman model, the ordered, sequential nature of photoreceptor recruitment depends on a feedback mechanism, whereby activation of Egfr activity also induces negative regulation of Egfr activity. Additional negative feedback regulatory mechanisms have since been discovered that likely have similar roles (reviewed e.g., in Shilo 2005).

Interesting recent work has explored cell biological mechanisms that ensure short-range rather than long-range Spi secretion during *Drosophila* eye development (Yogev et al. 2011), allowing for controlled, sequential photoreceptor recruitment. Following translation on ER-bound-ribosomes, the transmembrane Spi precursor protein (mSpi) is retained in the ER through COPI-dependent retrograde trafficking (Lee et al. 2001; Tsruya et al. 2002; Schlesinger et al. 2004). Secretion of the mature Spi ligand (sSpi) depends on its interaction with the chaperone Star. In most contexts in *Drosophila* in which Egfr signaling is important, Star mediates transport of mSpi to the late compartment of the secretory pathway (Lee et al. 2001; Tsruya et al. 2002), where the Rho-1 protease is localized. Rho-1 then carries out intramembrane cleavage of mSpi to release sSpi (Urban et al. 2001), which is subsequently secreted and affects development in a long-range fashion.

However, two factors present during *Drosophila* eye development change the dynamics of Spi secretion from long-range to short-range. The first factor is *rho-3/roughoid*, which is only expressed during eye development where limiting the range of Spi secretion is important (Wasserman et al. 2000). In contrast to Rho-1 protein, which is localized exclusively to the late secretory compartment, Rho-3 is also found in the ER (Yogev et al. 2008) where it comes into contact not only with Spi but also with the Star chaperone. Rho proteases are known to cleave and inactivate Star (Tsruya et al. 2007). Thus, in the *Drosophila* eye, Rho-3 inactivates Star, resulting in less Spi being trafficked to the late secretory compartment for cleavage by Rho-1 or Rho-3 (Yogev et al. 2008).

Second, the PLCγ encoded by the *small wing* (*sl*) gene (the only PLC γ encoded by the *Drosophila* genome) is required to prevent release of sSpi produced in the ER by Rho-3 (Thackeray et al. 1998; Schlesinger et al. 2004). The mechanism by which Sl accomplishes this is not completely clear. In addition to its catalytic domain, Sl contains a number of protein–protein interaction motifs, including SH2, SH3, and PH domains, and appears to function as a scaffolding protein in addition to its role as a PLC. Sl's scaffolding activity, but not its PLC catalytic activity, is required for its role in regulating Egfr activity during *Drosophila* eye development (Mankidy et al. 2003). In addition, it is known that Sl is a negative regulator of Egfr signaling that acts upstream of MAPK (Thackeray et al. 1998). Nevertheless, the data presented above make it clear that even prior to the secretion of the ligands, mechanisms exist to regulate Egfr activity and ensure controlled recruitment of ommatidial cells.

Limiting Egfr Signaling in the R8 Equivalence Group

Markers for Egfr activity, including phosphorylated MAPK (pERK) and the Egfr pathway targets *argos* and *pointed P1*, are first expressed above background levels in clusters of cells in the furrow that correspond to Ato intermediate groups. In conjunction with Ato expression, expression of pERK subsequently focuses down to just a few cells, which are likely to be the R8 equivalence group (Gabay et al. 1997; Golembo et al. 1996b; Kumar et al. 1998, 2003; Spencer et al. 1998; Chen and Chien 1999; Lesokhin et al. 1999; Wasserman et al. 2000; Rawlins et al. 2003; Spencer and Cagan 2003; Frankfort and Mardon 2004). Ato appears to play a critical role in activating Egfr signaling in these cells because in clones mutant for *ato* or *da* pERK expression is lost, and ectopic Ato also correlates with ectopic pERK (Chen and Chien 1999).

As mentioned before, R8 does not require Egfr signaling for its specification. Instead, high levels of Egfr activity in the intermediate groups may be required for their spacing, though this is controversial (reviewed in Doroquez and Rebay 2006; Bao 2010). Nevertheless, recent reports have identified mechanisms that keep Egfr activity levels from rising too high at these early stages and thereby disrupting development. Two groups have shown that the L1-like cell adhesion molecule Echinoid (Ed) reduces the intensity/duration of Egfr signaling in intermediate groups and R8 equivalence groups without altering the pattern, possibly by direct interaction between Ed and Egfr. This appears to be necessary to prevent Ato levels from rising too high in response to Egfr signaling, resulting in more than one equivalence group cell from developing as an R8 precursor (R8 "twinning"; Spencer et al. 1998; Rawlins et al. 2003; Spencer and Cagan 2003).

Based on expression of pERK expression and of the Egfr signaling pathway targets *aos* and *pnt* expression, in spite of the fact that R8 precursors themselves secrete Spi, they have low or no Egfr activity compared to their neighbors (Chen and Chien 1999; Lesokhin et al. 1999; Frankfort and Mardon 2004). Moreover, high levels of Egfr signaling in R8 precursors appear to interfere with their ability to differentiate as R8s and recruit other ommatidial cells (cf. Lesokhin et al. 1999; Yamada et al. 2003; Frankfort and Mardon 2004). Accordingly, two reports suggest mechanisms that prevent pre-R8s from responding to the Spi that they secrete. One of these involves EDL/MAE, a protein capable of interacting with the Ets transcription factor Pointed P2 (PntP2), which activates *pntP2* expression in response to Egfr signaling. EDL/MAE is expressed first in R8 and subsequently in R2/R5, and appears to prevent PntP2 transcriptional activation, suggesting that it subverts autocrine activation of Egfr by Spi (Yamada et al. 2003). The other mechanism for preventing R8 from responding to the Spi ligand it releases appears to be transcriptional repression of *pointed* by Sens (Frankfort and Mardon 2004; Fig. 4a, a').

Specifying the Identities of Different Ommatidial Cells

There are no strict cell lineage relationships among ommatidial cells, suggesting that cell fate determination occurs by induction (Ready et al. 1976; Lawrence and Green 1979; Tomlinson and Ready 1987b). As described before, Egfr is required for the recruitment of cells to each growing ommatidial cluster, but that does not appear to be sufficient to specify their identities. Instead, specification requires interplay between two signaling pathways: (1) Ras/MAPK via Egfr and the Sevenless (Sev) RTK and (2) Notch, as well as cell type-specific transcription factors, as described in more detail below.

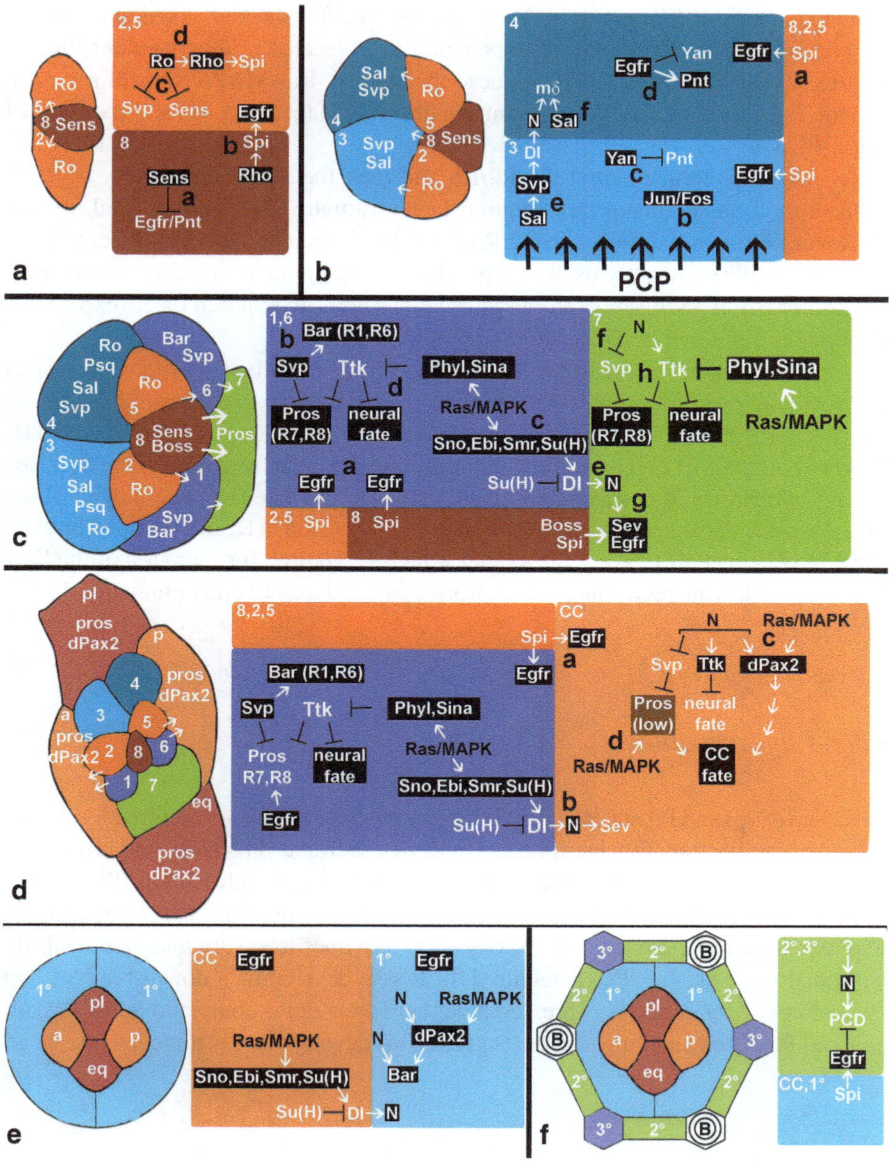

Fig. 4 Recruitment of ommatidial cells. Diagrams in (**a**–**e**) depict recruitment of **a** R2 and R5 photoreceptors; **b** R3 and R4 photoreceptors; **c** R1, R6, and R7 photoreceptors; **d** cone cells (CCs); **e** primary pigment cells (1°); **f** Roles of Notch and Egfr signaling in regulating programmed cell death (PCD) during recruitment of the secondary (2°) and tertiary (3°) pigment cells. See text for details. *B* bristle, *a* anterior cone cells, *p* posterior cone cells, *pl* polar cone cells, *eq* equatorial cone cells

Formation of the 5-Cell Precluster

At the time of specification of the R8 precursors (based on the point at which Ato becomes limited to a single cell), they form the center cells in the arc-/closed arc-stage ommatidial clusters (Brown et al. 2006). The R8 precursors are flanked in the arc by the presumptive R2/R5 photoreceptors, with R3/R4 located further out along the arc, next to R2 and R5, respectively. As described above, Ras/MAPK signaling via Spi and the Egfr are responsible for recruiting first R2/R5 (Fig. 4a, b') and then R3/R4 (Fig. 4b, a').

As they are recruited, first the R2/R5 and then the R3/R4 precursors begin to express the homeodomain transcription factor Rough (Ro; Kimmel et al. 1990). However, Ro is only required in R2/R5 for normal ommatidial development (Tomlinson et al. 1988). Exactly what activates Ro expression in these cells is not known, but Egfr or Hh signaling are both candidates since both are required for Ro expression (Domínguez et al. 1998; Domínguez 1999).

Ro appears to have at least three functions in R2/R5 cells: (1) it prevents R2/R5 from developing into R8s by directly binding to *sens* regulatory elements and repressing its expression in R2/R5 precursors (Dokucu et al. 1996; Frankfort et al. 2001; Pepple et al. 2008); (2) it is required to prevent R2/R5 cells from expressing Svp (see below), which would convert them to an R3/R4/R1/R6 fate (Heberlein et al. 1991; Fig. 4a, c'); and (3) it is required for *rhomboid* expression in R2/R5 (Tomlinson et al. 1988; Freeman et al. 1992a; Fig. 4a, d'), which is required for secretion of the Egfr ligand Spi, allowing recruitment of the R3/R4 pair and later of other ommatidial cells.

R3/R4 Specification

Seven-up (Svp) and Pipsqueak (Psq) are two transcription factors required for R3/R4 cell fate specification. Svp is expressed in R3/R4 as well as in R1/R6, and is required to prevent R3/R4 and R1/R6 precursors from developing as either R7 or R8 photoreceptors (Mlodzik et al. 1990; Miller et al. 2008). It is not currently clear what activates Svp expression in these cells, or exactly what Svp does to ensure that these cells do not transform their fate. Psq is required for R3/R4 development downstream of Svp (Weber et al. 1995), and is expressed at higher levels in R3 compared to R4 (Weber et al. 2008). The exact function of Psq in R3/R4 development is not currently clear.

Specification of R3 vs. R4 is Important for Planar Polarity

Unlike the R2/R5 and R1/R6 pairs, where there appears to be a single fate assumed by both cells, R3 and R4 are two different fates that lead to asymmetry within the ommatidium, as described earlier (Fig. 1b, c). Planar polarity signaling (PCP) is important for specification of the R3 vs. R4 fates (recently reviewed in Jenny 2010): PCP signaling is activated in R3, leading to activation of Jun/Fos, which are nuclear effectors of the PCP pathway (Weber et al. 2000, 2008; Fig. 4b, b'). In addition,

PCP signaling results in R3-specific transcription of the Notch ligand, *Dl*, and of *neuralized*, which encodes an E3 ubiquitin ligase required for Dl-Notch signaling, leading to activation of Notch signaling in R4 (Cooper and Bray 1999; Fanto and Mlodzik 1999; Tomlinson and Struhl 1999; del Alamo and Mlodzik 2006; Jenny 2010). A number of other factors that influence Notch signaling and the response to Notch signaling have also been shown to affect R3 vs. R4 fate specification (del Alamo and Mlodzik 2008; Zheng and Carthew 2008; Bhattacharya and Baker 2009; Singh and Mlodzik 2012).

Jun/Fos cooperate with the Ets family transcription factors Yan and Pnt to determine R3 vs. R4 fate. Jun/Fos and Yan are all required in the R3 precursor for the R3 fate. Yan functions to inhibit R4 fate, at least in part by inhibiting Pnt function (Fig. 4). In contrast, higher levels of Egfr signaling in the R4 precursor inactivate Yan and activate Pnt, promoting R4 fate (Weber et al. 2000, 2008; Fig. 4). It is unclear at this point what mechanisms trigger Egfr signaling to higher levels in the R4 vs. R3 precursor.

Svp and the zinc-finger transcription factors Spalt major and Spalt related (Salm and Salr), are also important for R3 vs. R4 cell fate. Sal appears to function upstream of Svp to mediate the interpretation of planar polarity signaling that leads to upregulation of Dl in R3, resulting in asymmetric distribution of PCP proteins (Fig. 4b, e'). In addition, Sal appears to be required in parallel to Notch signaling for expression of E(spl)mδ in R4 (Fig. 4b, f'; Fanto et al. 1998; Domingos et al. 2004a, b). It is not currently known what activates expression of Svp or Sal in R3/R4 cells, what the mechanism is by which Sal responds to PCP signaling, whether Svp functions directly to upregulate Dl, or what the targets of E(spl)mδ are during R4 cell fate specification.

After the 5-Cell Precluster

Cells not recruited into the 5-cell precluster undergo one more round of division during the second mitotic wave, and it is from this pool of cells that R1/R6, R7, the cone cells, and the pigment cells are recruited. As described before, based on mosaic analysis Rho-1, Rho-3, Star, and Spi are only required in R8, R2/R5, and possibly weakly in R3/R4, for development of a normal ommatidium. Nevertheless, recruitment of all the other ommatidial cells requires Egfr signaling. Presumably, this occurs via propagation of Spi outward from these three cells, which form the core of the nascent ommatidium.

R1/R6, R7, and cone cells form an equivalence group, based on the fact that their fates are interchangeable (see below). If Egfr is required for the recruitment of all these cells, then what provides the specificity? The answer appears to be that additional signals are required to distinguish among these cell types: Notch and the Sev RTK (recently reviewed in Mavromatakis and Tomlinson 2012a). Egfr and Sev signaling through the Ras/MAPK/Ets transcription factor pathway appear to be interchangeable (see below), so I will subsequently refer to signaling through either pathway as Ras/MAPK signaling.

Part of the interpretation of the information provided by various combinations of the Notch and Ras/MAPK pathways is provided by the RUNX family transcription factor Lozenge (Lz). Lz is initially expressed posterior to the furrow in all cells outside of the 5-cell precluster, and is later upregulated first in R1/R6, then R7, then the cone cells, and finally the pigment cells (Flores et al. 1998). Lz has a role in specification of all these cell types, although it does not appear to specify any particular fate per se (Batterham et al. 1996; Daga et al. 1996).

Lz is a direct target of Glass (Gl), which is a Zn-finger transcription factor cell autonomously required in all photoreceptors for their development (Moses et al. 1989; Moses and Rubin 1991), and of a transcriptional activation complex composed of two RD factors: Eya and So (Yan et al. 2003; Jemc and Rebay 2007; Morillo et al. 2012). Since Gl, Eya, and So are expressed in all cells posterior to the furrow (Moses and Rubin 1991; Bonini et al. 1993; Cheyette et al. 1994), it is not clear what prevents Lz expression in the 5-cell precluster. Yan, a repressor downstream in the Egfr signaling pathway, probably binds directly to the *lz* gene and keeps Lz levels low in uncommitted cells until they are recruited via Egfr signaling (Behan et al. 2002; Jackson Behan et al. 2005).

R1/R6

Following the second mitotic wave, the R1, R6, and R7 precursors are added to the 5-cell precluster (Ready et al. 1976; Tomlinson 1985). Based on neuronal markers, R1 and R6 begin to differentiate relatively quickly, while the onset of R7 differentiation occurs a few hours later (Tomlinson and Ready 1987b). As with the R3/R4 photoreceptor pair, recruitment of R1/R6 requires moderate levels of Ras/MAPK signaling, likely in response to Spi ligand emanating from adjacent R2/R5 and R8 cells (Fig. 4c, a').

Cell-specific transcription factors with known roles in R1/R6 development include Svp and BarH1/BarH2 (Bar). Svp appears to have a similar role in R1/R6 specification as it does in R3/R4: it is required to prevent R1/R6 precursors from developing as R7 or R8 photoreceptors (Mlodzik et al. 1990; Miller et al. 2008). The functionally redundant homeodomain transcription factors BarH1/BarH2 (Bar) are expressed in R1/R6 precursors and are autonomously required in these cells for their development (Higashijima et al. 1992a, b). In addition, expression of BarH1 in cone cell precursors can change at least a subset to an R1/R6 fate (Hayashi et al. 1998). However, so far it has not been investigated what happens to R1/R6 precursors in the absence of *BarH1/BarH2*, i.e., whether they fail to develop at all or whether they switch to some other photoreceptor type. Svp (Fig. 4c, b'), Gl, and Lz are all required for BarH1/BarH2 expression, and when expressed in all eye precursors Lz is capable of activating ectopic *BarH1/BarH2* expression (Higashijima et al. 1992a; Daga et al. 1996; Crew et al. 1997).

R7 and the Cone Cells Require Both Egfr and Notch Signaling

Unlike R1–R6, which appear to be recruited via a single signaling pathway—Egfr/Ras/MAPK signaling in response to Spi released from R8/R2/R5—recruitment of the R7 photoreceptor and the cone cells requires signaling through two pathways: Ras/MAPK and Notch (Cagan and Ready 1989a; Freeman 1996). The interplay between these two pathways determines what cell type will form, and occurs via two mechanisms, as described in more detail below: (1) Ras/MAPK signaling is responsible for inducing expression of the Notch ligand Dl in R1/R6 and R7 and (2) Ras/MAPK signaling relieves repression of photoreceptor development by Notch by leading to degradation of the Tramtrack (Ttk) repressor, which is a Notch target.

1. *Dl* transcription initiates in photoreceptors as they are recruited to nascent ommatidia (Parks et al. 1995). *Dl* transcription in photoreceptors is under control of the Egfr signaling pathway via two proteins called Ebi and Sno (Tsuda et al. 2002). Notch signaling is not required either to activate or repress Dl expression in photoreceptors. However, the Notch pathway transcription factor Su(H) represses Dl expression in this context. In the presence of Egfr signaling, Sno, Ebi, Su(H), and the *Drosophila* NCoR/SMRT corepressor homolog Smrter (Smr) form a complex or complexes that derepress Dl expression in a proteasome-dependent manner (Tsuda et al. 2002; Marygold et al. 2011). In addition to Egfr signaling, RBF/Rno regulate Ebi expression (Sukhanova et al. 2011).
   All of these proteins are conserved in vertebrates. Very little is known about the biochemical functions of either *Drosophila* Sno or its vertebrate orthologs (Coyle-Thompson and Banerjee 1993; Majumdar et al. 1997; Takano et al. 2011). Ebi, on the other hand, is the sole fly ortholog of vertebrate Transducin β-like protein 1 (TBL1) and TBL1-related protein (TBLR1), which are F box/WD40-containing factors. In vertebrates, TBL1 and TBLR1 form part of a corepressor silencing complex containing the NCoR/SMRT corepressor and HDAC3 (Guenther et al. 2000; Li et al. 2000; Zhang et al. 2002; Yoon et al. 2003; Tomita et al. 2004; Oberoi et al. 2011).
   Additional evidence suggests that upon phosphorylation, TBL1/TBLR1 are involved in the exchange of corepressors with coactivators for a number of signaling pathway transcription factors, probably by recruiting the ubiquitin-conjugating/19S proteasome complex via their F box domains (Perissi et al. 2004, 2010). One of the affected transcription factors is CBF1/RBPJ in the Notch signaling pathway, which is orthologous to the sole *Drosophila* CSL transcription factor Su(H) (Kao et al. 1998; Perissi et al. 2008, 2010).
   Thus, in both flies and vertebrates, Ebi/TBL1/TBLR1 interact with Su(H)/CBF1/RBPJ and the Smr/NCoR/SMRT corepressor, and in both contexts Ebi/TBL1/TBLR1 has been linked to transcriptional derepression that involves the proteasome. However, the vertebrate data do not clarify which kinase(s) are responsible in particular in vivo developmental contexts for phosphorylating TBL1/TBLR1, thus resulting in corepressor/coactivator exchange. In addition, what the targets of derepression may be in vivo are currently unknown.

In contrast, data from Tsuda et al. 2002 suggest that Egfr signaling is involved in promoting the Ebi-mediated derepression of Dl during *Drosophila* eye development. However, it has not been established in *Drosophila* that Ebi is a corepressor/coactivator exchange factor for Su(H), or how direct the effect is of Egfr signaling on Ebi function, or of Ebi/Su(H)/Smr on Dl derepression. Ascertaining whether MAPK or another kinase directly phosphorylates Ebi, and what the mechanisms are by which interactions among Ebi, Su(H), and Smr lead to Dl expression would simultaneously clarify the relationships between Egfr and Notch signaling, and would shed light on how the switch from corepression to coactivation occurs in vivo, as opposed to in tissue culture cells.
However, it should be noted that more recent experiments from *Drosophila* eye development have provided evidence that Ebi does not bind directly to *Dl*-regulatory elements (Tsuda et al. 2006). Instead, Su(H), Ebi, and Smr are all associated with regulatory elements in the *charlatan* (*chn*) gene, which encodes a C2H2-type zinc-finger protein with similarity to human NRSF/REST. Chn is likely a direct repressor of *Dl* during *Drosophila* eye development. Thus, the Ebi/Su(H)/Smr complex may instead be required to repress expression of *chn*, relieving repression of *Dl* by Chn (Tsuda et al. 2006). Future experiments are required to provide a resolution to this story.

2. The *tramtrack* (*ttk*) gene encodes two BTB-Zn finger proteins produced by alternative splicing: Ttk69 and Ttk88, which have the same N-termini but differ in their C-terminal DNA-binding domains (Read and Manley 1992; Xiong and Montell 1993). Loss-of-function mutations in *ttk* that affect the Ttk88 isoform result in a very mild phenotype that includes the formation of extra R7 cells in a few ommatidia (Xiong and Montell 1993; Lai et al. 1996; Yamamoto et al. 1996; Shi and Noll 2009). Loss of Ttk69 leads to defects in photoreceptor development, although these occur late in photoreceptor development (Tang et al. 1997; Lai and Li 1999). Conversely, ectopic Ttk69 or Ttk88 expression in photoreceptors inhibits their development (Li et al. 1997; Tang et al. 1997; Shi and Noll 2009).
Based on expression of an enhancer trap, *ttk* is transcribed in both photoreceptors and cone cells (CCs), but Ttk88 and Ttk69 proteins are expressed in CCs but not in photoreceptors (Lai et al. 1996, 1997). Thus, either Ttk translation is somehow prevented, or the Ttk protein is removed from photoreceptor precursors, allowing them to develop as neurons. For R1/R6/R7, this appears to be accomplished by the adaptor protein Phyllopod (Phyl), which links Ttk and the ubiquitin ligase Seven in absentia (Sina), targeting Ttk for degradation by the proteasome (Li et al. 1997, 2002; Tang et al. 1997). Ras/MAPK signaling activates Phyl expression during R1/R6 and R7 recruitment (Fig. 4c, d'; Chang et al. 1995; Dickson et al. 1995). Ebi, which functions downstream of Egfr signaling as described above, also appears to be involved in Ttk degradation (Dong et al. 1999; Boulton et al. 2000). A recent report (Tomlinson et al. 2011) has shown that Notch signaling regulates Ttk protein levels, though not *ttk* transcription. Ttk protein is lower when Notch activity is reduced, and ectopic N activation results in high levels of Ttk in R1/R6 and R7 and the loss of R1-/R6- and R7-specific markers. Currently, the mechanism by which this occurs is unknown.

Distinguishing the R7 Fate from Other Photoreceptor Types and Cone Cell Fate

The homeodomain transcription factor Prospero (Pros) is required for various aspects of R7 differentiation, and is important for promoting R7 vs. R8 differentiation (Kauffmann et al. 1996; Cook et al. 2003; Morey et al. 2008). Pros is initially expressed at low levels in both R7 and cone cell precursors, but is subsequently upregulated in R7 precursors (Kauffmann et al. 1996), and does not appear to play a role in cone cell specification (Cook et al. 2003).

A number of transcription factors have been shown to regulate *pros* transcription by binding directly to regulatory elements in the *pros* gene. These include Eya/So and Gl which as mentioned before are coexpressed in every cell posterior to the furrow. In addition, direct binding by Lz is required for activating *pros* transcription, thus limiting Pros expression to cells that do not join the 5-cell precluster (Hayashi et al. 2008). In addition, Ras/MAPK pathway transcription factors Yan and Pnt bind directly to the *pros* eye enhancer element (Xu et al. 2000).

Combinatorial regulation by the factors listed before appears to be sufficient to activate initial low levels of *pros* in R7 and cone cells, but does not explain why *pros* is not also expressed at low levels in R1/R6 or why it is upregulated in R7. However, as described before, Notch signaling is activated in R7 and the cone cells upon initiation of expression of the Notch transmembrane ligand Dl in the adjacent R1/R6 cells (Fig. 4c, e'; Parks et al. 1995; Cooper and Bray 2000; Tomlinson and Struhl 2001; Tsuda et al. 2002; Miller et al. 2009; Tomlinson et al. 2011). Notch is required to inhibit Svp expression, which is itself an inhibitor of *pros* expression (Fig. 4c, f'; Hayashi et al. 2008; Miller et al. 2008). Thus, Svp is expressed in R1/R6, where Notch signaling is low, and inhibits *pros* expression. In R7 (and also in cone cells), Notch signaling is higher, which leads to inactivation of Svp and allows low levels of *pros* expression in these cells.

However, as mentioned before, high levels of Notch signaling also lead to high levels of Ttk, which inhibits photoreceptor development and inhibits high levels of *pros* expression (Xu et al. 2000). How, then, do Pros levels increase in R7 precursors, and how are the R7 precursors distinguished from non-neuronal cone cells? As detailed below, this appears to occur via an additional boost in Ras/MAPK signaling via the Sev RTK, whose expression is activated by Notch (Fig. 4c, g') as well as by a contribution from the small GTPase Rap1 (Freeman 1996; Freeman 1997; Tomlinson et al. 2011; Mavromatakis and Tomlinson 2012a, b). High levels of Ras/MAPK and Notch signaling in the R7 precursor inhibit Svp and Ttk, leading to high levels of *pros* expression and to R7-specific differentiation (Fig. 4c, h'; Kauffmann et al. 1996; Xu et al. 2000).

The Sev RTK (Hafen et al. 1987; Basler and Hafen 1988; Bowtell et al. 1988) signals through a typical Sos/Ras/Raf/MAPK/Ets pathway (Rogge et al. 1991; Simon et al. 1991; Dickson et al. 1992b; Fortini et al. 1992; Lai and Rubin 1992; Biggs et al. 1994; Brunner et al. 1994a, b; O'Neill et al. 1994; Rebay and Rubin 1995). Sev is expressed in a surprisingly large group of cells given its role in specifying only

R7: in R3/R4 and the mystery cells at the precluster stage, and, following the second mitotic wave, at low levels in R1/R6 and at high levels in R7 and the cone cells (Tomlinson and Ready 1987a). Indeed, constitutive activation of Sev can convert mystery cells and cone cells to R7s (Basler et al. 1991).

Nevertheless, considerable evidence suggests that Sev function is only required in the R7 precursor for normal ommatidial development to occur. For instance, in the absence of Sev function, only R7 precursors are affected: they develop as cone cells instead (Tomlinson and Ready 1986). In addition, based on mosaic analysis, Sev is required only in the R7 precursor for its development as an R7; any other cell can be mutant and still give rise to a normal ommatidium (Tomlinson and Ready 1987a).

The ligand for the Sev RTK is a transmembrane protein called Bride of Sevenless (Boss), which at the stages at which photoreceptors are recruited is expressed and required solely in R8 (Reinke and Zipursky 1988; Hart et al. 1990; Krämer et al. 1991; Van Vactor et al. 1991). Several pieces of evidence suggest that Boss/Sev do not specify R7 fate per se, but instead provide high levels of Ras/MAPK signaling: (1) when R7 precursors misexpress Ro (R2/R5 fate, see previous discussion) they develop as outer photoreceptors, but Sev is still required for recruitment (Basler et al. 1990; Kimmel et al. 1990); (2) chimeras containing the Torso extracellular domain and the Sev intracellular domain are capable of replacing Sev function (Dickson et al. 1992a); and (3) high levels of signaling via the Egfr are capable of replacing Sev function (Freeman 1996).

Since Boss is a transmembrane ligand, its effect is limited to the cells adjacent to R8. Thus, the reason that Sev signaling is only activated in R7, even though Sev is also expressed in R3/R4 and the cone cells, appears to be that the R7 precursor is the only cell primed and in position to respond to the Boss ligand. For instance, the cone cell precursors are separated from R8 by intervening cells (Tomlinson 1985; Tomlinson and Ready 1987b; Wolff and Ready 1993), so that even though they express the Sev receptor, it never comes into contact with the Boss ligand on R8.

In contrast, R3/R4 precursors express Sev and are in contact with R8. The reason why R3/R4 precursors fail to develop as R7s appears to be that the timing is off, with R3/R4 recruitment occurring prior to initiation of Boss expression in R8 (Krämer et al. 1991; Van Vactor et al. 1991). Interestingly, R7 but not R3/R4 internalize Boss (Krämer et al. 1991), suggesting that some unknown R3-/R4-specific factor prevents the Sev–Boss interaction or the endocytosis of Sev–Boss in R3/R4, etc. Given its role in preventing R3/R4 from developing as R7 or R8, it is possible that Svp is somehow involved in making R3/R4 cells refractory to the Boss signal.

What about R1/R6? How do they become different from R7? After all, the trio of cells that comprise the future R1/R6 and R7 photoreceptors join the precluster at the same time and appear to form an equivalence group (Cooper and Bray 2000; Tomlinson and Struhl 2001; Miller et al. 2009). However, R1/R6 begin to differentiate earlier than R7, and thereby express Dl prior to the point at which Dl expression is initiated in the R7 precursors, and also at higher levels (Parks et al. 1995; Cooper and Bray 2000). Dl then activates Notch signaling in the R7 precursor. Although R7 eventually expresses Dl (Parks et al. 1995), and ectopic activation of Notch is certainly capable of converting R1/R6 to an R7 fate (Cooper and Bray 2000;

Tomlinson and Struhl 2001; Miller et al. 2009), R1/R6 appear to be inhibited from receiving the Dl signal from R7, possibly through cis-inhibition (Miller et al. 2009).

Ttk and dpax2 are Required for Cone Cell Development

The four cone cells are the first non-neuronal cells to be recruited into ommatidia. As with R7 precursors, cone cell precursors are subject to both Egfr and Notch signaling. Spi from R8, R2/R5 activates moderate levels of Egfr signaling, which is required for cone cell precursor recruitment (Fig. 4d, a'; Freeman 1996). In addition, Notch signaling is activated by Dl expressed in R1/R6, R7 (Fig. 4d, b'; Tsuda et al. 2002). As mentioned before, the cone cell precursors express Sev; however, they do not contact R8 and do not experience a boost in Ras/MAPK signaling as the R7 precursor does. Consequently, Ttk levels remain high in cone cell precursors and they develop as non-neuronal cells.

The *Drosophila* Pax2 homolog (dPax2; aka Sparkling (Spa)) is expressed in both cone cells and primary pigment cells and is required for proper development of both (Fu and Noll 1997). As with other ommatidial cell type markers, expression of dPax2 is under combinatorial control: in this case, the Egfr pathway Ets domain transcription factors (PntP2 and Yan) as well as the Notch pathway transcription factor Su(H) both bind directly to dPax2 *cis*-regulatory elements, reflecting the fact that both Egfr and Notch signaling are required for cone cell fate (Fig. 4d, c'). In addition, Lz binds directly to dPax2 *cis*-regulatory elements (Flores et al. 2000).

However, this does not explain why dPax2 is not expressed in the R7 precursor in which both Ras/MAPK and Notch signaling pathways are also active. In addition, although dPax2 is required for cone cell development, it is not capable of transforming photoreceptors to a cone cell fate (Fu and Noll 1997; Shi and Noll 2009). Thus, some other factor or factors must be involved both in preventing dPax2 expression in R7 and in determining whether a cell becomes a neuron or a cone cell.

The latter factor appears to be Ttk, which, as described before, is known to inhibit neuronal development. However, until recently it has not been clear what role Ttk might play in cone cell development: like dPax2, ectopic Ttk is incapable of converting photoreceptors to cone cells, and ttk^1 mutants (which affect only the Ttk88 isoform) have no apparent affect on cone cell development (Xiong and Montell 1993; Lai et al. 1996; Shi and Noll 2009). In contrast to either ttk^1 or *dPax2* single mutants, ttk^1; *dPax2* double mutants lack cone cells entirely. Moreover, ectopic coexpression of Ttk88 and dPax2 can transform photoreceptors into cone cells (Shi and Noll 2009). Thus, cone cell development requires the presence of two factors: Ttk and dPax2, with Ttk preventing neuronal development and dPax2 promoting cone cell development.

The E3 ubiquitin ligase complex, composed of Phyl, Sina, and Ebi, is required to prevent photoreceptors from developing as cone cells and is capable of transforming cone cells into photoreceptors (Chang et al. 1995; Dickson et al. 1995). Interestingly, this complex is not only required to degrade Ttk in photoreceptors, but also to prevent dPax2 in photoreceptors. The effects on dPax2 are at the transcriptional level, suggesting that the Phyl/Sina/Ebi complex targets a dPax2 transcriptional activator for degradation (Shi and Noll 2009). These authors speculate that Ebi's known role in mediating Su(H)/Smr regulation of gene expression (see previous discussion), but currently there is no direct evidence for this. In addition to these regulators, there is an additional unknown repressor of Pax2, which prevents its expression in R1/R6 and R7 (Swanson et al. 2010).

More About Pros and dpax2 in R7 and Cone Cell Development

Another recent report describes findings that suggest that Pros and dPax2 have a complex interaction with respect to R7 and cone cell development. Recall that Pros is expressed at low levels in cone cell precursors and at high levels in R7 precursors, while dPax2 is not expressed in R7, but is expressed in cone cell and primary pigment cell precursors. Whereas removal of either *pros* or *dPax2* function alone has only minor effects on cone cell *recruitment* (though dPax2 mutants affect cone cell *differentiation*; Fu and Noll 1997; Shi and Noll 2009; Charlton-Perkins et al. 2011), removal of both results in essentially no cone cells developing. On the other hand, whereas ectopic dPax2 can convert R7s to cone cells, in the absence of dPax2, ectopic Pros can convert cone cells to R7s (Charlton-Perkins et al. 2011). Thus, Pros and dPax2 function synergistically to promote cone cell recruitment, but function antagonistically for R7 versus cone cell development.

Importantly, although as described before *pros* and *dPax2* both require input from Egfr and Notch signaling pathways for their expression patterns, they also appear to function in feedback loops that regulate the signaling pathways. For instance, *pros* is a known direct target of Egfr signaling (Fig. 4d, d'; Kauffmann et al. 1996; Xu et al. 2000; Hayashi et al. 2008), but *pros* is also required for high levels of Egfr activity, and is capable of upregulating Egfr activity (Charlton-Perkins et al. 2011). Similarly, *dPax2* is a direct target of both Egfr and Notch signaling (Flores et al. 2000; Hayashi et al. 2008) but *dPax2* is also required to repress Dl transcription (Charlton-Perkins et al. 2011). At present, the mechanisms by which this occurs are unknown. Nonetheless, these data indicate that there is still much to be learned about signal integration and the interplay between signaling and cell type-specific transcription factors in the context of R7 versus cone cell specification.

Primary Pigment Cells

Ablation of cone cells, but not of photoreceptor cells, prevents formation of primary pigment cells, suggesting that signals from the cone cells are required for primary pigment cell fate (Miller and Cagan 1998). Studies have in fact linked both Egfr and

Notch signaling to primary pigment cell recruitment (Cagan and Ready 1989a; Parks et al. 1995; Freeman 1996; Miller and Cagan 1998).

As mentioned above, dPax2 is necessary for primary pigment cell development as well as cone cell development (Fu and Noll 1997). In addition, the homeodomain transcription factors BarH1/BarH2, which were described above as having a role in R1/R6 development, are also required for primary pigment cell development, and BarH1 misexpression can change a subset of cone cell precursors to primary pigment cell fate (Higashijima et al. 1992b; Hayashi et al. 1998). The function of dPax2 is required for BarH1/BarH2 expression in cone cells (Fu and Noll 1997).

Factors involved in dPax2 activation in primary pigment cells are similar to those in cone cells, with input from Egfr, Notch, and Lz. Egfr signaling leads to *Dl* transcription in cone cell precursors via a Sno-Ebi-dependent mechanism, leading to Notch activation in adjacent primary pigment cells. Notch signaling is required in primary pigment cell precursors for expression of BarH1. In contrast, Egfr signaling is *not* required in primary pigment cell precursors for BarH1 expression (Nagaraj and Banerjee 2007). Thus, primary pigment cell fate requires direct input from Notch signaling, while Egfr signaling is required indirectly for primary pigment cell fate via its role in activating Dl expression in cone cell precursors.

Secondaries and Tertiaries

More cells are generated during the proliferative phases of eye development than will actually end up in the ommatidial lattice. Following recruitment of photoreceptors, cone cells and 1° pigment cells, the remaining cells, referred to as interommatidial cells, either become 2° or 3° pigment cells or bristles, or undergo programmed cell death (PCD; Cagan and Ready 1989b; Wolff and Ready 1991a).

For 2° and 3° pigment cells, the decision appears to be made on the basis of whether a particular cell is in direct physical contact with 1° pigment cells: ablation studies have shown that 1° pigment cells and possibly cone cells are required to prevent 2° and 3° pigment cells from undergoing PCD. In contrast, ablation of photoreceptors did not have an effect on 2° and 3° pigment cell recruitment (Miller and Cagan 1998). These data suggest that factors intrinsic to and/or signaling among interommatidial cells promote PCD, and that a signal from the 1° pigment/cone cells allows survival and differentiation as a 2° or 3° pigment cell.

Notch promotes PCD of interommatidial cells, because reducing *Notch* function results in a loss of PCD (Cagan and Ready 1989a; Miller and Cagan 1998). Notch appears to function within the interommatidial cells to promote PCD (Miller and Cagan 1998). Accordingly, during the stage of 2° or 3° pigment cell recruitment Notch is expressed in all interommatidial cells, but apparently not in other ommatidial cells (Fehon et al. 1991; Kooh et al. 1993). However, it is not clear what ligand is responsible for activating Notch in this context: although Dl is expressed in a subset of interommatidial cells, reduction of Dl function does not appear to result in loss of PCD (Parks et al. 1995).

On the other side, high levels of Egfr signaling are required for development of 2° or 3° pigment cells and to prevent excessive PCD. Conversely, activation of the Egfr/Ras pathway during pupal eye development blocks PCD (Freeman 1996; Miller and Cagan 1998; Sawamoto et al. 1998). Consistent with Egfr signaling being the signal from 1° pigment/cone cells that allows interommatidial cell survival and 2° or 3° pigment cell differentiation, the Egfr ligand Spi is expressed at high levels in cone cells and at lower levels in 1° pigment cells, while Egfr is expressed at high levels in interommatidial cells (Miller and Cagan 1998).

The mechanism by which Egfr signaling inhibits PCD in interommatidial cells appears to be negative regulation of the PCD activator *head involution defective* (*hid*) both transcriptionally and posttranslationally (Grether et al. 1995; Bergmann et al. 1998; Kurada and White 1998; Wang et al. 1999; Goyal et al. 2000; Meier et al. 2000; Yu et al. 2002; Bao 2010). In contrast, the mechanism by which Notch signaling promotes PCD is not clear in this context. It is also not known how the expression patterns of Notch, Egfr, and their ligands are established. Although in some types of cancer cells Notch signaling has been shown to inhibit PCD (Axelson 2004; Leong and Karsan 2006), in other types of cancer cells activation of Notch signaling has been linked with PCD promotion (Greenblatt et al. 2007; Zheng et al. 2007; Platta et al. 2008; Ou et al. 2012). Thus, study of the pro-PCD effects of Notch signaling during *Drosophila* eye development will provide clues about the mechanisms by which Notch signaling can promote PCD in other contexts.

Unlike for other types of ommatidial cells, there are no known 2° or 3° pigment cell-specific transcription factors that define these cell types and drive their differentiation. Instead, 2° and 3° pigment cells appear to acquire their fates solely by cell sorting. *Drosophila* Neph1 and Nephrin homologs are expressed in complementary patterns during pupal eye development: Neph1 homologs Roughest (Rst) and Kin of irre (Kirre) are expressed in interommatidial cells, while the Nephrin homologs Hibris (Hbs) and Sticks and stones (Sns) are expressed in cone cells and 1° pigment cells (Ramos et al. 1993; Reiter et al. 1996; Bao and Cagan 2005). Heterophilic interactions between Neph1 and Nephrin homologs are required to enable contact between the cone/1° pigment cells and the interommatidial cells, leading to sorting of the interommatidial cells into a single row surrounding each ommatidium, and in maintaining separation between ommatidia (Wolff and Ready 1991a; Reiter et al. 1996; Bao and Cagan 2005; Bao 2010; Bao et al. 2010). It is currently not clear what regulates expression of these molecules in complementary patterns.

References

Artavanis-Tsakonas S, Muskavitch MA (2010) Notch: the past, the present, and the future. Curr Top Dev Biol 92:1–29

Axelson H (2004) Notch signaling and cancer: emerging complexity. Semin Cancer Biol 14:317–319

Baker NE (2000) Notch signaling in the nervous system Pieces still missing from the puzzle. Bioessays 22:264–273

Baker NE (2002) Notch and the patterning of ommatidial founder cells in the developing Drosophila eye. In: Moses K (ed) Results and problems in cell differentiation. Springer, Berlin, pp 35–58

Baker NE, Yu SY (1997) Proneural function of neurogenic genes in the developing Drosophila eye. Curr Biol 7:122–132

Baker NE, Zitron AE (1995) Drosophila eye development: Notch and Delta amplify a neurogenic pattern conferred on the morphogenetic furrow by scabrous. Mech Dev 49:173–189

Baker NE, Yu S, Han D (1996) Evolution of proneural atonal expression during distinct regulatory phases in the developing Drosophila eye. Curr Biol 6:1290–1301

Bang AG, Kintner C (2000) Rhomboid and Star facilitate presentation and processing of the Drosophila TGF-alpha homolog Spitz. Genes Dev 14:177–186

Bao S (2010) Two themes on the assembly of the Drosophila eye. Curr Top Dev Biol 93:85–127

Bao S, Cagan R (2005) Preferential adhesion mediated by Hibris and Roughest regulates morphogenesis and patterning in the Drosophila eye. Dev Cell 8:925–935

Bao S, Fischbach KF, Corbin V, Cagan RL (2010) Preferential adhesion maintains separation of ommatidia in the Drosophila eye. Dev Biol 344:948–956

Baonza A, Casci T, Freeman M (2001) A primary role for the epidermal growth factor receptor in ommatidial spacing in the Drosophila eye. Curr Biol 11:396–404

Baonza A, Freeman M (2001) Notch signaling and the initiation of neural development in the Drosophila eye. Development 128:3889–3898

Basler K, Christen B, Hafen E (1991) Ligand-independent activation of the sevenless receptor tyrosine kinase changes the fate of cells in the developing Drosophila eye. Cell 64:1069–1081

Basler K, Hafen E (1988) Sevenless and Drosophila eye development: a tyrosine kinase controls cell fate. Trends Genet 4:74–79

Basler K, Yen D, Tomlinson A, Hafen E (1990) Reprogramming cell fate in the developing Drosophila retina: transformation of R7 cells by ectopic expression of rough. Genes Dev 4:728–739

Batterham P, Crew JR, Sokac AM, Andrews JR, Pasquini GM, Davies AG, Stocker RF, Pollock JA (1996) Genetic analysis of the lozenge gene complex in Drosophila melanogaster: adult visual system phenotypes. J Neurogenet 10:193–220

Behan KJ, Nichols CD, Cheung TL, Farlow A, Hogan BM, Batterham P, Pollock JA (2002) Yan regulates Lozenge during Drosophila eye development. Dev Genes Evol 212:267–276

Bergmann A, Agapite J, McCall K, Steller H (1998) The Drosophila gene hid is a direct molecular target of Ras-dependent survival signaling. Cell 95:331–341

Bertrand N, Castro DS, Guillemot F (2002) Proneural genes and the specification of neural cell types. Nat Rev Neurosci 3:517–530

Bessa J, Gebelein B, Pichaud F, Casares F, Mann RS (2002) Combinatorial control of Drosophila eye development by eyeless, homothorax, and teashirt. Genes Dev 16:2415–2427

Bhattacharya A, Baker NE (2009) The HLH protein Extramacrochaetae is required for R7 cell and cone cell fates in the Drosophila eye. Dev Biol 327:288–300

Bhattacharya A, Baker NE (2011) A network of broadly expressed HLH genes regulates tissue-specific cell fates. Cell 147:881–892

Bhattacharya A, Baker NE (2012) The role of the bHLH protein hairy in morphogenetic furrow progression in the developing Drosophila eye. PLoS One 7:e47503

Biggs WH, Zavitz KH, Dickson B, van der Straten A, Brunner D, Hafen E, Zipursky SL (1994) The Drosophila rolled locus encodes a MAP kinase required in the sevenless signal transduction pathway. EMBO J 13:1628–1635

Bonini NM, Leiserson WM, Benzer S (1993) The eyes absent gene: genetic control of cell survival and differentiation in the developing Drosophila eye. Cell 72:379–395

Borod ER, Heberlein U (1998) Mutual regulation of decapentaplegic and hedgehog during the initiation of differentiation in the Drosophila retina. Dev Biol 197:187–197

Boulton SJ, Brook A, Staehling-Hampton K, Heitzler P, Dyson N (2000) A role for Ebi in neuronal cell cycle control. EMBO J 19:5376–5386

Bowtell DD, Simon MA, Rubin GM (1988) Nucleotide sequence and structure of the sevenless gene of Drosophila melanogaster. Genes Dev 2:620–634

Bray SJ (2006) Notch signaling: a simple pathway becomes complex. Nat Rev Mol Cell Biol 7:678–689

Brown KE, Baonza A, Freeman M (2006) Epithelial cell adhesion in the developing Drosophila retina is regulated by Atonal and the EGF receptor pathway. Dev Biol 300:710–721

Brown KE, Kerr M, Freeman M (2007) The EGFR ligands Spitz and Keren act cooperatively in the Drosophila eye. Dev Biol 307:105–113

Brown NL, Sattler CA, Paddock SW, Carroll SB (1995) Hairy and emc negatively regulate morphogenetic furrow progression in the Drosophila eye. Cell 80:879–887

Brown NL, Paddock SW, Sattler CA, Cronmiller C, Thomas BJ, Carroll SB (1996) Daughterless is required for Drosophila photoreceptor cell determination, eye morphogenesis, and cell cycle progression. Dev Biol 179:65–78

Brunner D, Dücker K, Oellers N, Hafen E, Scholz H, Klämbt C (1994a) The ETS domain protein pointed-P2 is a target of MAP kinase in the sevenless signal transduction pathway. Nature 370:386–389

Brunner D, Oellers N, Szabad J, Biggs WH, Zipursky SL, Hafen E (1994b) A gain-of-function mutation in Drosophila MAP kinase activates multiple receptor tyrosine kinase signaling pathways. Cell 76:875–888

Cagan RL, Ready DF (1989a) Notch is required for successive cell decisions in the developing Drosophila retina. Genes Dev 3:1099–1112

Cagan RL, Ready DF (1989b) The emergence of order in the Drosophila pupal retina. Dev Biol 136:346–362

Campuzano S (2001) Emc, a negative HLH regulator with multiple functions in Drosophila development. Oncogene 20:8299–8307

Carthew RW (2007) Pattern formation in the Drosophila eye. Curr Opin Genet Dev 17:309–313

Chang HC, Solomon NM, Wassarman DA, Karim FD, Therrien M, Rubin GM, Wolff T (1995) Phyllopod functions in the fate determination of a subset of photoreceptors in Drosophila. Cell 80:463–472

Charlton-Perkins M, Cook TA (2010) Building a fly eye: terminal differentiation events of the retina, corneal lens, and pigmented epithelia. Curr Top Dev Biol 93:129–173

Charlton-Perkins M, Whitaker SL, Fei Y, Xie B, Li-Kroeger D, Gebelein B, Cook T (2011) Prospero and Pax2 combinatorially control neural cell fate decisions by modulating Ras- and Notch-dependent signaling. Neural Dev 6:20

Chen CK, Chien CT (1999) Negative regulation of atonal in proneural cluster formation of Drosophila R8 photoreceptors. Proc Natl Acad Sci U S A 96:5055–5060

Cheyette BN, Green PJ, Martin K, Garren H, Hartenstein V, Zipursky SL (1994) The Drosophila sine oculis locus encodes a homeodomain-containing protein required for the development of the entire visual system. Neuron 12:977–996

Cook T, Pichaud F, Sonneville R, Papatsenko D, Desplan C (2003) Distinction between color photoreceptor cell fates is controlled by Prospero in Drosophila. Dev Cell 4:853–864

Cooper MT, Bray SJ (1999) Frizzled regulation of Notch signaling polarizes cell fate in the Drosophila eye. Nature 397:526–530

Cooper MT, Bray SJ (2000) R7 photoreceptor specification requires Notch activity. Curr Biol 10:1507–1510

Coyle-Thompson CA, Banerjee U (1993) The strawberry notch gene functions with Notch in common developmental pathways. Development 119:377–395

Crew JR, Batterham P, Pollock JA (1997) Developing compound eye in *lozenge* mutants of *Drosophila*: lozenge expression in the R7 equivalence group. Dev Genes Evol 206:481–493

Cubas P, Celis JF de, Campuzano S, Modolell J (1991) Proneural clusters of achaete-scute expression and the generation of sensory organs in the Drosophila imaginal wing disc. Genes Dev 5:996–1008

Cubas P, Modolell J (1992) The extramacrochaetae gene provides information for sensory organ patterning. EMBO J 11:3385–3393

Culí J, Modolell J (1998) Proneural gene self-stimulation in neural precursors: an essential mechanism for sense organ development that is regulated by Notch signaling. Genes Dev 12:2036–2047

Curtiss J, Mlodzik M (2000) Morphogenetic furrow initiation and progression during eye development in Drosophila: the roles of decapentaplegic, hedgehog and eyes absent. Development 127:1325–1336

Daga A, Karlovich CA, Dumstrei K, Banerjee U (1996) Patterning of cells in the Drosophila eye by Lozenge, which shares homologous domains with AML1. Genes Dev 10:1194–1205

del Alamo D, Mlodzik M (2006) Frizzled/PCP-dependent asymmetric neuralized expression determines R3/R4 fates in the Drosophila eye. Dev Cell 11:887–894

del Alamo D, Mlodzik M (2008) Self-modulation of Notch signaling during ommatidial development via the Roughened eye transcriptional repressor. Development 135:2895–2904

Dickson B, Sprenger F, Hafen E (1992a) Prepattern in the developing Drosophila eye revealed by an activated torso–sevenless chimeric receptor. Genes Dev 6:2327–2339

Dickson B, Sprenger F, Morrison D, Hafen E (1992b) Raf functions downstream of Ras1 in the Sevenless signal transduction pathway. Nature 360:600–603

Dickson BJ, Domínguez M, van der Straten A, Hafen E (1995) Control of Drosophila photoreceptor cell fates by phyllopod, a novel nuclear protein acting downstream of the Raf kinase. Cell 80:453–462

Distefano GM, Gangemi AJ, Khandelwal PJ, Saunders AJ, Marenda DR (2012) Drosophila lilliputian is required for proneural gene expression in retinal development. Dev Dyn 241:553–562

Dokucu ME, Zipursky SL, Cagan RL (1996) Atonal, rough and the resolution of proneural clusters in the developing Drosophila retina. Development 122:4139–4147

Domingos PM, Brown S, Barrio R, Ratnakumar K, Frankfort BJ, Mardon G, Steller H, Mollereau B (2004a) Regulation of R7 and R8 differentiation by the spalt genes. Dev Biol 273:121–133

Domingos PM, Mlodzik M, Mendes CS, Brown S, Steller H, Mollereau B (2004b) Spalt transcription factors are required for R3/R4 specification and establishment of planar cell polarity in the Drosophila eye. Development 131:5695–5702

Domínguez M (1999) Dual role for Hedgehog in the regulation of the proneural gene atonal during ommatidia development. Development 126:2345–2353

Domínguez M, Hafen E (1997) Hedgehog directly controls initiation and propagation of retinal differentiation in the Drosophila eye. Genes Dev 11:3254–3264

Domínguez M, Wasserman JD, Freeman M (1998) Multiple functions of the EGF receptor in Drosophila eye development. Curr Biol 8:1039–1048

Dong X, Tsuda L, Zavitz KH, Lin M, Li S, Carthew RW, Zipursky SL (1999) ebi regulates epidermal growth factor receptor signaling pathways in Drosophila. Genes Dev 13:954–965

Doroquez DB, Orr-Weaver TL, Rebay I (2007) Split ends antagonizes the Notch and potentiates the EGFR signaling pathways during Drosophila eye development. Mech Dev 124:792–806

Doroquez DB, Rebay I (2006) Signal integration during development: mechanisms of EGFR and Notch pathway function and cross-talk. Crit Rev Biochem Mol Biol 41:339–385

Ellis HM, Spann DR, Posakony JW (1990) Extramacrochaetae, a negative regulator of sensory organ development in Drosophila, defines a new class of helix-loop-helix proteins. Cell 61:27–38

Fanto M, Mayes CA, Mlodzik M (1998) Linking cell-fate specification to planar polarity: determination of the R3/R4 photoreceptors is a prerequisite for the interpretation of the Frizzled mediated polarity signal. Mech Dev 74:51–58

Fanto M, Mlodzik M (1999) Asymmetric Notch activation specifies photoreceptors R3 and R4 and planar polarity in the Drosophila eye. Nature 397:523–526

Fehon RG, Johansen K, Rebay I, Artavanis-Tsakonas S (1991) Complex cellular and subcellular regulation of notch expression during embryonic and imaginal development of Drosophila: implications for notch function. J Cell Biol 113:657–669
Firth LC, Baker NE (2009) Retinal determination genes as targets and possible effectors of extracellular signals. Dev Biol 327:366–375
Flores GV, Daga A, Kalhor HR, Banerjee U (1998) Lozenge is expressed in pluripotent precursor cells and patterns multiple cell types in the Drosophila eye through the control of cell-specific transcription factors. Development 125:3681–3687
Flores GV, Duan H, Yan H, Nagaraj R, Fu W, Zou Y, Noll M, Banerjee U (2000) Combinatorial signaling in the specification of unique cell fates. Cell 103:75–85
Fortini ME (2009) Notch signaling: the core pathway and its posttranslational regulation. Dev Cell 16:633–647
Fortini ME, Simon MA, Rubin GM (1992) Signaling by the sevenless protein tyrosine kinase is mimicked by Ras1 activation. Nature 355:559–561
Frankfort BJ, Mardon G (2002) R8 development in the Drosophila eye: a paradigm for neural selection and differentiation. Development 129:1295–1306
Frankfort BJ, Mardon G (2004) Senseless represses nuclear transduction of Egfr pathway activation. Development 131:563–570
Frankfort BJ, Nolo R, Zhang Z, Bellen H, Mardon G (2001) Senseless repression of rough is required for R8 photoreceptor differentiation in the developing Drosophila eye. Neuron 32:403–414
Frankfort BJ, Pepple KL, Mamlouk M, Rose MF, Mardon G (2004) Senseless is required for pupal retinal development in Drosophila. Genesis 38:182–194
Freeman M (1994a) Misexpression of the Drosophila argos gene, a secreted regulator of cell determination. Development 120:2297–2304
Freeman M (1994b) The spitz gene is required for photoreceptor determination in the Drosophila eye where it interacts with the EGF receptor. Mech Dev 48:25–33
Freeman M (1996) Reiterative use of the EGF receptor triggers differentiation of all cell types in the Drosophila eye. Cell 87:651–660
Freeman M (1997) Cell determination strategies in the Drosophila eye. Development 124:261–270
Freeman M (2002) A fly's eye view of EGF receptor signaling. EMBO J 21:6635–6642
Freeman M, Kimmel BE, Rubin GM (1992a) Identifying targets of the rough homeobox gene of Drosophila: evidence that rhomboid functions in eye development. Development 116:335–346
Freeman M, Klämbt C, Goodman CS, Rubin GM (1992b) The argos gene encodes a diffusible factor that regulates cell fate decisions in the Drosophila eye. Cell 69:963–975
Fu W, Baker NE (2003) Deciphering synergistic and redundant roles of Hedgehog, Decapentaplegic and Delta that drive the wave of differentiation in Drosophila eye development. Development 130:5229–5239
Fu W, Noll M (1997) The Pax2 homolog sparkling is required for development of cone and pigment cells in the Drosophila eye. Genes Dev 11:2066–2078
Gabay L, Scholz H, Golembo M, Klaes A, Shilo BZ, Klämbt C (1996) EGF receptor signaling induces pointed P1 transcription and inactivates Yan protein in the Drosophila embryonic ventral ectoderm. Development 122:3355–3362
Gabay L, Seger R, Shilo BZ (1997) In situ activation pattern of Drosophila EGF receptor pathway during development. Science 277:1103–1106
Garcàa-Bellido A, de Celis JF (2009) The complex tale of the achaete-scute complex: a paradigmatic case in the analysis of gene organization and function during development. Genetics 182:631–639
Garrell J, Modolell J (1990) The Drosophila extramacrochaetae locus, an antagonist of proneural genes that, like these genes, encodes a helix-loop-helix protein. Cell 61:39–48
Golembo M, Raz E, Shilo BZ (1996a) The Drosophila embryonic midline is the site of Spitz processing, and induces activation of the EGF receptor in the ventral ectoderm. Development 122:3363–3370

Golembo M, Schweitzer R, Freeman M, Shilo BZ (1996b) Argos transcription is induced by the Drosophila EGF receptor pathway to form an inhibitory feedback loop. Development 122:223–230

Goyal L, McCall K, Agapite J, Hartwieg E, Steller H (2000) Induction of apoptosis by Drosophila reaper, hid and grim through inhibition of IAP function. EMBO J 19:589–597

Greenblatt DY, Vaccaro AM, Jaskula-Sztul R, Ning L, Haymart M, Kunnimalaiyaan M, Chen H (2007) Valproic acid activates notch-1 signaling and regulates the neuroendocrine phenotype in carcinoid cancer cells. Oncologist 12:942–951

Greenwood S, Struhl G (1999) Progression of the morphogenetic furrow in the Drosophila eye: the roles of Hedgehog, Decapentaplegic and the Raf pathway. Development 126:5795–5808

Grether ME, Abrams JM, Agapite J, White K, Steller H (1995) The head involution defective gene of Drosophila melanogaster functions in programmed cell death. Genes Dev 9:1694–1708

Guenther MG, Lane WS, Fischle W, Verdin E, Lazar MA, Shiekhattar R (2000) A core SMRT corepressor complex containing HDAC3 and TBL1, a WD40-repeat protein linked to deafness. Genes Dev 14:1048–1057

Guichard A, Biehs B, Sturtevant MA, Wickline L, Chacko J, Howard K, Bier E (1999) rhomboid and Star interact synergistically to promote EGFR/MAPK signaling during Drosophila wing vein development. Development 126:2663–2676

Hafen E, Basler K, Edstroem JE, Rubin GM (1987) Sevenless, a cell-specific homeotic gene of Drosophila, encodes a putative transmembrane receptor with a tyrosine kinase domain. Science 236:55–63

Halder G, Callaerts P, Flister S, Walldorf U, Kloter U, Gehring WJ (1998) Eyeless initiates the expression of both sine oculis and eyes absent during Drosophila compound eye development. Development 125:2181–2191

Hart AC, Krämer H, Van Vactor DL, Paidhungat M, Zipursky SL (1990) Induction of cell fate in the Drosophila retina: the bride of sevenless protein is predicted to contain a large extracellular domain and seven transmembrane segments. Genes Dev 4:1835–1847

Hayashi T, Kojima T, Saigo K (1998) Specification of primary pigment cell and outer photoreceptor fates by BarH1 homeobox gene in the developing Drosophila eye. Dev Biol 200:131–145

Hayashi T, Xu C, Carthew RW (2008) Cell-type-specific transcription of prospero is controlled by combinatorial signaling in the Drosophila eye. Development 135:2787–2796

Heberlein U, Mlodzik M, Rubin GM (1991) Cell-fate determination in the developing Drosophila eye: role of the rough gene. Development 112:703–712

Heberlein U, Hariharan IK, Rubin GM (1993) Star is required for neuronal differentiation in the Drosophila retina and displays dosage-sensitive interactions with Ras1. Dev Biol 160:51–63

Heberlein U, Singh CM, Luk AY, Donohoe TJ (1995) Growth and differentiation in the Drosophila eye coordinated by hedgehog. Nature 373:709–711

Higashijima S, Kojima T, Michiue T, Ishimaru S, Emori Y, Saigo K (1992a) Dual Bar homeo box genes of Drosophila required in two photoreceptor cells, R1 and R6, and primary pigment cells for normal eye development. Genes Dev 6:50–60

Higashijima S, Michiue T, Emori Y, Saigo K (1992b) Subtype determination of Drosophila embryonic external sensory organs by redundant homeo box genes BarH1 and BarH2. Genes Dev 6:1005–1018

Hsiung F, Moses K (2002) Retinal development in Drosophila: specifying the first neuron. Hum Mol Genet 11:1207–1214

Jackson Behan K, Fair J, Singh S, Bogwitz M, Perry T, Grubor V, Cunningham F, Nichols CD, Cheung TL, Batterham P, Pollock JA (2005) Alternative splicing removes an Ets interaction domain from Lozenge during Drosophila eye development. Dev Genes Evol 215:423–435

Jarman AP, Grau Y, Jan LY, Jan YN (1993) Atonal is a proneural gene that directs chordotonal organ formation in the Drosophila peripheral nervous system. Cell 73:1307–1321

Jarman AP, Grell EH, Ackerman L, Jan LY, Jan YN (1994) Atonal is the proneural gene for Drosophila photoreceptors. Nature 369:398–400

Jarman AP, Sun Y, Jan LY, Jan YN (1995) Role of the proneural gene, atonal, in formation of Drosophila chordotonal organs and photoreceptors. Development 121:2019–2030

Jemc J, Rebay I (2007) Identification of transcriptional targets of the dual-function transcription factor/phosphatase eyes absent. Dev Biol 310:416–429

Jenny A (2010) Planar cell polarity signaling in the Drosophila eye. Curr Top Dev Biol 93:189–227

Kao HY, Ordentlich P, Koyano-Nakagawa N, Tang Z, Downes M, Kintner CR, Evans RM, Kadesch T (1998) A histone deacetylase corepressor complex regulates the Notch signal transduction pathway. Genes Dev 12:2269–2277

Kauffmann RC, Li S, Gallagher PA, Zhang J, Carthew RW (1996) Ras1 signaling and transcriptional competence in the R7 cell of Drosophila. Genes Dev 10:2167–2178

Kimmel BE, Heberlein U, Rubin GM (1990) The homeo domain protein rough is expressed in a subset of cells in the developing Drosophila eye where it can specify photoreceptor cell subtype. Genes Dev 4:712–727

Kolodkin AL, Pickup AT, Lin DM, Goodman CS, Banerjee U (1994) Characterization of Star and its interactions with sevenless and EGF receptor during photoreceptor cell development in Drosophila. Development 120:1731–1745

Kooh PJ, Fehon RG, Muskavitch MA (1993) Implications of dynamic patterns of Delta and Notch expression for cellular interactions during Drosophila development. Development 117:493–507

Kopan R, Ilagan MX (2009) The canonical Notch signaling pathway: unfolding the activation mechanism. Cell 137:216–233

Krämer H, Cagan RL, Zipursky SL (1991) Interaction of bride of sevenless membrane-bound ligand and the sevenless tyrosine-kinase receptor. Nature 352:207–212

Kumar JP (2009) The molecular circuitry governing retinal determination. Biochim Biophys Acta 1789:306–314

Kumar JP (2010) Retinal determination the beginning of eye development. Curr Top Dev Biol 93:1–28

Kumar JP (2011) My what big eyes you have: how the Drosophila retina grows. Dev Neurobiol 71:1133–1152

Kumar JP (2012) Building an ommatidium one cell at a time. Dev Dyn 241:136–149

Kumar JP, Tio M, Hsiung F, Akopyan S, Gabay L, Seger R, Shilo BZ, Moses K (1998) Dissecting the roles of the Drosophila EGF receptor in eye development and MAP kinase activation. Development 125:3875–3885

Kumar JP, Hsiung F, Powers MA, Moses K (2003) Nuclear translocation of activated MAP kinase is developmentally regulated in the developing Drosophila eye. Development 130:3703–3714

Kurada P, White K (1998) Ras promotes cell survival in Drosophila by downregulating hid expression. Cell 95:319–329

Lage P, Jan YN, Jarman AP (1997) Requirement for EGF receptor signaling in neural recruitment during formation of Drosophila chordotonal sense organ clusters. Curr Biol 7:166–175

Lai ZC, Harrison SD, Karim F, Li Y, Rubin GM (1996) Loss of tramtrack gene activity results in ectopic R7 cell formation, even in a sina mutant background. Proc Natl Acad Sci U S A 93:5025–5030

Lai, Z.-C., Fetchko, M, Li,Y (1997) Repression of Drosophila Photoreceptor Cell Fate Through Cooperative Action of Two Transcriptional Repressors Yan and Tramtrack. Genetics 147: 1131–1137

Lai ZC, Li Y (1999) Tramtrack69 is positively and autonomously required for Drosophila photoreceptor development. Genetics 152:299–305

Lai ZC, Rubin GM (1992) Negative control of photoreceptor development in Drosophila by the product of the yan gene, an ETS domain protein. Cell 70:609–620

Lawrence PA, Green SM (1979) Cell lineage in the developing retina of Drosophila. Dev Biol 71:142–152

Lee EC, Hu X, Yu SY, Baker NE (1996) The scabrous gene encodes a secreted glycoprotein dimer and regulates proneural development in Drosophila eyes. Mol Cell Biol 16:1179–1188

Lee JR, Urban S, Garvey CF, Freeman M (2001) Regulated intracellular ligand transport and proteolysis control EGF signal activation in Drosophila. Cell 107:161–171
Leong KG, Karsan A (2006) Recent insights into the role of Notch signaling in tumorigenesis. Blood 107:2223–2233
Lesokhin AM, Yu SY, Katz J, Baker NE (1999) Several levels of EGF receptor signaling during photoreceptor specification in wild-type, Ellipse, and null mutant Drosophila. Dev Biol 205:129–144
Li S, Li Y, Carthew RW, Lai ZC (1997) Photoreceptor cell differentiation requires regulated proteolysis of the transcriptional repressor Tramtrack. Cell 90:469–478
Li J, Wang J, Nawaz Z, Liu JM, Qin J, Wong J (2000) Both corepressor proteins SMRT and N-CoR exist in large protein complexes containing HDAC3. EMBO J 19:4342–4350
Li X, Perissi V, Liu F, Rose DW, Rosenfeld MG (2002) Tissue-specific regulation of retinal and pituitary precursor cell proliferation. Science 297:1180–1183
Li Y, Baker NE (2001) Proneural enhancement by Notch overcomes Suppressor-of-Hairless repressor function in the developing Drosophila eye. Curr Biol 11:330–338
Ligoxygakis P, Yu SY, Delidakis C, Baker NE (1998) A subset of notch functions during Drosophila eye development require Su(H) and the E(spl) gene complex. Development 125:2893–2900
Lim J, Choi KW (2003) Bar homeodomain proteins are anti-proneural in the Drosophila eye: transcriptional repression of atonal by Bar prevents ectopic retinal neurogenesis. Development 130:5965–5974
Lim J, Choi KW (2004) Induction and autoregulation of the anti-proneural gene Bar during retinal neurogenesis in Drosophila. Development 131:5573–5580
Lim J, Jafar-Nejad H, Hsu YC, Choi KW (2008) Novel function of the class I bHLH protein Daughterless in the negative regulation of proneural gene expression in the Drosophila eye. EMBO Rep 9:1128–1133
Majumdar A, Nagaraj R, Banerjee U (1997) Strawberry notch encodes a conserved nuclear protein that functions downstream of Notch and regulates gene expression along the developing wing margin of Drosophila. Genes Dev 11:1341–1353
Mankidy R, Hastings J, Thackeray JR (2003) Distinct phospholipase C-gamma-dependent signaling pathways in the Drosophila eye and wing are revealed by a new small wing allele. Genetics 164:553–563
Marygold SJ, Walker C, Orme M, Leevers S (2011) Genetic characterization of ebi reveals its critical role in Drosophila wing growth Fly (Austin) 5
Massari ME, Murre C (2000) Helix-loop-helix proteins: regulators of transcription in eucaryotic organisms. Mol Cell Biol 20:429–440
Mavromatakis YE, Tomlinson A (2012a) Stop and go: antagonistic signals in the specification of the Drosophila R7 photoreceptor viewed from an evolutionary perspective Fly (Austin) 6
Mavromatakis YE, Tomlinson A (2012b) The role of the small GTPase Rap in Drosophila R7 photoreceptor specification. Proc Natl Acad Sci U S A 109:3844–3849
Meier P, Silke J, Leevers SJ, Evan GI (2000) The Drosophila caspase DRONC is regulated by DIAP1. EMBO J 19:598–611
Melicharek D, Shah A, DiStefano G, Gangemi AJ, Orapallo A, Vrailas-Mortimer AD, Marenda DR (2008) Identification of novel regulators of atonal expression in the developing Drosophila retina. Genetics 180:2095–2110
Miller AC, Seymour H, King C, Herman TG (2008) Loss of seven-up from Drosophila R1/R6 photoreceptors reveals a stochastic fate choice that is normally biased by Notch. Development 135:707–715
Miller AC, Lyons EL, Herman TG (2009) cis-Inhibition of Notch by endogenous Delta biases the outcome of lateral inhibition. Curr Biol 19:1378–1383
Miller DT, Cagan RL (1998) Local induction of patterning and programmed cell death in the developing Drosophila retina. Development 125:2327–2335
Mlodzik M, Hiromi Y, Weber U, Goodman CS, Rubin GM (1990) The Drosophila seven-up gene, a member of the steroid receptor gene superfamily, controls photoreceptor cell fates. Cell 60:211–224

Morante J, Desplan C, Celik A (2007) Generating patterned arrays of photoreceptors. Curr Opin Genet Dev 17:314–319

Morey M, Yee SK, Herman T, Nern A, Blanco E, Zipursky SL (2008) Coordinate control of synaptic-layer specificity and rhodopsins in photoreceptor neurons. Nature 456:795–799

Morillo SA, Braid LR, Verheyen EM, Rebay I (2012) Nemo phosphorylates Eyes absent and enhances output from the Eya-Sine oculis transcriptional complex during Drosophila retinal determination. Dev Biol 365:267–276

Moses K, Ellis MC, Rubin GM (1989) The glass gene encodes a zinc-finger protein required by Drosophila photoreceptor cells. Nature 340:531–536

Moses K, Rubin GM (1991) Glass encodes a site-specific DNA-binding protein that is regulated in response to positional signals in the developing Drosophila eye. Genes Dev 5:583–593

Nagaraj R, Banerjee U (2007) Combinatorial signaling in the specification of primary pigment cells in the Drosophila eye. Development 134:825–831

Nolo R, Abbott LA, Bellen HJ (2000) Senseless, a Zn finger transcription factor, is necessary and sufficient for sensory organ development in Drosophila. Cell 102:349–362

O'Neill EM, Rebay I, Tjian R, Rubin GM (1994) The activities of two Ets-related transcription factors required for Drosophila eye development are modulated by the Ras/MAPK pathway. Cell 78:137–147

Oberoi J, Fairall L, Watson PJ, Yang JC, Czimmerer Z, Kampmann T, Goult BT, Greenwood JA, Gooch JT, Kallenberger BC, Nagy L, Neuhaus D, Schwabe JW (2011) Structural basis for the assembly of the SMRT/NCoR core transcriptional repression machinery. Nat Struct Mol Biol 18:177–184

Ohsako S, Hyer J, Panganiban G, Oliver I, Caudy M (1994) Hairy function as a DNA-binding helix-loop-helix repressor of Drosophila sensory organ formation. Genes Dev 8:2743–2755

Okabe M, Okano H (1997) Two-step induction of chordotonal organ precursors in Drosophila embryogenesis. Development 124:1045–1053

Ou JP, Lin HY, Su KY, Yu SL, Tseng IH, Chen CJ, Hsu HC, Chan DC, Sophia Chen YL (2012) Potential therapeutic role of Z-isochaihulactone in lung cancer through induction of apoptosis via Notch signaling. Evid Based Complement Alternat Med 2012:809204

Parks AL, Turner FR, Muskavitch MA (1995) Relationships between complex Delta expression and the specification of retinal cell fates during Drosophila eye development. Mech Dev 50:201–216

Pepple KL, Atkins M, Venken K, Wellnitz K, Harding M, Frankfort B, Mardon G (2008) Two-step selection of a single R8 photoreceptor: a bistable loop between senseless and rough locks in R8 fate. Development 135:4071–4079

Perissi V, Aggarwal A, Glass CK, Rose DW, Rosenfeld MG (2004) A corepressor/coactivator exchange complex required for transcriptional activation by nuclear receptors and other regulated transcription factors. Cell 116:511–526

Perissi V, Scafoglio C, Zhang J, Ohgi KA, Rose DW, Glass CK, Rosenfeld MG (2008) TBL1 and TBLR1 phosphorylation on regulated gene promoters overcomes dual CtBP and NCoR/SMRT transcriptional repression checkpoints. Mol Cell 29:755–766

Perissi V, Jepsen K, Glass CK, Rosenfeld MG (2010) Deconstructing repression: evolving models of co-repressor action. Nat Rev Genet 11:109–123

Pickup AT, Banerjee U (1999) The role of star in the production of an activated ligand for the EGF receptor signaling pathway. Dev Biol 205:254–259

Platta CS, Greenblatt DY, Kunnimalaiyaan M, Chen H (2008) Valproic acid induces Notch1 signaling in small cell lung cancer cells. J Surg Res 148:31–37

Powell LM, Jarman AP (2008) Context dependence of proneural bHLH proteins. Curr Opin Genet Dev 18:411–417

Quan XJ, Hassan BA (2005) From skin to nerve: flies, vertebrates and the first helix. Cell Mol Life Sci 62:2036–2049

Ramos RG, Igloi GL, Lichte B, Baumann U, Maier D, Schneider T, Brandstätter JH, Fröhlich A, Fischbach KF (1993) The irregular chiasm C-roughest locus of Drosophila, which affects axonal projections and programmed cell death, encodes a novel immunoglobulin-like protein. Genes Dev 7:2533–2547

Rawlins EL, White NM, Jarman AP (2003) Echinoid limits R8 photoreceptor specification by inhibiting inappropriate EGF receptor signaling within R8 equivalence groups. Development 130:3715–3724

Read D, Manley JL (1992) Alternatively spliced transcripts of the Drosophila tramtrack gene encode zinc finger proteins with distinct DNA binding specificities. EMBO J 11:1035–1044

Ready DF, Hanson TE, Benzer S (1976) Development of the Drosophila retina, a neurocrystalline lattice. Dev Biol 53:217–240

Rebay I, Rubin GM (1995) Yan functions as a general inhibitor of differentiation and is negatively regulated by activation of the Ras1/MAPK pathway. Cell 81:857–866

Reich A, Shilo BZ (2002) Keren, a new ligand of the Drosophila epidermal growth factor receptor, undergoes two modes of cleavage. EMBO J 21:4287–4296

Reinke R, Zipursky SL (1988) Cell-cell interaction in the Drosophila retina: the bride of sevenless gene is required in photoreceptor cell R8 for R7 cell development. Cell 55:321–330

Reiter C, Schimansky T, Nie Z, Fischbach KF (1996) Reorganization of membrane contacts prior to apoptosis in the Drosophila retina: the role of the IrreC-rst protein. Development 122:1931–1940

Rogge RD, Karlovich CA, Banerjee U (1991) Genetic dissection of a neurodevelopmental pathway: Son of sevenless functions downstream of the sevenless and EGF receptor tyrosine kinases. Cell 64:39–48

Roignant JY, Treisman JE (2009) Pattern formation in the Drosophila eye disc. Int J Dev Biol 53:795–804

Rosay P, Colas JF, Maroteaux L (1995) Dual organization of the Drosophila neuropeptide receptor NKD gene promoter. Mech Dev 51:329–339

Sawamoto K, Taguchi A, Hirota Y, Yamada C, Jin M, Okano H (1998) Argos induces programmed cell death in the developing Drosophila eye by inhibition of the Ras pathway. Cell Death Differ 5:548

Schlesinger A, Kiger A, Perrimon N, Shilo BZ (2004) Small wing PLCgamma is required for ER retention of cleaved Spitz during eye development in Drosophila. Dev Cell 7:535–545

Schweitzer R, Howes R, Smith R, Shilo BZ, Freeman M (1995a) Inhibition of Drosophila EGF receptor activation by the secreted protein Argos. Nature 376:699–702

Schweitzer R, Shaharabany M, Seger R, Shilo BZ (1995b) Secreted Spitz triggers the DER signaling pathway and is a limiting component in embryonic ventral ectoderm determination. Genes Dev 9:1518–1529

Shi Y, Noll M (2009) Determination of cell fates in the R7 equivalence group of the Drosophila eye by the concerted regulation of D-Pax2 and TTK88. Dev Biol 331:68–77

Shilo BZ (2005) Regulating the dynamics of EGF receptor signaling in space and time. Development 132:4017–4027

Silver SJ, Rebay I (2005) Signaling circuitries in development: insights from the retinal determination gene network. Development 132:3–13

Simon MA, Bowtell DD, Dodson GS, Laverty TR, Rubin GM (1991) Ras1 and a putative guanine nucleotide exchange factor perform crucial steps in signaling by the sevenless protein tyrosine kinase. Cell 67:701–716

Singh J, Mlodzik M (2012) Hibris, a Drosophila nephrin homolog, is required for presenilin-mediated Notch and APP-like cleavages. Dev Cell 23:82–96

Spencer SA, Cagan RL (2003) Echinoid is essential for regulation of Egfr signaling and R8 formation during Drosophila eye development. Development 130:3725–3733

Spencer SA, Powell PA, Miller DT, Cagan RL (1998) Regulation of EGF receptor signaling establishes pattern across the developing Drosophila retina. Development 125:4777–4790

Steele L, Sukhanova MJ, Xu J, Gordon GM, Huang Y, Yu L, Du W (2009) Retinoblastoma family protein promotes normal R8-photoreceptor differentiation in the absence of rhinoceros by inhibiting dE2F1 activity. Dev Biol 335:228–236

Sukhanova MJ, Steele LJ, Zhang T, Gordon GM, Du W (2011) RBF and Rno promote photoreceptor differentiation onset through modulating EGFR signaling in the Drosophila developing eye. Dev Biol 359:190–198
Sun Y, Jan LY, Jan YN (1998) Transcriptional regulation of atonal during development of the Drosophila peripheral nervous system. Development 125:3731–3740
Sundaram MV (2005) The love-hate relationship between Ras and Notch. Genes Dev 19:1825–1839
Suzuki T, Saigo K 2000 Transcriptional regulation of atonal required for Drosophila larval eye development by concerted action of eyes absent, sine oculis and hedgehog signaling independent of fused kinase and cubitus interruptus. Development 127:1531–1540
Swanson CI, Evans NC, Barolo S (2010) Structural rules and complex regulatory circuitry constrain expression of a Notch- and EGFR-regulated eye enhancer. Dev Cell 18:359–370
Takano A, Zochi R, Hibi M, Terashima T, Katsuyama Y (2011) Function of strawberry notch family genes in the zebrafish brain development. Kobe J Med Sci 56:E220–230
Tanaka-Matakatsu M, Du W (2008) Direct control of the proneural gene atonal by retinal determination factors during Drosophila eye development. Dev Biol 313:787–801
Tang AH, Neufeld TP, Kwan E, Rubin GM (1997) PHYL acts to down-regulate TTK88, a transcriptional repressor of neuronal cell fates, by a SINA-dependent mechanism. Cell 90:459–467
Tepass U, Harris KP (2007) Adherens junctions in Drosophila retinal morphogenesis. Trends Cell Biol 17:26–35
Thackeray JR, Gaines PC, Ebert P, Carlson JR (1998) Small wing encodes a phospholipase C-(gamma) that acts as a negative regulator of R7 development in Drosophila. Development 125:5033–5042
Tio M, Ma C, Moses K (1994) Spitz, a Drosophila homolog of transforming growth factor-alpha, is required in the founding photoreceptor cells of the compound eye facets. Mech Dev 48:13–23
Tio M, Moses K (1997) The Drosophila TGF alpha homolog Spitz acts in photoreceptor recruitment in the developing retina. Development 124:343–351
Tomita A, Buchholz DR, Shi YB (2004) Recruitment of N-CoR/SMRT-TBLR1 corepressor complex by unliganded thyroid hormone receptor for gene repression during frog development. Mol Cell Biol 24:3337–3346
Tomlinson A (1985) The cellular dynamics of pattern formation in the eye of Drosophila. J Embryol Exp Morphol 89:313–331
Tomlinson A, Kimmel BE, Rubin GM (1988) Rough, a Drosophila homeobox gene required in photoreceptors R2 and R5 for inductive interactions in the developing eye. Cell 55:771–784
Tomlinson A, Mavromatakis YE, Struhl G (2011) Three distinct roles for notch in Drosophila R7 photoreceptor specification. PLoS Biol 9:e1001132
Tomlinson A, Ready DF (1986) Sevenless: a cell-specific homeotic mutation of the Drosophila eye. Science 231:400–402
Tomlinson A, Ready DF (1987a) Cell fate in the Drosophila ommatidium. Dev Biol 123:264–275
Tomlinson A, Ready DF (1987b) Neuronal differentiation in Drosophila ommatidium. Dev Biol 120:366–376
Tomlinson A, Struhl G (1999) Decoding vectorial information from a gradient: sequential roles of the receptors Frizzled and Notch in establishing planar polarity in the Drosophila eye. Development 126:5725–5738
Tomlinson A, Struhl G (2001) Delta/Notch and Boss/Sevenless signals act combinatorially to specify the Drosophila R7 photoreceptor. Mol Cell 7:487–495
Tsruya R, Schlesinger A, Reich A, Gabay L, Sapir A, Shilo BZ (2002) Intracellular trafficking by Star regulates cleavage of the Drosophila EGF receptor ligand Spitz. Genes Dev 16:222–234
Tsruya R, Wojtalla A, Carmon S, Yogev S, Reich A, Bibi E, Merdes G, Schejter E, Shilo BZ (2007) Rhomboid cleaves Star to regulate the levels of secreted Spitz. EMBO J 26:1211–1220
Tsuda L, Nagaraj R, Zipursky SL, Banerjee U (2002) An EGFR/Ebi/Sno pathway promotes delta expression by inactivating Su(H)/SMRTER repression during inductive notch signaling. Cell 110:625–637

Tsuda L, Kaido M, Lim YM, Kato K, Aigaki T, Hayashi S (2006) An NRSF/REST-like repressor downstream of Ebi/SMRTER/Su(H) regulates eye development in Drosophila. EMBO J 25:3191–3202

Urban S, Lee JR, Freeman M (2001) Drosophila rhomboid-1 defines a family of putative intramembrane serine proteases. Cell 107:173–182

Urban S, Lee JR, Freeman M (2002) A family of Rhomboid intramembrane proteases activates all Drosophila membrane-tethered EGF ligands. EMBO J 21:4277–4286

Usui K, Goldstone C, Gibert JM, Simpson P (2008) Redundant mechanisms mediate bristle patterning on the Drosophila thorax. Proc Natl Acad Sci U S A 105:20112–20117

Van Doren M, Ellis HM, Posakony JW (1991) The Drosophila extramacrochaetae protein antagonizes sequence-specific DNA binding by daughterless/achaete-scute protein complexes. Development 113:245–255

Van Doren M, Bailey AM, Esnayra J, Ede K, Posakony JW (1994) Negative regulation of proneural gene activity: hairy is a direct transcriptional repressor of achaete. Genes Dev 8:2729–2742

Van Vactor DL, Cagan RL, Krämer H, Zipursky SL (1991) Induction in the developing compound eye of Drosophila: multiple mechanisms restrict R7 induction to a single retinal precursor cell. Cell 67:1145–1155

Voas MG, Rebay I (2003) The novel plant homeodomain protein rhinoceros antagonizes Ras signaling in the Drosophila eye. Genetics 165:1993–2006

Voas MG, Rebay I (2004) Signal integration during development: insights from the Drosophila eye. Dev Dyn 229:162–175

Wang SL, Hawkins CJ, Yoo SJ, Müller HA, Hay BA (1999) The Drosophila caspase inhibitor DIAP1 is essential for cell survival and is negatively regulated by HID. Cell 98:453–463

Wasserman JD, Urban S, Freeman M (2000) A family of rhomboid-like genes: Drosophila rhomboid-1 and roughoid/rhomboid-3 cooperate to activate EGF receptor signaling. Genes Dev 14:1651–1663

Weber U, Siegel V, Mlodzik M (1995) pipsqueak encodes a novel nuclear protein required downstream of seven-up for the development of photoreceptors R3 and R4. EMBO J 14:6247–6257

Weber U, Paricio N, Mlodzik M (2000) Jun mediates Frizzled-induced R3/R4 cell fate distinction and planar polarity determination in the Drosophila eye. Development 127:3619–3629

Weber U, Pataki C, Mihaly J, Mlodzik M (2008) Combinatorial signaling by the Frizzled/PCP and Egfr pathways during planar cell polarity establishment in the Drosophila eye. Dev Biol 316:110–123

Weinmaster G, Fischer JA (2011) Notch ligand ubiquitylation: what is it good for? Dev Cell 21:134–144

White NM, Jarman AP (2000) Drosophila atonal controls photoreceptor R8-specific properties and modulates both receptor tyrosine kinase and Hedgehog signaling. Development 127:1681–1689

Wolff T, Ready DF (1991a) Cell death in normal and rough eye mutants of Drosophila. Development 113:825–839

Wolff T, Ready DF (1991b) The beginning of pattern formation in the Drosophila compound eye: the morphogenetic furrow and the second mitotic wave. Development 113:841–850

Wolff T, Ready DF (1993) Pattern formation in the Drosophila retina. In: Bate M, Martinez-Arias A (eds) The development of Drosophila melanogaster. Cold Spring Harbor Laboratory Press, Cold Spring Harbor, pp 1277–1325

Xie B, Charlton-Perkins M, McDonald E, Gebelein B, Cook T (2007) Senseless functions as a molecular switch for color photoreceptor differentiation in Drosophila. Development 134:4243–4253

Xiong WC, Montell C (1993) tramtrack is a transcriptional repressor required for cell fate determination in the Drosophila eye. Genes Dev 7:1085–1096

Xu C, Kauffmann RC, Zhang J, Kladny S, Carthew RW (2000) Overlapping activators and repressors delimit transcriptional response to receptor tyrosine kinase signals in the Drosophila eye. Cell 103:87–97

Yamada T, Okabe M, Hiromi Y (2003) EDL/MAE regulates EGF-mediated induction by antagonizing Ets transcription factor Pointed. Development 130:4085–4096

Yamamoto D, Nihonmatsu I, Matsuo T, Miyamoto H, Kondo S, Hirata K, Ikegami Y (1996) Genetic interactions of *pokkuri* with seven in absentia, tramtrack and downstream components of the sevenless pathway in R7 photoreceptor induction in Drosophila melanogaster. Rouxs Arch Dev Biol 205:215–224

Yan H, Canon J, Banerjee U (2003) A transcriptional chain linking eye specification to terminal determination of cone cells in the Drosophila eye. Dev Biol 263:323–329

Yang L, Baker NE (2001) Role of the EGFR/Ras/Raf pathway in specification of photoreceptor cells in the Drosophila retina. Development 128:1183–1191

Yogev S, Schejter ED, Shilo BZ (2008) Drosophila EGFR signaling is modulated by differential compartmentalization of Rhomboid intramembrane proteases. EMBO J 27:1219–1230

Yogev S, Rousso T, Schejter ED, Shilo BZ (2011) Versatility of EGF receptor ligand processing in insects. Dev Biol 357:17–20

Yoon HG, Chan DW, Huang ZQ, Li J, Fondell JD, Qin J, Wong J (2003) Purification and functional characterization of the human N-CoR complex: the roles of HDAC3, TBL1 and TBLR1. EMBO J 22:1336–1346

Yu SY, Yoo SJ, Yang L, Zapata C, Srinivasan A, Hay BA, Baker NE (2002) A pathway of signals regulating effector and initiator caspases in the developing Drosophila eye. Development 129:3269–3278

Zhang J, Kalkum M, Chait BT, Roeder RG (2002) The N-CoR-HDAC3 nuclear receptor corepressor complex inhibits the JNK pathway through the integral subunit GPS2. Mol Cell 9:611–623

Zhang T, Ranade S, Cai CQ, Clouser C, Pignoni F (2006) Direct control of neurogenesis by selector factors in the fly eye: regulation of atonal by Ey and So. Development 133:4881–4889

Zheng L, Carthew RW (2008) Lola regulates cell fate by antagonizing Notch induction in the Drosophila eye. Mech Dev 125:18–29

Zheng Q, Qin H, Zhang H, Li J, Hou L, Wang H, Zhang X, Zhang S, Feng L, Liang Y, Han H, Yi D (2007) Notch signaling inhibits growth of the human lung adenocarcinoma cell line A549. Oncol Rep 17:847–852

Cell Polarity in *Drosophila* Retina

Sang-Chul Nam

Differentiation and Morphogenesis of *Drosophila* Retina

The compound eye of *Drosophila* is made up of about 800 ommatidia, each of which comprises a cluster of eight elongated columnar photoreceptor cells covered by a thin layer of pigment cells (Kumar and Ready 1995; Longley and Ready 1995). These clusters of eight photoreceptor cells (R1–R8) are made in the eye disc epithelium during the third-instar larval stage, before photoreceptor morphogenesis takes place. Along the length of each ommatidial column extends a light sensitive, tightly packed array of 60,000 microvilli called a rhabdomere (Kumar and Ready 1995; Longley and Ready 1995). At 37 % pupal development (PD) stage, the apical region of each of the photoreceptor cells is involuted by 90°, reorienting the apical domains toward the center of the cluster (Fig. 1; Kumar and Ready 1995; Longley and Ready 1995). At this time, the apical membrane domain, having been localized at the center of the photoreceptor cluster, is now surrounded immediately by adherens junctions (AJs), followed by the basolateral domains (Fig. 1; Izaddoost et al. 2002; Pellikka et al. 2002). The formation of the rhabdomere from the apical surface of the photoreceptor cells begins at 55 % PD and involves a series of complex cell–cell signaling interactions and the rapid expansion of the plasma membrane (Kumar and Ready 1995; Longley and Ready 1995). Because of the enormity of this growth/elongation and the rapidity with which it occurs, even small signaling defects can cause dramatic phenotypic consequences in the developing eye.

Apico-basal Cell Polarity Genes

The establishment and maintenance of cell polarity is an essential feature of all eukaryotic cells and is critical for the integrity of the organism. Recent studies have begun to reveal the molecular and genetic basis of apical–basal cell polarity by

S.-C. Nam (✉)
Department of Biology, Texas A&M International University
5201 University Blvd., Laredo, TX 78041, USA
e-mail: sang-chul.nam@tamiu.edu

A. Singh, M. Kango-Singh (eds.), *Molecular Genetics of Axial Patterning, Growth and Disease in the Drosophila Eye,* DOI 10.1007/978-1-4614-8232-1_5,

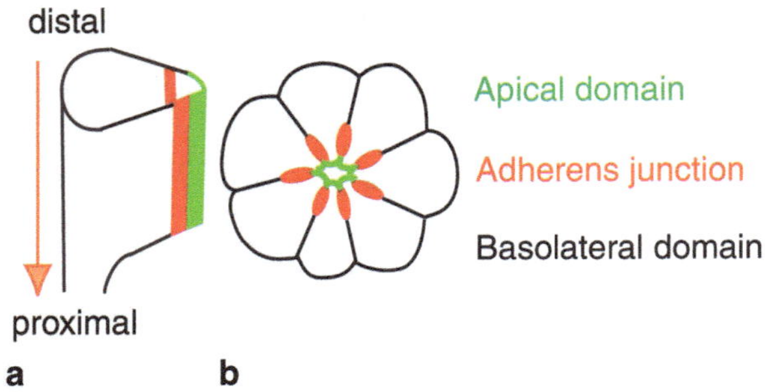

Fig. 1 Morphogenesis of *Drosophila* pupal photoreceptors. **a** Side view of developing photoreceptors at midstage of pupal development. The photoreceptors elongate from distal to proximal (*arrow*). **b** Cross-section of midstage pupal photoreceptors. Apical domain (*green*) localizes apical to AJ (*red*) in the center of a photoreceptor cluster

identifying important proteins involved in cell polarity determination and junction formation (Bilder 2001; Ohno 2001). Accumulating evidence suggests that important cues for the establishment of cell polarity are provided by the function of at least four evolutionarily conserved protein complexes. These are Crumbs (Crb) complex of Crb, Stardust (Sdt)/Pals-1-associated tight junction protein (Patj; Bachmann et al. 2001), partitioning-defective (Par) complex of Par-6/atypical protein kinase C (aPKC)/Par-3 (Ohno 2001), Scribble complex of Scrib/Dlg/Lgl (Bilder and Perrimon 2000), and Yurt complex of Yurt/Coracle/Neurexin-IV/Na-K-ATPase (Laprise et al. 2009). The Crb and Par complexes localize at the apical membrane domains or AJs, but the Scribble and Yurt complexes reside on the basolateral domains. Furthermore, there is the fifth complex of Par-1/Lkb1(Par-4)/AMPK. All five cell polarity complexes contribute to establish, maintain, and regulate the cell polarity through synergic and antagonistic collaborations (Laprise and Tepass 2011; Tepass 2012; Tepass et al. 2001; Table 1).

Crb/Sdt/Patj Complex in Retina Development

Crb complex including the Crb, Sdt, and Patj were discovered in *Drosophila* (Bhat et al. 1999; Tepass and Knust 1993). The *crb* and *sdt* genes were identified genetically as essential components for organizing apical–basal polarity and AJs in early embryonic epithelia (Bachmann et al. 2001; Bhat et al. 1999; Hong et al. 2001; Tepass et al. 1990). Genetic interaction studies suggested that *sdt* acts downstream of *crb* in the same pathway (Grawe et al. 1996; Tepass and Knust 1993). Molecular analysis of Crb and Sdt has shown that they are directly associated in the apical plasma membranes of epithelial cells (Bachmann et al. 2001; Hong et al. 2001). Crb is a transmembrane protein with a long extracellular domain and a short C-terminal cytoplasmic tail that recruits Sdt and Patj through direct biochemical interactions. (Hong et al. 2001; Roh et al. 2002) and now it is treated as a single protein complex (Fig. 2).

Table 1 Cell polarity genes and their regulators affecting retina morphogenesis

Drosophila	References	Vertebrate	References
crb	Chartier et al. 2012; Izaddoost et al. 2002; Johnson et al. 2002; Nam and Choi 2003; Pellikka et al. 2002; Pocha et al. 2011	*CRB1*	den Hollander et al. 1999; Mehalow et al. 2003; Pellikka et al. 2002; Zou et al. 2012
sdt	Hong et al. 2003; Nam and Choi 2003; Nam et al. 2007	*Pals1, nagie oko*	Cho et al. 2012; Park et al. 2011; Wei and Malicki 2002
patj	Nam and Choi 2006; Richard et al. 2006; Zhou and Hong 2012		
baz	Hong et al. 2003; Nam and Choi 2003; Nam et al. 2007; Walther and Pichaud 2010		
par-6	Nam and Choi 2003; Nam et al. 2007		
apkc	Nam and Choi 2003; Nam et al. 2007	*Heart and soul*	Horne-Badovinac et al. 2001
par-1	Nam et al. 2007		
pp2a	Nam et al. 2007		
lkb1	Amin et al. 2009		
ampk	Amin et al. 2009; Poels et al. 2012; Spasic et al. 2008		
moesin	Karagiosis and Ready 2004		
yurt	Laprise et al. 2006	*Mosaic eyes*	Hsu et al. 2006; Jensen and Westerfield 2004
pi3k	Pinal et al. 2006		
pten	Pinal et al. 2006		
cofilin	Pham et al. 2008		
abl	Xiong and Rebay 2011		
wasp	Zelhof and Hardy 2004		
myosin V	Li et al. 2007; Pocha et al. 2011		
rab11	Satoh et al. 2005; Wu et al. 2005		
sec6, sec8	Beronja et al. 2005		
a-spec	Chen et al. 2009		
kasrt	Chen et al. 2009		
β-spec	Chen et al. 2009		
spastin	Chen et al. 2010		
cnn	Chen et al. 2011		
kinesin-1	League and Nam 2011		
kinesin-2	Mukhopadhyay et al. 2010	*KIF3A*	Avasthi et al. 2009; Jimeno et al. 2006a; Jimeno et al. 2006b; Lopes et al. 2010
shot	Mui et al. 2011		
eyes shot	Husain et al. 2006; Zelhof et al. 2006	*Spacemaker*	Abd El-Aziz et al. 2008; Collin et al. 2008
prominin	Nie et al. 2012; Zelhof et al. 2006		Nie et al. 2012
chaoptin	Zelhof et al. 2006		

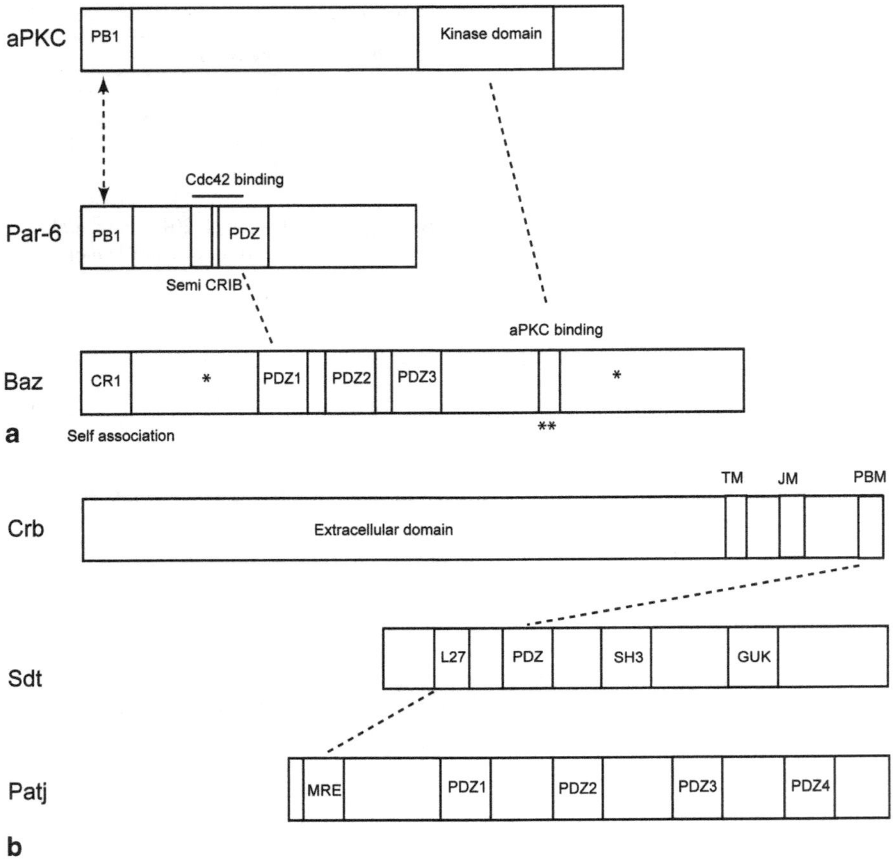

Fig. 2 The Par and Crb complexes and their interactions. **a** Domain organization of aPKC, PAR-6, and Baz. The Phagocyte oxidase/Bem1 (PB1) domain binds other PB1 domains. PSD-95, Discs large, and Zona occludens-1 (PDZ) domain binds other PDZ or PDZ-binding motif (PBM). CDC42/Rac-interactive-binding (CRIB) domain binds a small GTPase in GTP-bound state (the CRIB domain of PAR-6 is not sufficient enough, so is referred to as a "semi-CRIB" domain). An aPKC-binding domain in PAR3 is phosphorylated by the kinase. The amino terminal conserved region (CR1) is required for oligomerization of Baz. *Single asterisks* (*) indicate the Par-1 phosphorylation sites, and *double asterisk* (**) indicates the aPKC phosphorylation site in Baz. **b** Domain organization of Crb, Sdt, and Patj. Crb has a transmembrane domain (TM), juxtamembrane (JM), and PBM. Guanylate kinase (GUK) is a protein-binding domain. Lin2 and Lin7 (L27) domain bind MAGUK recruitment element (MRE) domain. *Dashed lines* indicate regions of the proteins that interact with one another

The apical–basal polarity is prominent in the photoreceptors due to the photosensitive organ, rhabdomere, formed on the apical surface of the cell. During pupal eye development, the apical domain of differentiating photoreceptors undergoes dynamic reorganization of the cell shape and size, resulting in the formation of rhabdomeres (Kumar and Ready 1995; Longley and Ready 1995). Recent studies have shown that Crb plays important roles in morphogenesis of the photoreceptor rhabdomere, providing evidence that at least some proteins involved in the apical–basal polarity of embryonic epithelia have essential roles in the organization of photoreceptors (Izaddoost et al. 2002; Pellikka et al. 2002). Crb is specifically localized to the rhabdomere stalk, a membrane domain that is juxtaposed apically to the emerging rhabdomere and basally to the AJ. Crb is required for positioning and growth of rhabdomere and AJ during the critical period of photoreceptor extension along the proximal–distal axis of the retina. Further analysis of Crb function has shown that the intracellular domain is necessary for the recruitment of AJ as well as localization of rhabdomere stalk (Izaddoost et al. 2002).

Importantly, Crb's mammalian homolog localizes to the region corresponding to the rhabdomere stalk membrane, that is, the inner segment between the outer segment (analogous to the rhabdomere) and the AJ of rod photoreceptors (Pellikka et al. 2002). Furthermore, mutations in *CRB1*, one of Crb homologs in human, cause severe retinal dystrophies such as retinitis pigmentosa type 12 (RP12; den Hollander et al. 1999) and Leber congenital amaurosis (den Hollander et al. 1999). These studies suggest that Crb and other cell polarity components involved in the specification of apical membrane of photoreceptors might be evolutionarily conserved. Crb's mammalian homolog, Crb1, was found to play an essential role in retinal differentiation in mice (den Hollander et al. 1999; Mehalow et al. 2003) and in zebrafish (Zou et al. 2012). Crb is also associated with light-induced retinal degeneration (Johnson et al. 2002) and superoxide-dependent retinal degeneration (Chartier et al. 2012). The Crb's degeneration protection was found to link with rhodopsin 1 trafficking (Pocha et al. 2011) and Rac1-NADPH oxidase complex activity (Chartier et al. 2012).

Sdt also shows colocalization with Crb specifically to the rhabdomere stalk region of the photoreceptor in pupal retina (Hong et al. 2003; Nam and Choi 2003). In the mutant of *sdt*, Crb and Patj were almost completely absent in rhabdomere stalks. In contrast, in the mutant of *crb*, Sdt was mislocalized together with Patj from the rhabdomere stalk (Nam and Choi 2003). These results provide an evidence of the interdependent function of Crb complex proteins in the developing retina. Std's vertebrate homolog, Pals1, was also reported to play important roles in retinal differentiation (Cho et al. 2012; Park et al. 2011) using a conditional knout-out technique. Previously, Sdt's zebrafish homolog, Nagie oko, was found to localize to the apical cell junctions of the retinal neuroepithelium and have an essential role in retinal differentiation (Wei and Malicki 2002).

Crb and Sdt are required for rhabdomere elongation and AJs during pupal photoreceptor morphogenesis (Hong et al. 2003; Izaddoost et al. 2002; Nam and Choi 2003; Pellikka et al. 2002). Patj binds Sdt to form a conserved heterotrimeric Crb complex (Roh et al. 2002). Recently, Patj's function in photoreceptor morphogenesis was clearly demonstrated. First, synthetic hypomorph of *patj* showed late-onset degeneration of photoreceptor cells in adult eye although the mutant eyes develop relatively normally (Nam and Choi 2006; Richard et al. 2006). Second, analysis of synthetic null mutant of *patj* null and $patj^{RNAi}$ demonstrated that Patj is essential for early development of the animal and for morphogenesis of AJ and apical membrane domains of photoreceptor cells during PD (Nam and Choi 2006). The role of patj in retina development was further demonstrated using a knock-out mutant of *patj* (Zhou and Hong 2012). In addition to Crb, Sdt and Patj were also associated with progressive light-induced retinal degeneration (Berger et al. 2007; Richard et al. 2006).

Par-3/Par-6/aPKC Complex in Cell Polarity

The Par-3/Par-6/aPKC complex is an evolutionarily conserved regulator of cell polarity that plays a central role in forming and maintaining cell junctions in early embryonic epithelial cells and in determining asymmetric cell division (Ohno 2001). The first discovery of Par-3 was done in the Par phenotype, which was manifested in the first embryo cell division in C. elegans (Kirby et al. 1990) . The normal *C. elegans* embryo goes through the asymmetric cell division to generate the differential cell fate of the two daughter cells. However, the *par* mutants show the symmetric cell division based on the loss of the polarized distribution of fate determinants. Based on this initial screening, several par mutant genes including *par-3* and *par-6* were isolated (Kirby et al. 1990). Later, through biochemical analysis, Par-3, Par-6, and aPKC were found to bind each other directly through the protein–protein interaction (Fig. 2); therefore, the single protein complex hypothesis was established (Macara 2004a, b). This single complex is called by a Par complex composed of Par-3(Baz)/Par-6/aPKC (Fig. 2). Baz contains three PDZ domains that can mediate protein–protein interactions with Par-6 (Petronczki and Knoblich 2001) and aPKC (Wodarz et al. 2000) and it is thought that these three proteins form a polarity complex (Macara 2004a, b; Fig. 2).

The cross-talk between the Par complex and Crb complex was identified through the direct protein–protein interactions. The Sdt–Par-6 (Hurd et al. 2003), Crb–aPKC (Sotillos et al. 2004), Sdt–Baz (Krahn et al. 2010; Sotillos et al. 2004), or Patj–Par-6 (Nam and Choi 2006) binding causes the physical interaction between the Crb and Par complex, and this interaction affects the tight collaboration between these two complexes to generate cell polarity or organ morphogenesis (Hurd et al. 2003; Nam and Choi 2006; Sotillos et al. 2004).

Par-3(Baz)/Par-6/aPKC in Retina Development

Par-6/aPKC colocalize with Crb complex proteins in photoreceptor cells during pupal stages. Both Par-6 and aPKC colocalized with Crb/Sdt/Patj at the rhabdomere stalks. In contrast, Baz localized at the AJ of photoreceptors basal to the rhabdomere stalk (Hong et al. 2003; Nam and Choi 2003). In *baz, par-6*, or *apkc* mutant photoreceptors, Crb/Sdt/Patj and AJ were strongly reduced and/or mislocalized (Hong et al. 2003; Nam and Choi 2003). These results demonstrate that Par-6/aPKC/Baz complex proteins are essential for proper localization of Crb complex and AJ components. In contrast, Par complex remained in the membrane although mislocalized, in the absence of Crb complex (Nam and Choi 2003). These data suggest that Par complex acts upstream to the Crb complex. The hierarchy among the Par complex of Baz/Par-6/aPKC was further examined and found that Baz acts upstream to Par-6/aPKC (Nam et al. 2007). Therefore, Baz is a nodal component for apical targeting of Par and Crb complexes (Nam et al. 2007).

The knowledge of Par complex in vertebrate eyes is very limited except the aPKC in zebrafish retina (Horne-Badovinac et al. 2001). The *heart and soul* mutation in

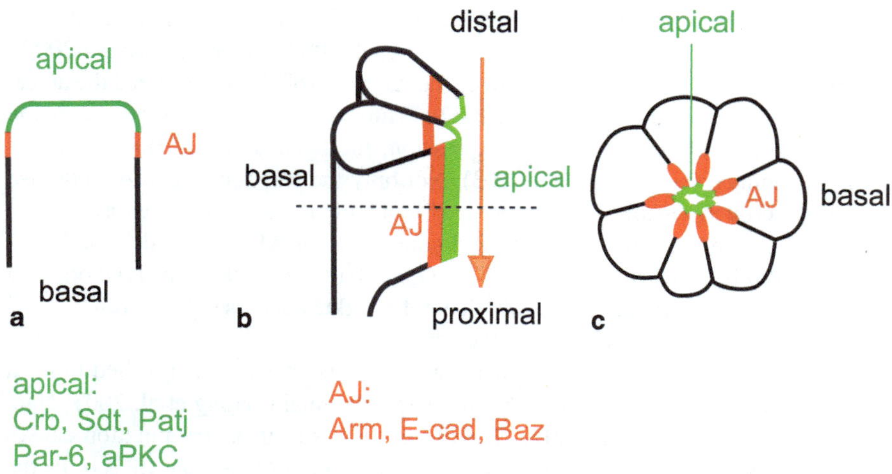

Fig. 3 Localization of Crb and Par complexes in *Drosophila* photoreceptors. **a** Schematic view of developing eye disc cell in third-instar larvae. Apical domains of photoreceptors (*green*) face the retinal surface and are held together by the AJ (*red*). **b** Longitudinal section of a photoreceptor cluster in midpupal stage. At this stage, the apical domain (*green*) and the AJ are oriented toward the center of the ommatidial cluster, as photoreceptors have rotated 90° inward during early pupal stage. **c** Tangential section was indicated by the *dashed line* in **b**. The tangential section of midpupal photoreceptors shows the apical domains (*green*) face into the center and surrounded by the AJ (*red*). All the Crb complex (Crb/Sdt/Patj) and Par-6/aPKC of Par complex localize at the apical domain (*green*), whereas Baz localizes at AJ (*red*) with Armadillo (Arm, β-Catenin homolog) and E-cadherin (E-cad)

aPKC lambda (an ortholog of aPKC) caused AJ defects in zebrafish retina (Horne-Badovinac et al. 2001). The localization of Par-3 in mouse retina was examined (Sottocornola et al. 2010), but the functional role of Par-3 in vertebrate retina is not known, yet.

Localization of Baz at AJs of Photoreceptors

Vertebrate Par-3, Baz homolog, localizes to the apical tight junction in vertebrate epithelial cells (Izumi et al. 1998; Suzuki et al. 2001). In most models, Baz, aPKC, and Par-6 form a complex to regulate epithelial polarity. In mammalian cells, this complex localizes above AJ at tight junctions (Nelson 2003). Considering aPKC and Par-6 can bind Baz (Hutterer et al. 2004; Wodarz et al. 2000), it was hypothesized that they might also localize to apical membrane domain in this context. Thus, the Par complex localization in the midstage of pupal developing eyes was examined (Hong et al. 2003; Nam and Choi 2003). Surprisingly, Baz localizes at the AJ (Hong et al. 2003; Nam and Choi 2003), but Par-6/aPKC localize at the apical membrane domain, *Drosophila* homolog of tight junction. Also, the Crb/Sdt/Patj, a Crb complex, localize at the apical domain (Nam and Choi 2003). Therefore, all the Crb and Par complex localize at the apical domain, except the Baz (Fig. 3).

Previously, it is reported that the apical domain and AJ domain are controlled by the intracellular domain of Crb (Izaddoost et al. 2002; Klebes and Knust 2000). Ectopic expression of Crb^{JM} mislocalized the AJ, but Crb^{PBM} mislocalized the apical domain, respectively (Izaddoost et al. 2002). Using this independent mislocalization of apical and AJ driven by Crb misexpression, the localization of Baz at AJ was further examined (Nam and Choi 2003). Not only the Baz localize at AJ, but also Baz was recruited together with AJ to ectopic membrane sites by misexpression of Crb^{JM}, suggesting that Baz is an integral component of AJ (Nam and Choi 2003). However, Baz is not recruited by Crb^{PBM}, whereas Par-6 and aPKC can be ectopically recruited by Crb^{PBM} rather than Crb^{JM}. Therefore, Baz appears to be recruited to AJ independently of Par-6/aPKC (Nam and Choi 2003).

Intriguingly, despite its specific localization to AJs, loss of Baz resulted in most severe disruption of AJ as well as the more apical domain (Hong et al. 2003; Nam and Choi 2003; Nam et al. 2007). Therefore, the result that *baz* mutation causes loss of Par-6/aPKC, Crb/Sdt/Patj, and AJ support the crucial role of Baz in the initial step of cell polarization (Nam and Choi 2003; Nam et al. 2007). However, the distinct localization of Baz from Par-6/aPKC in the photoreceptors suggested that Baz might be targeted to the membrane with Par-6 but be sorted out from Par-6 in subsequent steps of polarization to remain in the AJs (Choi et al. 2007; Nam and Choi 2003; Nam and Choi 2006). Furthermore, the localization of Baz at AJ is not in the specific pupal stage. The Baz localizes at the AJ in the early larval eye discs and does not overlap with the apical domain on which the other polarity components localize (Nam and Choi 2003).

The initial finding of Baz at AJ and its separate localization from Par-6/aPKC in the larval and pupal eyes discs (Hong et al. 2003; Nam and Choi 2003) was confirmed in early epithelia cells in *Drosophila* embryo where epithelial cells first form (Harris and Peifer 2004; Harris and Peifer 2005). Early in cellularization, Baz colocalizes with AJs rather than aPKC or Par-6. At gastrulating embryos, Baz continues to colocalize with AJ in the epidermis (Harris and Peifer 2004) and in the posterior midgut invagination. The aPKC remains apical to AJs in both tissues whereas PAR-6 becomes enriched above Baz, colocalizing with aPKC in the extreme apical domain. Thus, most cortical Baz remains segregated from aPKC and PAR-6 during gastrulation and retains close AJ association. Baz was previously found to localize above AJs at stage 14 when the epithelium is fully polarized (Wodarz et al. 2000). This was reexamined (Harris and Peifer 2004; Harris and Peifer 2005) and found some segregation at stages 11 and 12 that became more pronounced at stage 14 and later. At stage 14, segregation was most evident in the gut and in segmental furrows of the epidermis. In each case, Baz appears to localize just apical to AJs. However, PAR-6 continues to localize just apical to Baz. Thus, in late-stage epithelia, the apical domain is stratified into three regions; the apical, "midapical," and the AJ (Harris and Peifer 2004, 2005). Previous models placed AJs at the top of the epithelial polarity establishment hierarchy (Drubin and Nelson 1996). However, it was found that Baz establishes apical complexes along cellularization furrows in the absence of AJs, and that Baz is required for recruiting AJ into apical spot junctions. These results show that Baz acts upstream of AJs as epithelial polarity is established during *Drosophila* cellularization (Harris and Peifer 2004, 2005).

Localization of Baz is Controlled by Par-1 Kinase and PP2A Phosphatase

As Par-1 kinase activity is important for inducing Baz mislocalization and Baz is a biochemical substrate for Par-1 protein kinase (Benton and St Johnston 2003), it is possible that Par-1 phosphorylation of Baz may be responsible for the mislocalization of AJ and apical markers. To test this possibility, GFP-tagged wild-type Baz (Baz^{WT}) was expressed in differentiating retinal cells and examined whether the GFP-Baz proteins are normally localized to AJ or are recruited to ectopic positions in the photoreceptor cells (Choi et al. 2007; Nam et al. 2007). Most GFP-Baz^{WT} were severely displaced to apical or basolateral regions. The apical marker was also diffused and mislocalized basolaterally from the apical domain, implying the disruption of apical basal cell polarity (Choi et al. 2007; Nam et al. 2007). However, the unphosphorylatable Baz by Par-1 having mutations at the Par-1 phosphorylation sites, was predominantly localized to the normal AJ positions basal to the apical domain (Choi et al. 2007; Nam et al. 2007).

Furthermore, the phosphorylation of Baz by Par-1 was dephosphorylated by Protein phosphatase 2A (PP2A), therefore, the localization of Baz at AJ is controlled by Par-1 phosphorylation and PP2A dephosphorylation (Choi et al. 2007; Nam et al. 2007). The dephosphorylation of Baz by PP2A and its antagonistic function against the Par-1 was further confirmed in *Drosophila* neuroblasts polarity (Krahn et al. 2009).

Separation of Baz from Par-6/aPKC is Triggered by Crb and aPKC

Baz, PAR-6, and aPKC form a complex that plays a key role in the polarization of many cell types and cell polarity- dependent organ morphogenesis. In epithelial cells including the photoreceptors (Hong et al. 2003; Nam and Choi 2003) and embryo (Harris and Peifer 2004, 2005), however, Baz localizes at AJ below PAR-6 and aPKC. Recently, the molecular separation mechanism of Baz from Par-6/aPKC was discovered (Morais-de-Sa et al. 2010). The mechanism is that Baz is excluded from the apical Par-6/aPKC domain in epithelia by aPKC phosphorylation, which disrupts the Baz–aPKC interaction (Morais-de-Sa et al. 2010). Removal of Baz from the Par-6/aPKC complex also requires the Crb complex, which prevents the Baz–PAR-6 interaction. In the absence of Crb or aPKC phosphorylation of Baz, mislocalized Baz recruits AJ components apically, leading to a loss of the apical domain and an expansion of lateral (Morais-de-Sa et al. 2010; Fig. 2). This molecular sorting mechanism of Baz from Par-6/aPKC was also confirmed in developing pupal photoreceptor (Walther and Pichaud 2010). Furthermore, the molecular sorting mechanism of Baz was proposed as an apical boundary establishment during the photoreceptor polarity remodeling during the pupal eye development. However, the

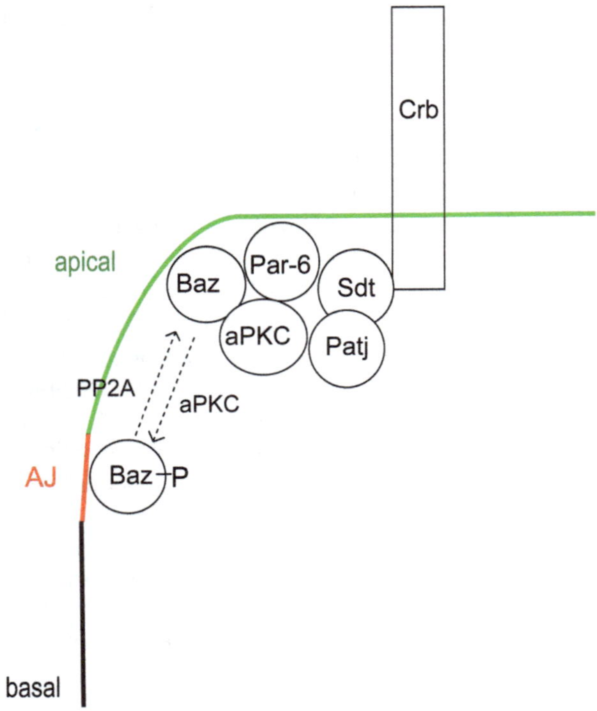

Fig. 4 Baz localization at AJ from Par-6/aPKC and Crb complexes. Par-6 and Crb complex (Crb/Sdt/Patj) proteins are targeted to the apical region of epithelia cells, except that Baz is localized to AJ (*red*) between the apical (*green*) and the basolateral domains. Baz protein phosphorylated by aPKC at the S980 aPKC site causes the Baz separation from the apical Baz/Par-6/aPKC complex, thereby the Baz is localized to AJ. The PP2A phosphatase antagonizes aPKC kinase function by dephosphorylating Baz

separation of Baz from Par-6/aPKC and Crb complex is not unique feature in pupal eyes, since the same separations were found in developing earlier larval eyes (Nam and Choi 2003), follicles cells (Morais-de-Sa et al. 2010) and embryos (Harris and Peifer 2004, 2005). There is a possibility of PP2A as a negative regulator against the aPKC on Baz, since the antagonistic relationship between aPKC and PP2A was found in *Drosophila* neuroblast (Chabu and Doe, 2009; Ogawa et al. 2009) and mammalian epithelia (Nunbhakdi-Craig et al. 2002; Figs. 2 and 4).

FERM (band 4.1, ezrin, radixin, moesin) Proteins in Retina

Crb has a highly evolutionary conserved short cytoplasmic tail that contains a PDZ domain-binding site (PBM) at its C-terminus and a juxtamembrane region (JM) that was predicted to act as a FERM domain-binding site (Klebes and Knust 2000; Fig. 2). Both PBM and JM are important for the function of Crb in retina (Izaddoost et al. 2002). However, the PBMs recruit the apical stalk membrane, in contrast, the JM control the AJ, respectively (Izaddoost et al. 2002). Based on the JM has a potential FERM-binding sequence, therefore, it was postulated that a FERM protein will control the AJ through the JM of Crb (Izaddoost et al. 2002). Two FERM proteins were suggested as potential candidates for mediating this interaction (Karagiosis and Ready 2004; Laprise et al. 2006). One of the FERM proteins is

Moesin. It localizes to rhabdomere base, and is essential for the apical membrane and rhabdomere (Karagiosis and Ready 2004). The other FERM protein is Yurt, which localizes at the basolateral membrane domain. It transiently localizes to the apical stalk membrane during a very late stage of pupal eye development. Yurt negatively regulates Crb since yurt mutants show an apical membrane expansion similar to Crb overexpression (Laprise et al. 2006). Although the roles of FERM proteins of Mosein and Yurt are clear on the Crb in the retinal development, the evidence of the Moesin/Yurt's role on Crb–AJ linkage is not clear, yet. The zebrafish gene *mosaic eyes* is a homolog of Yurt (Jensen and Westerfield 2004). The mosaic eyes mutation in zebrafish retina showed a retinal junction phenotype (Christensen and Jensen 2008; Hsu et al. 2006; Jensen and Westerfield 2004).

Role of LKB1 (Par-4), AMPK

LKB1 (Par-4) is essential for the correct distribution of polarity determinants during early embryo development in *C. elegans* (Morton et al. 1992; Watts et al. 2000) and *Drosophila* (Martin and St Johnston 2003). Recently, the role of LKB1 (Par-4) was found to have an essential role in apical junction and AJ regulation in the pupal retina morphogenesis (Amin et al. 2009). Especially, the loss of LKB1 caused the mixing of the apical domain and the basolateral domains, which is a true cell polarity defect (Amin et al. 2009).

Surprisingly, AMPK was found to be dispensable in retinal development (Amin et al. 2009; Spasic et al. 2008). Furthermore, it was found that LKB1 does not act primarily through AMPK to regulate cell polarity in the retina (Amin et al. 2009), although it was known that LKB1 acts primarily through the AMP kinase to establish and/or maintain cell polarity in other system or organ (Lee et al. 2007; Mirouse et al. 2007). But, it was found that AMPK was crucial to maintain the adult *Drosophila* photoreceptor from the neurodegeneration caused by light-induced excitation (Spasic et al. 2008). Furthermore, energy depletion in *ampk* mutants resulted in increase of autophagy and promotion of neurodegeneration in *Drosophila* retina (Poels et al. 2012).

Role of Phosphatidinositol Lipids in Retina Morphogenesis

The apical and basolateral surfaces of the cell have completely different protein and lipid compositions and so the cell has mechanisms to specifically sort these components to one surface or the other. The lipids phosphatidylinositol bisphosphase (PIP2) and phosphatidylinositole triphosphate (PIP3) are crucial determinants of the identities and formation of the apical and basolateral surfaces, respectively. PIP2 localizes at the apical membrane domain and PIP3 localizes at the basolateral membrane domain, respectively (Bryant and Mostov 2008; Martin-Belmonte and Mostov 2007). Phosphatidylinositol 3-kinase (PI3K) convert PIP2 to PIP3 and phosphatase and tensin homolog (PTEN) convert PIP3 to PIP2 (Di Paolo and De Camilli 2006; Gassama-Diagne et al. 2006; Martin-Belmonte and Mostov 2007).

In developing midpupal photoreceptors, PIP3 was enriched in the whole apical membrane. The PIP3 was later restricted into the rhabdomere in the late pupal photoreceptors (Pinal et al. 2006). In contrast, PIP2 was mainly localized in the AJ as well as the apical and basolateral domains at lower levels (Pinal et al. 2006). Therefore, the appropriate control of the PIP3 levels in the cell membrane may be required to define the apical rhabdomere area (Pinal et al. 2006).

Trafficking and Secretion During Retina Morphogenesis

During retina development in pupal stage, a massive trafficking from cytoplasm to apical membrane domains is essential. Therefore, cellular-trafficking components were postulated to involve the rhabdomere elongation and growth. Rab11, a small GTPase implicated in membrane traffic, in the trans-Golgi network, cytoplasmic vesicles, and the rhabdomere base (Satoh et al. 2005). When Rab11 activity is reduced, rhabdomere morphogenesis was inhibited. Then, it was proposed that Rab11 has a role in the post-Golgi transport to the rhabdomeric membranes of photoreceptors. Furthermore, other exocytosis genes of Sec6, Sec8, and Sec15 were identified to be involved in the rhabdomere morphogenesis (Beronja et al. 2005; Li et al. 2007; Wu et al. 2005). However, cell polarity protein targeting was not affected in the absence of the Sec6 (Beronja et al. 2005). Therefore, the targeting of the cell polarity proteins to the membrane may be independent from the exocytosis.

Role of Spectrins in Membrane Domain Modulations

Spectrins are major proteins in the cytoskeletal network of most cells. In *Drosophila*, βHeavy-Spectrin encoded by *karst* gene functions together with Crb during photoreceptor morphogenesis (Pellikka et al. 2002). Recently, it has been shown that Karst colocalizes with Crb at the rhabdomere stalk and interacts with Crb (Medina et al. 2002; Pellikka et al. 2002). The rhabdomere stalk length is reduced in photoreceptors of *karst* mutant adult eyes. This phenotype was enhanced by the presence of one copy of *crb* mutation, indicating that *karst* and *crb* genetically cooperate for rhabdomere stalk maintenance. The rhabdomere stalk localization of Karst depends on Crb (Pellikka et al. 2002). While Karst localizes apically, β-Spectrin is preferentially distributed in the basolateral region (Chen et al. 2009). Overexpression of the basolateral β-Spectrin caused a strong shrinkage of apical membrane domains, and loss of the β-Spectrin causes an expansion of apical domains, implying an antagonistic relationship between β-Spectrin and Karst. These results indicate that Spectrins are required for controlling photoreceptor morphogenesis through the modulations of cell membrane domains (Chen et al. 2009).

Function of Actin Cytoskeleton in Retina Morphogenesis

Drosophila photoreceptors undergo massive elongation during PD (Fig. 1). This elongation includes the rhabdomere elongation. The rhabdomere is a stabilized actin cytoskeleton, therefore, it was postulated that the actin cytoskeleton might provide an important role in this elongation step. Recently, cofilin/actin-depolymerzing factor was found to be required for this process (Pham et al. 2008). Furthermore, Abelson (Abl) tyorisine kinase, a regulator of actin cytoskeleton, was found to be an essential role for this rhabdomere elongation step (Xiong and Rebay 2011). However, there is no direct evidence between the actin cytoskeleton and the cell polarity genes, yet. Wiskott–Alrich syndrome protein (WASP) and CYFIP/Sra-1 (a member of the WAVE/SCAR complex and regulator of Actin remodeling) were suggested and identified as regulators of the actin-based rhabdomeres biogenesis (Galy et al. 2011; Zelhof and Hardy 2004). Furthermore, Myosin V was found to involve the material trafficking for the rhabdomere components (Li et al. 2007) and to be stabilized by the Crb for the rhodopsin trafficking (Pocha et al. 2011).

Stable/Acetylated Microtubules in *Drosophila* Retina

In animal photoreceptor cells, the surface membrane is enlarged for the storage of opsin photopigment. Insect eyes use an actin-based structure for surface membrane enlargement, but mammalian eyes use microtubule-based structure (Land and Nilsson 2002). Previously, the microtubules in developing early *Drosophila* photoreceptors were reported in early developing eye discs during the third-instar larval stage (Lei and Warrior 2000; Mosley-Bishop et al. 1999; Whited et al. 2004). Nuclear positioning or migration functions were defects in the larval eye discs with the microtubule-dependent genes including *klarsicht* (Mosley-Bishop et al. 1999), *dynactin* (Whited et al. 2004), and *lissencephay1* (Lei and Warrior 2000). However, the functional role of microtubules in rhabdomere morphogenesis was not reported. Previously, a subcellular localization of a microtubule structure at the base of the rhabdomere was identified in the pupal photoreceptors (Fan and Ready 1997). Furthermore, the microtubules at the rhabdomere base were recently identified as stable/acetylated microtubules (Chen et al. 2010). Given the specific localization of stable microtubules in developing pupal photoreceptors (Fig. 5), these subcellular structures might provide a functional role for photoreceptor morphogenesis.

Role of Spastin in *Drosophila* Retina Elongation

Spastin is a microtubule-severing AAA ATPase involved in constructing neuronal and noncentrosomal microtubule arrays (Lumb et al. 2011; Roll-Mecak and McNally 2009; Salinas et al. 2007). In mammals, Spastin has been shown to modulate the

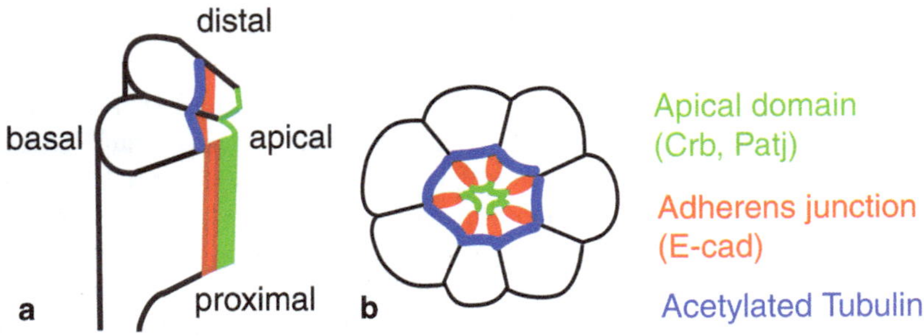

Fig. 5 Schematic diagram of the localization of stabilized microtubules in pupal photoreceptors. The apical markers (Crb) localize at the apical domain (*green*). The E-cad localizes at AJ (*red*), which are more basal to the apical domain. The acetylated tubulin (*blue*) localizes at the outside from the AJs (*red*)

microtubule cytoskeleton (Errico et al. 2002). The *spastin* mutation in developing pupal eyes causes a mild mislocalization of the apical membrane domain at the distal section, but the apical domain was dramatically reduced at the proximal section of the developing pupal eye (Chen et al. 2010). Since the rhabdomeres in developing pupal eyes grow from distal to proximal (Izaddoost et al. 2002), this phenotype strongly suggests that *spastin* is required for apical domain maintenance during rhabdomere elongation. This role of *spastin* in apical domain modulation was further supported by *spastin's* gain-of-function phenotype. Spastin overexpression in photoreceptors caused the expansion of the apical membrane domain from apical to basolateral in the developing photoreceptor (Chen et al. 2010). Although the localizations of the apical domain and AJs were severely expanded, there were no defects in cell polarity. These results strongly suggest that Spastin is essential for apical domain biogenesis during rhabdomere elongation in *Drosophila* photoreceptor morphogenesis.

Role of Centrosomin (Cnn) in *Drosophila* Retina Morphogenesis

Cnn is a core protein for centrosome, which is a major microtubule-organizing center. The effects of the *cnn* mutation on developing eyes were recently reported (Chen et al. 2011). Photoreceptors deficient in Cnn displayed dramatic morphogenesis defects including the mislocalization of Crb and Baz during midstage pupal eye development, suggesting that Cnn is required for photoreceptor morphogenesis during pupal eye development. This role of Cnn in apical domain modulation was further supported by Cnn's gain-of-function phenotype. Cnn overexpression in photoreceptors caused the expansion of the apical Crb membrane domain, Baz and AJs (Chen et al. 2011). These results strongly suggest that the interaction of Baz and Cnn is essential for apical domain and AJ modulation during photoreceptor morphogenesis.

Role of Kinesin Motors in *Drosophila* Retina Morphogenesis

Crb, a cell polarity gene, has been shown to provide a positional cue for the extension of the apical membrane domain, AJ, and rhabdomere along the growing proximal–distal axis during *Drosophila* photoreceptor morphogenesis (Izaddoost et al. 2002; Pellikka et al. 2002). In developing *Drosophila* photoreceptors, a stabilized microtubule structure was discovered (Chen et al. 2010; Fan and Ready 1997) and its presence was linked to polarity protein localization (Chen et al. 2010). It was therefore proposed that the microtubules may provide trafficking routes for the polarity proteins during photoreceptor morphogenesis (League and Nam 2011). Recently, Crb localization was examined in the developing photoreceptors of *kinesin-1* mutants (League and Nam 2011). The *kinesin-1* mutant photoreceptors showed a range of abnormalities in the apical membrane domain depending on the position along the proximal–distal axis in pupal photoreceptors (League and Nam 2011). The *kinesin-1* mutant showed a progressive mislocalization in the apical domain along the distal–proximal axis during rhabdomere elongation (League and Nam 2011). The *kinesin-1* mutation also led to a similar progressive defect in the stabilized microtubule structures, strongly suggesting that Kinesin-1 motor is essential for microtubule structure and Crb localization during distal to proximal rhabdomere elongation in pupal morphogenesis. This role of Kinesin-1 in apical domain control was further supported by *kinesin-1*'s dominant-negative mutation phenotypes, which showed disruption of the apical membrane domain and the stabilized microtubules in the developing photoreceptors (League and Nam 2011). These phenotypes suggest that Kinesin-1 is essential for the microtubule structures and apical membrane domains during the distal–proximal elongation of photoreceptors, but is dispensable for early eye development.

Another Kinesin motor, Kinesin-2, was analyzed in photoreceptor development (Mukhopadhyay et al. 2010). It turned out that Kinesin-2 is essential for viability of developing photoreceptors and localization of junctional proteins during early eye development of eye disc differentiation (Mukhopadhyay et al. 2010). The early *kinesin-2* mutants have abnormal nuclear position in differentiating photoreceptors. These cells eventually die in the pupal stage, indicating Kinesin-2's role in cell viability. Furthermore, Kinesin-2 was essential for Baz localization to the AJ in pupal photoreceptors (Mukhopadhyay et al. 2010). These findings suggest that Kinesin-2 motor plays a primary role in the localization of AJ and cell polarity proteins in the developing retina (Mukhopadhyay et al. 2010). In contrast, the Kinesin-1 motor participates in a more specific step of apical domain elongation during the rhabdomere morphogenesis (League and Nam 2011). Kinesin-2's function in mouse retina was reported. The absence of Kinesin-2 caused optin-trafficking defects and followed by cell death (Avasthi et al. 2009; Jimeno et al. 2006a, b; Lopes et al. 2010).

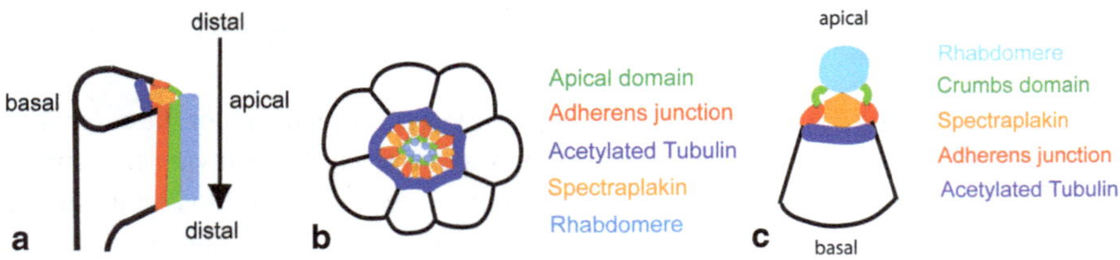

Fig. 6 Schematic representation of midpupal photoreceptor and localization of Shot. Shot (*orange*) localizes in between AJ (*red*), at the basal side of the apical Crumbs domain (*green*), at the apical side of the stable microtubule (*blue*), and at the basal side of the rhabdomere (*light blue*)

Role of Spectraplakin, an Actin-Microtubule Linker, in *Drosophila* Retina Morphogenesis

Coordinated interactions between microtubule and actin cytoskeletons are involved in many polarized cellular processes. Since Spectraplakin is able to bind both microtubule and actin cytoskeletons, the role of Short stop (Shot, *Drosophila* homolog of Spectraplakin; Lee et al. 2000; Lee and Kolodziej 2002) was analyzed in the regulations of apical Crb domain in developing *Drosophila* photoreceptors (Mui et al. 2011). The localization pattern of Shot in developing pupal photoreceptors showed a unique intracellular distribution. Shot localized at rhabdomere terminal web (Ready 2002; Xia and Ready 2011), which is at the basal side of the apical Crb or rhabdomere, and in between the AJs (Fig. 6). The rhabdomere terminal web, where the Shot localizes, may be the interface where the stable microtubules and F-actins of rhabdomere meet together. Since Shot has an actin-microtubule cross-linking activity, Shot might cross-link the two cytoskeletons of actin and microtubules at the rhabdomere terminal web. The *shot* mutant photoreceptors showed dramatic mislocalizations of Crb, AJs, and the stable microtubules (Mui et al. 2011). This role of Shot in Crb and AJ regulation was further supported by *shot*'s gain-of-function phenotype (Mui et al. 2011). Shot overexpression in photoreceptors caused a cell polarity defect including dramatic mislocalization of Crb, AJs, and the stable microtubules in the developing photoreceptors. These data suggest that Shot, an actin-microtubule cross-linker, is essential in the apical and adherens junction controls during the photoreceptors morphogenesis.

Inter-Retina Space Formation During *Drosophila* Retina Morphogenesis

Drosophila has an open rhabdom system in which the rhabdomeres of each ommatidium are separated from each other (Land and Nilsson 2002). This system evolved from the ancestral insect eye that has fused rhabdoms. Recently, several genes involving in this rhabdomere separation were identified (Husain et al. 2006; Zelhof et al. 2006). Eyes shut (spacemaker), prominin, and chaoptin are responsible to generate inter-rhabdomere space, which is an extracellular lumen. Although the eyes shut mutation do not affect the cell polarity defects in retina, the Eyes shut is secreted to the inter-retina space through the rhabdomere stalk (Husain et al. 2006), which is controlled by cell polarity genes. Therefore, there is a potential possibility of cell polarity genes' role in Eyes shut secretion and then inter-retina space formation.

Surprisingly, several studies identified mutations in a human ortholog of *Drosophila* eyes shut is responsible to cause retinitis pigmentosa (Abd El-Aziz et al. 2008; Collin et al. 2008). These two fundamentally different types of photoreceptors use totally different materials to increase the surface areas for housing photopigments. Vertebrate eyes utilize a microtubule-based cilia, but *Drosophila* eyes use a actin-based rhabdomere. However, they use the same way to make inter-retina space in the retina. A similar conserved of Prominin was also identified between *Drosophila* and mouse (Nie et al. 2012).

Significance and Perspective

Evolutionary conservation in the structure and function of polarity genes makes the *Drosophila* retina an excellent model for studying the genetic and molecular basis of retinal cell organization and retinal diseases resulting from mutations in polarity genes (den Hollander et al. 2001; Izaddoost et al. 2002; Pellikka et al. 2002). For example, mutations in human Crb homolog cause retinal diseases such as a late-onset retina degeneration of retinis pigmentosa (den Hollander et al. 1999) and an early-onset retina degeneration of Leber Congenital Amaurosis (den Hollander et al. 2001). Searching for new genes-interacting Crb will help in the search for new therapy targets for these eye diseases. This study may broaden our knowledge of the *Drosophila* photoreceptor, one of the best genetic model systems, and expand its usefulness as a model system for human retina diseases. Based on the strict conservation of genes and cellular structures between *Drosophila* and human retinas, narrowly defined classic "homology"-based approaches might not be enough to understand deeper similarities between the two. The new concept of "deep homology" deals with homology in contexts in which structures are not homologous in the classical sense (Shubin et al. 2009). The new concept of "deep homology" could be an appropriate way to fully understand the deeply conserved mechanisms between these two eyes.

References

Abd El-Aziz MM, Barragan I, O'Driscoll CA, Goodstadt L, Prigmore E, Borrego S, Mena M, Pieras JI, El-Ashry MF, Safieh LA et al (2008) EYS, encoding an ortholog of Drosophila spacemaker, is mutated in autosomal recessive retinitis pigmentosa. Nat Genet 40:1285–1287

Amin N, Khan A, St Johnston D, Tomlinson I, Martin S, Brenman J, McNeill H (2009) LKB1 regulates polarity remodeling and adherens junction formation in the Drosophila eye. Proc Natl Acad Sci U S A 106:8941–8946

Avasthi P, Watt CB, Williams DS, Le YZ, Li S, Chen CK, Marc RE, Frederick JM, Baehr W (2009) Trafficking of membrane proteins to cone but not rod outer segments is dependent on heterotrimeric kinesin-II. J Neurosci 29:14287–14298
Bachmann A, Schneider M, Theilenberg E, Grawe F, Knust E (2001) Drosophila stardust is a partner of crumbs in the control of epithelial cell polarity. Nature 414:638–643
Benton R, St Johnston D (2003) Drosophila PAR-1 and 14-3-3 inhibit Bazooka/PAR-3 to establish complementary cortical domains in polarized cells. Cell 115:691–704
Berger S, Bulgakova NA, Grawe F, Johnson K, Knust E (2007) Unraveling the genetic complexity of Drosophila stardust during photoreceptor morphogenesis and prevention of light-induced degeneration. Genetics 176:2189–2200
Beronja S, Laprise P, Papoulas O, Pellikka M, Sisson J, Tepass U (2005) Essential function of Drosophila Sec6 in apical exocytosis of epithelial photoreceptor cells. J Cell Biol 169:635–646
Bhat MA, Izaddoost S, Lu Y, Cho KO, Choi KW, Bellen HJ (1999) Discs lost, a novel multi-PDZ domain protein, establishes and maintains epithelial polarity. Cell 96:833–845
Bilder D, Perrimon N (2000) Localization of apical epithelial determinants by the basolateral PDZ protein Scribble. Nature 403:676–680
Bilder D (2001) Cell polarity: squaring the circle. Curr Biol 11:R132–135
Bryant DM, Mostov KE (2008) From cells to organs: building polarized tissue. Nat Rev Mol Cell Biol 9:887–901
Chabu C, Doe CQ (2009) Twins/PP2A regulates aPKC to control neuroblast cell polarity and self-renewal. Dev Biol 330:399–405
Chartier FJ, Hardy EJ, Laprise P (2012) Crumbs limits oxidase-dependent signaling to maintain epithelial integrity and prevent photoreceptor cell death. J Cell Biol 198:991–998
Chen TW, Chen G, Funkhouser LJ, Nam SC (2009) Membrane domain modulation by spectrins in Drosophila photoreceptor morphogenesis. Genesis 47:744–750
Chen G, League GP, Nam SC (2010) Role of spastin in apical domain control along the rhabdomere elongation in Drosophila photoreceptor. PLoS One 5:e9480
Chen G, Rogers AK, League GP, Nam SC (2011) Genetic interaction of centrosomin and bazooka in apical domain regulation in Drosophila photoreceptor. PLoS One 6:e16127
Choi KW, Nam SC, Mukhopadhyay B (2007) Par-1 and PP2A: Yin-Yang of bazooka localization. Fly (Austin) 1:235–237
Cho SH, Kim JY, Simons DL, Song JY, Le JH, Swindell EC, Jamrich M, Wu SM, Kim S (2012) Genetic ablation of Pals1 in retinal progenitor cells models the retinal pathology of Leber congenital amaurosis. Hum Mol Genet 21:2663–2676
Christensen AK, Jensen AM (2008) Tissue-specific requirements for specific domains in the FERM protein Moe/Epb4.115 during early zebrafish development. BMC Dev Biol 8:3
Collin RW, Littink KW, Klevering BJ, den Born LI van, Koenekoop RK, Zonneveld MN, Blokland EA, Strom TM, Hoyng CB, den Hollander AI, Cremers FP (2008) Identification of a 2 Mb human ortholog of Drosophila eyes shut/spacemaker that is mutated in patients with retinitis pigmentosa. Am J Hum Genet 83:594–603
den Hollander AI, ten Brink JB, Kok YJ de, Soest S van, den Born LI van, Driel MA van, Pol DJ van de, Payne AM, Bhattacharya SS, Kellner U et al (1999) Mutations in a human homologue of Drosophila crumbs cause retinitis pigmentosa (RP12). Nat Genet 23:217–221
den Hollander AI, Heckenlively JR, den Born LI van, Kok YJ de, Velde-Visser SD van der, Kellner U, Jurklies B, Schooneveld MJ van, Blankenagel A, Rohrschneider K et al (2001) Leber congenital amaurosis and retinitis pigmentosa with Coats-like exudative vasculopathy are associated with mutations in the crumbs homologue 1 (CRB1) gene. Am J Hum Genet 69:198–203
Di Paolo G, De Camilli P (2006) Phosphoinositides in cell regulation and membrane dynamics. Nature 443:651–657
Drubin DG, Nelson WJ (1996) Origins of cell polarity. Cell 84:335–344
Errico A, Ballabio A, Rugarli EI (2002) Spastin, the protein mutated in autosomal dominant hereditary spastic paraplegia, is involved in microtubule dynamics. Hum Mol Genet 11:153–163
Fan SS, Ready DF (1997) Glued participates in distinct microtubule-based activities in Drosophila eye development. Development 124:1497–1507

Galy A, Schenck A, Sahin HB, Qurashi A, Sahel JA, Diebold C, Giangrande A (2011) CYFIP dependent actin remodeling controls specific aspects of Drosophila eye morphogenesis. Dev Biol 359:37–46

Gassama-Diagne A, Yu W, ter Beest M, Martin-Belmonte F, Kierbel A, Engel J, Mostov K (2006) Phosphatidylinositol-3,4,5-trisphosphate regulates the formation of the basolateral plasma membrane in epithelial cells. Nat Cell Biol 8:963–970

Grawe F, Wodarz A, Lee B, Knust E, Skaer H (1996) The Drosophila genes crumbs and stardust are involved in the biogenesis of adherens junctions. Development 122:951–959

Harris TJ, Peifer M (2004) Adherens junction-dependent and -independent steps in the establishment of epithelial cell polarity in Drosophila. J Cell Biol 167:135–147

Harris TJ, Peifer M (2005) The positioning and segregation of apical cues during epithelial polarity establishment in Drosophila. J Cell Biol 170:813–823

Hong Y, Stronach B, Perrimon N, Jan LY, Jan YN (2001) Drosophila stardust interacts with crumbs to control polarity of epithelia but not neuroblasts. Nature 414:634–638

Hong Y, Ackerman L, Jan LY, Jan YN (2003) Distinct roles of Bazooka and Stardust in the specification of Drosophila photoreceptor membrane architecture. Proc Natl Acad Sci U S A 100:12712–12717

Horne-Badovinac S, Lin D, Waldron S, Schwarz M, Mbamalu G, Pawson T, Jan Y, Stainier DY, Abdelilah-Seyfried S (2001) Positional cloning of heart and soul reveals multiple roles for PKC lambda in zebrafish organogenesis. Curr Biol 11:1492–1502

Hsu YC, Willoughby JJ, Christensen AK, Jensen AM (2006) Mosaic eyes is a novel component of the crumbs complex and negatively regulates photoreceptor apical size. Development 133:4849–4859

Hurd TW, Gao L, Roh MH, Macara IG, Margolis B (2003) Direct interaction of two polarity complexes implicated in epithelial tight junction assembly. Nat Cell Biol 5:137–142

Husain N, Pellikka M, Hong H, Klimentova T, Choe KM, Clandinin TR, Tepass U (2006) The agrin/perlecan-related protein eyes shut is essential for epithelial lumen formation in the Drosophila retina. Dev Cell 11:483–493

Hutterer A, Betschinger J, Petronczki M, Knoblich JA (2004) Sequential roles of Cdc42, Par-6, aPKC, and Lgl in the establishment of epithelial polarity during Drosophila embryogenesis. Dev Cell 6:845–854

Izaddoost S, Nam SC, Bhat MA, Bellen HJ, Choi KW (2002) Drosophila crumbs is a positional cue in photoreceptor adherens junctions and rhabdomeres. Nature 416:178–183

Izumi Y, Hirose T, Tamai Y, Hirai S, Nagashima Y, Fujimoto T, Tabuse Y, Kemphues KJ, Ohno S (1998) An atypical PKC directly associates and colocalizes at the epithelial tight junction with ASIP, a mammalian homologue of Caenorhabditis elegans polarity protein PAR-3. J Cell Biol 143:95–106

Jensen AM, Westerfield M (2004) Zebrafish mosaic eyes is a novel FERM protein required for retinal lamination and retinal pigmented epithelial tight junction formation. Curr Biol 14:711–717

Jimeno D, Feiner L, Lillo C, Teofilo K, Goldstein LS, Pierce EA, Williams DS (2006a) Analysis of kinesin-2 function in photoreceptor cells using synchronous Cre-loxP knockout of Kif3a with RHO-Cre. Invest Ophthalmol Vis Sci 47:5039–5046

Jimeno D, Lillo C, Roberts EA, Goldstein LS, Williams DS (2006b) Kinesin-2 and photoreceptor cell death: requirement of motor subunits. Exp Eye Res 82:351–353

Johnson K, Grawe F, Grzeschik N, Knust E (2002) Drosophila crumbs is required to inhibit light-induced photoreceptor degeneration. Curr Biol 12:1675–1680

Karagiosis SA, Ready DF (2004) Moesin contributes an essential structural role in Drosophila photoreceptor morphogenesis. Development 131:725–732

Kirby C, Kusch M, Kemphues K (1990) Mutations in the par genes of Caenorhabditis elegans affect cytoplasmic reorganization during the first cell cycle. Dev Biol 142:203–215

Klebes A, Knust E (2000) A conserved motif in crumbs is required for E-cadherin localisation and zonula adherens formation in Drosophila. Curr Biol 10:76–85

Krahn MP, Egger-Adam D, Wodarz A (2009) PP2A antagonizes phosphorylation of Bazooka by PAR-1 to control apical-basal polarity in dividing embryonic neuroblasts. Dev Cell 16:901–908

Krahn MP, Klopfenstein DR, Fischer N, Wodarz A (2010) Membrane targeting of Bazooka/PAR-3 is mediated by direct binding to phosphoinositide lipids. Curr Biol 20:636–642

Kumar JP, Ready DF (1995) Rhodopsin plays an essential structural role in Drosophila photoreceptor development. Development 121:4359–4370

Land MF, Nilsson DE (2002) Animal Eyes. Oxford University Press, Oxford

Laprise P, Tepass U (2011) Novel insights into epithelial polarity proteins in Drosophila. Trends Cell Biol 21:401–408

Laprise P, Beronja S, Silva-Gagliardi NF, Pellikka M, Jensen AM, McGlade CJ, Tepass U (2006) The FERM protein Yurt is a negative regulatory component of the crumbs complex that controls epithelial polarity and apical membrane size. Dev Cell 11:363–374

Laprise P, Lau KM, Harris KP, Silva-Gagliardi NF, Paul SM, Beronja S, Beitel GJ, McGlade CJ, Tepass U (2009) Yurt, Coracle, Neurexin IV and the Na(+),K(+)-ATPase form a novel group of epithelial polarity proteins. Nature 459:1141–1145

League GP, Nam SC (2011) Role of kinesin heavy chain in crumbs localization along the rhabdomere elongation in Drosophila photoreceptor. PLoS One 6:e21218

Lee S, Kolodziej PA (2002) Short stop provides an essential link between F-actin and microtubules during axon extension. Development 129:1195–1204

Lee S, Harris KL, Whitington PM, Kolodziej PA (2000) Short stop is allelic to kakapo, and encodes rod-like cytoskeletal-associated proteins required for axon extension. J Neurosci 20:1096–1108

Lee JH, Koh H, Kim M, Kim Y, Lee SY, Karess RE, Lee SH, Shong M, Kim JM, Kim J, Chung J (2007) Energy-dependent regulation of cell structure by AMP-activated protein kinase. Nature 447:1017–1020

Lei Y, Warrior R (2000) The Drosophila Lissencephaly1 (DLis1) gene is required for nuclear migration. Dev Biol 226:57–72

Li BX, Satoh AK, Ready DF (2007) Myosin V, Rab11, and dRip11 direct apical secretion and cellular morphogenesis in developing Drosophila photoreceptors. J Cell Biol 177:659–669

Longley RL Jr, Ready DF (1995) Integrins and the development of three-dimensional structure in the Drosophila compound eye. Dev Biol 171:415–433

Lopes VS, Jimeno D, Khanobdee K, Song X, Chen B, Nusinowitz S, Williams DS (2010) Dysfunction of heterotrimeric kinesin-2 in rod photoreceptor cells and the role of opsin mislocalization in rapid cell death. Mol Biol Cell 21:4076–4088

Lumb JH, Connell JW, Allison R, Reid E (2011) The AAA ATPase spastin links microtubule severing to membrane modelling. Biochim Biophys Acta 1823:192–197

Macara IG (2004a) Par proteins: partners in polarization. Curr Biol 14:R160–R162

Macara IG (2004b) Parsing the polarity code. Nat Rev Mol Cell Biol 5:220–231

Martin SG, St Johnston D (2003) A role for Drosophila LKB1 in anterior-posterior axis formation and epithelial polarity. Nature 421:379–384

Martin-Belmonte F, Mostov K (2007) Phosphoinositides control epithelial development. Cell Cycle 6:1957–1961

Medina E, Williams J, Klipfell E, Zarnescu D, Thomas G, Le Bivic A (2002) Crumbs interacts with moesin and beta (heavy)-spectrin in the apical membrane skeleton of Drosophila. J Cell Biol 158:941–951

Mehalow AK, Kameya S, Smith RS, Hawes NL, Denegre JM, Young JA, Bechtold L, Haider NB, Tepass U, Heckenlively JR et al (2003) CRB1 is essential for external limiting membrane integrity and photoreceptor morphogenesis in the mammalian retina. Hum Mol Genet 12:2179–2189

Mirouse V, Swick LL, Kazgan N, St Johnston D, Brenman JE (2007) LKB1 and AMPK maintain epithelial cell polarity under energetic stress. J Cell Biol 177:387–392

Morais-de-Sa E, Mirouse V, St Johnston D (2010) aPKC phosphorylation of Bazooka defines the apical/lateral border in Drosophila epithelial cells. Cell 141:509–523

Morton DG, Roos JM, Kemphues KJ (1992) Par-4, a gene required for cytoplasmic localization and determination of specific cell types in Caenorhabditis elegans embryogenesis. Genetics 130:771–790

Mosley-Bishop KL, Li Q, Patterson L, Fischer JA (1999) Molecular analysis of the klarsicht gene and its role in nuclear migration within differentiating cells of the Drosophila eye. Curr Biol 9:1211–1220
Mui UN, Lubczyk CM, Nam SC (2011) Role of spectraplakin in Drosophila photoreceptor morphogenesis. PLoS One 6:e25965
Mukhopadhyay B, Nam SC, Choi KW (2010) Kinesin II is required for cell survival and adherens junction positioning in Drosophila photoreceptors. Genesis 48:522–530
Nam SC, Choi KW (2003) Interaction of Par-6 and crumbs complexes is essential for photoreceptor morphogenesis in Drosophila. Development 130:4363–4372
Nam SC, Choi KW (2006) Domain-specific early and late function of Dpatj in Drosophila photoreceptor cells. Dev Dyn 235:1501–1507
Nam SC, Mukhopadhyay B, Choi KW (2007) Antagonistic functions of Par-1 kinase and protein phosphatase 2A are required for localization of Bazooka and photoreceptor morphogenesis in Drosophila. Dev Biol 306:624–635
Nelson WJ (2003) Adaptation of core mechanisms to generate cell polarity. Nature 422:766–774
Nie J, Mahato S, Mustill W, Tipping C, Bhattacharya SS, Zelhof AC (2012) Cross species analysis of Prominin reveals a conserved cellular role in invertebrate and vertebrate photoreceptor cells. Dev Biol 371(2):312–320
Nunbhakdi-Craig V, Machleidt T, Ogris E, Bellotto D, White CL 3rd, Sontag E (2002) Protein phosphatase 2A associates with and regulates atypical PKC and the epithelial tight junction complex. J Cell Biol 158:967–978
Ogawa H, Ohta N, Moon W, Matsuzaki F (2009) Protein phosphatase 2A negatively regulates aPKC signaling by modulating phosphorylation of Par-6 in Drosophila neuroblast asymmetric divisions. J Cell Sci 122:3242–3249
Ohno S (2001) Intercellular junctions and cellular polarity: the PAR-aPKC complex, a conserved core cassette playing fundamental roles in cell polarity. Curr Opin Cell Biol 13:641–648
Park B, Alves CH, Lundvig DM, Tanimoto N, Beck SC, Huber G, Richard F, Klooster J, Andlauer TF, Swindell EC et al (2011) PALS1 is essential for retinal pigment epithelium structure and neural retina stratification. J Neurosci 31:17230–17241
Pellikka M, Tanentzapf G, Pinto M, Smith C, McGlade CJ, Ready DF, Tepass U (2002) Crumbs, the Drosophila homologue of human CRB1/RP12, is essential for photoreceptor morphogenesis. Nature 416:143–149
Petronczki M, Knoblich JA (2001) DmPAR-6 directs epithelial polarity and asymmetric cell division of neuroblasts in Drosophila. Nat Cell Biol 3:43–49
Pham H, Yu H, Laski FA (2008) Cofilin/ADF is required for retinal elongation and morphogenesis of the Drosophila rhabdomere. Dev Biol 318:82–91
Pinal N, Goberdhan DC, Collinson L, Fujita Y, Cox IM, Wilson C, Pichaud F (2006) Regulated and polarized PtdIns (3,4,5) P3 accumulation is essential for apical membrane morphogenesis in photoreceptor epithelial cells. Curr Biol 16:140–149
Pocha SM, Shevchenko A, Knust E (2011) Crumbs regulates rhodopsin transport by interacting with and stabilizing myosin V. J Cell Biol 195:827–838
Poels J, Spasic MR, Gistelinck M, Mutert J, Schellens A, Callaerts P, Norga KK (2012) Autophagy and phagocytosis-like cell cannibalism exert opposing effects on cellular survival during metabolic stress. Cell Death Differ 19:1590–1601
Ready DF (2002) Drosophila compound eye morphogenesis: blind mechanical engineers? In: Moses K (ed) Drosophila eye development (results and problems in cell differentiation). Springer, Heidelberg, pp 191–204
Richard M, Grawe F, Knust E (2006) DPATJ plays a role in retinal morphogenesis and protects against light-dependent degeneration of photoreceptor cells in the Drosophila eye. Dev Dyn 235:895–907
Roh MH, Makarova O, Liu CJ, Shin K, Lee S, Laurinec S, Goyal M, Wiggins R, Margolis B (2002) The Maguk protein, Pals1, functions as an adapter, linking mammalian homologues of crumbs and discs lost. J Cell Biol 157:161–172

Roll-Mecak A, McNally FJ (2009) Microtubule-severing enzymes. Curr Opin Cell Biol 22:96–103
Salinas S, Carazo-Salas RE, Proukakis C, Schiavo G, Warner TT (2007) Spastin and microtubules: functions in health and disease. J Neurosci Res 85:2778–2782
Satoh AK, O'Tousa JE, Ozaki K, Ready DF (2005) Rab11 mediates post-Golgi trafficking of rhodopsin to the photosensitive apical membrane of Drosophila photoreceptors. Development 132:1487–1497
Shubin N, Tabin C, Carroll S (2009) Deep homology and the origins of evolutionary novelty. Nature 457:818–823
Sotillos S, Diaz-Meco MT, Caminero E, Moscat J, Campuzano S (2004) DaPKC-dependent phosphorylation of crumbs is required for epithelial cell polarity in Drosophila. J Cell Biol 166:549–557
Sottocornola R, Royer C, Vives V, Tordella L, Zhong S, Wang Y, Ratnayaka I, Shipman M, Cheung A, Gaston-Massuet C et al (2010) ASPP2 binds Par-3 and controls the polarity and proliferation of neural progenitors during CNS development. Dev Cell 19:126–137
Spasic MR, Callaerts P, Norga KK (2008) Drosophila alicorn is a neuronal maintenance factor protecting against activity-induced retinal degeneration. J Neurosci 28:6419–6429
Suzuki A, Yamanaka T, Hirose T, Manabe N, Mizuno K, Shimizu M, Akimoto K, Izumi Y, Ohnishi T, Ohno S (2001) Atypical protein kinase C is involved in the evolutionarily conserved par protein complex and plays a critical role in establishing epithelia-specific junctional structures. J Cell Biol 152:1183–1196
Tepass U (2012) The apical polarity protein network in Drosophila epithelial cells: regulation of polarity, junctions, morphogenesis, cell growth, and survival. Annu Rev Cell Dev Biol 28:655–685
Tepass U, Knust E (1993) Crumbs and stardust act in a genetic pathway that controls the organization of epithelia in Drosophila melanogaster. Dev Biol 159:311–326
Tepass U, Tanentzapf G, Ward R, Fehon R (2001) Epithelial cell polarity and cell junctions in Drosophila. Annu Rev Genet 35:747–784
Tepass U, Theres C, Knust, E (1990) Crumbs encodes an EGF-like protein expressed on apical membranes of Drosophila epithelial cells and required for organization of epithelia. Cell 61:787–799
Walther RF, Pichaud F (2010) Crumbs/DaPKC-dependent apical exclusion of bazooka promotes photoreceptor polarity remodeling. Curr Biol 20:1065–1074
Watts JL, Morton DG, Bestman J, Kemphues KJ (2000) The C. elegans par-4 gene encodes a putative serine-threonine kinase required for establishing embryonic asymmetry. Development 127:1467–1475
Wei X, Malicki J (2002) Nagie oko, encoding a MAGUK-family protein, is essential for cellular patterning of the retina. Nat Genet 31:150–157
Whited JL, Cassell A, Brouillette M, Garrity PA (2004) Dynactin is required to maintain nuclear position within postmitotic Drosophila photoreceptor neurons. Development 131:4677–4686
Wodarz A, Ramrath A, Grimm A, Knust E (2000) Drosophila atypical protein kinase C associates with bazooka and controls polarity of epithelia and neuroblasts. J Cell Biol 150:1361–1374
Wu S, Mehta SQ, Pichaud F, Bellen HJ, Quiocho FA (2005) Sec15 interacts with Rab11 via a novel domain and affects Rab11 localization in vivo. Nat Struct Mol Biol 12:879–885
Xia H, Ready DF (2011) Ectoplasm, ghost in the R cell machine? Dev Neurobiol 71:1246–1257
Xiong W, Rebay I (2011) Abelson tyrosine kinase is required for Drosophila photoreceptor morphogenesis and retinal epithelial patterning. Dev Dyn 240:1745–1755
Zelhof AC, Hardy RW (2004) WASp is required for the correct temporal morphogenesis of rhabdomere microvilli. J Cell Biol 164:417–426
Zelhof AC, Hardy RW, Becker A, Zuker CS (2006) Transforming the architecture of compound eyes. Nature 443:696–699
Zhou W, Hong Y (2012) Drosophila Patj plays a supporting role in apical-basal polarity but is essential for viability. Development 139:2891–2896
Zou J, Wang X, Wei X (2012) Crb apical polarity proteins maintain zebrafish retinal cone mosaics via intercellular binding of their extracellular domains. Dev Cell 22:1261–1274

Negative Regulation for Neural Patterning in the *Drosophila* eye

Kwang-Wook Choi

Introduction

An adult compound eye consists of about 800 unit eyes called ommatidia. The ommatidia are organized in a highly ordered structure and provide an ideal sensory system for genetic dissection of neural development and cellular pattern formation. Each ommatidium contains eight photoreceptor neurons and a dozen accessory cells including cone cells, pigment cells, and bristles that are formed in a stereotypic pattern (Ready et al. 1976).

The adult eye develops from eye imaginal disc, an epithelial primordium for eye proper and the surrounding head tissues. In the early phase of development, the eye disc grows by cell proliferation without retinal differentiation. This growth phase continues until retinal differentiation is initiated at the early third-instar larval stage. Among several different cell types in the adult eye, photoreceptor neurons are the first kind to be generated in the eye disc. The initial pattern of these cells provides the structural foundation for the subsequent patterning events to establish the adult eye. Thus, how these initial events are organized at the cell and molecular levels is an important question in eye development.

Following the growth phase of eye disc development, retinal neurogenesis is initiated in the morphogenetic furrow (the furrow in short), a groove formed along the dorso-ventral axis of the eye disc. This furrow is first generated at the posterior margin of the eye disc and progresses anteriorly during neurogenesis. Importantly, columns of neuronal cell clusters are generated immediately posterior to the furrow (Ready et al. 1976). Thus, the furrow is an important site for retinal organization where multiple cell signaling pathways are coordinated to specify the founder cells for photoreceptor neurons.

K.-W. Choi (✉)
Department of Biological Sciences, Korea Advanced Institute of Science and Technology, 291 Daehak-ro, Yuseong-gu, Daejeon 305–701, Rep. of Korea
e-mail: kchoi100@kaist.ac.kr

Graduate School of Nanoscience and Technology, Korea Advanced Institute of Science and Technology, 291 Daehak-ro, Yuseong-gu, Daejeon 305–701, Rep. of Korea

A. Singh, M. Kango-Singh (eds.), *Molecular Genetics of Axial Patterning, Growth and Disease in the Drosophila Eye,* DOI 10.1007/978-1-4614-8232-1_6,

A critical event for retinal neurogenesis in the furrow is to induce the expression of proneural proteins like Atonal (Ato) that promotes neural differentiation. Ato expression in the furrow is dynamically induced by positive genetic factors expressed within the furrow or adjacent regions anterior or posterior to it. Secreted signaling molecules like Decapentaplegic (Dpp) and Hedgehog (Hh) act as positive factors to induce Ato expression (Baker and Yu 1997; Borod and Heberlein 1998; Greenwood and Struhl 1999). While these factors act as positive regulators of retinal differentiation, spatial patterning of neurogenesis is also dependent on the function of specific negative regulators. The ommatidial pattern consists of repetitive arrays of photoreceptor clusters. Each of these clusters is surrounded by nonneuronal interommatidial cells, resulting in the formation of a precise polka dot pattern in the eye (Fig. 1a). Thus, the function of negative factors in neurogenesis is not only essential for the establishment of the ommatidial pattern but also to provide necessary conditions for subsequent differentiation of nonneuronal interommatidial cells in the eye.

One of the key negative regulators of Ato induction is Notch. Notch is a transmembrane protein that functions as a receptor for the membrane-bound ligands, Delta (Dl) and Serrate (Ser). Notch signaling is required at multiple steps during eye development (Cagan and Ready 1989). Notch initially promotes neural differentiation but later antagonizes it by lateral inhibition in the cells surrounding the Ato-expressing cells (Baker and Yu 1997; Baonza and Freeman 2001). In addition, other negative regulators such as EGF receptor (EGFR) and Bar transcription factors have been identified as inhibitory factors for retinal neurogenesis within or behind the furrow. In this chapter, the role of a few key positive factors involved in the retinal fate induction will be briefly introduced. Next, the function of negative regulators will be discussed in more details to illustrate how the interaction of these positive and negative factors leads to the generation of the initial ommatidial pattern in the eye disc.

Retinal Neurogenesis: Positive Regulation of Ato Expression

Retinal differentiation begins with the formation of the morphogenetic furrow from the posterior margin of an eye imaginal disc. As the furrow progresses anteriorly, columns of R8 founder neurons are generated from the posterior part of the furrow (Wolff and Ready 1991). Subsequently, additional photoreceptor cells are recruited to R8 by specific cell–cell interactions to generate the remaining seven photoreceptor cells in the order of R2/R5, R3/R4, and R1/R6/R7. This process of sequential induction of photoreceptor cell fates illustrates the critical role of the initial R8 selection in retinal neurogenesis.

Neurogenesis is promoted by a group of factors called proneural genes that encodes bHLH family transcription factors. Generation of the R8 founder cells from undifferentiated cells requires the function of proneural gene *ato* at the furrow (Jarman et al. 1994). Ato protein is a homolog of mammalian Ato7 (also called MATH5; Brown et al. 1998). The expression of Ato protein is transient and dynamically regulated at the furrow, suggesting that *ato* is subject to negative regulation. Ato

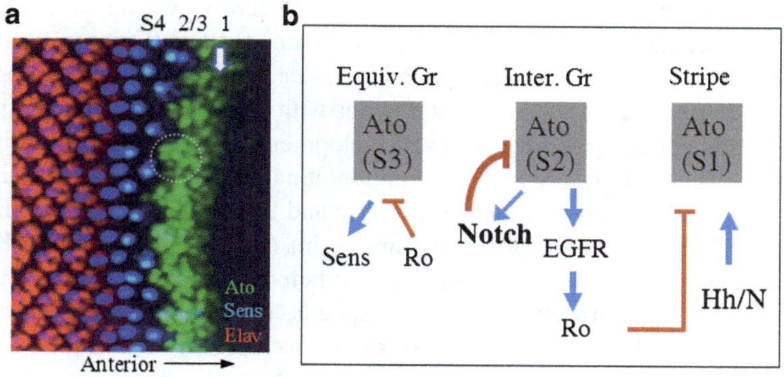

Fig. 1 Regulation of Ato expression. (**a**) Ato expression pattern in eye disc. Ato is expressed initially in a stripe pattern at and immediately anterior to the furrow (stage 1). The stripe pattern is resolved into the intermediate groups (stage 2) and the equivalence groups (stage 3). Later, Ato is expressed only in R8 cells (stage 4). Ato expression in the stages 1 and 2–4 are controlled by the 3′ and 5′ regulatory region, respectively. (**b**) A simplified diagram for gene functions involved in early stages of neurogenesis. The initial Ato expression (S1) is regulated by positive factors including Hh, Dpp, and N, which leads to autoactivated *ato* expression in the stage S2. The S2 Ato induces epidermal growth factor receptor/mitogen-activated protein kinase (EGFR/MAPK) signaling, which inhibits the S1 stage *ato* expression resulting in evenly spaced intermediate groups. One cell from each R8 equivalence group maintains Ato expression to become an R8 while Ato expression in other cells is repressed by Ro

expression can be divided into four stages based on the distinct pattern of expression (Fig. 1): (1) first expression in a stripe pattern across the disc in the most anterior region of the furrow, (2) expression in about 10 cell clusters called intermediate (or proneural) groups just posterior to the stripe, (3) expression in two to three cells of an R8 equivalence group, and (4) selected expression in a single R8 founder cell from each equivalence group (Frankfort and Mardon 2002). Transient expression and sequential restriction of Ato expression in the furrow indicate that *ato* expression is regulated by specific spatial and temporal regulatory factors. Analysis of *ato* regulation has identified two *cis*-regulatory regions responsible for *ato* expression at the furrow (Sun et al. 1998). The eye-specific *ato 3′ cis*-regulatory region controls the early stripe pattern (stage 1) and contains binding sites for transcription factors for retinal determination (RD; Tanaka-Matakatsu and Du 2008; Zhang et al. 2006) while the *5′* regulatory region is responsible for the rest of *ato* expression posterior to the stripe (stages 2–4).

Accumulated evidence indicates that the diffusible factor Hh and transmembrane protein Notch (N) provide positive signaling to activate *ato* transcription (Baker and Yu 1997; Dominguez and Hafen 1997; Frankfort and Mardon 2002; Fu and Baker 2003; Hsiung and Moses 2002). Hh is expressed in all photoreceptors and secreted to act on more anterior cells in the furrow. The stage 2 Ato expression (Fig. 1) induces the expression of Rhomboid family proteins that activate the TGFα family ligand Spitz (Spi) for EGFR signaling in the adjacent cells.

While Hh and Dpp are secreted factors that promote *ato* expression, eventually *ato* gene must be activated by specific transcription factors. Eyeless (Ey) is a *Drosophila* homolog of Pax-6 that is considered to be a master regulator of eye development (Halder et al. 1995). It is a transcription factor with a homeodomain and a paired domain, and is not only required for eye development but also sufficient to induce ectopic eyes when ectopically expressed in nonretinal tissues. Ey is expressed very early in the embryonic primordium for eye disc and is later detected in the anterior region to the furrow as retinal differentiation begins (Baker and Firth 2011; Kumar and Moses 2001). Since Ey is necessary and sufficient for retinal induction, it may directly activate *ato* transcription in an eye-specific manner. However, *ato* acts as a common proneural gene in the initial stage of development of multiple segment-specific sensory organs, not only the eye but also the auditory organs and stretch receptors. Hence, it has been proposed that these sensory organs may have the same origins (Niwa et al. 2004). In this view, Ato is a common transcription factor necessary for all these three sensory organs rather than the retina-specific proneural factor. The 3′ region of *ato* is required for Ato expression not only in the eye but also in the Johnston's auditory organ in the antenna and the chordotonal organs (Sun et al. 1998). In fact, protosensory organs are formed by Dpp-dependent Ato expression. Furthermore, two Mad-binding sites were identified that are essential for Ato expression in all three sensory organs. This supported that the target sites for the Dpp signal are conserved during the diversification of sensory organs. Interestingly, ectopic eye formation induced by Ey depends on the presence of Dpp and other signaling molecules, and Ato can be expressed in *ey* mutant if cell death is prevented. These findings led to a model that Ey functions as a downstream or parallel component of Ato rather than as an upstream master control factor for Ato expression and retinal differentiation (Niwa et al. 2004; Treisman 2004).

In contrast to this model, further analysis of *ato 3′ cis*-regulatory sequences have provided evidence that Ato expression in the eye disc is directly regulated by Ey and other so- genes" like Sine oculis (So) and Dacshund (Dac; Zhang et al. 2006). Hence, the main role of Dpp in Ato induction in the furrow is to regulate the RD gene expression level (Tanaka-Matakatsu and Du 2008). It is interesting to note that the 3′ *cis*-regulatory region of *ato* consists of multiple modules. Hence, the activation of *ato* expression for sensory organs in different imaginal discs appears to be determined by the modular organization of *ato*-regulatory region instead of a common regulatory region for Dpp signaling.

Role of Notch for Lateral Inhibition

The initial stripe pattern of Ato expression is sequentially restricted to proneural groups that are separated by ato-negative cells in the interommatidial space. Notch is a conserved key factor involved in lateral inhibition in diverse developmental processes, especially in neurogenesis (Beatus and Lendahl 1998; Cabrera 1990). In the absence of N function, the lack of lateral inhibition results in the formation of excess number of R8 cells, resulting in more compact spacing of ommatidia (Roignant and Treisman 2009).

Scabrous (Sca) is one of the first genes that were found to be required for proper spacing of R8 founder neurons. Sca is a secreted factor that is released from the R8 cells. Sca is a protein related to Fibrinogen that acts as a lateral inhibitor of the R8 differentiation (Baker et al. 1990; Ellis et al. 1994). Sca is expressed early in the furrow and is known to be required for lateral inhibition by N-Dl interaction at the level of intermediate clusters. In the absence of Sca, excess cells are selected to become R8 cells, thus disrupting the normal pattern of R8 spacing. Evidence suggests that Sca functions together with Gp150, a target of protein phosphatase DPTP10D. Loss of Gp150 shows more R8 cells as in *sca* mutant eye. Moreover, double mutants show similar phenotypes, and both proteins are localized together in the endosomes (Fetchko et al. 2002; Li et al. 2003), suggesting that they work in the same pathway for N signaling. Gp150 acts downstream to Sca in the cells that respond to secreted Sca protein. It has been proposed that N activity is downregulated in the neuronal cells by an endosomal pathway, and Sca and Gp150 are indirectly involved in the activation of N signaling in nonneuronal cells by blocking the endosomal pathway. Gp150 is required for all Sca functions identified so far, indicating that Sca-Gp150 pathway is a conserved step for N regulation in diverse developmental events (Li et al. 2003). It remains to be studied whether Gp150 function in this process is regulated by the protein phosphatase activity of DPTP10D.

Additional evidence supports the role of endosomal trafficking for N signaling. Clathrin adaptor protein complex-1 (AP-1) plays a role in sorting of membrane proteins in the Golgi network and endosomes (Kametaka et al. 2012). Loss of AP-1 or its accessory protein like *Drosophila* Aftiphilin results in an intracellular accumulation of Sca and downregulation of N, since N degradation is promoted in the lysosomes. Sca protein secreted from R8 cells directly interacts with the N extracellular domain to stabilize it (Powell et al. 2001). Knockdown of AP-1 or Aftiphilin by RNAi causes abnormal pattern of R8 initiation, consistent with the role of AP-1 function of N regulation during R8 specification. Notch signaling is also regulated by endocytosis of the Notch ligand Dl. This process is mediated by ubiquitination of Dl by Neuralized (Neur), a ring finger domain protein that functions as an ubiquitin ligase (Lai et al. 2001; Pavlopoulos et al. 2001; Weinmaster and Fischer 2011; Yeh et al. 2001). Monoubiquitinated Dl is internalized for endocytosis in the signal sending cells. For an unknown mechanism, this endocytosis of Dl is necessary for Notch signaling.

N is a type I single-pass transmembrane protein. When activated by its ligands, it is cleaved by the N-secretase complex, resulting in the formation of the Notch intracellular domain (NICD) that enters the cell nucleus to promote transcription of downstream target genes (Kopan and Goate 2002). The γ-secretase is a multi-subunit complex containing Presenilin (Psn) as the catalytic subunit. Since the catalytic activity of Psn depends on the maturation of a holoprotein by proteolytic cleavage (Annaert and De Strooper 1999), the regulation of Psn maturation is important for N signaling. Loss of Psn causes the lack of lateral inhibition in the furrow, indicating the importance of N processing by Psn for proper ommatidial spacing (Ye et al. 1999).

Cis-Inhibition and Ligand-Independent Notch Signaling

In addition to its function in R8 selection, Notch is also involved in the following recruitment of other photoreceptor cells. Posterior to the furrow, emerging photoreceptor clusters in the eye disc are assembled in an asymmetric trapezoidal pattern. These clusters have two opposite forms of chirality in the dorsal and ventral halves of an eye, thus showing a mirror symmetric planar cell polarity (PCP) about the dorsoventral midline (Choi et al. 1996; Ready et al. 1976; Singh et al. 2012). Generation of these chiral forms depends on the specification of two photoreceptors, R3 and R4, from R3/4 equivalent precursor cells. One of these two cells is located closer to the equator and has higher Fz signaling than the polar cell. Interestingly, different Fz activity levels in these two cells lead to the asymmetric activation of N signaling, leading to the generation of R3 and R4 cells with low and high N activation, respectively (Cooper and Bray 1999; Fanto and Mlodzik 1999).

The R8 selection and the subsequent R3/4 specification are distinct processes. However, the asymmetric activation of N in R3 and R4 cells has a similarity to the situation of lateral inhibition in the furrow between the R8 founder cell and the surrounding cells. It is worth noting that Hibris (Hbs), initially found as a protein interacting with Roughest (Rst), a cell adhesion factor, is involved in N signaling not only for R3/4 planar polarity but also proneural patterning in the furrow. A recent study has identified Hbs as a new factor that functionally and physically interacts with Psn and Nicastrin (Ncs), another component of the γ-secretase complex (Singh and Mlodzik 2012).

Studies on the R3/4 specification led to the finding of new mechanisms for N signaling. In the conventional mechanism, N activation is mediated by specific ligands such as Dl and Ser. Notch activation leads to transcriptional repression of Dl and Neur in the signal receiving cells, whereas Dl and Neur are upregulated in the signal sending cells (Cooper and Bray 1999; del Alamo and Mlodzik 2006; Fanto and Mlodzik 1999; Tomlinson & Struhl 1999). Alternatively, Fz-Dishevelled (Dsh) complex may inhibit N activation in R3 cells. A recent study has found that Ral, a small Ras-like GTPase, is upregulated in response to Fz activation in the equatorial cells, and the upregulated Ral activity represses Notch activation in a ligand-independent manner (Fig. 2). In this mechanism, Ral GTPase activity may interfere with the ligand-independent Notch activation by regulating Notch trafficking to the lysosome, generation of NICD, or nuclear translocation (Cho and Fischer 2011, 2012).

N signaling is asymmetrically transmitted with directionality because Dl ligand is downregulated in the signal receiving cells by a negative feedback loop (Heitzler and Simpson 1991; Rooke and Xu 1998). In addition to this traditional feedback system, *cis*-inhibition has been proposed as an alternative mechanism for directional activation of N signaling. In photoreceptor recruitment, R1, R6, and R7 cells are the last cells to be specified. R1 and R6 cells are recruited together and express Dl to activate the R7 fate in the neighbor. In this process, Dl ligand expressed in R1 and R6 cells *cis*-inhibits N in the same cell, which prevents inappropriate N activation in R1/R6 by Dl ligand from the R7 cell. Such *cis*-inhibition of N signaling by Dl in

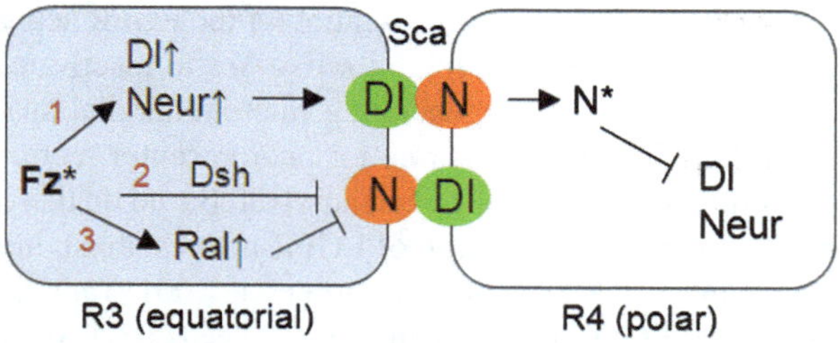

Fig. 2 Regulation of asymmetric Notch signaling. R3/R4 precursor cells are initially equivalent. The cell located closer to the equator has more Fz signaling that increases the level of Dl and Neur expression. This cell having more Dl becomes R3. In contrast, the other cell on the polar side becomes R4 with more N signaling. This asymmetric signaling takes place by increased expression of Dl and Neur by Fz (step 1), inactivation of Notch by Fz/Dsh complex (step 2), and/or Fz-dependent upregulation of Ral that inhibits N signaling (step 3). (Adapted from Cho and Fischer 2011)

the signal sending cell may also function in R3/R4 recruitment. It is an intriguing question whether similar *cis*-inhibition and ligand-independent regulation of Notch signaling plays a role in asymmetric N signaling for neural induction at the furrow.

Epidermal Growth Factor Signaling in Ommatidial Spacing

EGFR signaling is another important mechanism that functions throughout different stages of eye development. EGFR is activated by the positive ligand Spi, a TGFα homolog, while it is inactivated by the antagonist, Argos (Freeman 1994; Rutledge et al. 1992; Schweitzer et al. 1995). Similar to the vertebrate EGFR, *Drosophila* EGFR is also dimerized upon binding of the Spi ligand and activated by autophosphorylation of the dimer. Activated EGFR triggers the conserved intracellular signaling pathway that involves Ras and MAP kinase (Kumar et al. 1998). Ato expression in R8 cells induces Spi ligand expression that activates EGFR in the neighboring cells (Dominguez et al. 1998). Thus, one of the major functions of EGFR signaling is to activate photoreceptor precursor cells to initiate the retinal differentiation in all photoreceptors except the R8 neuron, the source of EGFR ligand (Dominguez et al. 1998; Yang and Baker 2001).

The role of EGFR in ommatidial spacing during neurogenesis was first implicated by abnormal spacing of photoreceptor clusters caused by *Ellipse* dominant mutations. Ellipse mutations turned out to be alleles of *EGFR* and it was suggested that EGFR may be important for controlling the ommatidial spacing (Baker and Rubin 1989). The role of EGFR for ommatidial spacing was supported by an analysis of EGFR-dependent Rough (Ro) expression. Ro is a homeobox transcription factor induced by EGFR signaling (Kimmel et al. 1990) and negatively regulates the initial *ato* transcription (Fig. 1b), thus generating the spacing between intermediate groups (Dokucu et al. 1996; Pepple et al. 2008). Analysis of *EGFR* mutant clones supports that EGFR is required for ommatidial spacing (Dominguez et al. 1998). Rhomboid-1 expression induced by Ato in R8 cells is essential for the EGFR activation in nascent ommatidia, which secrete a negative inhibitor like Sca to inactivate Ato expression in the neighboring cells, therefore generating interommatidial spacing. Although EGFR is essential for cell survival, its role for photoreceptor recruitment is largely independent of its function in promoting cell survival. Based on these findings, it has been proposed that the primary function of EGFR is to establish the spatial pattern of ommatidia by regulating R8 spacing (Baonza et al. 2001; Yang and Baker 2001).

However, the role of EGFR in R8 spacing has not been clearly demonstrated probably due to the multiple functions of the EGFR signaling in eye development and different experimental methods used for functional analysis. For instance, an analysis of EGFR function using a temperature-sensitive allele did not support its role for ommatidial spacing. In this approach, a temperature-sensitive allele of *EGFR* called $EGFR^{tsla}$ was used to minimize the defects in cell proliferation. $EGFR^{tsla}$ encodes a mutant protein that becomes quickly inactive or functionally null at the restrictive temperature. Analysis of $EGFR^{tsla}$ mutant clones at the restrictive temperature showed normal rate of furrow progression and normal spacing of R8 cells (Rodrigues et al. 2005), which seems to be contradicting with the previous clonal analysis (Baonza et al. 2001; Yang and Baker 2001).

In the earlier clonal analysis with the *EGFR* null allele, *EGFR* loss-of-function (LOF) clones were generated using a *Minute* (*M*) mutation to generate larger mutant clones. Larger $M^{+/+}$ mutant clones can be generated because $M^{+/+}$ cells have growth advantage compared to the neighboring $M^{+/-}$ cells. Importantly, it was found that the $M^{+/+}$ twin spot has strong noncell autonomous effects on the $EGFR^{-}$ mutant cells. Thus, it was concluded that the defects in Ato expression and ommatidial spacing in the mosaic clone experiments may be due to effects of the *M* rather than the effects of *EGFR* mutation. However, it is still possible that $EGFR^{tsla}$ mutant clones at the restrictive temperature may have a very low but sufficient level of functional EGFR protein for normal R8 spacing. Interestingly, two EGFR ligands, Spi and Keren, are redundant for EGFR signaling, but loss of both ligands causes abnormal R8 spacing (Brown et al. 2007). Although this supports the requirements of EGFR signaling for normal R8 spacing, additional studies are needed to draw a definitive conclusion about the role of EGFR signaling for the lateral inhibition of Ato expression and R8 spacing.

Antiproneural Function of Bar

Ato is transiently expressed in the selected R8 cells and turned off soon in several hours. The inhibition of Ato expression behind the furrow is important to prevent the formation of ectopic photoreceptors while maintaining the ommatidial spacing. This repression is mediated by the *Bar* genes that are expressed posterior to the furrow (Lim and Choi 2003). The first Bar mutation (Bar^1) was found as a dominant allele that reduces the eye size (Steinberg and Abramowitz 1938). Bar^1 is a duplication of the *Bar* gene suggesting that abnormal overexpression of Bar results in reduction

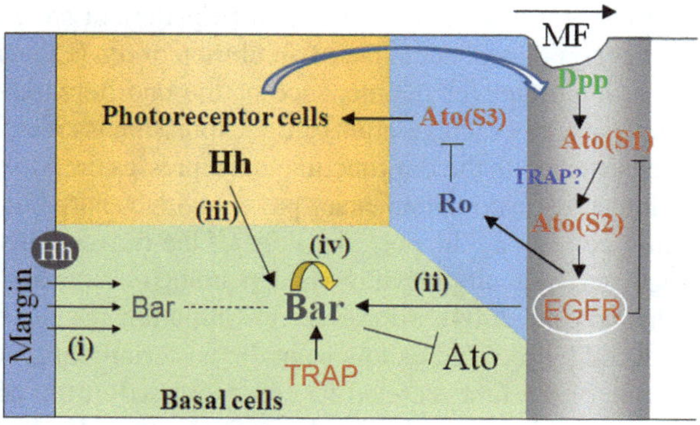

Fig. 3 Bar regulation during retinal neurogenesis.Ato expression in the morphogenetic furrow (MF) is activated by Hh produced by photoreceptor cells and initiates the generation of photoreceptor neurons. Bar proteins are expressed in basal undifferentiated cells behind the furrow (*green region*) by several mechanisms. Positive and inhibitory relationships indicated by *arrows* may be indirect: (i) at the time of furrow initiation, Bar expression in the basal undifferentiated cells is induced by a secreted signaling factor, Hh, from the posterior margin (*yellow region*), (ii) during furrow migration, Bar expression near the furrow is induced by Ato from the furrow. EGFR signaling may partially mediate nonautonomous effects of Ato on Bar expression, (iii) Hh produced in photoreceptor cells induces Dpp expression and may also contribute to Bar expression during furrow migration, and finally, and (iv) Bar is autoregulated to maintain its expression. (Adapted from Lim and Choi 2004)

of the eye. Further analysis has shown that *dpp* expression in the furrow is strongly reduced in Bar^1. Since Dpp is required for furrow progression, it was suggested that Bar^1 mutation causes premature furrow stop (Chanut and Heberlein 1997; Curtiss and Mlodzik 2000; Heberlein et al. 1993).

Bar encodes two related and functionally redundant homeodomain proteins, *BarH1* and *BarH2* (hereafter abbreviated as "Bar") that is present in a tandem repeat (Akimaru and Saigo 1991; Higashijima et al. 1992). Expression of Bar is regulated dynamically during eye development. In the eye disc, it is specifically expressed in the nuclei of R1 and R6 photoreceptors and later in primary pigment cells. Consistent with this expression pattern, Bar is important for differentiation of R1, R6, and primary pigment cells (Higashijima et al. 1992).

In addition to these cells, Bar is also expressed in all undifferentiated retinal precursors posterior to the furrow (Higashijima et al. 1992), which can be distinguished by the position of their nuclei. Because the nuclei of differentiating photoreceptors migrate apically while those of undifferentiated cells stay in the basal region (Tomlinson and Ready 1987), these undifferentiated cells posterior to the furrow are referred to as the "basal cells" (Fig. 3). Bar expression in these basal cells was shown to be crucial for regulating the neural patterning in early steps of eye development (Lim and Choi 2003). *Bar* LOF mutant clones showed ectopic *ato* induction at the transcription level, indicating that Bar acts as a transcriptional repressor of *ato*. It is important to note that loss of Bar has little effect on the spacing of intermediate groups and the selection of R8 from the equivalence group in the endogenous furrow.

Hence, Bar is required for repressing the early stripe pattern of *ato* expression, but it may have little effect on N-dependent lateral inhibition of *ato* (Lim and Choi 2003).

Because Bar is important for maintenance of the undifferentiated state of the basal cells, spatial and temporal regulation of Bar expression is crucial for proper eye development. Consistent with the Bar function that represses the Ato expression, Bar and Ato expression shows a complementary pattern with a sharp boundary between the Bar^+ and Ato^+ cells along the posterior edge of the furrow. This pattern of Bar expression is regulated by multiple pathways depending on time and position in the disc (Fig. 3; Lim and Choi 2004). Prior to furrow initiation and Ato expression, Bar appears to be induced by secreted factors from the posterior margin of the disc. For example, Hh is one of the first secreted factors expressed in the posterior margin at the time of furrow initiation, and it has been shown that Hh signaling is in part responsible for initial Bar induction (Lim and Choi 2004). Evidence also suggests that Bar is induced by several factors functioning in the furrow. Immediately behind the furrow, Bar expression depends on EGFR signaling, which is induced by Ato expression in the intermediate groups (Fig. 3).

An important function of Ato is to regulate its own expression. Ato protein induced by activation of the 3′ regulatory region can turn on itself by binding to its own 5′ regulatory region (Sun et al. 1998). Interestingly, Thyroid hormone receptor associated proteins (TRAP)/Mediator complex is involved in the regulation of *ato* expression in the proneural groups. The TRAP complex acts as a coactivator for a variety of transcriptional activators (Ito and Roeder 2001; Malik and Roeder 2000). Among many Mediator complex proteins, two *Drosophila* TRAPs, Kohtalo (Kto, TRAP230) and Blind spot (Bli, also called Skuld, TRAP240), have been extensively studied for their roles in retinal neurogenesis. In *TRAP* mutant clones, Ato is ectopically induced behind the furrow (Treisman 2001).

In contrast, TRAPs are required for *ato* expression in the intermediate groups. Because Ato expression in the intermediate groups is dependent on Ato itself, it is possible that TRAP complex might act as coactivator for Ato. Indeed, in the absence of TRAP, Ato fails to induce EGFR signaling and Sca expression that are necessary for lateral inhibition, thus resulting in ectopic Ato expression. Similarly, Kto and Skd are also required for positive Ato functions to induce Ato targets such as Ato itself and Senseless (Sens) within the proneural clusters. Hence, TRAP complex is required for Ato expression and other Ato target genes such as *sca*, *sens*, and *rho* in the intermediate groups (Fig. 4; Lim et al. 2007).

These studies suggest that proneural and anti-proneural genes function in a negative feedback network. EGFR signaling activated by Ato is necessary for nonautonomous Bar expression near the furrow. In turn, Bar represses *ato* in the basal cells behind the furrow (see a model in Fig. 3). In developing wing disc, TRAP230/240 proteins are necessary for the activation of Wnt signaling and Notch target genes (Carrera et al. 2008; Janody and Treisman 2011). Wg is expressed in the dorsal and ventral margin of an eye disc where it antagonizes the Dpp function for furrow progression. It would be interesting to see whether the TRAP mediator complex also plays a role for Wg regulation and N signaling in the eye.

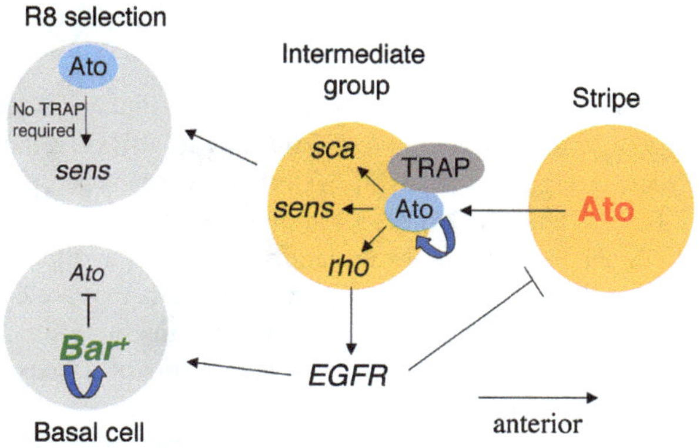

Fig. 4 A model of TRAP-mediated Ato activation in early retinal neurogenesis. Kto (TRAP230) and Skd (TRAP240) function as coactivators for Ato in the proneural clusters and are therefore required for expression of Ato target genes such as Ato itself, Sens, Sca, and for activation of EGFR signaling. Ato and Sens are required for selection and differentiation of the R8 founder neurons. In contrast, EGFR signaling and Sca are involved in repressing Ato expression in cells between proneural clusters. Kto/Skd may not be required for Ato activation in the R8 cells in which Ato activation may depend on other coactivators. In the basal cells, Bar induced by EGFR signaling represses Ato expression. A *blue arrow* indicates autoregulation. (Adapted from Lim et al. 2007)

As described above, Bar and Ato are expressed in a complementary pattern and are antagonistic to each other. It is worth noting that Bar expression is also complementary to Dpp that is expressed in the furrow. Dpp is critical for the initial steps of eye morphogenesis such as furrow initiation, progression, and *ato* activation (Borod and Heberlein 1998; Greenwood and Struhl 1999; Heberlein and Moses 1995). Because *dpp* transcription is induced in the furrow, it provides a marker for the boundary between undifferentiated cells in the anterior domain and differentiating cells in the posterior domain of eye disc. Similarly, *dpp*-expressing cells in limb discs mark the boundary between the anterior-posterior (A/P) compartments (Raftery et al. 1991).

It has been suggested that localized expression of *dpp* at the A/P boundary is largely under negative regulation because subfragments of *dpp*-regulatory region fused to *lacZ* reporter constructs typically result in ectopic *lacZ* expression rather than its loss (Sanicola et al. 1995). Consistent with this idea, Engrailed (En) homeodomain protein, which acts as the selector for posterior compartments in limb discs (Blair 1992; Kornberg 1981; Morata and Lawrence 1975), is a direct repressor of *dpp*, thus defining the posterior boundary of the *dpp* stripe. In eye development, Dpp is an upstream factor for *ato* induction in the furrow (Fu and Baker 2003; Greenwood and Struhl 1999). Ato expression results in the activation of Bar expression in the region posterior to the furrow indirectly by a combination of EGFR and Hh signaling (Lim and Choi 2004). Since ectopically expressed Bar can repress *dpp-lacZ* in eye and other discs (Lim and Choi 2003), it is possible that Bar expression in the basal cells may be important for preventing ectopic Dpp expression behind the furrow. This negative regulation between *dpp* and *Bar* may play a role in defining the A/P boundary in developing eye.

Dual Function of Daughterless (Da) in Ato Regulation

In addition to Ato, there are other bHLH family proteins that are required for specification of sensory organs. *Achaete-Scute Complex* (*ASC*) bHLH genes are expressed with spatially regulated pattern to specify external sensory organ precursors (SOPs; Cubas et al. 1991; Ghysen and Dambly-Chaudiere 1989; Romani et al. 1989; Skeath and Carroll 1991), and *amos* for multiple dendritic neurons in olfactory sense organs (Goulding et al. 2000; Huang et al. 2000). These proteins that are expressed in specific tissues are categorized as class II bHLH family transcription factors in contrast to the class I bHLH factors expressed in a broader range of tissues. Tissue-specific class II bHLH proteins form heterodimeric complexes with class I bHLH proteins and directly bind to E-box consensus sequences of target genes through their basic domains.

Da, initially identified as an important factor for sex determination, is the only known neural class I bHLH protein in *Drosophila* (Caudy et al. 1988b). Like other class I proteins, Da has been thought to be expressed ubiquitously in a broad range of tissues and involved in diverse developmental processes including neurogenesis, depending on its class II bHLH-binding partners (Brown et al. 1996; Caudy et al. 1988a, 1988b; Cronmiller et al. 1988). Since each tissue-specific class II bHLH protein requires Da to form a functional heterodimeric complex, both class I Da and class II bHLH proteins are important for their proneural function.

Interestingly, although Da protein is ubiquitously expressed in the eye disc, it is selectively upregulated in the morphogenetic furrow (Brown et al. 1996). Further analysis has revealed that there are two distinct patterns of Da upregulation in the furrow: a broad low-level upregulation and a stronger Da expression in the nonneural cells surrounding the R8 cells between proneural clusters (Fig. 5; Lim et al. 2008). This pattern indicates that Da expression may be regulated in coordination with the process of neurogenesis in the furrow. Indeed, Da expression is dynamically regulated in the furrow by multiple mechanisms including Hh and Dpp signaling pathways.

Because Ato is known to form a dimer with the type I bHLH factor Da to function as an active transcription factor, the upregulation of Da in the nonneural cells between proneural clusters is unexpected. Remarkably, loss of Da in the furrow leads to an expansion of Ato expression in mutant clones, indicating that Da acts as a negative factor for Ato expression. Despite the expanded Ato expression, there is no retinal differentiation within *da* LOF clones because Ato target genes like *sens* necessary for retinal differentiation are not induced in the absence of Da. In contrast, overexpression of Da results in the repression of Ato expression in the furrow. These results, together with the specific upregulation of Da between proneural groups, suggest that Da has both proneural and antiproneural functions depending on the expression level and cell types in the furrow (Lim et al. 2008).

Because Da functions as a negative regulator of Ato expression in the furrow, it is an intriguing question whether this Da function is related to the lateral inhibition by N signaling. Notch-dependent lateral inhibition is mediated by E(spl), another bHLH family transcription factor. Clonal loss of Da in the region covering the furrow leads to

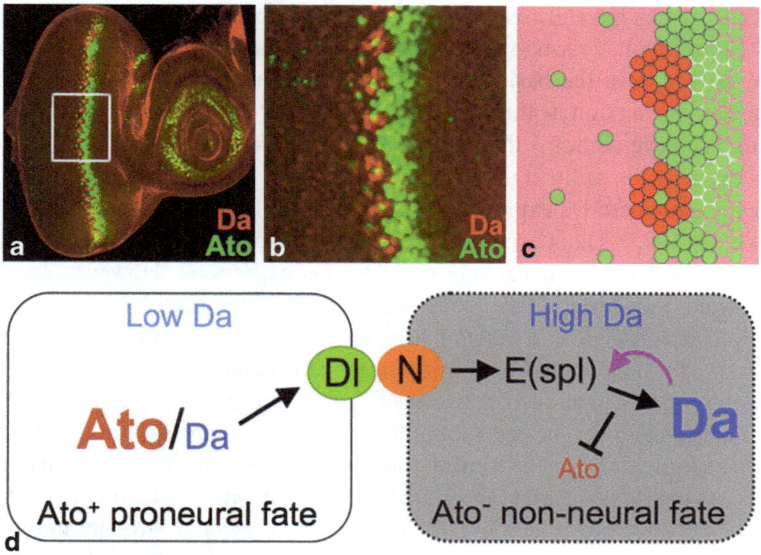

Fig. 5 Antiproneural function of Da. (**a–c**) Expression pattern of Da. Third-instar eye disc stained with antibodies against Da and Ato. An area around the furrow (*rectangle*) in (**a**) is magnified in (**b**). (**c**) Is a schematic of (**b**). In the furrow region, Da is expressed with a relatively low level in all Ato-expressing cells (*green*), but it is highly expressed in the cells surrounding singled-out Ato-positive R8 cells just behind the furrow (*red*). Outside the furrow, Da is expressed broadly at a low-level anterior and posterior to the furrow region (*pink*). (**d**) A model for Da function during retinal neurogenesis. Da has dual functions as a proneural and an antiproneural factor depending on expression level during early retinal neurogenesis. In Ato-positive proneural cells, a low level of Da forms a heterodimer with Ato to function as a proneural factor. In neighboring cells, Da is further upregulated by N-E(spl) pathway. A positive feedback regulation between E(spl) and Da represses Ato expression to antagonize neural specification. (Adapted from Lim et al. 2008)

a loss of E(spl) with concomitant upregulation of Ato. Moreover, ectopic expression of Da causes strong reduction of E(spl) expression. Interestingly, loss of N or E(spl) also results in reduction of Da expression. Thus, Da promotes Notch signaling to activate E(spl) expression, and both Da and Notch signaling cooperatively repress Ato expression to refine single R8 cell selection. In this model, a high level of Ato induced in the proneural group cells activate Dl, resulting in the Notch activation in the adjacent cells. Consequently, E(spl) expressed in these cells induces Da expression, resulting in a high-level Da expression and repression of Ato in these cells. The high-level Da in these cells also activate E(spl) expression by a feedback regulation, thus strengthen the difference between the proneural and nonneural cells (Fig. 5; Lim et al. 2008). The dual function of Da was also found from a genetic screen for *ato* modifiers. One of the dominant enhancers turned out to be LOF alleles of *da*. LOF *da* clones showed expanded 3′ *ato-lacZ* in the furrow while loss of 5′ *ato-lacZ* expression posterior to the furrow, consistent with the positive and negative regulation of *ato*, using 5′ and 3′ *ato* genomic enhancers, respectively (Melicharek et al. 2008).

Recent studies have also shown that an interaction of broadly expressed type I bHLH genes regulate tissue-specific cell fates. The Id family HLH proteins, called type V, do not have the basic domain so that the heterodimers of type II and V protein cannot function as transcription factors due to their inability to bind DNA. Extramacrochaete (Emc) is the only type V HLH protein in *Drosophila*. Although Emc is expressed broadly in most tissues, it is expressed at a low level in the furrow of an eye disc where Da expression is upregulated (Bhattacharya and Baker 2011; Brown et al. 1995). Da is expressed highly in the clones of LOF *emc* mutant cells. This indicates that Emc negatively regulates Da expression. However, loss of Da causes strong reduction of Emc, indicating that Da is required for Emc expression. Thus, Emc and Da appear to function in an unusual negative feedback loop where Da is necessary for the expression of its inhibitor Emc (Bhattacharya and Baker 2012).

It was proposed that such network of type 1 and 5 HLH proteins might be a general mechanism for the regulation of type II HLH expression in developmental decision makings during neurogenesis, not only in *Drosophila* but also in mammalian systems. As mentioned earlier, there are two levels of Da upregulation in the furrow: a general weak Da upregulation in the furrow and a selective stronger upregulation between the proneural groups in the furrow. In contrast to the high levels of Da in the furrow, Emc is low. Thus, it would be interesting to see whether Emc may be expressed dynamically within the furrow to cross-talk with different levels of Da in the furrow.

Concluding Remarks

Neural differentiation involves a series of inductive events to generate neurons from an undifferentiated epithelium. Neurogenesis is initiated in the morphogenetic furrow by secreted factors like Dpp and Hh as well as proneural transcription factors like Ato. Generation of ommatidial pattern in the developing eye is established by interaction of these positive factors and various negative regulators.

Several key negative factors involved in Ato repression and ommatidial spacing were discussed in this chapter. Notch is a major negative regulator of Ato expression and is essential for spatial patterning as well as fate specification of photoreceptor cells. Lateral inhibition by N is activated by ligands that may also be mediated by ligand-independent and *cis*-inhibition mechanisms.

Ato expression is also regulated negatively by Da. Da is an essential partner of Ato for its proneural function, but its high level around the proneural groups antagonizes Ato expression to generate interommatidial space. This antiproneural function of Da is positively regulated by E(spl), a target of Notch signaling. Da and E(spl) form a feedback loop to promote their expression, which probably reinforces the lateral inhibition of Ato expression.

Bar homeodomain proteins are major negative regulator of Ato expression behind the furrow. Bar-expressing undifferentiated cells near the furrow can repress Ato expression, thus preventing ectopic formation of photoreceptors. Negative feedback regulation between Ato and Bar is important for ommatidial patterning. It remains to be seen whether homologs for Ato and Bar have similar functional relationship in developing vertebrate eyes.

Acknowledgments I am thankful to Dr. Janghoo Lim for his work on Bar, and Drs. Kyung-Ok Cho and Amit Singh for helpful comments. I apologize to many investigators whose work might not have been cited due to space limitation or my ignorance. This work was supported by grants from the World Class University Program (R31-2008-000-10071-0) and National Research Laboratory grant through the National Research Foundation of Korea funded by the Ministry of Education, Science & Technology.

References

Akimaru H, Saigo K (1991) DNA binding activity of the BarH1 homeodomain of Drosophila. Nucleic Acids Symp Ser (25):29–30

Annaert W, De Strooper B (1999) Presenilins: molecular switches between proteolysis and signal transduction. Trends Neurosci 22:439–443

Baker NE, Firth LC (2011) Retinal determination genes function along with cell–cell signals to regulate Drosophila eye development: examples of multi-layered regulation by master regulators. Bioessays 33:538–546

Baker NE, Mlodzik M, Rubin GM (1990) Spacing differentiation in the developing Drosophila eye: a fibrinogen-related lateral inhibitor encoded by scabrous. Science 250:1370–1377

Baker NE, Rubin GM (1989) Effect on eye development of dominant mutations in Drosophila homologue of the EGF receptor. Nature 340:150–153

Baker NE, Yu SY (1997) Proneural function of neurogenic genes in the developing Drosophila eye. Curr Biol 7:122–132

Baonza A, Casci T, Freeman M (2001) A primary role for the epidermal growth factor receptor in ommatidial spacing in the Drosophila eye. Curr Biol 11:396–404

Baonza A, Freeman M (2001) Notch signalling and the initiation of neural development in the Drosophila eye. Development 128:3889–3898

Beatus P, Lendahl U (1998) Notch and neurogenesis. J Neurosci Res 54:125–136

Bhattacharya A, Baker NE (2011) A network of broadly expressed HLH genes regulates tissue-specific cell fates. Cell 147:881–892

Bhattacharya A, Baker NE (2012) The Role of the bHLH Protein Hairy in Morphogenetic Furrow Progression in the Developing Drosophila Eye. PLoS One 7:e47503

Blair SS (1992) Engrailed expression in the anterior lineage compartment of the developing wing blade of Drosophila. Development 115:21–33

Borod ER, Heberlein U (1998) Mutual regulation of decapentaplegic and hedgehog during the initiation of differentiation in the Drosophila retina. Dev Biol 197:187–197

Brown KE, Kerr M, Freeman M (2007) The EGFR ligands Spitz and Keren act cooperatively in the Drosophila eye. Dev Biol 307:105–113

Brown NL, Kanekar S, Vetter ML, Tucker PK, Gemza DL, Glaser T (1998) Math5 encodes a murine basic helix-loop-helix transcription factor expressed during early stages of retinal neurogenesis. Development 125:4821–4833

Brown NL, Paddock SW, Sattler CA, Cronmiller C, Thomas BJ, Carroll SB (1996) daughterless is required for Drosophila photoreceptor cell determination, eye morphogenesis, and cell cycle progression. Dev Biol 179:65–78

Brown NL, Sattler CA, Paddock SW, Carroll SB (1995) Hairy and emc negatively regulate morphogenetic furrow progression in the Drosophila eye. Cell 80:879–887

Cabrera CV (1990) Lateral inhibition and cell fate during neurogenesis in Drosophila: the interactions between scute, Notch and Delta. Development 110:733–742

Cagan RL, Ready DF (1989) Notch is required for successive cell decisions in the developing Drosophila retina. Genes Dev 3:1099–1112
Carrera I, Janody F, Leeds N, Duveau F, Treisman JE (2008) Pygopus activates Wingless target gene transcription through the mediator complex subunits Med12 and Med13. Proc Natl Acad Sci U S A 105:6644–6649
Caudy M, Grell EH, Dambly-Chaudiere C, Ghysen A, Jan LY, Jan YN (1988a) The maternal sex determination gene daughterless has zygotic activity necessary for the formation of peripheral neurons in Drosophila. Genes Dev 2:843–852
Caudy M, Vassin H, Brand M, Tuma R, Jan LY, Jan YN (1988b) daughterless, a Drosophila gene essential for both neurogenesis and sex determination, has sequence similarities to myc and the achaete-scute complex. Cell 55:1061–1067
Chanut F, Heberlein U (1997) Role of decapentaplegic in initiation and progression of the morphogenetic furrow in the developing Drosophila retina. Development 124:559–567
Cho B, Fischer JA (2011) Ral GTPase promotes asymmetric Notch activation in the Drosophila eye in response to Frizzled/PCP signaling by repressing ligand-independent receptor activation. Development 138:1349–1359
Cho B, Fischer JA (2012) Ral inhibits ligand-independent Notch signaling in Drosophila. Small GTPases 3:186–191
Choi KW, Mozer B, Benzer S (1996) Independent determination of symmetry and polarity in the Drosophila eye. Proc Natl Acad Sci U S A 93:5737–5741
Cooper MT, Bray SJ (1999) Frizzled regulation of Notch signalling polarizes cell fate in the Drosophila eye. Nature 397:526–530
Cronmiller C, Schedl P, Cline TW (1988) Molecular characterization of daughterless, a Drosophila sex determination gene with multiple roles in development. Genes Dev 2:1666–1676
Cubas P, Celis JF de, Campuzano S, Modolell J (1991) Proneural clusters of achaete-scute expression and the generation of sensory organs in the Drosophila imaginal wing disc. Genes Dev 5:996–1008
Curtiss J, Mlodzik M (2000) Morphogenetic furrow initiation and progression during eye development in Drosophila: the roles of decapentaplegic, hedgehog and eyes absent. Development 127:1325–1336
del Alamo D, Mlodzik M (2006) Frizzled/PCP-dependent asymmetric neuralized expression determines R3/R4 fates in the Drosophila eye. Dev Cell 11:887–894
Dokucu ME, Zipursky SL, Cagan RL (1996) Atonal, rough and the resolution of proneural clusters in the developing Drosophila retina. Development 122:139–4147
Dominguez M, Hafen E (1997) Hedgehog directly controls initiation and propagation of retinal differentiation in the Drosophila eye. Genes Dev 11:3254–3264
Dominguez M, Wasserman JD, Freeman M (1998) Multiple functions of the EGF receptor in Drosophila eye development. Curr Biol 8:1039–1048
Ellis MC, Weber U, Wiersdorff V, Mlodzik M (1994) Confrontation of scabrous expressing and non-expressing cells is essential for normal ommatidial spacing in the Drosophila eye. Development 120:1959–1969
Fanto M, Mlodzik M (1999) Asymmetric Notch activation specifies photoreceptors R3 and R4 and planar polarity in the Drosophila eye. Nature 397:523–526
Fetchko M, Huang W, Li Y, Lai ZC (2002) Drosophila Gp150 is required for early ommatidial development through modulation of Notch signaling. EMBO J 21:1074–1083
Frankfort BJ, Mardon G (2002) R8 development in the Drosophila eye: a paradigm for neural selection and differentiation. Development 129:1295–1306
Freeman M (1994) The spitz gene is required for photoreceptor determination in the Drosophila eye where it interacts with the EGF receptor. Mech Dev 48:25–33
Fu W, Baker NE (2003) Deciphering synergistic and redundant roles of Hedgehog, Decapentaplegic and Delta that drive the wave of differentiation in Drosophila eye development. Development 130:5229–5239

Ghysen A, Dambly-Chaudiere C (1989) Genesis of the Drosophila peripheral nervous system. Trends Genet 5:251–255

Goulding SE, zur Lage P, Jarman AP (2000) amos, a proneural gene for Drosophila olfactory sense organs that is regulated by lozenge. Neuron 25:69–78

Greenwood S, Struhl G (1999) Progression of the morphogenetic furrow in the Drosophila eye: the roles of Hedgehog, Decapentaplegic and the Raf pathway. Development 126:5795–5808

Halder G, Callaerts P, Gehring WJ (1995) Induction of ectopic eyes by targeted expression of the eyeless gene in Drosophila. Science 267:1788–1792

Heberlein U, Moses K (1995) Mechanisms of Drosophila retinal morphogenesis: the virtues of being progressive. Cell 81:987–990

Heberlein U, Wolff T, Rubin GM (1993) The TGF beta homolog dpp and the segment polarity gene hedgehog are required for propagation of a morphogenetic wave in the Drosophila retina. Cell 75:913–926

Heitzler P, Simpson P (1991) The choice of cell fate in the epidermis of Drosophila. Cell 64:1083–1092

Higashijima S, Kojima T, Michiue T, Ishimaru S, Emori Y, Saigo K (1992) Dual Bar homeo box genes of Drosophila required in two photoreceptor cells, R1 and R6, and primary pigment cells for normal eye development. Genes Dev 6:50–60

Hsiung F, Moses K (2002) Retinal development in Drosophila: specifying the first neuron. Hum Mol Genet 11:1207–1214

Huang ML, Hsu CH, Chien CT (2000) The proneural gene amos promotes multiple dendritic neuron formation in the Drosophila peripheral nervous system. Neuron 25:57–67

Ito M, Roeder RG (2001) The TRAP/SMCC/Mediator complex and thyroid hormone receptor function. Trends Endocrinol Metab 12:127–134

Janody F, Treisman JE (2011) Requirements for mediator complex subunits distinguish three classes of notch target genes at the Drosophila wing margin Dev Dyn 240:2051–2059

Jarman AP, Grell EH, Ackerman L, Jan LY, Jan YN (1994) Atonal is the proneural gene for Drosophila photoreceptors. Nature 369:398–400

Kametaka S, Kametaka A, Yonekura S, Haruta M, Takenoshita S, Goto S, Waguri S (2012) AP-1 clathrin adaptor and CG8538/Aftiphilin are involved in Notch signaling during eye development in Drosophila melanogaster. J Cell Sci 125:634–648

Kimmel BE, Heberlein U, Rubin GM (1990) The homeo domain protein rough is expressed in a subset of cells in the developing Drosophila eye where it can specify photoreceptor cell subtype. Genes Dev 4:712–727

Kopan R, Goate A (2002) Aph-2/Nicastrin: an essential component of gamma-secretase and regulator of Notch signaling and Presenilin localization. Neuron 33:321–324

Kornberg T (1981) Engrailed: a gene controlling compartment and segment formation in Drosophila. Proc Natl Acad Sci U S A 78:1095–1099

Kumar JP, Moses K (2001) EGF receptor and Notch signaling act upstream of Eyeless/Pax6 to control eye specification. Cell 104:687–697

Kumar JP, Tio M, Hsiung F, Akopyan S, Gabay L, Seger R, Shilo BZ, Moses K (1998) Dissecting the roles of the Drosophila EGF receptor in eye development and MAP kinase activation. Development 125:3875–3885

Lai EC, Deblandre GA, Kintner C, Rubin GM (2001) Drosophila neuralized is a ubiquitin ligase that promotes the internalization and degradation of delta. Dev Cell 1:783–794

Li Y, Fetchko M, Lai ZC, Baker NE (2003) Scabrous and Gp150 are endosomal proteins that regulate Notch activity. Development 130:2819–2827

Lim J, Choi KW (2003) Bar homeodomain proteins are anti-proneural in the Drosophila eye: transcriptional repression of atonal by Bar prevents ectopic retinal neurogenesis. Development 130:5965–5974

Lim J, Choi KW (2004) Induction and autoregulation of the anti-proneural gene Bar during retinal neurogenesis in Drosophila. Development 131:5573–5580

Lim J, Jafar-Nejad H, Hsu YC, Choi KW (2008) Novel function of the class I bHLH protein Daughterless in the negative regulation of proneural gene expression in the Drosophila eye. EMBO Rep 9:1128–1133

Lim J, Lee OK, Hsu YC, Singh A, Choi KW (2007) Drosophila TRAP230/240 are essential coactivators for Atonal in retinal neurogenesis. Dev Biol 308:322–330

Malik S, Roeder RG (2000) Transcriptional regulation through Mediator-like coactivators in yeast and metazoan cells. Trends Biochem Sci 25:277–283

Melicharek D, Shah A, DiStefano G, Gangemi AJ, Orapallo A, Vrailas-Mortimer AD, Marenda DR (2008) Identification of novel regulators of atonal expression in the developing Drosophila retina. Genetics 180:2095–2110

Morata G, Lawrence PA (1975) Control of compartment development by the engrailed gene in Drosophila. Nature 255:614–617

Niwa N, Hiromi Y, Okabe M (2004) A conserved developmental program for sensory organ formation in Drosophila melanogaster. Nat Genet 36:293–297

Pavlopoulos E, Pitsouli C, Klueg KM, Muskavitch MA, Moschonas NK, Delidakis C (2001) Neuralized Encodes a peripheral membrane protein involved in delta signaling and endocytosis. Dev Cell 1:807–816

Pepple KL, Atkins M, Venken K, Wellnitz K, Harding M, Frankfort B, Mardon G (2008) Two-step selection of a single R8 photoreceptor: a bistable loop between senseless and rough locks in R8 fate. Development 135:4071–4079

Powell PA, Wesley C, Spencer S, Cagan RL (2001) Scabrous complexes with Notch to mediate boundary formation. Nature 409:626–630

Raftery LA, Sanicola M, Blackman RK, Gelbart WM (1991) The relationship of decapentaplegic and engrailed expression in Drosophila imaginal disks: do these genes mark the anterior– posterior compartment boundary?. Development 113:27–33

Ready DF, Hanson TE, Benzer S (1976) Development of the Drosophila retina, a neurocrystalline lattice. Dev Biol 53:217–240

Rodrigues AB, Werner E, Moses K (2005) Genetic and biochemical analysis of the role of EGFR in the morphogenetic furrow of the developing Drosophila eye. Development 132:4697–4707

Roignant JY, Treisman JE (2009) Pattern formation in the Drosophila eye disc. Int J Dev Biol 53:795–804

Romani S, Campuzano S, Macagno ER, Modolell J (1989) Expression of achaete and scute genes in Drosophila imaginal discs and their function in sensory organ development. Genes Dev 3:997–1007

Rooke JE, Xu T (1998) Positive and negative signals between interacting cells for establishing neural fate. Bioessays 20:209–214

Rutledge BJ, Zhang K, Bier E, Jan YN, Perrimon N (1992) The Drosophila spitz gene encodes a putative EGF-like growth factor involved in dorsal–ventral axis formation and neurogenesis. Genes Dev 6:1503–1517

Sanicola M, Sekelsky J, Elson S, Gelbart WM (1995) Drawing a stripe in Drosophila imaginal disks: negative regulation of decapentaplegic and patched expression by engrailed. Genetics 139:745–756

Schweitzer R, Howes R, Smith R, Shilo BZ, Freeman M (1995) Inhibition of Drosophila EGF receptor activation by the secreted protein Argos. Nature 376:699–702

Singh A, Tare M, Puli OR, Kango-Singh M (2012) A glimpse into dorso-ventral patterning of the Drosophila eye. Dev Dyn 241:69–84

Singh J, Mlodzik M (2012) Hibris, a Drosophila nephrin homolog, is required for presenilin-mediated Notch and APP-like cleavages. Dev Cell 23:82–96

Skeath JB, Carroll SB (1991) Regulation of achaete-scute gene expression and sensory organ pattern formation in the Drosophila wing. Genes Dev 5:984–995

Steinberg AG, Abramowitz M (1938) The Bar "Locus" and the v reaction in Drosophila melanogaster. Proc Natl Acad Sci U S A 24:107–111

Sun Y, Jan LY, Jan YN (1998) Transcriptional regulation of atonal during development of the Drosophila peripheral nervous system. Development 125:3731–3740
Tanaka-Matakatsu M, Du W (2008) Direct control of the proneural gene atonal by retinal determination factors during Drosophila eye development. Dev Biol 313:787–801
Tomlinson A, Ready DF (1987) Neuronal differentiation in Drosophila ommatidium. Dev Biol 120:366–376
Tomlinson A, Struhl G (1999) Decoding vectorial information from a gradient: sequential roles of the receptors Frizzled and Notch in establishing planar polarity in the Drosophila eye. Development 126:5725–5738
Treisman J (2001) Drosophila homologues of the transcriptional coactivation complex subunits TRAP240 and TRAP230 are required for identical processes in eye-antennal disc development. Development 128:603–615
Treisman JE (2004) Coming to our senses. Bioessays 26:825–828
Weinmaster G, Fischer JA (2011) Notch ligand ubiquitylation: what is it good for?. Dev Cell 21:134–144
Wolff T, Ready DF (1991) The beginning of pattern formation in the Drosophila compound eye: the morphogenetic furrow and the second mitotic wave. Development 113:841–850
Yang L, Baker NE (2001) Role of the EGFR/Ras/Raf pathway in specification of photoreceptor cells in the Drosophila retina. Development 128:1183–1191
Ye Y, Lukinova N, Fortini ME (1999) Neurogenic phenotypes and altered Notch processing in Drosophila Presenilin mutants. Nature 398:525–529
Yeh E, Dermer M, Commisso C, Zhou L, McGlade CJ, Boulianne GL (2001) Neuralized functions as an E3 ubiquitin ligase during Drosophila development. Curr Biol 11:1675–1679
Zhang T, Ranade S, Cai CQ, Clouser C, Pignoni F (2006) Direct control of neurogenesis by selector factors in the fly eye: regulation of atonal by Ey and So. Development 133:4881–4889

Cell Adhesion During *Drosophila* Eye Development

Ruth I. Johnson

Introduction

Over a brief few days, thousands of cells in the *Drosophila* eye are organized to generate a precisely patterned functional organ. Eye morphogenesis requires coordinated cell fate specification and differentiation, local cell movements, niche acquisition, and apoptosis to remove surplus cells. The eye has provided a superb model tissue for studies of the molecular bases of these events and the past decade has been punctuated with studies on the adhesion molecules at play as the fly eye develops. Because of its structure—a neuroepithelium composed of several discrete and easily discernable cell types—the eye provides unique opportunities to examine the roles of adhesion between cells as a complex organ is generated. Indeed, dynamic adhesion plays a significant role in orchestrating, regulating, and driving eye morphogenesis.

A Descriptive Overview of Eye Development

The mature fly eye is marked by a neatly patterned honeycomb array of about 750 unit eyes (ommatidia) that are separated by an interweaving cell lattice. The precise arrangement of cells into this simple repeating pattern is striking and easily observed from the second day of pupation (Fig. 1d, e). Each mature ommatidium is composed of a bundle of eight photoreceptors capped by four lens-secreting cone cells and surrounded by two primary pigment cells (1°s) that are surrounded by a hexagonal lattice of secondary (2°) and tertiary (3°) pigment cells. Three mechanosensory bristle organules are positioned at alternate vertices of each hexagon.

The selection and arrangement of ommatidia into this striking hexagonal pattern begins in the late third larval instar eye disc (Fig. 1a, c). As soon as the morphogenetic

R. I. Johnson (✉)
Biology Department, Wesleyan University, Middletown, CT 06459, USA
e-mail: rijohnson@wesleyan.edu

A. Singh, M. Kango-Singh (eds.), *Molecular Genetics of Axial Patterning, Growth and Disease in the Drosophila Eye*, DOI 10.1007/978-1-4614-8232-1_7,

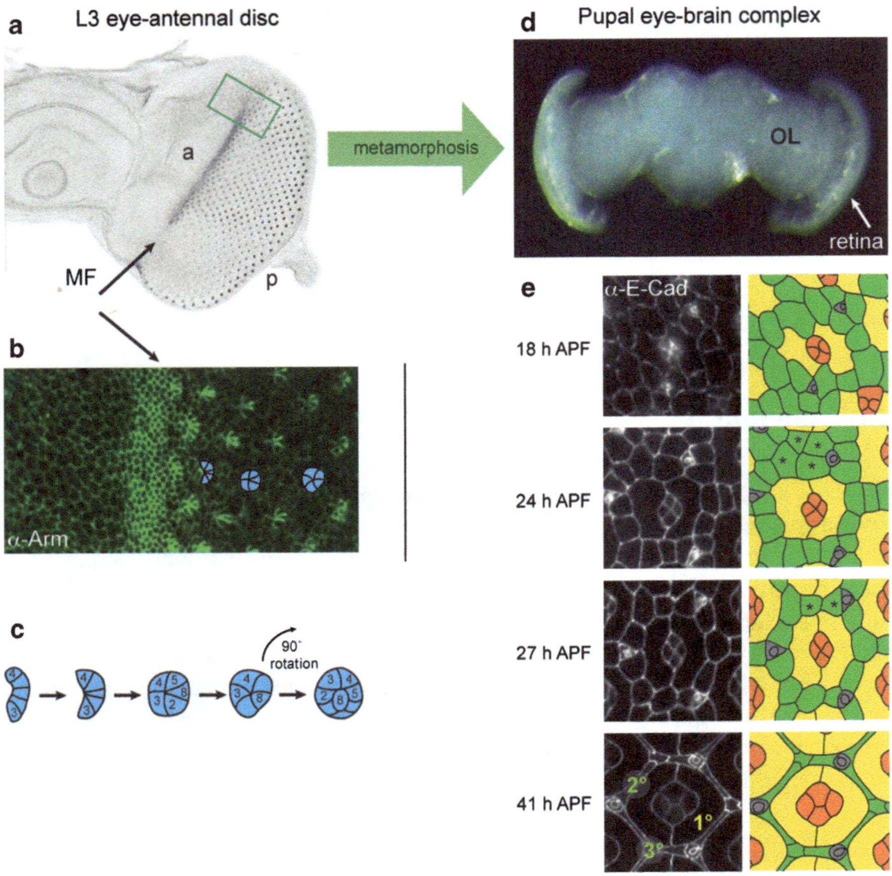

Fig. 1 Morphogenesis of the eye neuroepithelium during larval and pupal stages. **a** An eye-antennal disc dissected from a wandering L3 larva. The *green box* approximately corresponds to **b**. Ommatidial clusters emerge as the morphogenetic furrow (*MF*) travels from posterior (*p*, *right*) to anterior (*a*), illustrated in **c**. Once the first five photoreceptor precursors have been recruited (R8, R2 and R5, R3, and R4, colored *blue*), ommatidia begin to rotate. Recruitment of the remaining three R-cells completes each ommatidium. **d** Pupal eye–brain complex dissected 40 h after pupation (APF). The two retinas attach to the optic lobes (*OL*). **e** Small regions of retina dissected at 18, 24, 27, and 28 h APF, illustrated in panels on *right*. A central ommatidium is pictured in each panel. Cone cells (*orange*), primary pigment cells (1°s, *yellow*), and bristle groups (*grey*) occupy specific niches and the interommatidial cells (IPCs, *green*) rearrange into single file (*asterisks*). Excess IPCs are removed by apoptosis leaving three tertiary (3°) and six secondary (2°) pigment cells about each ommatidium (one of each of these cells is labeled in **e**, *bottom* panel). (Image in **b** is adapted from Escudero et al. 2007; **e** is adapted from Johnson et al. 2011)

furrow (MF) has passed, sequential recruitment of photoreceptor precursor cells (R-cells) begins as a wave across the eye disc. The photoreceptor bundles will form the core of each ommatidium and, remarkably, each bundle rotates as a unit a full 90° surrounded by a sea of relatively motionless undifferentiated cells—rather like a swimmer turning 90° in a swimming pool. Signals emitting from the R-cells then initialize recruitment of four cone cells per ommatidium from the surrounding pool of undifferentiated cells and the organism enters pupation.

In the early pupa, the eye undergoes dramatic morphogenesis to emerge inverted and cupping the optic lobe: photoreceptor axons projecting from each ommatidium synapse at discrete layers of the medulla within the optic lobe, attaching the two tissues. Apart from regularly spaced fledgling ommatidia (photoreceptor plus cone cell groups) the eye neuroepithelium is still relatively disorganized. Two cells are then recruited as 1° cells, usually those immediately adjoining the anterior and posterior boundaries of each ommatidial group, and patterning of the lattice commences. In this last step, the remaining sea of interommatidial pigment precursor cells (IPCs) is reduced to a well-ordered single-file lattice as cells move and compete for specific niches and excess cells are targeted for apoptosis (Fig. 1e). The product is a precisely ordered interweaving cell lattice that optically insulates each ommatidium from its neighbor in the pigmented adult eye.

The eight photoreceptors (R-cells) now lie below the tissue surface, encased by the cone and 1° cell layer (Fig. 2). The six outer R-cells (R1–R6) surround the two inner R-cells (the R7 and R8, the latter lies beneath the R7). The outer R-cells arrange in an asymmetrical trapezoidal shape around the R7/R8 group: this asymmetry is established in the larval disc when differential Notch activity regulated by planar cell polarity (PCP) distinguishes the R4 precursor from the R3 (Cooper and Bray 1999; Fanto and Mlodzik 1999). During pupal development, the eight photoreceptor precursors undergo extensive morphogenesis. Each R-cell bends, stooping a full 90° to orient its apical surface toward the ommatidial core and the apical membrane is elaborated into a mass of microvilli that form each light-sensing rhabdomere (Fig. 2b; Knust 2007).

To appropriately arrange and shape the photoreceptors, ommatidia, and interweaving pigment cell lattice, adhesive junctions must be carefully regulated. In this chapter, we explore the dynamic and cell-type-specific changes in adhesion that shape the fly eye. We focus on the adhesive forces that drive morphogenesis of the epithelial cells as the eye is patterned. Mechanisms that drive rhabdomere morphogenesis and mediate interactions between photoreceptor axons projecting from the retina to synapse in the medulla are not discussed (for recent reviews, see Knust 2007; Schwabe et al. 2009).

Junctions in the Fly Eye

In *Drosophila*, the adhesive adherens junction (AJ) is the most apical junction in epithelial cells (Muller 2000; Tepass et al. 2001). Occluding septate junctions (SJ) that share some similarity with vertebrate tight junctions lie below AJs (Furuse and Tsukita 2006). The apical outlines of epithelial cells of the fly eye are easily revealed by detecting *Drosophila* orthologs of E-Cadherin (E-Cad, encoded by *shotgun* (*shg*)) and the associated Catenins (Fig. 1). The eye is a stratified epithelium (Fig. 2). Tall columnar epithelial cells are exposed apically until lens tissue is secreted (during late pupal stages) and maintain a foothold on the underlying basal membrane that is

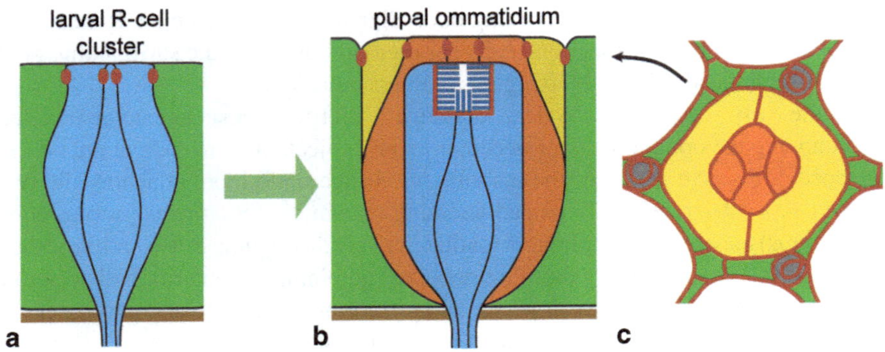

Fig. 2 Morphology and adherens junctions (AJs) in the fly retina. **a** Longitudinal illustration of larval photoreceptor precursor cells (R-cells, *blue*) clustered into a fledgling ommatidium and surrounded by undifferentiated epithelial cells (*green*). Apical AJs are depicted in *red*. **b** By the pupal stage, R-cells are surrounded by cone (*orange*) and primary pigment cells (1°s, *yellow*) and interommatidial pigment precursor cells (IPCs, *green*) form an insulating barrier between ommatidia. The apical membranes of photoreceptor cells elaborate to form rhabdomeres (*dark blue lines*) flanked by AJs (*red*). The basement membrane is *brown*. **c** Apical view of a single pupal ommatidium, *arrow* indicates corresponding plane. Bristle groups are colored *gray*. (Illustrations are inspired and adapted from Tepass and Harris 2007)

pierced periodically by photoreceptor axons. In contrast, the R-cells become buried within ommatidia and bend 90° to reorient the AJs and apical R-cell surface that is remodeled to generate rhabdomeres (Fig. 2b; Knust 2007).

Homophilic interactions between Cadherins generate the adhesive backbone of AJs. E-Cad is expressed in all cells of the eye neuroepithelium but N-Cadherin (N-Cad) in only select cell types. This differential expression of Cadherins that preferentially generate homophilic (rather than heterophilic) interactions generates a mechanism to aggregate like cell types during eye patterning that follows a model of "differential adhesion" proposed many years ago (Duguay et al. 2003; Steinberg 1970, 2007; Steinberg and Takeichi 1994). The patterning of cone cells that express *N-Cad* exemplifies differential adhesion.

In addition to the Cadherins, the fly eye requires several other adhesive transmembrane proteins for its correct development. Many locate at or close to AJs and are also implicated in signal transduction, highlighting the importance of the AJ as a location for these important processes. The current list includes several atypical Cadherins (Flamingo (Fmi), Dachsous (Ds), and Fat (Ft)), the Nectins Echinoid (Ed) and Friend of Echinoid (Fred), and *Drosophila* orthologs of the Nephrin/Neph proteins. The latter group interact heterophilically and are expressed in complementary groups of cells in the eye exemplifying a second model—of preferential heterophilic adhesion—that drives tissue patterning (Bao and Cagan 2005).

A clearly visible belt of F-actin tracks the AJs of neuroepithelial cells of the fly eye (see, e.g., Chu et al. 2012; Johnson et al. 2008). Cadherin–cytoskeletal interactions mediated by Catenins are considered to stabilize junctions—though the mechanism has been challenged in recent years (Weis and Nelson 2006). β-Catenin (encoded by *armadillo* (*arm*)) binds the cytoplasmic Cadherin domain to promote Cadherin stability (Huber and Weis 2001). α-Catenin interacts with β-catenin or Actin (Drees et al. 2005; Weis and Nelson 2006; Yamada et al. 2005). Binding of p120-catenin to E-Cad is proposed to stabilize E-Cad, preventing its endocytosis as new AJs are generated (Miyashita and Ozawa 2007; Thoreson et al. 2000; Xiao et al. 2005).

The eye neuroepithelium is dramatically remodeled during development: ommatidial clusters rotate, individual cells make small local movements to generate simple patterns, and photoreceptors and support IPCs change shape. The morphologies of these events have been well described but the molecular mechanisms that dynamically remodel AJs and the cytoskeleton without compromising tissue integrity are poorly understood. In the eye, these mechanisms may include prudent regulation of Cadherins and the associated α- and β-catenins that indirectly link Cadherins to the Actin cytoskeleton. Mechanisms may include regulation by the Rho GTPases, Catenin and Cadherin phosphorylation, clustering or dispersal of Cadherins, Cadherin trafficking, and regulation by tight junction components including ZO-1 (Gumbiner 2005; Nelson 2008; Yap et al. 2007). Similar mechanisms may regulate those non-Cadherin adhesion molecules required for eye development that also link indirectly to the Actin cytoskeleton. As discussed further, many of these non-Cadherins have a more important role than the Cadherins in particular events during eye morphogenesis.

The Larval Eye: Photoreceptor Groups that Rotate as Units

A wave of morphogenesis passes across the larval eye disc anteriorly, the front marked by a physical indentation aptly named the morphogenetic furrow (Fig. 1a, b). Eight R-cells are recruited in a series of sequential steps to form the core of each ommatidium (Fig. 1c; reviewed in Kumar 2012). Periodically spaced founding R8 cells emerge after the MF passes. Next, the R2/R5 and R3/R4 are recruited as pairs and then after a short pause the photoreceptor set is completed with recruitment of the R1/R6 and R7 precursors. The five-R-cell preclusters are initially arranged as distinctive arcs (with R3 and R4 cells on either end; Fig. 1c) but quickly condense into tight clusters (R3 and R4 cells now neighbors) that rotate 90° while the R1/R6/R7 cells are recruited. The direction of rotation is dictated by a set of PCP proteins that together polarize the eye field in a horizontal axis from the center (equator) toward each dorsal and ventral pole (reviewed in Goodrich and Strutt 2011; Maung and Jenny 2011). When functioning correctly, PCP signaling dictates that Notch activity is amplified in the most polar cell of each R3/R4 cluster. Clusters then rotate to move the R3/R4 toward the appropriate pole; movement is anticlockwise in the ventral hemisphere and clockwise in the dorsal hemisphere of the eye.

Packing R-Cells Tightly Together

As photoreceptor clusters rotate relative to the surrounding pool of undifferentiated cells, the R-cell precursors are tightly attached to each other via Cadherin-based AJs (Brown et al. 2006; Mirkovic and Mlodzik 2006 Figs. 1b and 3a). Dense localization of E-Cad is observed at boundaries between R2, R8, and R5 cells and dense

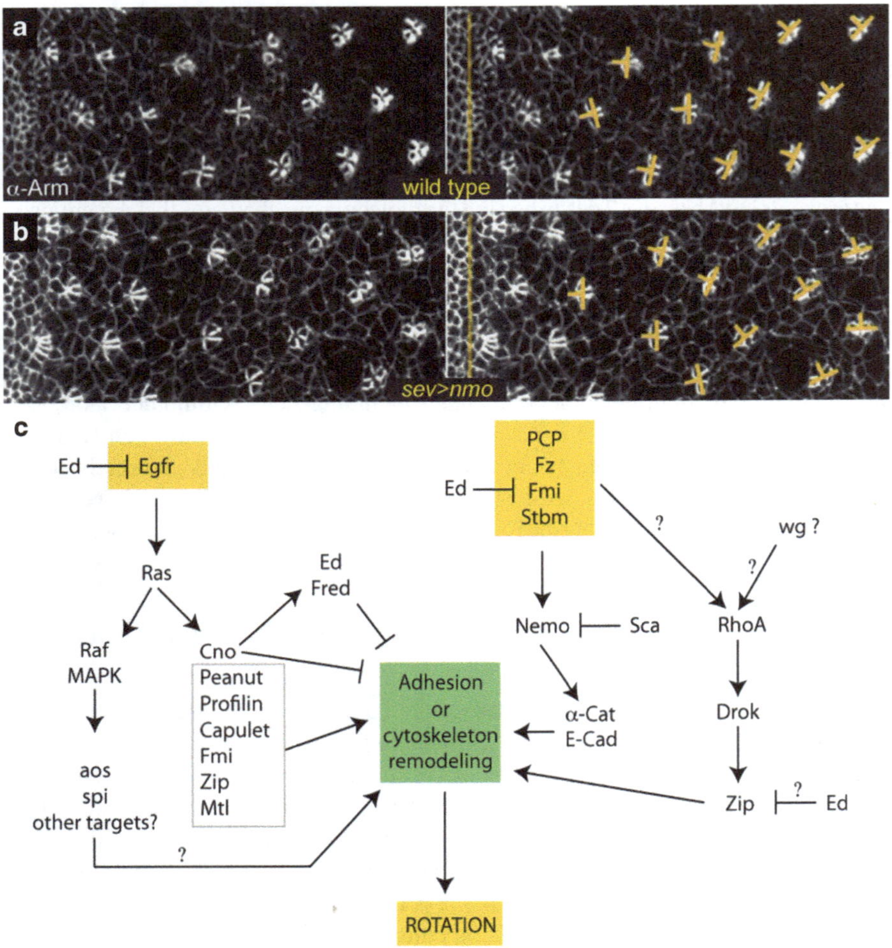

Fig. 3 Regulation of ommatidial rotation. **a** Rotation of wild-type ommatidia begins shortly after the five photoreceptor precursors cluster together. Rotation is emphasized in *right-hand* panels (*yellow*), a *vertical line* indicates the morphogenetic furrow. **b** Overexpression of Nemo accelerates rotation. **c** A model integrating factors that regulate ommatidial rotation, see text for details. (Panels **a** and **b** are from Mirkovic et al. 2011)

N-Cad is detected between the R3 and R4 (Mirkovic and Mlodzik 2006). Accumulation of E-Cad but not N-Cad requires the activity of the Ras GTPase Rap1 (O'Keefe et al. 2009). Cytoplasmic AJ-associated proteins such as β-catenin /Arm accumulate at all adjoining R-cell membranes. These observations correlate with a model of tight adhesion that maintains each photoreceptor precursor cluster. The mechanisms that closely pack R-cells together and fortify these specific AJs are not well understood though several signal transduction pathways have been implicated. Epidermal growth factor receptor (Egfr) signaling is proposed to promote *E-Cad* translation or stabilize junctions to promote R-cell adhesion (Brown et al. 2006). Activity of the fibroblast growth factor receptor Breathless (Btl) is also required: photoreceptor precursors that lack the Btl ligand Branchless (Bnl) fail to pack into tight clusters though expression of E-Cad at AJs appears normal. In addition, dynamic Myosin II and remodeling of the Actin cytoskeleton is required for correct photoreceptor cluster morphogenesis (Chu et al. 2012; Escudero et al. 2007).

Regulating Adhesion During Rotation: Cadherins

In contrast to that observed at boundaries between R-cells, lower levels of E-Cad are detected about the circumference of each R-cell cluster. In fact, E-Cad levels match that detected at AJs of surrounding undifferentiated epithelial cells. This turns out to be crucial—hypomorphic mutations of the E-Cad locus *shg* reduced cluster rotation (Mirkovic and Mlodzik 2006). In contrast, ectopic expression of N-Cad opposed rotation and ommatidia lacking *N-Cad* rotated at a faster rate (Mirkovic and Mlodzik 2006). These data lead to a model of opposing E-Cad and N-Cad function to carefully regulate ommatidial rotation. The mechanism of rotation remains unclear but may be mediated by Cadherin interactions similar to those proposed to mediate movement of border cell clusters within the Drosophila ovary. Like ommatidia, border cell clusters also display lower levels of E-Cad on the outer cluster surface. The presence of E-Cad on this outer surface is nonetheless absolutely required to facilitate correct migration of clusters between nurse cells in the ovary (Niewiadomska et al. 1999). An E-Cad/α-catenin fusion protein that lacks the cytoplasmic tail of E-Cad did not restore migration to *shg* mutant border cell clusters, which suggested that migration is independent of AJ–cytoskeleton linkage (Pacquelet and Rorth 2005). Instead migration may be mediated by repeated catch-and-release interactions between E-Cad molecules on opposing border cell and nurse cell membranes. This mechanism may include adenomatous polyposis coli (Apc), which appears to mediate turnover of AJs in migrating border cells (to mediate "release" of E-Cad interactions) (De Graeve et al. 2012).

Determining Direction: PCP Signaling

PCP protein complexes that determine the direction of rotation assemble at the apical boundaries between the R3 and R4 cells (reviewed by Goodrich and Strutt 2011; Maung and Jenny 2011). The seven-pass atypical Cadherin Fmi (also known as Starry night, Stan) is present in both cells and generates homophilic interactions across the R3/R4 boundary. The transmembrane receptor Frizzled (Fz) and its cytoplasmic transducers Dishevelled (Dsh) and Diego (Dgo) accumulate in the R3 cell at the R3/R4 cell boundary. Meanwhile, the tetraspanin protein Strabismus (Stbm, also known as Van Gogh, Vang) and its cytoplasmic interactor Prickle (Pk) accumulate on the R4 side of the R3/R4 cell border. Asymmetric localization of Fz and Stbm/Vang complexes at the R3/R4 boundary is established immediately before the photoreceptor clusters begin to rotate. Asymmetry is dependent on inhibitory interactions between these complexes and disrupted in PCP mutants. Since the direction and degree of ommatidial rotation is frequently incorrect when components of PCP complexes are mutated or ectopically expressed, it is reasonable to speculate that factors that implement rotation may be asymmetrically localized or activated across the R3/R4 boundary.

The MAP kinase Nemo (Nmo) was among the first proteins to be implicated in regulating ommatidial rotation (Choi and Benzer 1994). Recently, Nmo was shown to be bound by Stbm/Vang and localized to the R4 side of the R3/R4 boundary (Mirkovic et al. 2011). *nmo*^{null} ommatidia failed to rotate and ectopic Nemo (but not a kinase-dead Nmo) caused overrotation (Fig. 3b), a phenotype rescued in *stbm* heterozygotes. In addition to binding Stbm/Vang, Nemo also bound E-Cad and targeted β-catenin/Arm for phosphorylation in vitro. Genetic data indicated that Nemo may promote ommatidial rotation via β-catenin phosphorylation: in larvae raised at 18 °C, expression of either *armS10* (an isoform of β-cat that is resistant to degradation) or *nmo* generated no defects in eye tissue but when *armS10* and *nmo* were coexpressed, many overrotated ommatidia were observed. This interaction was obliterated when a kinase-dead *nmo* or an *armS10*^{AAA} construct (with the serine–threonine targets of Nmo mutated) was used. These data lead to a model of asymmetrically regulated adhesion that facilitates rotation: phosphorylation of β-catenin by Nmo modulates E-Cad–β-catenin complex activity to modify adhesion and junction–cytoskeleton interactions at the R3/R4 (Mirkovic et al. 2011). Although genetic analysis did not indicate an absolute requirement for Nmo in the R4, stabilizing Nmo may provide a mechanism that contributes to linking PCP signaling to junction regulation—though how asymmetric Nemo then directs rotation in the correct orientation remains unclear. In PCP mutations that disrupt Stbm/Vang localization, Nmo's function would also be disrupted, accounting for the high frequency of rotation errors observed. Observations from several years prior provide clues for applying the brakes to ommatidial rotation: the secreted glycoprotein Scabrous (Sca) suppresses Nmo activity (via an unknown mechanism) to prevent overrotation (Chou and Chien 2002). In an interesting twist, Sca is generated by cells within the MF and delivered to rotating ommatidia by long cellular extensions (Chou and Chien 2002).

Multiple Roles: Egfr Signaling

Careful analysis of the locus *roulette*—initially described alongside Nemo for its role in ommatidial rotation (Choi and Benzer 1994)—revealed this to be an allele of Argos, an antagonist of Egfr signaling (Brown and Freeman 2003; Gaengel and Mlodzik 2003; Strutt and Strutt 2003). Extensive genetic data reflect that Egfr activity affects rotation via a variety of mechanisms including modulating cytoskeletal or junction-associated proteins including the Afadin Canoe (Cno), the Septin Peanut, Profilin, Capulet, Zip, Rho GTPase Mtl, and the core PCP protein Fmi (Escudero et al. 2007; Gaengel and Mlodzik 2003; Munoz-Soriano et al. 2011). These mechanisms are independent of Raf/MAPK activity, but transcriptional targets of Egfr may also be involved (Gaengel and Mlodzik 2003). When it comes to understanding the role of Egfr activity as a multitasking regulator of ommatidial rotation, we have merely scratched the surface.

Myosin II: Generating Force

Though Zip was isolated in a screen for modifiers of an Nmo-overexpression phenotype (*zip* suppressed ectopic Nmo), subsequent genetic assays and the observation that Zip was unaffected in *nmo*$^{-}$ clones lead the authors to conclude that Zip and Nmo functioned independently (Fiehler and Wolff 2007). Zip accumulates together with F-actin at the interface between rotating ommatidia and the surrounding undifferentiated IPCs and decreasing Zip activity reduced the rate of rotation (Escudero et al. 2007; Fiehler and Wolff, 2007). The Actomyosin band likely generates forces that power ommatidial rotation.

Two current models go toward explaining how the band of Actin and Myosin is generated around ommatidia. In the first, RhoA and its effector Drok (Rho-associated protein kinase) activate the regulatory Myosin II light chain (encoded by *spaghetti squash, sqh)* that targets Myosin II to the apical cell cortex (Winter et al. 2001). This RhoA-Drok-Myosin II pathway is proposed to function downstream of signaling pathways including PCP (Verdier et al. 2006; Winter et al. 2001), wingless (Fiehler and Wolff 2007), and Egfr signaling (Escudero et al. 2007). A second (untested) model arises from detailed analyses of the interface between *ed* mutant and wild-type cells in epithelia: an Actomyosin cable was detected at boundaries of *ed*$^{-}$ and wild-type cells, where Ed (a Nectin-like protein) failed to accumulate (Chang et al. 2011; Wei et al. 2005). These data implicated Ed in repressing Actomyosin assembly. Consistent with these observations, lower levels of Ed are detected at AJs that line the boundaries between ommatidial clusters and the surrounding IPCs (Fetting et al. 2009; Ho et al. 2010). This is precisely where Zip accumulation is observed (Fiehler and Wolff 2007).

Regulating Adhesion During Rotation: Ed and Fred

Homophilic adhesions between Ed and a second Nectin-like protein Fred are also important in facilitating and then slowing ommatidial rotation. During the initial stages of rotation, *ed* and *fred* expression are highest in IPCs and R-cells, respectively. Homophilic interactions are proposed to segregate R-cells from IPCs and generate a field of differential adhesion that promotes ommatidial clustering and rotation. At the same time, Ed binds AP-2 that is suggested to promote endocytosis of Fmi and Egfr in IPCs and further enhance ommatidial rotation (Ho et al. 2010). Then, when the R1/R6/R7 precursors, that contain high levels of both Ed and Fred, are recruited, homophilic adhesion is proposed to prevent ommatidia from rotating beyond 90°—by increasing adhesion between ommatidia and the surrounding IPCs (Fetting et al. 2009). The fly Afadin Cno is likely to further stabilize these ommatidial "brakes" by linking Ed/Fred to the cytoskeleton: Afadins are cytoplasmic adaptor proteins that interact with Nectins and the Actin-binding protein α-actinin to stabilize adhesion (Ooshio et al. 2007; Takahashi et al. 1999). As expected, *cno* mutant ommatidia overrotate (Fetting et al. 2009; Gaengel and Mlodzik 2003). Cno is an effector of the small GTPase Rap1 as well as Egfr activity (O'Keefe et al. 2009) which regulates ommatidial rotation independently of Nmo (Brown and Freeman 2003; Gaengel and Mlodzik 2003; Strutt and Strutt 2003).

A complex picture of interacting molecular mechanisms that together regulate ommatidial rotation is beginning to take shape. In Fig. 3c, I present a tentative interaction map of factors regulating ommatidial rotation. These include PCP and Egfr signaling, dynamic adhesion, and cytoskeletal regulation. Understanding how these elements are cohesively integrated is a major challenge.

The Pupal Eye: Generating a Distinctive Pattern

The non-neuronal epithelial cells rearrange to occupy precise niches during pupal development (Fig. 1e; Cagan and Ready 1989). Following recruitment of a pair of 1° cells that encircle each ommatidium, the remaining sea of undifferentiated cells is rearranged. Directed movements drive IPCs from multiple into single rows that separate each ommatidium. This directed cell intercalation, coupled with apoptotic removal of surplus cells, leaves nine cells and three bristle groups organized as a hexagon about each ommatidium. The apical surfaces of these tall columnar epithelial cells adopt distinctive shapes. The corner 3° cells become hexagonal and 2° cells stretch to occupy rectangular territories between each hexagon vertex (Fig. 1e). In the very center, the four cone cells form a stereotypical arrangement surrounded by two banana-shaped 1°s. The mechanisms that drive cells into specific niches and shapes have garnered significant attention in recent years. It turns out that several differentially expressed transmembrane adhesion proteins are essential to correctly organize cells into specific niches, excess cells are removed by apoptosis, and the principles of minimal free energy can explain many cell shapes. Many components that regulate these events are, however, still unclear.

IgSF Proteins: Preferential Adhesion Between Different Receptors

Of primary importance are several *Drosophila* immunoglobulin super family (IgSF) adhesion proteins, orthologs of mammalian Nephrin, and Neph1. These include Hibris (Hbs) and Roughest (Rst) and, in a less dominant role, Sticks and Stones (Sns) and Kin-of-Irre (Kirre; Bao and Cagan 2005; Bao et al. 2010; Reiter et al.

1996). When Hbs–Rst and Sns–Kirre complexes are disrupted, the eye is incorrectly patterned.

The pupal eye is readily available for live imaging and in one such study *rst*$^{-}$ IPCs were observed to move (slowly) within the unpatterned epithelium but not reach or maintain the appropriate niches (Fig. 4a, b; Larson et al. 2008). In one explanatory model, expansion of the cone and 1° cell core pushes IPCs into single file. In a second model, intercalation of grouped IPCs is induced by core expansion as well as engagement of the Actin remodeling machinery that generates powerful cell extensions that probe between neighboring cells (Fig. 4; Johnson et al. 2011). These extensions are generated apically and directed toward target 1°s (Cagan 2009; Cagan and Ready 1989). If successful in reaching the target, the cell extension is secured by rapidly generated IgSF complexes that spread laterally to extend and secure the new cell niche and drive cell intercalation to completion. Indeed, reducing Rst or Hbs leaves many IPCs in multiple rows and when either Sns or Kirre are reduced in addition, the effect is greatly enhanced (Bao and Cagan 2005; Bao et al. 2010). At present, we believe that Rst/Hbs complexes are stabilized and maintained by cytoplasmic proteins including Cindr and PICK1 that couple junctions to the cytoskeleton or regulate junction turnover (Hohne et al. 2011; Johnson et al. 2012).

Two key mechanisms drive the central role of IgSF proteins in stabilizing intercalating cells: *trans* heterophilic interactions are favored over homophilic IgSF interactions, and the expression patterns of partner IgSFs evolve into complementary domains (Fig. 4c, d; Bao and Cagan 2005; Bao et al. 2010; Johnson et al. 2012; Reiter et al. 1996). Unpatterned IPCs (pre-24 h APF) express all IgSFs but coinciding with the period of cell intercalation, Rst and Kirre are removed from 1° cells while expression in IPCs is retained. At the same time, Hbs and Sns are removed from IPCs and retained in 1°s. The result is differential expression of IgSF partners and complementary localization at apical IPC–1° cell boundaries to form heterophilic junctions that drive adhesion between the different cell types (coined preferential adhesion; Bao and Cagan 2005). These mechanisms also account for separation of ommatidia by a row of IPCs—in *IgSF* mutant retina the IPCs are frequently excluded from neighboring ommatidia: 1° cells, now equally competent to bind IPCs and 1°s, frequently interact directly with 1°s of neighboring ommatidia (Bao et al. 2010).

The morphology of IPCs in the hours immediately following intercalation further demonstrates a preference for heterophilic interactions between IgSF proteins. The IPCs form rounded or scalloped contacts with neighboring 1°s that are proposed to maximize cell contacts (Fig. 1e; Bao et al. 2010). In the absence of either Rst or Hbs, cells fail to scallop. Conversely, ectopic expression of either IgSF in single IPCs induces that cell to expand its niche, invading the territory of neighboring cells (Fig. 4f; Bao and Cagan 2005; Bao et al. 2010).

As the eye matures, scalloping is reduced to render smooth boundaries between IPCs and 1° cells and the apical cell profiles adopt characteristic shapes. Several mechanisms likely interact to generate the mature pattern. First, gradual expansion of the apical volume of 1° cells would contribute to the round shape of the ommatidial core (Larson et al. 2010). At the same time, an increasingly dense and well-organized supporting Actin cytoskeleton traverses the "width" of 1° and interommatidial cells

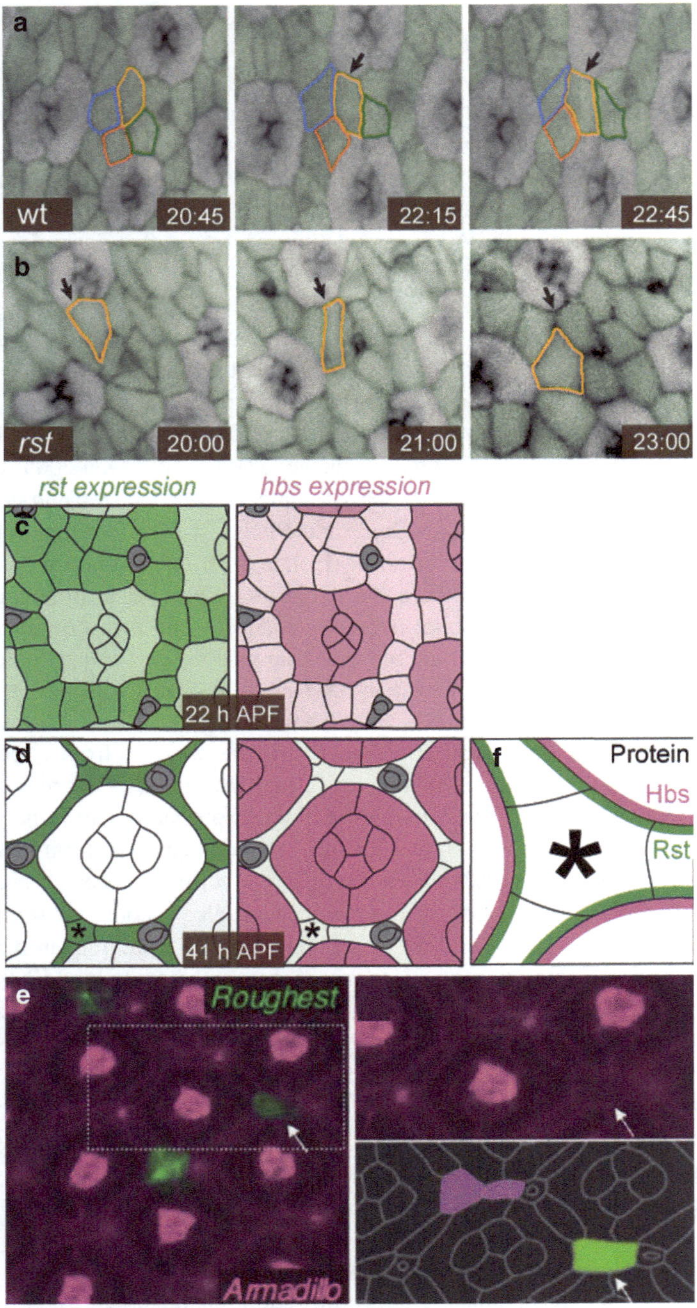

Fig. 4 Interactions between Rst and Hbs drive intercalation and cell stabilization. **a** Selected frames from a movie of wild-type retina imaged from ~20 h after pupation (APF), during the process of interommatidial pigment precursor cell (IPC) intercalation. Time points (APF) are indicated on each frame. IPCs are *pseudocolored green* and select cells outlined in color for emphasis: note intercalation of the *yellow* and *green* cells and immediate lateral widening of the *yellow* cell on

and accumulates densely at the zonula adherens (Johnson et al. 2008, 2011). Additionally, forces generated by interacting Myosin may "lengthen" IPC–1° boundaries (this hypothesis is untested, but scalloping is easily observed in images of mature *zip*$^{-}$ retina; Fiehler and Wolff 2007). Finally, the landscape of differentially expressed IgSFs is underlain by uniform localization of E-Cad at all cell contacts. The amount of E-Cad detected at AJs increases mildly corresponding with the release of cell scalloping—homophilically interacting E-Cad may generate stronger adhesive forces to override those generated by IgSF complexes .

The IgSF proteins must be carefully regulated for the correctly patterned eye to emerge. Complementary display of partner IgSFs on the surface of neighboring cell types evolves as IPCs intercalate. This is likely achieved through temporally regulated gene transcription and degradation of perduring proteins. Our limited understanding of these events is informed by two separate studies. First, we have observed that excessively motile and unstable cells that move from IPC to 1° cell niches or vice versa following cell intercalation continue to express the IgSF characteristic of the former niche even late in development (Johnson et al. 2012). These observations suggest that the instructions that limit IgSF expression are delivered during a discrete time period. The nature and origin of these instructions have not yet been described, though *hbs* is a target of Notch activity (Krejci et al. 2009) likely stimulated by neighboring cone cells that express *Delta* to recruit two 1° cells (Nagaraj and Banerjee 2007) 6–8 h before *hbs* expression is restricted from IPCs. It is conceivable that once the 1°s have surrounded the cone cells, the remaining IPCs are excluded from Delta (Dl) signaling. The mechanism that restricts *rst* expression to IPCs is not known, though decapentaplegic (Dpp) activity is required for Rst (Cordero et al. 2007). Second, when *Hbs* was expressed in IPCs, numerous Rst–Hbs complexes were detected in cytoplasmic puncta likely to be endocytic vesicles (Cagan 2011). This observation suggested that *cis*-interacting Rst–Hbs complexes are removed from cell membranes, leaving only IgSF complexes that have formed in *trans*. These events provide an efficient mechanism for removing perdurant complexes after transcription is resolved to adjacent complementary cell types.

How are mature IgSF proteins and complexes regulated? In tissue mutant for the adaptor protein *cindr*, IgSF complexes remain at IPC–IPC membranes even while transcription is restricted into the correct complementary patterns (Johnson et al. 2008, 2012). Similar phenotypes were observed when endocytosis was inhibited or when the formation of IgSF *trans* complexes was inhibited (by reducing expression

reaching its target ommatidium. **b** Cells in rst mutant retina successfully generate cell extensions (note *yellow* cell) to effect cell intercalation, but fail to maintain stable adhesions with ommatidia (compare *arrows* with those in panel **a**) and fail to maintain stable niches. **c, d** Illustrations depicting *rst* and *hbs* expression at 24 and 41 h APF. **e** Illustration of localization of Rst and Hbs at 41 h APF at primary pigment cell (1°)–IPC cell boundaries. *Asterisk* labels a tertiary pigment cell (3°). **f** Ectopic expression of Rst in a single IPC enables that cell to maximize its adhesion to neighboring 1°s and expand its niche. Panels on *right* show one of these cells at higher magnification. (Panel **a** is from Johnson et al. 2011; **b** from Larson et al. 2008; **c–e** from Johnson et al. 2012; **f** from Bao and Cagan 2005) of partner IgSF proteins). Cindr interacted with both Rst and Hbs, but interactions with Sns and Kirre were not convincingly observed. While not conclusive, the data suggested that Cindr functioned like its mammalian counterpart Cd2ap to mediate IgSF-complex interactions with the Actin cytoskeleton (reviewed by Faul et al. 2007) while also recapitulating the function of the ortholog Cin85 in targeting Rst and Hbs to the endocytic pathway (Tossidou et al. 2010).

Several other cytoplasmic factors that regulate the IgSFs have been identified. These include a BAR domain protein PICK1 and the fly ortholog of ZO-1 (Polychaetoid, Pyd). PICK1 has been implicated in stabilizing the Rst ortholog Neph1 at the plasma membrane or recycling endocytosed Neph1: reducing *dPICK1* during fly eye development reduced IgSF accumulation at AJs (Hohne et al. 2011). Pyd has been implicated in inhibiting accumulation of several AJ proteins including the IgSFs (Seppa et al. 2008). Reducing either *dPICK1* or *pyd* disrupted IPC patterning. These data underscore the importance of mechanisms that stabilize IgSFs at AJs for correct tissue patterning.

Cadherins: Adhesion Segregates and Patterns Cone Cells

Unlike E-Cad, which is ubiquitously detected at all AJs in the eye neuroepithelium, N-Cad is specific to the four cone cells and particularly important for their assembly into their final arrangement (Hayashi and Carthew 2004). Both E-Cad and N-Cad generate homophilic adhesion complexes across neighboring membranes, exemplified by localization of N-Cad only at cone cell–cone cell boundaries (Fig. 5a; Hayashi and Carthew 2004). This differential adhesion segregates the cone cells from the 1°s. The shapes and arrangement of mature cone cells are very distinctive (Fig. 5b). Mathematical models that explain this arrangement consider that strong homophilic adhesions work to model the elastic cone cell membranes into shapes that minimize their surface area and free energy and maximally expand favorable homophilic cell–cell interfaces (Hilgenfeldt et al. 2008; Kafer et al. 2007). The process is similar to the behavior of groups of soap bubbles that minimize surface contact with the surrounding water but maximize contacts between neighboring bubbles. Accordingly, cone cell–cone cell boundaries are straight and this configuration determines the shape of the remaining apical circumference of the cell, which is also constrained by the amount of the remaining apical membrane. This analogy was confirmed by genetically manipulating the cone cell number: in each case cone cells arranged in configurations that matched the arrangement of soap bubbles (Fig. 5c; Hayashi and Carthew 2004). So removing N-Cad from one cone cell reduces adhesion affinity and the length of those straight junctions that still form at the interface with wild-type cone cell neighbors. At the same time, the remaining apical membrane—the interface between the *N-Cad*$^-$ cone cell and neighboring 1° cell—expands (Fig. 5e; Hayashi and Carthew 2004). The Carthew group have also considered the contribution of N-Cad dynamics to cone cell geometries and included in their modeling parameters that took into account: (1) recycling of unbound N-Cad back to cone cell–cone cell

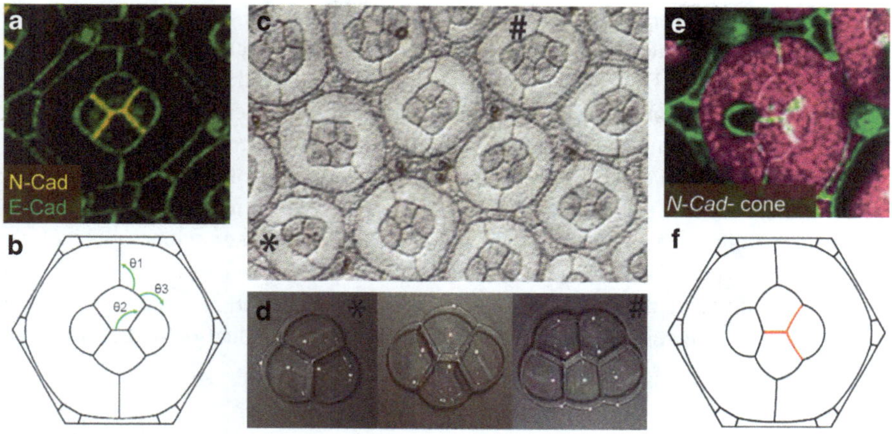

Fig. 5 N-Cadherin is essential for cone cell patterning. **a** A single ommatidium, stained for E-Cad (*green*) and N-Cad (*red*, colocalization results in *orange* color). **b** The primary pigment cell (1°) and cone cells arrange in distinctive geometries. Hilgenfeldt et al. (2008) found the average angles of θ1, θ2, and θ3 to be 109 ± 6°, 118 ± 6°, and 130 ± 9°, respectively. **c, d** These angles are recapitulated in four adherent soap bubbles and changed when the number of bubbles or cone cells is modified (* and #). Tissue in **c** is stained with cobalt sulfide. **e** An ommatidia with one N-Cad mutant cell (lack of *magenta*). **f** The configuration of this cone cell is modified, as correctly predicted by a simulation that included parameters predicting recycling of unbound N-Cad back to the membranes labeled in *red*. (Images **a** and **b** are modified from Hilgenfeldt et al. 2008; **c** and **d** are from Hayashi and Carthew 2004; **e** and **f** are from Gemp et al. 2011)

boundaries or (2) degradation of unbound N-Cad (Gemp et al. 2011). The geometries of actual mosaic *N-Cad*$^-$ cone cell quartets were more like those predicted by simulations that included recycling parameters suggesting that this dynamic process contributes to cone cell arrangement (Fig. 5f) .

These beautiful computer models explain the final configuration of cone cells and their relationship to the surrounding 1°s. But the story is more complex. In early pupal stages, the anterior and posterior cone cells, rather than the polar and equatorial, are in direct contact (see Fig. 1e). Precisely what orchestrates the switching of these cone cell contacts is still unknown, though the IgSF Hbs plays a role (Grillo-Hill and Wolff 2009).

Adhesive Junctions Moonlighting as Signaling Centers

In this chapter, I have considered the contribution of the AJs to the morphogenesis of the eye. However, many adhesion receptors and proteins that localize to AJs have roles in signaling. The IgSF Hbs, for example, interacts with Presenilin (Psn) to promote Notch cleavage and signaling in a multitude of tissues (Singh and Mlodzik 2012). In the eye, this relationship is important for photoreceptor differentiation and Hbs was observed at R-cell membranes surrounding the rotating ommatidia; whether

Hbs modifies Notch signaling to affect pupal eye patterning is likely but has not yet been studied. Notch activity is also regulated by Pyd (ZO-1), which binds the E3 ligase Suppressor of Deltex (Su(dx); Djiane et al. 2011). The regulatory nature of this relationship appears to be tissue specific but was not studied in the eye, though in a separate study Pyd inhibited the accumulation of AJ components and may similarly disrupt Notch (Seppa et al. 2008). These two are examples of "active" interactions between junction and signaling proteins, but the interactions may also be "passive." Apart from Notch, other signaling molecules residing apically or at junctions include the Egfr and atypical Cadherins Ds and Ft that affect planar polarity and also signal to the Hippo pathway to regulate organ size (Staley and Irvine 2012). It stands to reason that when adhesive cell junctions are disrupted, signal transduction and correct eye patterning would be too .

Perspective

We are now in an age when using *Drosophila* as a tool for translational research is gaining attention and acceptance. Studying defects, especially in the adult eye, provides an easily scorable assay for a variety of genetic and pharmacological manipulations (e.g., Vidal et al. 2005). However, the underlying biology that generates these visible defects is varied. Dissection of larval or pupal stages can reveal a plethora of developmental defects that we do not yet fully understand. Many of these defects are indicative of disrupted adhesion and cell movement, and disrupted signaling consequent to these. Dissecting the molecular nature of these phenotypes will provide a more informative context for increasingly popular translational studies.

We do not yet fully understand how activity at the AJs determines the pattern and shape of the fly eye. The list of transmembrane adhesion proteins that are instructive in shaping the eye—E-Cad, N-Cad, Rst, Hbs, Ed, Fred—may yet be incomplete. The next decade will undoubtedly prove fruitful in elucidating the signals that regulate adhesion proteins, the cytoplasmic proteins with which they interact, and interactions with the dynamic cytoskeleton.

Acknowledgments I am grateful to Mark Hellerman, Ursula Weber, and Jun Wu for very helpful comments on this chapter.

References

Bao S, Cagan R (2005) Preferential adhesion mediated by Hibris and Roughest regulates morphogenesis and patterning in the *Drosophila* eye. Dev Cell 8:925–935

Bao S, Fischbach KF, Corbin V, Cagan RL (2010) Preferential adhesion maintains separation of ommatidia in the *Drosophila* eye. Dev Biol 344:948–956

Brown KE, Freeman M (2003) Egfr signalling defines a protective function for ommatidial orientation in the *Drosophila* eye. Development 130:5401–5412

Brown KE, Baonza A, Freeman M (2006) Epithelial cell adhesion in the developing *Drosophila* retina is regulated by Atonal and the EGF receptor pathway. Dev Biol 300:710–721
Cagan R (2009) Principles of *Drosophila* eye differentiation. Curr Top Dev Biol 89:115–135
Cagan RL (2011) The *Drosophila* nephrocyte. Curr Opin Nephrol Hypertens 20:409–415
Cagan RL, Ready DF (1989) The emergence of order in the *Drosophila* pupal retina. Dev Biol 136:346–362
Chang LH, Chen P, Lien MT, Ho YH, Lin CM, Pan YT, Wei SY, Hsu JC (2011) Differential adhesion and actomyosin cable collaborate to drive Echinoid-mediated cell sorting. Development 138:3803–3812
Choi KW, Benzer S (1994) Rotation of photoreceptor clusters in the developing *Drosophila* eye requires the nemo gene. Cell 78:125–136
Chou YH, Chien CT (2002) Scabrous controls ommatidial rotation in the *Drosophila* compound eye. Dev Cell 3:839–850
Chu D, Pan H, Wan P, Wu J, Luo J, Zhu H, Chen J (2012) AIP1 acts with cofilin to control actin dynamics during epithelial morphogenesis. Development 139:3561–3571
Cooper MT, Bray SJ (1999) Frizzled regulation of Notch signalling polarizes cell fate in the *Drosophila* eye. Nature 397:526–530
Cordero JB, Larson DE, Craig CR, Hays R, Cagan R (2007) Dynamic Decapentaplegic signaling regulates patterning and adhesion in the *Drosophila* pupal retina. Development 134:1861–1871
De Graeve FM, Van de Bor V, Ghiglione C, Cerezo D, Jouandin P, Ueda R, Shashidhara LS, Noselli S (2012) *Drosophila* apc regulates delamination of invasive epithelial clusters. Dev Biol 368:76–85
Djiane A, Shimizu H, Wilkin M, Mazleyrat S, Jennings MD, Avis J, Bray S, Baron M (2011) Su(dx) E3 ubiquitin ligase-dependent and -independent functions of polychaetoid, the *Drosophila* ZO-1 homologue. J Cell Biol 192:189–200
Drees F, Pokutta S, Yamada S, Nelson WJ, Weis WI (2005) Alpha-catenin is a molecular switch that binds E-cadherin-beta-catenin and regulates actin-filament assembly. Cell 123:903–915
Duguay D, Foty RA, Steinberg MS (2003) Cadherin-mediated cell adhesion and tissue segregation: qualitative and quantitative determinants. Dev Biol 253:309–323
Escudero LM, Bischoff M, Freeman M (2007) Myosin II regulates complex cellular arrangement and epithelial architecture in *Drosophila*. Dev Cell 13:717–729
Fanto M, Mlodzik M (1999) Asymmetric Notch activation specifies photoreceptors R3 and R4 and planar polarity in the *Drosophila* eye. Nature 397:523–526
Faul C, Asanuma K, Yanagida-Asanuma E, Kim K, Mundel P (2007) Actin up: regulation of podocyte structure and function by components of the actin cytoskeleton. Trends Cell Biol 17:428–437
Fetting JL, Spencer SA, Wolff T (2009) The cell adhesion molecules Echinoid and Friend of Echinoid coordinate cell adhesion and cell signaling to regulate the fidelity of ommatidial rotation in the *Drosophila* eye. Development 136:3323–3333
Fiehler RW, Wolff T (2007) *Drosophila* Myosin II, Zipper, is essential for ommatidial rotation. Dev Biol 310:348–362
Furuse M, Tsukita S (2006) Claudins in occluding junctions of humans and flies. Trends Cell Biol 16:181–188
Gaengel K, Mlodzik M (2003) Egfr signaling regulates ommatidial rotation and cell motility in the *Drosophila* eye via MAPK/Pnt signaling and the Ras effector Canoe/AF6. Development 130:5413–5423
Gemp IM, Carthew RW, Hilgenfeldt S (2011) Cadherin-dependent cell morphology in an epithelium: constructing a quantitative dynamical model. PLoS Comput Biol 7:e1002115
Goodrich LV, Strutt D (2011) Principles of planar polarity in animal development. Development 138:1877–1892
Grillo-Hill BK, Wolff T (2009) Dynamic cell shapes and contacts in the developing *Drosophila* retina are regulated by the Ig cell adhesion protein hibris. Dev Dyn 238:2223–2234

Gumbiner BM (2005) Regulation of cadherin-mediated adhesion in morphogenesis. Nat Rev Mol Cell Biol 6:622–634

Hayashi T, Carthew RW (2004) Surface mechanics mediate pattern formation in the developing retina. Nature 431:647–652

Hilgenfeldt S, Erisken S, Carthew RW (2008) Physical modeling of cell geometric order in an epithelial tissue. Proc Natl Acad Sci U S A 105:907–911

Ho YH, Lien MT, Lin CM, Wei SY, Chang LH, Hsu JC (2010) Echinoid regulates Flamingo endocytosis to control ommatidial rotation in the *Drosophila* eye. Development 137:745–754

Hohne M, Lorscheider J, Bardeleben A von, Dufner M, Scharf MA, Godel M, Helmstadter M, Schurek EM, Zank S, Gerke P et al (2011) The BAR domain protein PICK1 regulates cell recognition and morphogenesis by interacting with Neph proteins. Mol Cell Biol 31:3241–3251

Huber AH, Weis WI (2001) The structure of the beta-catenin/E-cadherin complex and the molecular basis of diverse ligand recognition by beta-catenin. Cell 105:391–402

Johnson RI, Seppa MJ, Cagan RL (2008) The *Drosophila* CD2AP/CIN85 orthologue Cindr regulates junctions and cytoskeleton dynamics during tissue patterning. J Cell Biol 180:1191–1204

Johnson RI, Sedgwick A, D'Souza-Schorey C, Cagan RL (2011) Role for a Cindr-Arf6 axis in patterning emerging epithelia. Mol Biol Cell 22:4513–4526

Johnson RI, Bao S, Cagan RL (2012) Interactions between *Drosophila* IgCAM adhesion receptors and Cindr, the Cd2ap/Cin85 ortholog. Dev Dyn 241:1933–1943

Kafer J, Hayashi T, Maree AF, Carthew RW, Graner F (2007) Cell adhesion and cortex contractility determine cell patterning in the *Drosophila* retina. Proc Natl Acad Sci U S A 104:18549–18554

Knust E (2007) Photoreceptor morphogenesis and retinal degeneration: lessons from *Drosophila*. Curr Opin Neurobiol 17:541–547

Krejci A, Bernard F, Housden BE, Collins S, Bray SJ (2009) Direct response to Notch activation: signaling crosstalk and incoherent logic. Sci Signal 2:ra1

Kumar JP (2012) Building an ommatidium one cell at a time. Dev Dyn 241:136–149

Larson DE, Liberman Z, Cagan RL (2008) Cellular behavior in the developing *Drosophila* pupal retina. Mech Dev 125:223–232

Larson DE, Johnson RI, Swat M, Cordero JB, Glazier JA, Cagan RL (2010) Computer simulation of cellular patterning within the *Drosophila* pupal eye. PLoS Comput Biol 6:e1000841

Maung SM, Jenny A (2011) Planar cell polarity in *Drosophila*. Organogenesis 7:165–179

Mirkovic I, Mlodzik M (2006) Cooperative activities of drosophila DE-cadherin and DN-cadherin regulate the cell motility process of ommatidial rotation. Development 133:3283–3293

Mirkovic I, Gault WJ, Rahnama M, Jenny A, Gaengel K, Bessette D, Gottardi CJ, Verheyen EM, Mlodzik M (2011) Nemo kinase phosphorylates beta-catenin to promote ommatidial rotation and connects core PCP factors to E-cadherin-beta-catenin. Nat Struct Mol Biol 18:665–672

Miyashita Y, Ozawa M (2007) Increased internalization of p120-uncoupled E-cadherin and a requirement for a dileucine motif in the cytoplasmic domain for endocytosis of the protein. J Biol Chem282:11540–11548

Muller HA (2000) Genetic control of epithelial cell polarity: lessons from *Drosophila*. Dev Dyn 218:52–67

Munoz-Soriano V, Belacortu Y, Durupt FC, Munoz-Descalzo S, Paricio N (2011) Mtl interacts with members of Egfr signaling and cell adhesion genes in the *Drosophila* eye. Fly (Austin) 5:88–101

Nagaraj R, Banerjee U (2007) Combinatorial signaling in the specification of primary pigment cells in the *Drosophila* eye. Development 134:825–831

Nelson WJ (2008) Regulation of cell-cell adhesion by the cadherin-catenin complex. Biochem Soc Trans 36:149–155

Niewiadomska P, Godt D, Tepass U (1999) DE-Cadherin is required for intercellular motility during *Drosophila* oogenesis. J Cell Biol 144:533–547

O'Keefe DD, Gonzalez-Nino E, Burnett M, Dylla L, Lambeth SM, Licon E, Amesoli C, Edgar BA, Curtiss J (2009) Rap1 maintains adhesion between cells to affect Egfr signaling and planar cell polarity in *Drosophila*. Dev Biol 333:143–160

Ooshio T, Fujita N, Yamada A, Sato T, Kitagawa Y, Okamoto R, Nakata S, Miki A, Irie K, Takai Y (2007) Cooperative roles of Par-3 and afadin in the formation of adherens and tight junctions. J Cell Sci 120:2352–2365

Pacquelet A, Rorth P (2005) Regulatory mechanisms required for DE-cadherin function in cell migration and other types of adhesion. J Cell Biol 170:803–812

Reiter C, Schimansky T, Nie Z, Fischbach KF (1996) Reorganization of membrane contacts prior to apoptosis in the *Drosophila* retina: the role of the IrreC-rst protein. Development 122:1931–1940

Schwabe T, Gontang AC, Clandinin TR (2009) More than just glue: the diverse roles of cell adhesion molecules in the *Drosophila* nervous system. Cell Adh Migr 3:36–42

Seppa MJ, Johnson RI, Bao S, Cagan RL (2008) Polychaetoid controls patterning by modulating adhesion in the *Drosophila* pupal retina. Dev Biol 318:1–16

Singh J, Mlodzik M (2012) Hibris, a *Drosophila* nephrin homolog, is required for presenilin-mediated Notch and APP-like cleavages. Dev Cell 23:82–96

Staley BK, Irvine KD (2012) Hippo signaling in *Drosophila*: recent advances and insights. Dev Dyn 241:3–15

Steinberg MS (1970) Does differential adhesion govern self-assembly processes in histogenesis? Equilibrium configurations and the emergence of a hierarchy among populations of embryonic cells. J Exp Zool 173:395–433

Steinberg MS (2007) Differential adhesion in morphogenesis: a modern view. Curr Opin Genet Dev 17:281–286

Steinberg MS, Takeichi M (1994) Experimental specification of cell sorting, tissue spreading, and specific spatial patterning by quantitative differences in cadherin expression. Proc Natl Acad Sci U S A 91:206–209

Strutt H, Strutt D (2003) EGF signaling and ommatidial rotation in the *Drosophila* eye. Curr Biol 13:1451–1457

Takahashi K, Nakanishi H, Miyahara M, Mandai K, Satoh K, Satoh A, Nishioka H, Aoki J, Nomoto A, Mizoguchi A et al (1999) Nectin/PRR: an immunoglobulin-like cell adhesion molecule recruited to cadherin-based adherens junctions through interaction with Afadin, a PDZ domain-containing protein. J Cell Biol 145:539–549

Tepass U, Harris KP (2007) Adherens junctions in *Drosophila* retinal morphogenesis. Trends Cell Biol 17(1):26–35

Tepass U, Tanentzapf G, Ward R, Fehon R (2001) Epithelial cell polarity and cell junctions in *Drosophila*. Annu Rev Genet 35:747–784

Thoreson MA, Anastasiadis PZ, Daniel JM, Ireton RC, Wheelock MJ, Johnson KR, Hummingbird DK, Reynolds AB (2000) Selective uncoupling of p120(ctn) from E-cadherin disrupts strong adhesion. J Cell Biol 148:189–202

Tossidou I, Teng B, Drobot L, Meyer-Schwesinger C, Worthmann K, Haller H, Schiffer M (2010) CIN85/RukL is a novel binding partner of nephrin and podocin and mediates slit diaphragm turnover in podocytes. J Biol Chem 285:25285–25295

Verdier V, Guang Chao C, Settleman J (2006) Rho-kinase regulates tissue morphogenesis via non-muscle myosin and LIM-kinase during *Drosophila* development. BMC Dev Biol 6:38

Vidal M, Wells S, Ryan A, Cagan R (2005) ZD6474 suppresses oncogenic RET isoforms in a *Drosophila* model for type 2 multiple endocrine neoplasia syndromes and papillary thyroid carcinoma. Cancer Res 65:3538–3541

Wei SY, Escudero LM, Yu F, Chang LH, Chen LY, Ho YH, Lin CM, Chou CS, Chia W, Modolell J et al (2005) Echinoid is a component of adherens junctions that cooperates with DE-Cadherin to mediate cell adhesion. Dev Cell 8:493–504

Weis WI, Nelson WJ (2006) Re-solving the cadherin-catenin-actin conundrum. J Biol Chem 281:35593–35597

Winter CG, Wang B, Ballew A, Royou A, Karess R, Axelrod JD, Luo L (2001) *Drosophila* Rho-associated kinase (Drok) links Frizzled-mediated planar cell polarity signaling to the actin cytoskeleton. Cell 105:81–91

Xiao K, Garner J, Buckley KM, Vincent PA, Chiasson CM, Dejana E, Faundez V, Kowalczyk AP (2005) p120-Catenin regulates clathrin-dependent endocytosis of VE-cadherin. Mol Biol Cell16:5141–5151

Yamada S, Pokutta S, Drees F, Weis WI, Nelson WJ (2005) Deconstructing the cadherin-catenin-actin complex. Cell 123:889–901

Yap AS, Crampton MS, Hardin J (2007) Making and breaking contacts: the cellular biology of cadherin regulation. Curr Opin Cell Biol 19:508–514

Modulation of Developmental Signaling by the Proteostasis Network

Kristin D. Patterson and Janice A. Fischer

Introduction

Interplay between a number of cell communication pathways controls cell behaviors as the *Drosophila* larval eye disc develops into the ordered array of ommatidia seen in the adult eye (Kumar 2012; Roignant and Treisman 2009; Tsachaki and Sprecher 2012). Disruption of signaling by a change in either the levels or timing of expression of signaling pathway components, often leads to dramatic effects: the lack of eye tissue altogether, tumor growth, misdetermination of cell types, altered patterning, or degeneration (Artavanis-Tsakonas and Muskavitch 2010; Baker 2007; Herranz and Milán 2008). A limited number of signaling pathways used iteratively during the course of eye development can elicit different responses, depending on context. Organisms have evolved mechanisms to ensure that cells receive and interpret instructions about their environments, enabling them to activate a normal developmental plan. Regulation of transcription is one way by which cells modulate signaling. However, proteomic studies have shown that the levels of particular mRNAs in a cell are related only partially to the levels of protein encoded by those mRNAs; the proteome is also controlled at translation and posttranslationally (Vogel and Marcotte 2012). Signaling can be affected by posttranslational protein quality control, or proteostasis, which includes modulation of protein folding, chemical modification, trafficking, and degradation by the proteostasis network (Powers et al. 2009).

Cells maintain sets of proteins, their proteomes, to fulfill their functions. Most proteins achieve functional status only after folding into complex three-dimensional structures and in vitro biophysical studies have shown that proteins navigate a complex thermodynamic landscape as they explore different conformations in search of a low energy, stable structure (Hartl et al. 2011). A denatured polypeptide is able to

K. D. Patterson (✉) · J. A. Fischer
Section of Molecular Cell and Developmental Biology,
The University of Texas at Austin, Austin, TX 78712, USA
e-mail: kpatterson@austin.utexas.edu

J. A. Fischer
e-mail: janicefischer@austin.utexas.edu

A. Singh, M. Kango-Singh (eds.), *Molecular Genetics of Axial Patterning, Growth and Disease in the Drosophila Eye*, DOI 10.1007/978-1-4614-8232-1_8,

fold into its functional form without outside assistance under laboratory conditions. However, protein folding in the cell is normally more challenging, especially under conditions of stress. For one, translation is error prone, which means that on average, 15 % of proteins have at least one amino acid substitution, and most such errors affect folding. Second, the crowded space inside a cell provides opportunities for proteins to aggregate in nonproductive interactions instead of assuming a functional structure. In addition, protein function can depend on the ability to alter conformation in response to posttranslational modification or interactions with binding partners, and therefore, protein activity can require trafficking to particular cellular compartments. Exposure to stress conditions such as heat, changes in pH, and reactive oxygen species (ROS) affects protein folding as well (Gidalevitz et al. 2011; Powers et al. 2009). Given all the challenges to the attainment of functional protein structure, it is not surprising that organisms invest in tools that ensure protein quality, the proteostasis network. Signaling protein homeostasis is achieved in large part by two processes: (1) protein folding assisted by endoplasmic reticulum (ER) chaperones and modification enzymes and (2) Ubiquitin-directed trafficking and degradation.

Chaperones

Molecular chaperones are proteins that assist other proteins to achieve their native fold without themselves becoming part of the structure (Vabulas et al. 2010). Many intercellular signaling proteins are synthesized, folded, and trafficked in the ER secretory system. A protein translated in the ER is retained by chaperones to provide time for posttranslational modifications, such as glycosylation and disulfide linkages, that influence folding of many secreted proteins. Because accumulation of misfolded proteins in the ER (ER stress) is detrimental to cell function, proteins in the ER monitor folding. Persistent misfolding results in retrotranslocation to the cytoplasm and degradation by the proteasome in a process called ER-associated degradation (ERAD). ER chaperones, modifying enzymes, and ERAD reduce accumulation of unfolded proteins in the ER by ensuring that ER proteins are either folded or degraded. When the capacity of the ER to secrete or degrade proteins is overloaded, the unfolded protein response (UPR) is activated. The UPR is a cellular adaptation that decreases stress by increasing protein folding, trafficking, and degradation capability, and decreasing protein synthesis. Under long-term ER stress, a proapoptotic pathway may be activated (Araki and Nagata 2012; Liu and Ye 2011).

Ubiquitin

The Ubiquitin-Proteasome system, along with a contribution from the lysosome and autophagy, is the proteolysis arm of the protein homeostasis machinery. Ubiquitination is a posttranslational modification in which the small protein Ubiquitin (Ub) is attached covalently to a lysine residue in a substrate protein through a three-step enzymatic process: (1) E1-activating enzyme charges an E2-conjugating enzyme with Ub; (2) an E3 ligase enzyme binds to substrate protein and serves as an adapter between the E2 and substrate; (3) the Ub moiety is transferred directly or indirectly through the E3, to the ε-amino group of the substrate lysine (Varshavsky 2012). Ubiquitination is a highly flexible modification for three reasons. First, large numbers of E2 and E3 enzymes allow for developmental regulation and substrate specificity. In the fly genome, there are at least 34 E2s and 207 E3s (Du et al. 2011). Second, the ubiquitination reaction is reversible through the action of substrate-specific deubiquitinases (DUBs). Finally, Ub can modify a protein in several ways: a protein can be tagged with one Ub on one lysine residue (monoubiquitination), several substrate lysines can be modified with one Ub each (multi-monoubiquitination), or a single substrate lysine can have a chain of Ub moieties attached through lysines found in the Ub sequence (polyubiquitination). Each type of Ub linkage varies in structure and may be recognized by particular versions of protein motifs such as the Ubiquitin interacting motif (UIM) and the Ubiquitin-binding domain (UBD). Some Ub linkages target proteins to the proteasome for degradation. However, not all Ub modification leads to proteasomal degradation (Chen and Sun 2009; Kulathu and Komander 2012). Signaling activity in the fly eye is influenced by Ub-mediated endocytosis and trafficking of many signaling molecules.

Proteostasis in Specialized Cells

Studies of stressed cells, for example, during aging and disease, have highlighted the importance of protein homeostasis in maintaining a healthy proteome (Taylor and Dillin 2011). In a developing multicellular context, two additional factors must be considered. First, specialized cells must maintain specialized proteomes, suggesting that the various mechanisms of protein quality control are differentially active in different cell types. Furthermore, during early stages of development, cells change their protein composition depending on time and location as they become determined and eventually differentiate into a specific cell type. The almost crystalline regularity of the adult eye and its function in vision is the result of fine regulation of and interaction between a few well-studied and evolutionarily conserved signaling pathways, including the Notch, epidermal growth factor receptor (EGFR), and rhodopsin (Rh) pathways. Detailed summaries of the function of each of these signaling pathways in eye development can be found elsewhere (Fortini 2009; Kumar 2012; Nagaraj and Banerjee 2004; Roignant and Treisman 2009; Shieh 2011; Sundaram 2005; Tsachaki and Sprecher 2012). The proteostasis machinery is an efficient mechanism for altering the proteome in response to developmental signals. We review evidence from the Notch, EGFR, and Rh pathways that protein homeostasis plays a unique and substantial role in enabling the iterative use of signaling pathways in the confines of the developing *Drosophila* eye and in adult vision.

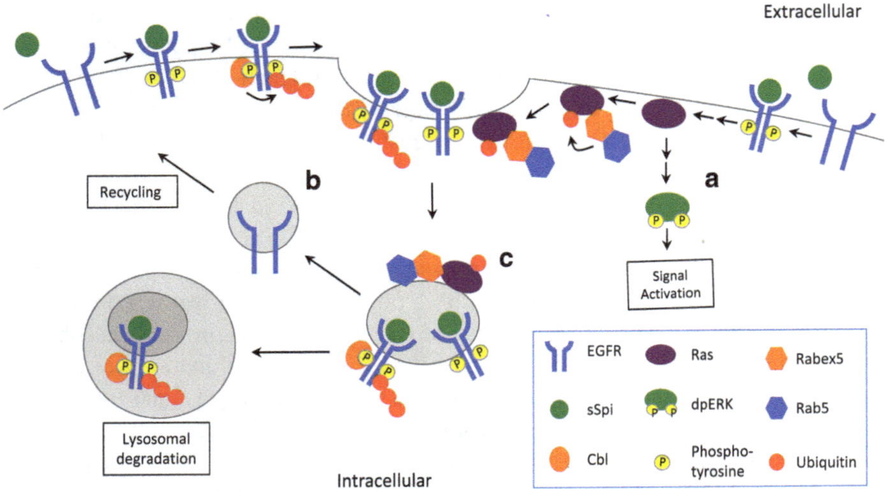

Fig. 1 Downregulation of EGFR signaling by Ub-mediated endocytosis and endosomal sorting. **a** The EGFR is a receptor tyrosine kinase. Upon activation by sSpi, signal is transduced through Ras to double phosphorylate ERK resulting in transcription modulation. *Double arrows* indicate that some steps in the pathway are not shown. **b** Desensitization of the EGFR is achieved through endocytosis and endosomal sorting. The Ub E3 ligase Cbl binds to EGFR at the cell surface. Ubiquitination of EGFR by Cbl leads to endosomal sorting in favor of lysosomal degradation. Receptors without Ub are recycled to the cell surface. **c** Ubiquitination of Ras by the E3 ligase Rabex-5 results in movement of Ras from the cell surface to endosomes and decreased Ras activity. The next steps of ubiquitinated Ras trafficking are unknown

The EGFR Signaling Cycle

Binding of the EGFR, a receptor tyrosine kinase, by the ligand Spitz (sSpi) activates the Ras signal transduction pathway. The Ras GTPase cycles between active and inactive forms through the action of guanine nucleotide exchange factors (GEFs) that catalyze the association of Ras with GTP and GTPase-activation proteins (GAPs) that return Ras to the inactive GDP-bound form. Downstream of active Ras, the MAP kinase Erk (rolled in *Drosophila*) becomes active through double phosphorylation (dpErk), and the function of dpErk results in transcriptional regulation through degradation of the repressor Yan and activation of the positive regulator Pointed (Pnt; Brunner et al. 1994; O'Neill et al. 1994; Fig. 1a).

Iterative activation of the EGFR is required for the proliferation, survival, and differentiation of all the cell types that make up a facet—photoreceptors (except R8), cone cells, and pigment cells (Freeman 1996). The morphogenetic furrow is a visible boundary between an undifferentiated epithelium in the anterior portion of the larval eye disc and differentiating tissue to the posterior. The first cell recruited to the developing facet posterior to the furrow is the R8 photoreceptor. The R8 begins to produce the secreted EGFR ligand sSpi. Cells adjacent to R8 transduce high levels of EGFR signal through Ras and begin to differentiate to make a five-cell cluster, including R8 and R2-R5. As cells differentiate, they produce more sSpi that diffuses farther and recruits more cells to differentiate in an outwardly expanding pattern. While high levels of EGFR activation encourage differentiation, lower levels are important in cell proliferation and survival. Cells further from R8 than R2–R5 do not initially activate the EGFR; instead, they enter a round of mitosis that is essential to form enough cells for each facet. Later, a low level of EGFR activation promotes cell division and survival of the additional cells (Baker and Yu 2001; Yang and Baker 2003). Therefore, normal eye development depends on restriction of the EGFR signal in space and the ability to control the level of signal in time. How do eye disc cells remain responsive to an EGFR signal, yet control the level and timing of response? Three proteostatic mechanisms for fine-tuning the activation and repression of EGFR signaling are discussed below.

EGFR Signal Activation by Ligand Processing

A flexible system of ligand processing, dependent on a family of Rhomboid proteases and the ER chaperone Star, activates the EGFR pathway. EGFR and Spi proteins are expressed widely in the developing eye. Spi is produced as a transmembrane protein with a single EGF repeat in the extracellular domain. Immature, full-length Spi is retained in the ER (Schweitzer et al. 1995). Intramembrane cleavage of Spi by Rhomboid proteases leads to secretion of the C-terminal half of the protein, the functional EGFR ligand (sSpi; Schweitzer et al. 1995; Urban et al. 2001). However, unprocessed Spi and the processing protease do not reside in the same subcellular compartment; Spi is found primarily in the ER and Rhomboid-1 (Rho-1), the defining member of the Rhomboid family, is localized to a late-endosome (Lee et al. 2001; Tsruya et al. 2002). Therefore, many eye disc cells are prepared to activate EGFR signaling, but compartmentalization of the ligand and protease prevents them from doing so.

To produce active EGFR ligand, Spi is trafficked by Star from the ER to Rho-1-containing endosomes (Lee et al. 2001; Yogev et al. 2010; Fig. 2a). Star was identified originally based on a haploinsufficient rough eye phenotype due to a reduction in photoreceptor number (Brown and Freeman 2003; Heberlein and Rubin 1991; Tsruya et al. 2007). Conversely, Star overexpression results in increased production of sSpi and extra photoreceptors, consistent with an expansion in EGFR signaling (Hsiung et al. 2001; Tsruya et al. 2002). Star is an example of an ER chaperone that is the limiting factor in production of a ligand and therefore activation of a signaling pathway. Therefore, it is significant that Star is also a substrate for Rhomboid proteases and neither part of cleaved Star functions as a Spi chaperone (Tsruya et al. 2007). Cleavage of Star by Rho-1 attenuates EGFR signaling by reducing the number of Star molecules available to chaperone Spi.

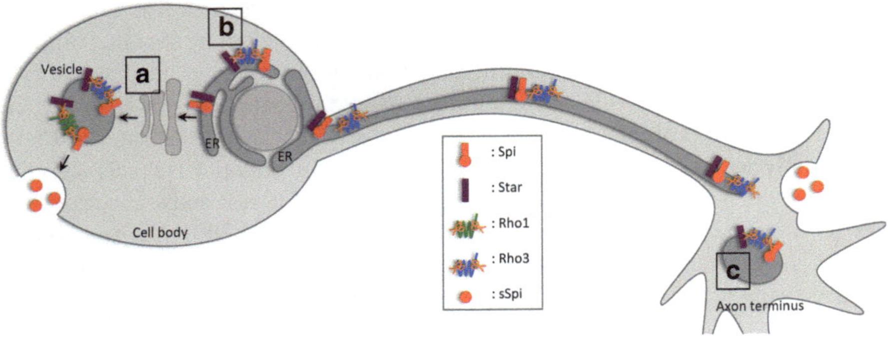

Fig. 2 Level and location of sSpi production depends on ER chaperone and protease activity. **a** Spi is trafficked from the ER to vesicles by the chaperone Star. Spi and Star are cleaved by Rho-1 and Rho-3 in the vesicle to produce sSpi and inactivate Star. **b** Rho-3 in the ER results in premature processing and retention of some Spi and Star. **c** Extension of the ER into the photoreceptor axon allows trafficking of Spi and Rho-3 to the axon terminus and production of sSpi in a second location

ER Rho-3 Reduces sSpi Secretion

Star is haploinsufficient only in the eye. The extreme sensitivity of eye development to Star activity can be explained by the presence in the developing eye of a second protease, Rhomboid-3 (Rho-3) (also known as Roughoid; Wasserman et al. 2000). Like Rho-1, Rho-3 processes Spi to its secreted form. Rho-1 and Rho-3 colocalize to the late endosome, but Rho-3 is also in the ER. Rho-3 is able to process Spi and cleave Star in the ER, before trafficking to the late-endosome (Fig. 2b). sSpi produced in the ER is retained in the ER; it is not secreted (Schlesinger et al. 2004). In addition, Rho-3 cleavage of Star in the ER reduces the amount of functional Star available to traffic full-length Spi to the late-endosomal compartment for processing and secretion (Yogev et al. 2008). Therefore, Rho-3 in the ER attenuates signaling through EGFR in the eye by reducing secretion of sSpi. In contrast, loss of Rho-3 in the eye results in excess recruitment of photoreceptors into facets, as would be expected if the EGFR were overactive (Yogev et al. 2008). Rho-3-dependent signal reduction enables a short range signal to recruit just a few cells to each ommatidium at any given time-point.

Spi Trafficking and Polarized Signaling

Spi trafficking in the ER before activation by proteolytic processing also provides a mechanism for localized signaling in polarized cells. After the outer photoreceptor neurons are determined, they extend axons basally through the optic stalk to the lamina of the brain. The photoreceptors continue to produce sSpi in the cell body for recruitment of cells to the facet and they begin to produce sSpi at the axon terminus where it is required for neurogenesis in the brain lamina. Yogev and colleagues (2010) have shown that in eye discs lacking Rho-3, photoreceptor neurogenesis appears normal (except that too many photoreceptors are produced), but sSpi is not produced at the axon terminus and neurogenesis in the lamina is blocked. The ER compartment extends throughout the axon in these cells and the ER localization of Spi and Rho-3, along with Star, provides a mechanism for the proteins to be trafficked into the axon (Fig. 2c). At the axon terminus, the proteins exit the ER and join the same type of late-endosomal compartment found in the cell body for ligand processing (Yogev et al. 2010). In this case, chaperone-mediated trafficking of Spi has two opposing effects in the polarized photoreceptor neuron: it attenuates the Rho-1-based production of sSpi in the cell body, and it permits the production of sSpi at the axon terminus.

Desensitization of the EGFR

Ligand binding results in activation of EGFR signaling followed immediately by desensitization, a temporary decrease in response to signal due to a transient decrease in the number of receptors on the cell surface. A series of studies have shown that EGFR desensitization is affected by endocytosis and endosomal sorting that leads to segregation of EGFR receptors into either of two pathways: recycling to the plasma membrane or degradation in the lysosome (Sorkin and Goh 2008). The role of Ub modification on EGFR endocytosis is unclear at this time (Haglund and Dikic 2012; Sorkin and Goh 2008), but there is substantial evidence that the EGFR is ubiquitinated by the proto-oncogene product Cbl and that Ub influences endosomal trafficking of the EGFR in favor of lysosomal degradation (Fig. 1b).

Cbl was first identified as a retroviral gene product (v-Cbl) involved in B cell lymphoma and myeloid leukemia (Langdon et al. 1989). A single Cbl-encoding gene exists in *Drosophila* (*d-cbl*; Hime et al. 1997; Meisner et al. 1997). A genetic screen conducted in *C. elegans* identified the homolog of *cbl*, called *sli-1*, as a suppressor of EGFR signaling, but the mechanism of action was unknown (Jongeward et al. 1995; Yoon et al. 1995). Subsequent work using in vitro reconstitution assays and mammalian cell culture models dissected the role of Cbl in EGFR regulation. EGFR is a substrate of Cbl-Ub ligase activity (Joazeiro et al. 1999; Levkowitz et al. 1999). EGFR remains in complex with ligand and Cbl during endosomal sorting allowing the receptor to become either multi-monoubiquitinated or polyubiquitinated with K^{63}-linked chains (Haglund et al. 2003; Umebayashi et al. 2008). Many of the proteins involved in endosomal sorting, including the hepatocyte growth factor-regulated tyrosine kinase substrate (Hrs) and the endosomal sorting complex required for transport (ESCRT) complexes, have UBDs and are involved in recruiting ubiquitinated substrates to the multivesicular body (MVB) and ultimately to the lysosome for degradation (Haglund and Dikic 2012). In mammalian cell models, ubiquitinated EGFRs are targeted to the lysosome while EGFRs without Ub modification are recycled to the plasma membrane (Levkowitz et al. 1998; Peschard and Park 2003; Umebayashi et al. 2008). Disrupting the interaction between Cbl and EGFR during endosomal sorting increases EGFR signaling and oncogenic Cbl proteins are either unable to bind EGFR or lack Ub ligase activity (Peschard and Park 2003; Umebayashi et al. 2008). Taken together, this information suggests that Cbl-mediated ubiquitination of EGFR leads to lysosomal degradation of EGFR, which limits activation of the EGFR pathway and cell growth.

D-cbl Isoforms Regulate EGFR and Notch

D-cbl modulates several signaling pathways during eye development. Genetic interaction studies suggest that D-Cbl is a negative regulator of EGFR signaling during all stages of eye development (Meisner et al. 1997; Wang et al. 2008). Mosaic analysis supports a role for D-cbl in developmental decisions controlled by EGFR signaling. For example, the ommatidia in clones lacking D-cbl function have increased numbers of photoreceptors, cone cells, and pigment cells, mimicking the over-recruitment defects seen in eye discs with overactive EGFR signaling. Also, molecular markers of EGFR activity are inversely related to D-cbl levels. At pupal stages, EGFR signaling inhibits apoptosis and apoptosis is increased in *D-cbl*$^{-}$ clones (Wang et al. 2008). More detailed analysis in eye discs showed that D-cbl mutant ommatidia were overpopulated with R7 photoreceptors and cone cells and that removing one copy of *egfr+* or *Notch+* were both able to suppress the recruitment phenotype. Mosaic analysis in the wing disc shows that loss of D-cbl upregulates Notch signaling (Wang et al. 2010). D-cbl is produced in two isoforms, D-cblS and D-cblL, respectively. Both proteins have E3 Ub ligase and EGFR-binding domains (Pai et al. 2006). In eye discs, overexpression of D-cblL downregulates EGFR signaling, but not Notch signaling. In contrast, D-cblS antagonizes Notch, but not EGFR (Wang et al. 2008; Wang et al. 2010). Therefore, differential expression of D-cbl isoforms modulates EGFR and Notch signaling. Future work in flies is needed to connect the *d-cbl*$^{-}$ phenotype with protein ubiquitination and degradation.

Downregulation of Ras by Endosomal Trafficking

Recent studies, using cell culture and the *Drosophila* eye as models, describe a mechanism of Ras downregulation by ubiquitination and endosomal sorting. Hypomorphic mutation of the *Drosophila* E1 Ub-activating enzyme (Uba1) results in an overgrowth of eye tissue due to a decrease in cell death and an increase in proliferation. Clones of eye cells lacking Uba1 have increased levels of dpERK and Pnt, both markers of EGFR activity. However, decreased EGFR levels do not rescue the *Uba1*-defects. Instead, partial rescue of cell death and proliferation abnormalities is seen when the level of Ras is reduced, suggesting that Uba1 may affect cell signaling by ubiquitinating a protein downstream of EGFR, maybe even Ras (Pfleger et al. 2007; Yan et al. 2009). Indeed, experiments in S2 cells showed that mono- and di-ubiquitinated species of Ras are present in *Drosophila* cells (Yan et al. 2009).

Rabex-5 Ubiquitinates Ras

Rabex-5 (also known as RABGEF1) is an E3 ligase that places Ub on Ras. Rabex-5 was originally identified as a GEF for Rab5, a molecule known to function in endosomal trafficking (Horiuchi et al. 1997). Rabex-5 also has Ub ligase activity through an atypical A20-type zinc finger (A20_ZF) domain (Lee et al. 2006; Mattera et al. 2006). In mammalian cells, the level of Ras ubiquitination is directly related to the level of Rabex-5 and only deleted versions of Rabex-5 proteins that contain the A20_ZF domain led to Ras ubiquitination (Xu et al. 2010). In *Drosophila* S2 cells, overexpression of Rabex led to increased Ras ubiquitination (Yan et al. 2010). Finally, Rabex-5 is able to ubiquitinate Ras in vitro, suggesting that the interaction is direct (Xu et al. 2010).

Ubiquitination of Ras changes its subcellular localization and attenuates Ras activity (Jura et al. 2006; Fig. 1c). In mammalian cells, increased Rabex-5 activity leads to increased trafficking of Ras to endosomes. The converse is also true—less Ras is associated with endosomes in cells with less Rabex-5. Furthermore, the translocation of Ras from the plasma membrane to the endosome is not dependent on the Rab-GEF activity of Rabex-5 (Xu et al. 2010). Flies in which Rabex-5 has been deleted or reduced using RNAi constructs have an impressive mutant phenotype. Deletion alleles result in the death of abnormally large larvae with melanotic tumors. Flies expressing reduced Rabex-5 survive to adulthood, but are larger than wild-type flies. Conditional downregulation of Rabex-5 accomplished by expressing RNAi in the eye results in large eyes, small eyes, and extra antennae due to cell-fate defects. All these phenotypes are consistent with loss of Rabex-5 leading to an increase in Ras signaling. In fact, molecular markers show that Ras is overactive in flies with lower Rabex-5 function and reduction of Ras led to rescue of the defects due to loss of Rabex-5 activity. Expression of Rabex-5 variants showed that the Ubiquitin ligase activity of Rabex-5 is required for its function in regulating Ras (Yan et al. 2010). The identification of Rabex-5-mediated Ras ubiquitination highlights the role of proteostasis in adjusting developmental signaling to allow seemingly opposing outcomes such as growth, survival, and differentiation.

The Notch Signaling Cycle

Canonical Notch signaling depends on interaction between the Notch receptor and a ligand, either Delta (Dl) or Serrate (Ser) in the fly eye, on the adjacent cell. Unlike other signaling pathways, Notch is not activated by enzymatic signal transduction. Instead, removal of an extracellular portion of the receptor (the ectodomain), usually triggered by ligand binding, renders the receptor susceptible to proteolytic cleavages by the metalloprotease Kuzbanian and γ-secretase. Cleavage releases a portion of the Notch intracellular domain (NICD) that, upon transport to the nucleus, interacts with Suppressor of Hairless (SuH) and Mastermind to activate transcription of downstream effectors (Artavanis-Tsakonas and Muskavitch 2010; Fig. 3a). The

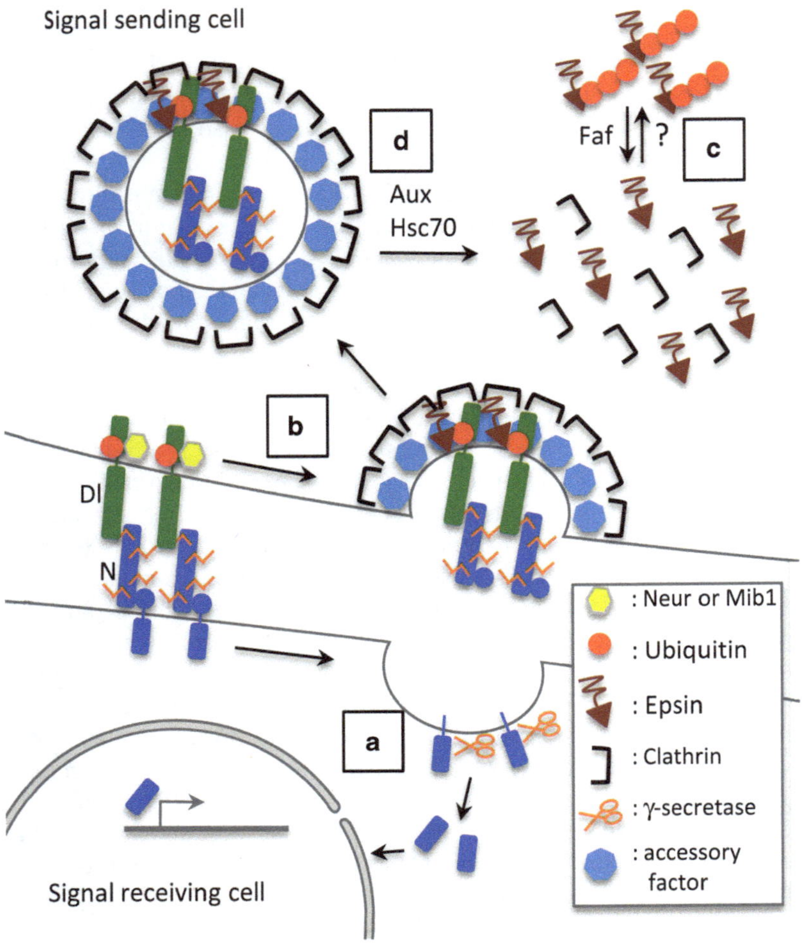

Fig. 3 Ligand-dependent Notch signaling requires adequate pools of Epsin and Clathrin for activation of receptor by ligand endocytosis. **a** Notch receptor is cleaved to release the N^{ICD}, a transcription regulator. **b** A population of Dl is ubiquitinated by Neur or Mib1 and recruited to specialized Clathrin-coated endocytic vesicles by Epsin. **c** Epsin is inactive when ubiquitinated. In the eye, the DUB Faf is required to increase the level of active Epsin by removing Ubiquitin from Epsin. **d** Epsin and Clathrin levels are increased by Aux and Hsc70 activity in uncoating Clathrin-coated vesicles

nonenzymatic nature of Notch signaling renders the pathway more sensitive than other signaling mechanisms to levels of receptor activation because there is a direct relationship between the number of activated receptors and the number of N^{ICD} molecules available to activate transcription.

Notch activation is used repeatedly for patterning the fly eye. It is required for proliferation of early eye progenitors (Amore and Casares 2010; Dominguez et al. 2004; Kenyon et al. 2003), origin of the morphogenetic furrow (Kumar and Moses 2001), establishment of the boundary between dorsal and ventral compartments

(Cho and Choi 1998; Domínguez and de Celis 1998; Singh et al. 2012), selection of R8 photoreceptors (Baker et al. 1996; Cagan and Ready 1989), ommatidial polarity by distinguishing R3 from R4 (Cooper and Bray 1999; Fanto and Mlodzik 1999; Tomlinson and Struhl 1999), selecting R7 from the R1/R6/R7 equivalence group (Cooper and Bray 2000; Tomlinson and Struhl 2001; Tomlinson et al. 2011), and the identity of non-neural ommatidial cells (Nagaraj and Banerjee 2004). Because Notch signaling occurs between two adjacent cells, the main function of Notch in all these developmental decisions is to make one cell different from its neighbor. The outcome of many of these cell-fate decisions relies on context; the downstream effects of Notch activation are different depending on the coordinated action of several pathways and the production of other regulatory molecules depends on developmental history (Nagaraj and Banerjee 2004). In addition, Notch pathway proteins are modulated by proteostatic mechanisms. In the fly eye, adjustment of the quantity and quality of Notch signaling through ER modification and Ub-mediated degradation is essential for specific developmental outcomes.

Adjusting Notch Quality via ER Function

The Notch receptor is a single-pass transmembrane protein with two functional domains in the extracellular region: a series of 36 EGF-like repeats and the negative regulatory region (NRR). EGR-like repeats 11 and 12 are involved in ligand binding (Rebay et al. 1991). The NRR is made up of three Lin12/Notch repeats (LNRs) and a heterodimerization region. The structure of the NRR blocks access to the proteolytic cleavage site and prevents Notch activation in the absence of ligand. Casual interaction between Notch and ligand is not sufficient to activate signaling. Instead, current models propose that tight receptor ligand binding followed by endocytosis of both into their respective cells results in sufficient force to alter the structure of the LNR domain and expose the site for proteolytic cleavage (Musse et al. 2012). Both ligand-binding and proteolytic cleavage depend on proper folding and modification of Notch in the ER. As Notch is transported through the secretory pathway, it is modified with *O*-linked fucose and glucose moieties on the EGF repeats and characteristic disulfide bonds are introduced in the EGF repeats and the LNRs. Regulation of ER function is important for Notch signaling during eye development for two reasons: (1) retention of improperly folded receptor prevents spurious Notch activation and (2) regulation of at least one ER protein results in tissue-specific regulation of Notch signaling.

Ero1L

Ero1L is a thiol oxidase involved indirectly in disulfide bond formation. In many tissues, loss of Ero1L function causes *Notch*-like phenotypes. Genetic interactions and mosaic clone analysis indicate that Ero1L function is Notch pathway-specific and required in signal-receiving cells. Experiments in S2 cells and an in vitro oxidation assay show that Ero1L is required for disulfide bond formation in Notch LNRs, but not in EGF repeats. Cells lacking Ero1L retain Notch in the ER and activate the UPR. Notch is likely a major Ero1L target because the UPR is much less robust in Ero1L mutant cells that also lack Notch (Tien et al. 2008). The disulfide bonds formed in the LNR region are critical for regulating activation by proteolytic processing. Therefore, it is proposed that ER retention of improperly folded Notch in *Ero1L*$^-$ cells prevents detrimental Notch activation by ligand-independent cleavage.

Glycosylation by OFUT1 and Fng

O-fucosyltransferase-1 (OFUT1) is a soluble ER enzyme that modifies proteins by placing an *O*-fucose moiety on a consensus sequence found in most EGF repeats, including those in *Drosophila* Notch. *OFUT1*$^-$ animals display an array of *Notch*-like defects in the eye and elsewhere, implicating OFUT1 in diverse Notch signaling events. Further inquiry into OFUT1 function was carried out using the wing or S2 cells as models. In cells lacking OFUT1, the Notch receptor fails to reach the plasma membrane and instead accumulates in the ER. Overexpression of either wild-type OFUT1 or OFUT1 that is enzymatically inactive rescues Notch trafficking defects caused by loss of OFUT function. Therefore, independent of its function as a glycosylase, OFUT1 probably has a chaperone function required for the folding or trafficking of Notch from the ER to the plasma membrane (Okajima et al. 2005; Okajima et al. 2008).

O-fucosylated Notch is also a substrate for further glycosylation by Fringe (Fng). Notch glycosylated by Fng has increased affinity for Dl, but cannot be activated by Ser (Brückner et al. 2000; Okajima and Irvine 2002; Okajima et al. 2003; Panin et al. 1997; Sasamura et al. 2003; Xu et al. 2007). Differential ligand binding is particularly important early in eye development when the dorsal–ventral (D/V) boundary is established. At that time, Notch and Dl are ubiquitous while Ser and Fng are localized to the ventral half of the eye disc. Notch receptors in ventral cells are glycosylated by Fng and are not activated by Ser. Instead, Notch in one row of cells on the ventral side of the D/V border is activated by Dl on the surface of cells on the dorsal side of the border. The converse is also true: Notch that is not glycosylated by Fng is not activated by Dl in the dorsal compartment, but is activated at the D/V boundary by ventral cells expressing Ser. Therefore, although Notch and its ligands are widely expressed in the early eye disc, modification in the ER by Fng results in tissue-specific activation of signaling to form a stable D-V border.

Glucosylation by Rumi

Loss of the ER protein Rumi causes temperature-sensitive *Notch*-like defects in all tissues, including the eye. Mosaic analysis indicates that Rumi is required in signal-receiving cells. Rumi is an *O*-glucosyltransferase that attaches a glucose moiety to multiple EGF repeats in the Notch receptor; the glucose serves as substrate for further modification by xylosyltransferases (Rana and Haltiwanger 2011). Notch receptors isolated from cells with reduced Rumi have lower levels of glucosylation. In Rumi-depleted wing disc clones, Notch accumulates throughout the cell and only a fraction of the receptor is trafficked to the plasma membrane. The ER, however, appears normal, the UPR is not activated, and Notch is able to bind ligand. Notch signaling is rescued in *rumi* mutants by overexpression of an active form of Notch (N^{ECN}), indicating that Rumi functions upstream of Notch receptor activation (proteolytic processing). These results, combined with the temperature-sensitive nature of the *rumi* mutant phenotype, lead to the hypothesis that glucosylation by Rumi assists the Notch receptor to fold into a structure that can be activated by proteolytic cleavage (Acar et al. 2008; Leonardi et al. 2011). Notch receptor maturation requires Rumi generally; it remains to be determined if glucosylation results in tissue-specific Notch regulation in the eye.

Notch Ligand Endocytosis Depends on Epsin Homeostasis

Notch ligands are perpetually placed on the cell surface and removed by endocytosis to maintain homeostasis. Unlike other signaling pathways, activation of the Notch receptor requires endocytosis of ligand into the signal-sending cell. How does the cell internalize Dl constitutively, yet avoid persistent, high levels of Notch activation? Not all ligand endocytosis leads to signal activation. Signaling competent Notch ligand endocytosis depends on Neuralized (Neur), Mind-bomb (Mib1 in *Drosophila*), and Epsin (Weinmaster and Fischer 2011). Neur and Mib1 are E3 Ub ligases for Notch ligands and their abilities to bind and ubiquitinate Dl are correlated with Notch activation in flies (Daskalaki et al. 2011; Weinmaster and Fischer 2011 and references therein; Fig. 3b). Epsin is an endocytic protein implicated in recruitment of ubiquitinated cargo to endocytic vesicles in many contexts, including the fly eye (Overstreet et al. 2003; Overstreet et al. 2004; Sen et al. 2012; Tian et al. 2004; Wang and Struhl 2004). The requirement for Neur or Mib1 and Epsin activities suggests that a specialized endocytic pathway, dependent on Notch ligand ubiquitination, leads to signaling.

Ubiquitination of Epsin by Faf

In the *Drosophila* eye, regulation of Epsin homeostasis by Ub and chaperones is important for a subset of Notch signaling events. Epsin proteins in eye discs are ubiquitinated by an unknown E3 ligase and deubiquitinated by Faf (Fig. 3c). In eye discs lacking Faf, a greater fraction of Epsin is ubiquitinated and that translates to at least a twofold reduction in Epsin levels (Chen et al. 2002). Loss of Faf results in ommatidia with extra photoreceptors, indicative of a Notch signaling defect posterior to the morphogenetic furrow (Fischer et al. 1997; Fischer-Vize et al. 1992). Genetic interaction and mosaic analysis show that Faf function is required in a subset of signal-sending cells to facilitate Dl endocytosis (Cadavid et al. 2000; Overstreet et al. 2004). Therefore, deubiquitination of Epsin by Faf increases the level of active Epsin available for Dl endocytosis. The requirement for Faf in only a subset of Notch signaling events suggests that Epsin is not always limiting. Two mechanisms are proposed to explain the inactivity of ubiquitinated Epsin: ubiquitinated Epsin may be targeted for proteasomal degradation, or perhaps Ub moieties on Epsin interfere with Epsin's ability to recruit Dl to endocytic vesicles by blocking the Epsin UIMs.

Auxilin and Hsc70

The level of available Epsin protein is also controlled by cytoplasmic chaperone activity. A mutagenesis screen for genes that interact with *lqf* uncovered an interaction between *lqf* and *auxilin* (aux; Eun et al. 2007). Auxilin is a co-chaperone that binds to Clathrin and recruits Hsc70 to Clathrin-coated endocytic vesicles (Ungewickell et al. 1995). Hsc70 activity at each of three Clathrin arms causes a change in Clathrin conformation and results in vesicle uncoating (Rothnie et al. 2011; Xing et al. 2010). In flies, loss of Auxilin causes widespread Notch-like developmental defects. Mosaic analysis in the eye demonstrates that Auxilin is required in signal-sending cells during all Notch signaling events. Overexpressing Clathrin or Epsin in the fly eye partially rescues the eye phenotype and overexpressing both proteins together in the eye compensates for the lack of Auxilin almost completely (Eun et al. 2008; Kandachar et al. 2008). The implication is that at least in the eye, the functions of Auxilin and Hsc70 in uncoating Clathrin-coated vesicles are crucial for maintenance of an adequate pool of Clathrin and Epsin for use in Dl endocytosis (Fig. 3c).

Control of Ligand-Independent Notch Activation During Endocytic Trafficking

The population of Notch on the cell surface is not static; endocytosis internalizes many Notch receptors whether or not they are bound to ligand. Therefore, cells have a heterogeneous mix of endocytic vesicles that contain full-length and processed Notch. Full-length Notch can be processed during endosomal trafficking leading to ligand-independent signaling. A study of Notch receptor trafficking in eye discs indicates that most Notch receptors are internalized, trafficked through the early endosome, and either recycled to the plasma membrane or directed to the MVB for further trafficking to the lysosome for degradation (Vaccari et al. 2008). The mechanistic details concerning Notch receptor regulation after endocytosis are not well understood, but the E3 Ub ligases Deltex (Dx) and Suppressor of Deltex [Su(dx)] have been implicated in stimulation and inhibition of ligand-independent Notch signaling through direct interactions with the Notch receptor during trafficking

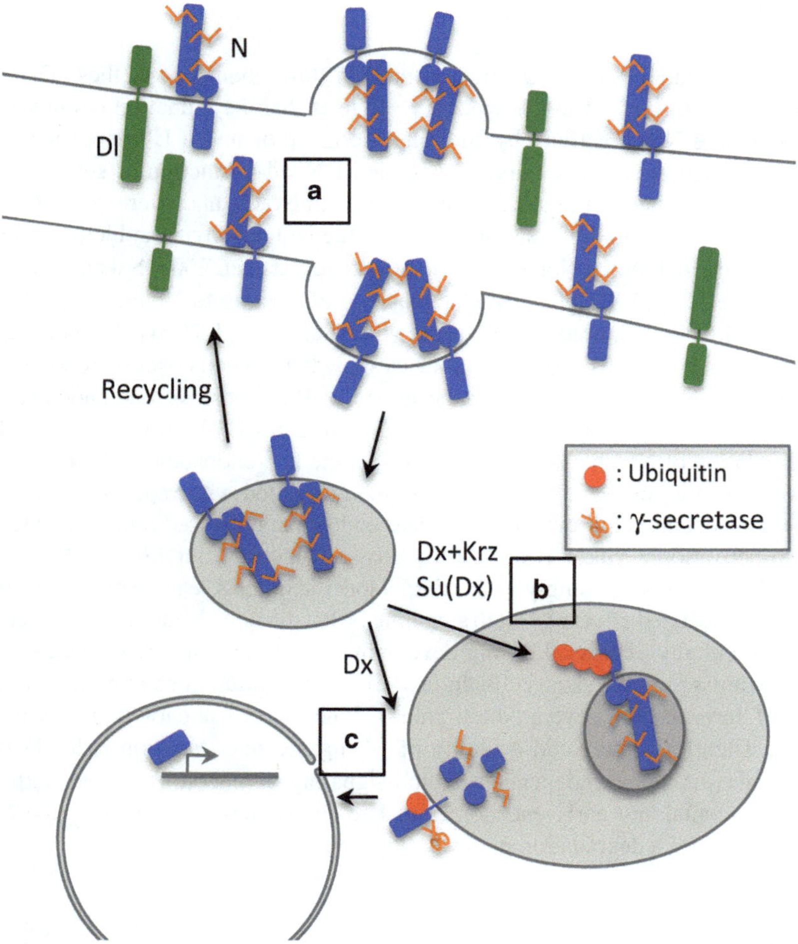

Fig. 4 Some robust cell-fate decisions depend on ligand-independent Notch signaling. **a** Adjacent cells with equal levels of Notch and Dl on the cell surface are shown. Internalization of Notch into either cell can lead to ligand-independent signal activation. Trafficking after endocytosis is shown in one cell, but ligand-independent signaling is controlled in either cell or in both cells. **b** Su(Dx) and interaction between Dx and Krz encourage polyubiquitination of full-length Notch followed by degradation in the lysosomal lumen to prevent ligand-independent Notch activation. **c** In contrast, Dx monoubiquitinates Notch resulting in trafficking to the lysosomal membrane. Subsequent partial degradation of the ectodomain in the lysosome permits production of N^{ICD} and increases ligand-independent Notch signaling

(Mukherjee et al. 2005; Sakata et al. 2004). In addition, Ral GTPase plays a role in preventing ligand-independent Notch signaling (Cho and Fischer 2011). In some developmental contexts in the eye, modulation of ligand-independent signaling is critical for robust cell-fate choices (Fig. 4a).

Nedd-4 E3 Ligases

The Nedd-4 family of E3 Ub ligases include three members in flies, DNedd-4, Su(dx), and Dsmurf. Available evidence suggests that they function redundantly to downregulate Notch activity by shuttling the receptor into a Ub-dependent endocytic degradation pathway. First, Su(dx) and DNedd-4 function as suppressors of Notch activity; they both interact genetically with Notch, and overexpression of either protein results in defects typical of too little Notch activity and less expression of molecular markers of Notch activation (Mazaleyrat et al. 2003; Sakata et al. 2004; Wilkin et al. 2008). Both proteins are capable of binding to Notch and, at least in the case of DNedd-4, ubiquitinating it in vitro (Sakata et al. 2004). Antibody uptake experiments in S2 cells and imaginal discs show that, in cells overexpressing Notch and Su(dx), Notch is trafficked to the lumen of Rab7-positive late-endosomes. In contrast, overexpression of a Su(dx) variant that lacks E3 ligase activity results in accumulation of Notch nearby, but not inside, the late-endosomes. The localization of Notch to the lumen of the late-endosome correlated with reduced Notch activity, while accumulation outside of the late-endosome increased ligand-independent Notch activation (Wilkin et al. 2008). Loss-of-function mutations in Su(dx) and DNedd-4 lead to surprisingly weak, tissue-specific defects. One explanation for the weak and localized effect on Notch signaling is that the three Nedd-4 family members are differentially expressed during development and they have redundant function when expressed in the same cells. In fact, flies having mutations in both *Su(dx)* and *Dsmurf* have a more severe Notch gain-of-function eye phenotype (Wilkin et al. 2008). Therefore, the Nedd-4 family of Ub ligases may function redundantly as antagonists of ligand-independent Notch signaling by encouraging the trafficking of Notch-containing early-endosomes to late-endosomes and directing the Notch receptor toward a degradation pathway (Fig. 4b).

Dx and Ral

Targeted regulation of ligand-independent Notch activation is important in eye development for interpretation of planar cell polarity (PCP) signals, including Frizzled (Fz) that is expressed in a gradient from the D/V boundary. On either side of the boundary, facets develop with opposite chirality defined by the asymmetric positioning of the R3 and R4 photoreceptors. The presumptive R3 cell is closer to the boundary so it has a higher level of Fz activation than the presumptive R4. Asymmetric development depends on activation of Notch in R4 by Dl in R3 in response to Fz (Cooper and Bray 1999; del Alamo and Mlodzik 2006; Fanto and Mlodzik 1999; Tomlinson and Struhl 1999). However, it appears that ligand-dependent activation of Notch in R4 is not sufficient to generate a robust response to PCP signals. The E3 Ub ligase Dx and the GTPase Ral may amplify R3/R4 asymmetry because of Fz in different ways.

Dx is required at several points for the endocytosis and subsequent trafficking of the Notch receptor (Yamada et al. 2011). Increased Dx, unlike the Nedd-4 family described previously, prolongs the time Notch spends in the endosomal pathway and increases activation of molecular markers downstream of Notch signaling (Hori et al. 2004). Wilkin et al. (2008) extended the analysis of trafficking in Notch signaling to show that Dx supports the trafficking of Notch to the limiting membrane of the late-endosome/lysosome, where the ectodomain of Notch is degraded and cleavage by γ-secretase releases the N^{ICD} to activate signaling. Therefore, Dx diverts Notch from a degradation path to support a lysosomal- and γ -secretase-dependent mechanism of ligand-independent Notch activation (Fig. 4c). The ability of Dx to stimulate ligand-independent Notch activation is in turn regulated by the nonvisual β-arrestin Kurtz (Krz) and the ESCRTIII component Shrub (Matsuno et al. 2002; Mukherjee et al. 2005). A model emerges from these studies whereby Dx alone enhances ligand-independent activation by monoubiquitinating Notch and directing it to the lysosomal membrane. In contrast, Notch receptors bound to Dx and Krz become polyubiquitinated, interact with ESCRTIII, and are translocated into the lysosomal lumen for degradation (Mukherjee et al. 2005; Fig. 4). In flies lacking Dx, the eyes develop almost normally, but a small percentage of facets are symmetrical and have two R3 cells (Fuwa et al. 2006). This makes sense if Dx-mediated ligand-independent Notch activation in R4 assists PCP signals in breaking Notch symmetry during the R3/R4 cell-fate decision. However, it is not known if Dx is required in R4 and the regulation of Dx activity by PCP signals has not been studied in the eye.

Hypomorphic alleles of the Ral GTPase result in ommatidia with *Notch*-like defects, including many that are symmetrical and misrotated. More detailed study of the chirality defects in *Ral*$^{-}$ mosaic eyes determined that Ral activity in pre-R3 cells results in increased Notch activation in pre-R4 cells. Furthermore, *Ral* transcription is upregulated in the pre-R3 cell in response to Fz signaling and Ral exerts its effect by downregulating ligand-independent Notch activation in the pre-R3 cell. When Ral is equal in both cells or over expressed in R4, only a small fraction of facets are symmetrical or have reversed chirality suggesting that Ral activity is one of several mechanisms needed to bias facet polarity (Cho and Fischer 2011). Ral has been implicated in membrane trafficking in other systems and it is localized to internal vesicles in the fly eye, suggesting that it may regulate the trafficking of Notch (Cho and Fischer 2011; Wu et al. 2008). Further experiments are needed to determine how Ral downregulates ligand-independent Notch activation. Regardless of mechanism, it is interesting that control of ligand-independent Notch signaling in R3 by Ral is required for reliable facet asymmetry.

Modulation of Visual Signal Transduction

Light is detected in the fly eye by the Rh family of G-protein-coupled receptors. Each Rh is composed of a transmembrane Opsin protein and the chromophore 3-hydroxy 11-*cis*-retinal in *Drosophila*. Isomerization of the chromophore upon activation by light results in signal transduction through the G-protein Gq and phospholipase C (PLC) to ultimately open the transient receptor potential (TRP) and TRP-like (TRPL) cation channels (Montell 2012; Stamnes et al. 1991). The visual response in flies is one of the fastest signaling cascades known, ten times faster than signaling in human rod photoreceptors. In addition, fly photoreceptors successfully adapt to a wide-range of light intensities (Hardie and Raghu 2001). The speed and adaptability of the fly visual system is important for flies to detect and fly away from movement in variable light conditions, thereby increasing survival (Hardie 2007; Mishra et al. 2007). Proteostasis contributes to the unusual features of *Drosophila* visual signaling in several ways. First, many Rh molecules are folded in the ER and trafficked with TRP channels to the specialized membrane of the rhabdomere. Second, phototransduction components are organized and retained in the rhabdomere by the regulatory protein scaffold inactivation-no-afterpotential D (INAD). And third, adaptation to light levels depends on continued protein trafficking into and out of the rhabdomere.

Folding and Cotrafficking of Rh and TRP

The six outer photoreceptors express a single Rh, Rh1. Each photoreceptor cell contains about 100 million Rh1 molecules. Because Rh1 is synthesized in massive quantities, environmental variation or mutations that affect Rh1 folding in the ER or trafficking to the rhabdomere lead to dramatic cellular consequences, including the induction of a strong UPR and neurodegeneration (Mendes et al. 2005; Montell 2012; Stamnes et al. 1991). Studies of Rh1 in flies have identified three ER chaperones that assist Rh1 folding and trafficking: neither activation nor afterpotential A (NinaA), Calnexin (Cnx), and Exit Protein of Rh and TRP (XPORT). NinaA is a cyclophilin, a protein with peptidyl prolyl isomerase activity. NinaA may catalyze protein folding, however, NinaA also has noncatalytic chaperone activity in Rh-1 maturation (Colley et al. 1995). Cnx binds to glycoproteins, including Rh1, to decrease aggregation of unfolded proteins and retain them in the ER (Molinari et al. 2004; Rosenbaum et al. 2006; Ware et al. 1995). XPORT interacts not only with Rh1 and TRP, but also with the chaperones Hsp27 and Hsp90, suggesting a role for XPORT as a co-chaperone that recruits protein-folding and/or -trafficking molecules to a complex with Rh1 and TRP. Rh1 and TRP are both trafficked to the rhabdomere through Rab11-containing endosomes, suggesting the possibility that XPORT functions in a process that ensures the maturation and cotrafficking of two important members of the phototransduction pathway (Rosenbaum et al. 2011). Photoreceptors lacking NinaA, Cnx, or XPORT fail to produce and transport normal quantities of mature Rh1 to the rhabdomere. Instead, Rh1 is retained in the ER and eventually degraded by activation of the UPR. In some cases, the UPR cannot relieve ER stress and a cell death pathway leads to neurodegeneration (Colley et al. 1991; Rosenbaum et al. 2006; Rosenbaum et al. 2011).

Control of Signal Termination by INAD

Organization of visual signal transduction components in the rhabdomere contributes to the rapid kinetics of the visual response (Montell 2012; Stamnes et al. 1991). Phototransduction components are trafficked to and physically coupled in the rhabdomere by the scaffold protein INAD (Chevesich et al. 1997; Tsunoda et al. 1997). INAD has five PDZ (Postsynaptic density 95, Discs Large, Zona Occludens 1) protein-binding domains. The core components of the INAD complex include PLC, TRP, and eye-PKC (an eye-specific protein kinase C involved in negative feedback regulation; Li and Montell 2000). Other phototransducing molecules, such as TRPL and Rh1, may bind INAD as well (Chevesich et al. 1997; Xu et al. 1998). Single-cell patch-clamp recordings indicate that cells without INAD have a slow, weak response to light (Scott and Zucker 1998). Therefore, coupling of signal activators by INAD explains the quick, robust visual response in flies.

INAD is not only a passive scaffold protein; the PDZ5 domain also functions as a redox-switch in the regulation of signal termination (Hardie 2007; Liu et al. 2011; Mishra et al. 2007; Montell 2007). In the dark, PDZ5 assumes one conformation with a functional binding domain. Upon stimulation by light, two cysteines in PDZ5 transiently form a disulfide bond, resulting in a structure change that eliminates the binding domain. Studies have suggested that the alternate PDZ5 structure prevents binding to PLC and the TRP channel; therefore, light exposure uncouples signal transduction from membrane depolarization (Liu et al. 2011; Tsunoda et al. 1997). Recent work provides evidence that the acidic environment produced during signal activation favors the conformation change that terminates signaling (Liu et al. 2011). A point mutation in INAD that prevents disulfide bond formation results in prolonged signal activation and defective escape behavior in flies (Mishra et al. 2007).

Cellular Adaptation to Light Is Mediated by Protein Trafficking

The subcellular localization of phototransduction proteins is a dynamic process. Movement of proteins into and out of the rhabdomere in response to light explains the ability of flies to adapt to very dim or very bright light. When flies are kept in the dark, signal-activating proteins such as Rh1, TRP, TRPL, and Gq are primarily located in the rhabdomere. In contrast, the visual arrestins, Arr1 and Arr2, proteins involved in adaptive signal attenuation, are located in the cell body. When exposed to light, Arr1 and Arr2 move into the rhabdomere and interact with Rh1. Arr1 facilitates Rh1 endocytosis and colocalizes with a fraction of Rh1 in MVBs (Elsaesser et al. 2010; Lee et al. 2003; Satoh and Ready 2005; Satoh et al. 2010). Light exposure also leads to translocation of the signal-enhancing proteins TRPL and Gq from the rhabdomere to the cell body. TRPL is transported in two phases; the first occurs within minutes of light activation and is probably due to diffusion from the rhabdomere to the nearby membrane, the second requires hours and may involve Rab5- and RabX4-dependent endocytic trafficking (Bahner et al. 2002; Cronin et al. 2006; Lieu et al. 2012; Oberegelsbacher et al. 2011). Translocation of Gq to the cell body requires from minutes to hours and is readily reversible upon removal of light (Cronin et al. 2004; Frechter et al. 2007; Kosloff et al. 2003). For the most part, the mechanisms involved in protein transport during visual adaptation are undefined.

Conclusion

We have provided examples that show the extent to which protein homeostasis regulates signaling in developmental and functional contexts in the fly eye. *Drosophila* eye development depends on the iterative use of a few signaling pathways as well as cells' ability to control the timing and level of response to signaling. In most cases, widespread synthesis of signaling and transduction molecules primes many cells to respond quickly to signal. However, the level of activation is controlled both positively and negatively by several proteostatic mechanisms, including folding and modification in the ER, subcellular localization, and proteolytic processing or degradation. Throughout eye development, proteostasis mechanisms turn signaling pathways on and off, allowing iterative use of the same signal in the restricted confines of the eye disc. Posttranslational protein processing and modification also results in qualitatively different signals from the same pathway. The result is conversion of an undifferentiated epithelium into a highly ordered and functional organ comprising many different cell types. Visual signal transduction in the adult eye continues to depend on proteostatic mechanisms to respond quickly and adapt to different light levels.

Acknowledgments KDP is supported by a re-entry supplement to NIH Grant No. HD30680 awarded to Janice A. Fischer.

References

Acar M, Jafar-Nejad H, Takeuchi H et al (2008) Rumi is a CAP10 domain glycosyltransferase that modifies Notch and is required for Notch signaling. Cell 132:247–258

Amore G, Casares F (2010) Size matters: the contribution of cell proliferation to the progression of the specification Drosophila eye gene regulatory network. Dev Biol 344:569–577

Araki K, Nagata K (2012) Protein folding and quality control in the ER. Cold Spring Harb Perspect Biol 4:a015438

Artavanis-Tsakonas S, Muskavitch MA (2010) Notch: the past, the present, and the future. Curr Top Dev Biol 92:1–29

Bahner M, Frechter S, Da Silva N et al (2002) Light-regulated subcellular translocation of *Drosophila* TRPL channels induces long-term adaptation and modifies the light induced current. Neuron 34:83–93

Baker NE (2007) Patterning signals and proliferation in Drosophila imaginal discs. Curr Opin Genet Dev 17:287–293

Baker NE, Yu S, Han D (1996) Evolution of proneural atonal expression during distinct regulatory phases in the developing Drosophila eye. Curr Biol 6:1290–1301

Baker NE, Yu SY (2001) The EGF receptor defines domains of cell cycle progression and survival to regulate cell number in the developing Drosophila eye. Cell 104:699–708

Brown KE, Freeman M (2003) Egfr signaling defines a protective function for ommatidial orientation in the Drosophila eye. Development 130:5401–5412

Brunner D, Oellers N, Szabad J et al (1994) A gain-of-function mutation in Drosophila MAP kinase activates multiple receptor tyrosine kinase signaling pathways. Cell 76:875–888

Brückner K, Perez L, Clausen H et al (2000) Glycosyltransferase activity of Fringe modulates Notch-Delta interactions. Nature 406:411–415

Cadavid AL, Ginzel A, Fischer JA (2000) The function of the Drosophila fat facets deubiquitinating enzyme in limiting photoreceptor cell number is intimately associated with endocytosis. Development 127:1727–1736

Cagan RL, Ready DF (1989) Notch is required for successive cell decisions in the developing Drosophila retina. Genes Dev 3:1099–1112

Chen X, Zhang B, Fischer JA (2002) A specific protein substrate for a deubiquitinating enzyme: liquid facets is the substrate of Fat facets. Genes Dev 16:289–294

Chen ZJ, Sun LJ (2009) Nonproteolytic functions of ubiquitin in cell signaling. Mol Cell 33:275–286

Chevesich J, Kreuz AJ, Montell C (1997) Requirement for the PDZ domain protein, INAD, for localization of the TRP store-operated channel to a signaling complex. Neuron 18:95–105

Cho B, Fischer JA (2011) Ral GTPase promotes asymmetric Notch activation in the Drosophila eye in response to Frizzled/PCP signaling by repressing ligand-independent receptor activation. Development 138:1349–1359

Cho KO, Choi KW (1998) Fringe is essential for mirror symmetry and morphogenesis in the Drosophila eye. Nature 396:272–276

Colley NJ, Baker EK, Stamnes MA et al (1991) The cyclophilin homolog ninaA is required in the secretory pathway. Cell 67:255–263

Colley NJ, Cassill JA, Baker EK et al (1995) Defective intracellular transport is the molecular basis of rhodopsin-dependent dominant retinal degeneration. Proc Natl Acad Sci USA 92:3070–3074

Cooper MT, Bray SJ (1999) Frizzled regulation of Notch signaling polarizes cell fate in the Drosophila eye. Nature 397:526–530

Cooper MT, Bray SJ (2000) R7 photoreceptor specification requires Notch activity. Curr Biol 10:1507–1510

Cronin MA, Diao F, Tsunoda S (2004) Light-dependent subcellular translocation of $G_q\alpha$ in *Drosophila* photoreceptors is facilitated by the photoreceptor-specific myosin III NINAC. J Cell Sci 117:4797–4806

Cronin MA, Lieu MH, Tsunoda S (2006) Two stages of light-dependent TRPL-channel translocation in *Drosophila* photoreceptors. J Cell Sci 119:2935–2944

Daskalaki A, Shalaby NA, Kux K et al (2011) Distinct intracellular motifs of Delta mediate its ubiquitylation and activation by Mindbomb1 and Neuralized. J Cell Biol 195:1017–1031

del Alamo D, Mlodzik M (2006) Frizzled/PCP-dependent asymmetric neuralized expression determines R3/R4 fates in the Drosophila eye. Dev Cell 11:887–894

Dominguez M, Ferres-Marco D, Gutierrez-Avio FJ et al (2004) Growth and specification of the eye are controlled independently by Eyegone and Eyeless in Drosophila melanogaster. Nat Genet 36:31–39

Domànguez M, de Celis JF(1998) A dorsal/ventral boundary established by Notch controls growth and polarity in the Drosophila eye. Nature 396:276–278

Du J, Zhang J, Su Y et al (2011) In vivo RNAi screen reveals neddylation genes as novel regulators of Hedgehog signaling. PLoS One 6:e24168

Eun SH, Lea K, Overstreet E et al (2007) Identification of genes that interact with Drosophila liquid facets. Genetics 175:1163–1174

Elsaesser R, Kalra D, Li R et al (2010) Light-induced translocation of *Drosophila* visual Arrestin2 depends on Rac2. Proc Nati Acad Sci USA 107:4740–4745

Eun SH, Banks SM, Fischer JA (2008) Auxilin is essential for Delta signaling. Development 135:1089–1095

Fanto M, Mlodzik M (1999) Asymmetric Notch activation specifies photoreceptors R3 and R4 and planar polarity in the Drosophila eye. Nature 397:523–526

Fischer JA, Leavell SK, Li Q (1997) Mutagenesis screens for interacting genes reveal three roles for fat facets during Drosophila eye development. Dev Genet 21:167–174

Fischer-Vize JA, Rubin GM, Lehmann R (1992) The fat facets gene is required for Drosophila eye and embryo development. Development 116:985–1000

Fortini ME (2009) Notch signaling: the core pathway and its posttranslational regulation. Dev Cell 16:633–647

Frechter S, Elia N, Tzarfaty V et al (2007) Translocation of $G_q\alpha$ mediates long-term adaptation in *Drosophila* photoreceptors. J Neurosci 27:5571–5583
Freeman M (1996) Reiterative use of the EGF receptor triggers differentiation of all cell types in the Drosophila eye. Cell 87:651–660
Fuwa TJ, Hori K, Sasamura T et al (2006) The first deltex null mutant indicates tissue-specific deltex-dependent Notch signaling in Drosophila. Mol Genet Genomics 275:251–263
Gidalevitz T, Prahlad V, Morimoto RI (2012) The stress of protein misfolding: from single cells to multicellular organisms. Cold Spring Harb Perspect Biol 3. pii:a009704, doi:10.1101/cshperspect.a009704
Haglund K, Dikic I (2012) The role of ubiquitylation in receptor endocytosis and endosomal sorting. J Cell Sci 125:265–275
Haglund K, Sigismund S, Polo S et al (2003) Multiple monoubiquitination of RTKs is sufficient for their endocytosis and degradation. Nat Cell Biol 5:461–466
Hardie RC (2007) Vision: dynamic platforms. Nature 450:37–39
Hardie RC, Raghu P (2001) Visual transduction in Drosophila. Nature 413:186–193
Hartl FU, Bracher A, Hayer-Hartl M (2011) Molecular chaperones in protein folding and proteostasis. Nature 475:324–332
Heberlein U, Rubin GM (1991) Star is required in a subset of photoreceptor cells in the developing Drosophila retina and displays dosage sensitive interactions with rough. Dev Biol 144:353–361
Herranz H, Milán M (2008) Signaling molecules, growth regulators and cell cycle control in Drosophila. Cell Cycle 7:3335–3337
Hime GR, Dhungat MP, Ng A et al (1997) D-Cbl, the Drosophila homologue of the c-Cbl proto-oncogene, interacts with the Drosophila EGF receptor in vivo, despite lacking C-terminal adaptor binding sites. Oncogene 14:2709–2719
Hori K, Fostier M, Ito M et al (2004) Drosophila deltex mediates suppressor of Hairless-independent and late-endosomal activation of Notch signaling. Development 131:5527–5537
Horiuchi H, Lippé R, McBride HM et al (1997) A novel Rab5 GDP/GTP exchange factor complexed to Rabaptin-5 links nucleotide exchange to effector recruitment and function. Cell 90:1149–1159
Hsiung F, Griffis ER, Pickup A et al (2001) Function of the Drosophila TGF-alpha homolog Spitz is controlled by Star and interacts directly with Star. Mech Dev 107:13–23
Joazeiro CA, Wing SS, Huang H et al (1999) The tyrosine kinase negative regulator c-Cbl as a RING-type, E2-dependent Ubiquitin-protein ligase. Science 286:309–312
Jongeward GD, Clandinin TR, Sternberg PW (1995) sli-1, a negative regulator of let-23-mediated signaling in C. elegans. Genetics 139:1553–1566
Jura N, Scotto-Lavino E, Sobczyk A et al (2006) Differential modification of Ras proteins by ubiquitination. Mol Cell 21:679–687
Kandachar V, Bai T, Chang HC (2008) The clathrin-binding motif and the J-domain of Drosophila Auxilin are essential for facilitating Notch ligand endocytosis. BMC Dev Biol 8:50
Kenyon KL, Ranade SS, Curtiss J et al (2003) Coordinating proliferation and tissue specification to promote regional identity in the Drosophila head. Dev Cell 5:403–414
Kosloff M, Elia N, joel-Almagor T et al (2003) Regulation of light-dependent $G_q\alpha$ translocation and morphological changes in fly photoreceptors. EMBO J 22:459–468
Kulathu Y, Komander D (2012) Atypical ubiquitylation—the unexplored world of polyubiquitin beyond Lys48 and Lys63 linkages. Nat Rev Mol Cell Biol 13:508–523
Kumar JP (2012) Building an ommatidium one cell at a time. Dev Dyn 241:136–149
Kumar JP, Moses K (2001) The EGF receptor and notch signaling pathways control the initiation of the morphogenetic furrow during Drosophila eye development. Development 128:2689–2697
Langdon WY, Hartley JW, Klinken SP et al (1989) v-cbl, an oncogene from a dual-recombinant murine retrovirus that induces early B-lineage lymphomas. Proc Natl Acad Sci USA 86:1168–1172
Lee JR, Urban S, Garvey CF et al (2001) Regulated intracellular ligand transport and proteolysis control EGF signal activation in Drosophila. Cell 107:161–171
Lee SJ, Xu H, Kang LW, et al (2003) Light adaptation through phosphoinositide-regulated translocation of Drosophila visual arrestin. Neuron 39:121–132

Lee S, Tsai YC, Mattera R et al (2006) Structural basis for Ubiquitin recognition and autoubiquitination by Rabex-5. Nat Struct Mol Biol 13:264–271

Leonardi J, Fernandez-Valdivia R, Li YD et al (2011) Multiple O-glucosylation sites on Notch function as a buffer against temperature-dependent loss of signaling. Development 138:3569–3578

Levkowitz G, Waterman H, Zamir E et al (1998) c-Cbl/Sli-1 regulates endocytic sorting and ubiquitination of the epidermal growth factor receptor. Genes Dev 12:3663–3674

Levkowitz G, Waterman H, Ettenberg SA et al (1999) Ubiquitin ligase activity and tyrosine phosphorylation underlie suppression of growth factor signaling by c-Cbl/Sli-1. Mol Cell 4:1029–1040

Li HS, Montell C (2000) TRP and the PDZ protein, INAD, form the core complex required for retention of the signalplex in Drosophila photoreceptor cells. J Cell Biol 150:1411–1422

Liu Y, Ye Y (2011) Proteostasis regulation at the endoplasmic reticulum: a new perturbation site for targeted cancer therapy. Cell Res 21:867–883

Lieu MH, Vallejos MJ, Michael E et al (2012) Mechanisms underlying stage-1 TRPL channel translocation in *Drosophila* photoreceptors. PLoS One 7:e31622

Matsuno K, Ito M, Hori K et al (2002) Involvement of a proline-rich motif and RING-H2 finger of Deltex in the regulation of Notch signaling. Development 129:1049–1059

Mattera R, Tsai YC, Weissman AM et al (2006) The Rab5 guanine nucleotide exchange factor Rabex-5 binds Ubiquitin (Ub) and functions as a Ub ligase through an atypical Ub-interacting motif and a zinc finger domain. J Biol Chem 281:6874–6883

Mazaleyrat SL, Fostier M, Wilkin MB et al (2003) Down-regulation of Notch target gene expression by Suppressor of deltex. Dev Biol 255:363–372

Meisner H, Daga A, Buxton J et al (1997) Interactions of Drosophila Cbl with epidermal growth factor receptors and role of Cbl in R7 photoreceptor cell development. Mol Cell Biol 17:2217–2225

Mendes HF, van der Spuy J, Chapple JP et al (2005) Mechanisms of cell death in rhodopsin retinitis pigmentosa: implications for therapy. Trends Mol Med 11:177–185

Mishra P, Socolich M, Wall MA et al (2007) Dynamic scaffolding in a G protein-coupled signaling system. Cell 131:80–92

Molinari M, Eriksson KK, Calanca V et al (2004) Contrasting functions of calreticulin and calnexin in glycoprotein folding and ER quality control. Mol Cell 13:125–135

Montell C (2007) Dynamic regulation of the INAD signaling scaffold becomes crystal clear. Cell 131:19–21

Montell C (2012) Drosophila visual transduction. Trends Neurosci 35:356–363

Mukherjee A, Veraksa A, Bauer A et al (2005) Regulation of Notch signaling by non-visual beta-arrestin. Nat Cell Biol 7:1191–1201

Musse AA, Meloty-Kapella L, Weinmaster G (2012) Notch ligand endocytosis: mechanistic basis of signaling activity. Semin Cell Dev Biol 23:429–436

Nagaraj R, Banerjee U (2004) The little R cell that could. Int J Dev Biol 48:755–760

Okajima T, Irvine KD (2002) Regulation of notch signaling by o-linked fucose. Cell 111:893–904

Oberegelsbacher C, Schneidler C, Voolstra O et al (2011) The *Drosophila* TRPL ion channel shares a Rab-dependent translocation pathway with rhodopsin. Eur J Cell Biol 90:620–630

Okajima T, Xu A, Irvine KD (2003) Modulation of notch-ligand binding by protein O-fucosyltransferase 1 and fringe. J Biol Chem 278:42340–42345

Okajima T, Xu A, Lei L et al (2005) Chaperone activity of protein O-fucosyltransferase 1 promotes notch receptor folding. Science 307:1599–1603

Okajima T, Reddy B, Matsuda T et al (2008) Contributions of chaperone and glycosyltransferase activities of O-fucosyltransferase 1 to Notch signaling. BMC Biol 6:1

O'Neill EM, Rebay I, Tjian R et al (1994) The activities of two Ets-related transcription factors required for Drosophila eye development are modulated by the Ras/MAPK pathway. Cell 78:137–147

Overstreet E, Chen X, Wendland B et al (2003) Either part of a Drosophila epsin protein, divided after the ENTH domain, functions in endocytosis of delta in the developing eye. Curr Biol 13:854–860

Overstreet E, Fitch E, Fischer JA (2004) Fat facets and Liquid facets promote Delta endocytosis and Delta signaling in the signaling cells. Development 131:5355–5366

Pai LM, Wang PY, Chen SR et al (2006) Differential effects of Cbl isoforms on Egfr signaling in Drosophila. Mech Dev 123:450–462

Panin VM, Papayannopoulos V, Wilson R et al (1997) Fringe modulates Notch-ligand interactions. Nature 387:908–912

Peschard P, Park M (2003) Escape from Cbl-mediated downregulation: a recurrent theme for oncogenic deregulation of receptor tyrosine kinases. Cancer Cell 3:519–523

Pfleger CM, Harvey KF, Yan H et al (2007) Mutation of the gene encoding the Ubiquitin activating enzyme uba1 causes tissue overgrowth in Drosophila. Fly (Austin) 1:95–105

Powers ET, Morimoto RI, Dillin A et al (2009) Biological and chemical approaches to diseases of proteostasis deficiency. Annu Rev Biochem 78:959–991

Rana NA, Haltiwanger RS (2011) Fringe benefits: Functional and structural impacts of O-glycosylation on the extracellular domain of Notch receptors. Curr Opin Struct Biol 21:583–589

Rebay I, Fleming RJ, Fehon RG et al (1991) Specific EGF repeats of Notch mediate interactions with Delta and Serrate: implications for Notch as a multifunctional receptor. Cell 67:687–699

Roignant JY, Treisman JE (2009) Pattern formation in the Drosophila eye disc. Int J Dev Biol 53:795–804

Rosenbaum EE, Hardie RC, Colley NJ (2006) Calnexin is essential for rhodopsin maturation, Ca2+ regulation, and photoreceptor cell survival. Neuron 49:229–241

Rosenbaum EE, Brehm KS, Vasiljevic E et al (2011) XPORT-dependent transport of TRP and rhodopsin. Neuron 72:602–615

Rothnie A, Clarke AR, Kuzmic P et al (2011) A sequential mechanism for clathrin cage disassembly by 70-kDa heat-shock cognate protein (Hsc70) and auxilin. Proc Natl Acad Sci U S A 108:6927–6932

Sakata T, Sakaguchi H, Tsuda L et al (2004) Drosophila Nedd4 regulates endocytosis of notch and suppresses its ligand-independent activation. Curr Biol 14:2228–2236

Sasamura T, Sasaki N, Miyashita F et al (2003) neurotic, a novel maternal neurogenic gene, encodes an O-fucosyltransferase that is essential for Notch-Delta interactions. Development 130:4785–4795

Satoh AK, Ready DF (2005) Arrestin1 mediates light-dependent rhodopsin endocytosis and cell survival. Curr Biol 15:1722–1733

Satoh AK, Xia H, Yan L et al (2010) Arrestin translocation is stoichiometric to rhodopsin isomerization and accelerated by phototransduction in *Drosophila* photoreceptors. Neuron 67:997–1008

Schlesinger A, Kiger A, Perrimon N et al (2004) Small wing PLCγ is required for ER retention of cleaved Spitz during eye development in Drosophila. Dev Cell 7:535–545

Schweitzer R, Shaharabany M, Seger R et al (1995) Secreted Spitz triggers the DER signaling pathway and is a limiting component in embryonic ventral ectoderm determination. Genes Dev 9:1518–1529

Scott K, Zuker CS (1998) Assembly of the *Drosophila* phototransduction cascade into a signalling complex shapes elementary responses. Nature 395:805–808

Sen A, Madhivanan K, Mukherjee D et al (2012) The epsin protein family: coordinators of endocytosis and signaling. Biomol Concepts 3:117–126

Shieh B-H (2011) Molecular genetics of retinal degeneration: A Drosophila perspective. Fly 5: 356-368.

Singh A, Tare M, Puli OR et al (2012) A glimpse into dorso-ventral patterning of the Drosophila eye. Dev Dyn 241:69–84

Sorkin A, Goh LK (2008) Endocytosis and intracellular trafficking of ErbBs. Exp Cell Res 314:3093–3106

Stamnes MA, Shieh BH, Chuman L et al (1991) The cyclophilin homolog ninaA is a tissue-specific integral membrane protein required for the proper synthesis of a subset of Drosophila rhodopsins. Cell 65:219–227

Sundaram MV (2005) The love-hate relationship between Ras and Notch. Genes Dev 19:1825–1839

Taylor RC, Dillin A (2012) Aging as an event of proteostasis collapse. Cold Spring Harb Perspect Biol 3. pii:a004440. doi:10.1101/cshperspect.a004440

Tian X, Hansen D, Schedl T et al (2004) Epsin potentiates Notch pathway activity in Drosophila and C. elegans. Development 131:5807–5815

Tien AC, Rajan A, Schulze KL et al (2008) Ero1L, a thiol oxidase, is required for Notch signaling through cysteine bridge formation of the Lin12-Notch repeats in Drosophila melanogaster. J Cell Biol 182:1113–1125

Tomlinson A, Struhl G (1999) Decoding vectorial information from a gradient: sequential roles of the receptors Frizzled and Notch in establishing planar polarity in the Drosophila eye. Development 126:5725–5738

Tomlinson A, Struhl G (2001) Delta/Notch and Boss/Sevenless signals act combinatorially to specify the Drosophila R7 photoreceptor. Mol Cell 7:487–495

Tomlinson A, Mavromatakis YE, Struhl G (2011) Three distinct roles for notch in Drosophila R7 photoreceptor specification. PLoS Biol 9:e1001132

Tsachaki M, Sprecher SG (2012) Genetic and developmental mechanisms underlying the formation of the Drosophila compound eye. Dev Dyn 241:40–56

Tsruya R, Schlesinger A, Reich A et al (2002) Intracellular trafficking by Star regulates cleavage of the Drosophila EGF receptor ligand Spitz. Genes Dev 16:222–234

Tsruya R, Wojtalla A, Carmon S et al (2007) Rhomboid cleaves Star to regulate the levels of secreted Spitz. EMBO J 26:1211–1220

Tsunoda S, Sierralta J, Sun Y et al (1997) A multivalent PDZ-domain protein assembles signalling complexes in a G-protein-coupled cascade. Nature 388:243–243

Umebayashi K, Stenmark H, Yoshimori T (2008) Ubc4/5 and c-Cbl continue to ubiquitinate EGF receptor after internalization to facilitate polyubiquitination and degradation. Mol Biol Cell 19:3454–3462

Ungewickell E, Ungewickell H, Holstein SE et al (1995) Role of auxilin in uncoating clathrin-coated vesicles. Nature 378:632–635

Urban S, Lee JR, Freeman M (2001) Drosophila rhomboid-1 defines a family of putative intramembrane serine proteases. Cell 107:173–182

Vabulas RM, Raychaudhuri S, Hayer-Hartl M et al (2010) Protein folding in the cytoplasm and the heat shock response. Cold Spring Harb Perspect Biol 2:a004390

Vaccari T, Lu H, Kanwar R et al (2008) Endosomal entry regulates Notch receptor activation in Drosophila melanogaster. J Cell Biol 180:755–762

Varshavsky A (2012) The Ubiquitin system, an immense realm. Annu Rev Biochem 81:167–176

Vogel C, Marcotte EM (2012) Insights into the regulation of protein abundance from proteomic and transcriptomic analyses. Nat Rev Genet 13:227–232

Wang W, Struhl G (2004) Drosophila Epsin mediates a select endocytic pathway that DSL ligands must enter to activate Notch. Development 131:5367–5380

Wang Y, Werz C, Xu D et al (2008) Drosophila cbl is essential for control of cell death and cell differentiation during eye development. PLoS One 3:e1447

Wang Y, Chen Z, Bergmann A (2010) Regulation of EGFR and Notch signaling by distinct isoforms of D-cbl during Drosophila development. Dev Biol 342:1–10

Ware FE, Vassilakos A, Peterson PA et al (1995) The molecular chaperone calnexin binds Glc1Man9GlcNAc2 oligosaccharide as an initial step in recognizing unfolded glycoproteins. J Biol Chem 270:4697–4704

Wasserman JD, Urban S, Freeman M (2000) A family of rhomboid-like genes: Drosophila rhomboid-1 and roughoid/rhomboid-3 cooperate to activate EGF receptor signaling. Genes Dev 14:1651–1663

Weinmaster G, Fischer JA (2011) Notch ligand ubiquitylation: what is it good for? Dev Cell 21:134–144

Wilkin MB, Carbery AM, Fostier M et al (2004) Regulation of notch endosomal sorting and signaling by Drosophila Nedd4 family proteins. Curr Biol 14:2237–2244
Wilkin M, Tongngok P, Gensch N et al (2008) *Drosophila* HOPS and AP-3 complex genes are required for a Deltex-regulated activation of Notch in the endosomal trafficking pathway. Dev Cell 15:762–772
Wu H, Rossi G, Brennwald P (2008) The ghost in the machine: small GTPases as spatial regulators of exocytosis. Trends Cell Biol 18:397–404
Xing Y, Böcking T, Wolf M et al (2010) Structure of clathrin coat with bound Hsc70 and auxilin: mechanism of Hsc70-facilitated disassembly. EMBO J 29:655–665
Xu A, Haines N, Dlugosz M et al (2007) In vitro reconstitution of the modulation of Drosophila Notch-ligand binding by Fringe. J Biol Chem 282:35153–35162
Xu L, Lubkov V, Taylor LJ et al (2010) Feedback regulation of Ras signaling by Rabex-5-mediated ubiquitination. Curr Biol 20:1372–1377
Xu XZ, Choudhury A, Li X et al (1998) Coordination of an array of signaling proteins through homo- and heteromeric interactions between PDZ domains and target proteins. J Cell Biol 142:545–555
Yamada K, Fuwa TJ, Ayukawa T et al (2011) Roles of Drosophila deltex in Notch receptor endocytic trafficking and activation. Genes Cells 16:261–272
Yan H, Chin ML, Horvath EA et al (2009) Impairment of ubiquitylation by mutation in Drosophila E1 promotes both cell-autonomous and non-cell-autonomous Ras-ERK activation in vivo. J Cell Sci 122:1461–1470
Yan H, Jahanshahi M, Horvath EA et al (2010) Rabex-5 Ubiquitin ligase activity restricts Ras signaling to establish pathway homeostasis in Drosophila. Curr Biol 20:1378–1382
Yang L, Baker NE (2003) Cell cycle withdrawal, progression, and cell survival regulation by EGFR and its effectors in the differentiating Drosophila eye. Dev Cell 4:359–369
Yogev S, Schejter ED, Shilo BZ (2008) Drosophila EGFR signaling is modulated by differential compartmentalization of Rhomboid intramembrane proteases. EMBO J 27:1219–1230
Yogev S, Schejter ED, Shilo BZ (2010) Polarized secretion of Drosophila EGFR ligand from photoreceptor neurons is controlled by ER localization of the ligand-processing machinery. PLoS Biol 8. pii:e1000505. doi:10.1371/journal.pbio.1000505
Yoon CH, Lee J, Jongeward GD et al (1995) Similarity of sli-1, a regulator of vulval development in C. elegans, to the mammalian proto-oncogene c-cbl. Science 269:1102–1105

Drosophila Eye as a Model to Study Regulation of Growth Control: The Discovery of Size Control Pathways

Shilpi Verghese, Indrayani Waghmare, Shree Ram Singh and Madhuri Kango-Singh

Introduction

In the biological sense, the term growth has intricate ramifications that we have only started to comprehend. Growth is the overall increase in cell mass or size of a tissue or organism (Conlon and Raff 1999; Cook and Tyers 2007; Edgar 1999; Raff 1996). Growth may be due to increase in cell number resulting from cell division (cell proliferation), increase in cellular mass without cell division (cell enlargement), or due to release of more extracellular matrix (cell accretion). These processes are intimately linked and it is clear that if coordinated growth has to occur in an organism, it is necessary for various biological pathways to interact and relay appropriate signals to proper cell types. Growth regulation is precisely controlled and affected by several intrinsic and extrinsic factors (Cooper 2004; Crickmore and Mann 2008; Grebien et al. 2005; Johnston and Gallant 2002). The intrinsic factors mainly involve synthesis and secretion of signals or ligands, which bind to their cognate receptors to relay downstream signals. These signals consist of variety of molecules such as hormones, mitogens, apoptosis-inducing signals, patterning and axis determining signals, etc.

Shilpi Verghese and Indrayani Waghmare have contributed equally to this work.

M. Kango-Singh (✉)
Center for Tissue Regeneration and Engineering at Dayton (TREND),
Department of Biology SC342, University of Dayton,
300 College Park, Dayton, OH 45469, USA
e-mail: mkangosingh1@udayton.edu

Premedical Program, University of Dayton,
300 College Park, Dayton, OH 45469, USA

S. Verghese · I. Waghmare
Department of Biology SC342, University of Dayton,
300 College Park, Dayton, OH 45469, USA

S. R. Singh
Mouse Cancer Genetics Program,
National Cancer Institute, Frederick, MD 21702, USA

A. Singh, M. Kango-Singh (eds.), *Molecular Genetics of Axial Patterning, Growth and Disease in the Drosophila Eye,* DOI 10.1007/978-1-4614-8232-1_9,

which eventually determine organ size and tissue homeostasis (Johnston and Gallant 2002; Mitchison et al. 1997; Montagne 2000; Tumaneng et al. 2012a). Growth of a tissue or organ is impacted not only by cell division but also by regulated cell death (apoptosis or programmed cell death; Bangs and White 2000; Jacobson et al. 1997; Martin et al. 2009; Oldham et al. 2000a; Richardson and Kumar 2002; Rusconi et al. 2000).

In this chapter, we will focus on growth regulation in imaginal discs (epithelial sacs that are precursors of adult appendages) in *D. melanogaster* (Bergantinos et al. 2010; Bryant 1978; Bryant 1987; Bryant 2001; Bryant and Schmidt 1990). The obvious advantages that *Drosophila* has to offer as a model organism include short life cycle, high fecundity, low cost maintenance, and lack of redundancy in genome (Bier 2005; Blair 2003; Boutros and Ahringer 2008; Pagliarini et al. 2003; St. Johnston 2002; Vidal and Cagan 2006). Furthermore, the sophisticated tools available in fly genetics provide great deal of versatility in terms of designing experiments. The plethora of knowledge thus generated through exhausting efforts of scientists has not only revealed to us the classic information about how growth occurs but has also lead to better understanding of growth-related diseases such as cancer.

Drosophila Eye as a Model to Study Regulation of Growth

The compound eyes of *Drosophila* arise from the eye-antennal imaginal discs, a monolayer epithelial sheet of cells that is responsible for the development of the eyes, the antennae, the ocelli, and a major part of the adult head cuticle. Each eye of the adult fruit fly on an average consists of about 800 ommatidia (Wolff and Ready 1993). Ommatidia arise from a set of 19 precursor cells that are generated by spatially and temporally coordinated cellular processes such as cell-proliferation, cell-differentiation, and cell-death in the eye imaginal discs. Eighteen of these cells contribute to the eye per se, whereas the nineteenth cell gives rise to a sensory bristle (Cagan 1993). A key feature that distinguishes eye from the rest of the organs is its ability to perceive light and relay the signal to distinct areas in the brain called the optic lobes. The eye imaginal discs arise from about 50 primordial cells that express the *Drosophila* PAX 6 gene *eyeless* (*ey*) during mid-to-late embryogenesis. Two such discs develop in each larva and differentiate into two compound eyes, antennae, ocelli, and the head cuticle in the adult.

Much is known about the regulation of growth and differentiation of the eye-antennal imaginal discs (Baker 2001; Cagan 1993; Dominguez and Casares 2005; Hafen 1991; Kramer and Cagan 1994; Kumar 2001). Until the second larval instar of development, the cells of the eye-antennal discs proliferate without differentiation (Baker 2001; Wolff and Ready 1993). During the second instar stage, a unique process of cell differentiation begins in the eye-antennal disc that paves the way for formation of photoreceptor neurons in the posterior region of the eye-antennal imaginal disc (Wolff and Ready 1993). The differentiation occurs in the wake of a so-called "morphogenetic furrow"—a front marked by apical constriction of epithelial cells in response to complex developmental signaling from the Hedgehog (Hh), Decapentaplegic (Dpp), Wingless (Wg), and Epidermal growth factor receptor (EGFR) pathways (Acquisti et al. 2009; Chen and Chien 1999; Firth et al. 2010; Harvey et al. 2001; Kango-Singh et al. 2003; Penton et al. 1997). Posterior to the morphogenetic furrow, the cells begin to acquire particular photoreceptor cell fates and organize into ommatidial clusters.

Anterior to the furrow, the cells divide asynchronously and do not differentiate, however, in the morphogenetic furrows, cells arrest in the G1 phase of the cell cycle, synchronize, and either start to differentiate into photoreceptor cells as they leave the furrow or undergo one additional round of cell division, referred to as the second mitotic wave (SMW) before differentiating into the remaining photoreceptor, cone, pigment, and bristle cells (Baker 2001; Dickson and Hafen 1993; Wolff and Ready 1993). The cells posterior to the morphogenetic furrow enter G1 arrest caused by Dpp (*decapentaplegic*) signaling that is maintained by the *roughex* (*rux*) gene, which negatively regulates G1-S transition. The cells that are temporarily trapped in the G1 phase begin differentiation with specification of the R8 (photoreceptor) cell due to expression of the proneural protein Atonal (Ato) (Baker et al. 1996; Chen and Chien 1999; Daniel et al. 1999; Dominguez 1999; Greenwood and Struhl 1999; Jarman et al. 1994). R8 recruits other photoreceptor cells-R2, R3, R4, and R5 to form a cluster of five photoreceptor precursors. Once specified, these cells never enter cell cycle or cell division again. All other nonspecified cells re-enter cell cycle only once at the SMW (Baker 2001; de Nooij and Hariharan 1995). Cells in the SMW undergo G2/M phase that is mediated through local signaling from Spitz (Spi). Binding of Spi to its cognate receptor EGFR in precursor cells causes activation of downstream *string* (*stg*) that completes the G2-M transition during mitosis. Local Spi-EGFR signaling also plays an important role limiting the progression of SMW. For instance, on an average the Spi signal from one precluster can span to a length of seven cells only causing these cells to divide whereas the remaining cells remain arrested in G2 phase and fail to divide (Baker 2001; Brumby and Richardson 2003; de Nooij and Hariharan 1995; Jarman et al. 1994; Price et al. 2002; Wolff and Ready 1991). The progression of the morphogenetic furrow is complete by the mid-third instar of larval development, and the eye-antennal disc is fully grown to about 50,000 cells (Kumar 2009; Kumar and Moses 2000; Kumar and Moses 2001; Sun 2007).

Following development in larval stages, supernumerary cells are eliminated via apoptosis during pupal development. This event is mediated through Notch signaling (Bonini and Fortini 1999; Burke and Basler 1997; Sawamoto and Okano 1996; Treisman and Heberlein 1998; Zipursky 1989). By contrast, survival of pupal cells is brought about by EGFR expression that mediates its cell survival function through suppressing the transcriptional activity of the proapoptotic gene *head involution defective* (*hid*) (Bonini and Fortini 1999). In addition, survival signals emanating from cone or primary pigment cells in each ommatidium play a role in survival and proliferation of secondary and tertiary pigment cells, and secondary bristle organs (Cagan 1993, 2009; Rubin 1989; Singh et al. 2012; Tsachaki and Sprecher 2012; Yamamoto 1993). During metamorphosis, the two eye-antennal imaginal discs fuse at the dorsal midline to form the fly head with three ocelli, two antennae, and compound eyes.

Thus, the eye-antennal disc is ideal for the study of organogenesis, morphogenesis, pattern formation, and several cell biological processes including the regulation of cell cycle, cell death, cell junctions and adhesion, transport of molecules, cell signaling, and metabolism. Recently, the eye discs have been used as an experimental system for genetic screens to discover postembryonic lethality, and for screening small molecule inhibitors in chemical and drug screens.

The Mosaic Analysis Systems and the *Drosophila* Eye

Mutagenesis screens are a very well-established tool for gene discovery in flies (for review, see Bellen et al. 1989, 2011; Blair 2003; Pfeiffer et al. 2010; St. Johnston 2002; Venken and Bellen 2012; Xu and Rubin 1993). Over the years, the *mosaic techniques* have evolved to include the Flipase(*FLP*)-Flipase recognition target (*FRT*), *eyGAL4 UASFlp EGUF*, Flp-out clones, and Mosaic Analysis with Repressible Cell Marker MARCM (for review, see Blair 2003; St. Johnston 2002). One of the first tissue-specific mosaic systems was developed in the eye-antennal discs where the mosaic clones were restricted to the eye-antennal discs by virtue of expression of the Flippase gene under the control of the Eyeless Promoter (commonly referred to as the '*ey-FLP* system', Newsome et al. 2000). This tissue-specific system was further refined by the development of the "*cell-lethal*" system, where effects of loss of function of a gene could be surveyed more clearly because the wild-type twin-clones are eliminated due to the presence of *cell-lethal* mutations (the *cell-lethal FLP-FRT* system; Newsome et al. 2000). We focus on the genetic screens performed about 10–12 years ago (simultaneously in many labs) that led to the identification of many new genes that were shown to belong to the two major growth regulatory networks: the Hippo pathway and the Tuberous Sclerosis Complex/Target of Rapamycin—TSC-TOR pathway.

Genetic Screens for Genes That Regulate Growth: The "Big-Head" and "Pin-Head" Mutations

Barry Dickson's group (Newsome et al. 2000) improved the traditional FLP-*FRT* approach developed in the Rubin Lab (Xu and Rubin 1993), to allow generation of essentially mutant eye discs by eliminating the wild-type twin clone via a *cell-lethal* mutation (the *cell-lethal* FLP-*FRT* system) (Fig. 1a). This so-called "*cell-lethal*" approach allows the mutant clones to grow to their highest potential due to elimination of competitive interactions between the mutant cells and their wild-type neighbors. Using this system, several groups carried out mutagenesis screens in flies (on the X, 2L, 2R, 3L, 3R chromosomes) and found mutations that affected patterning, growth, cell death, and differentiation (for review, see St. Johnston 2002).

Of special interest were the genes mutations which caused a remarkable effect on growth without disrupting the patterning process (Conlon and Raff 1999; Johnston and Gallant 2002; Mitchison et al. 1997; Oldham et al. 2000a; Raff 1996; Su and

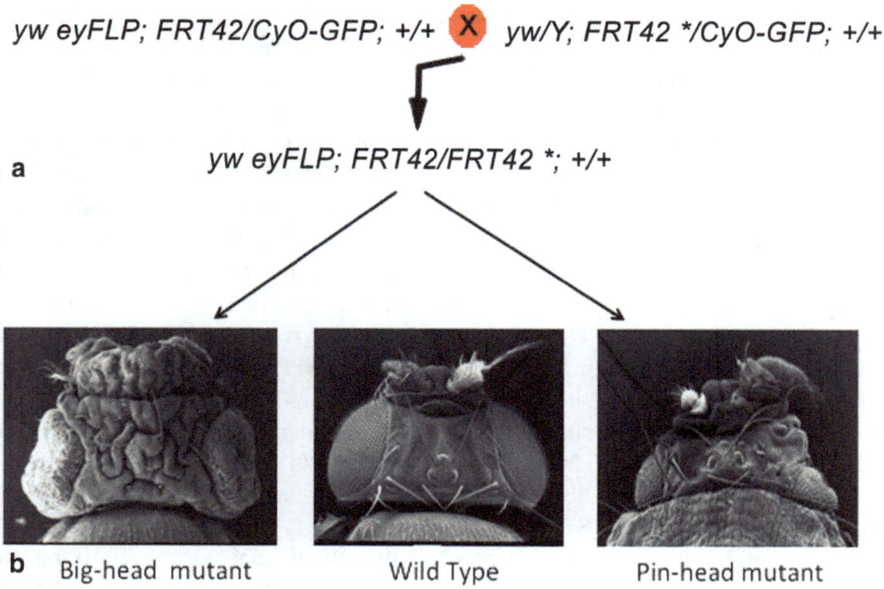

Fig. 1 Genetic screens for eye-specific mosaics lead to the identification of several Hippo and TSC-TOR pathway mutants. **a** Modified mutagenesis scheme where * represents a point mutation induced by a mutagen (e.g., Ethyl Methane Sulphonate (EMS)), **b** Typical phenotypes of Hippo and TSC-TOR pathway mutant from the mutagenesis screen

O'Farrell 1998; Tumaneng et al. 2012a). Characterization of these mutants revealed the mechanisms that regulate growth and tissue size by controlling cell number (Hippo pathway) (Zhao et al. 2011b) or cell size (InR/TSC-TOR pathway) (Kim and Guan 2011; Loewith 2011; Montagne 2000; Potter et al. 2003; Soulard et al. 2009) in a developing organ. Typically, loss of function mutations in positive regulators of these pathways caused development of enlarged heads that showed overgrowth—referred to as the "big head" mutations (Fig. 1b) (Hafen 2004; Oldham and Hafen 2003; Pan 2007, 2010). In contrast, loss of function of negative regulators of these pathways caused reduction in head size and development of smaller organs, which may be due to cell death or reduction in cell size, and were referred to as the "pin head" mutations (Fig. 1b).

The Hippo Signaling Pathway

The Hippo signaling pathway was first discovered in flies following characterization of "big-head" mutants identified from genetic screens (for review, see Edgar 2006; Pan 2007; Saucedo and Edgar 2007). Analysis of the loss of function phenotypes revealed that a fundamental function of the Hippo pathway was the regulation of organ size (Boggiano and Fehon 2012; Harvey and Hariharan 2012; Schroeder and Halder 2012; Staley and Irvine 2012). Interestingly, the pathway received its name just after some growth regulatory genes (*warts* (*wts*), *salvador* (*sav*, aka *shar-pie, shrp*)) were characterized. Warts (*wts*) was named based on the bumpy "warts-like" phenotype of the mutant cells in mitotic (mosaic) clones on the body of the adult flies that were reminiscent of the warts on toads (Justice et al. 1995). Another group led by Xu et al. (1995) also independently found *warts* in the initial FLP/*FRT*-based screen and named it *large tumor suppressor* (*LATS*) (Xu et al. 1995). Two independent groups identified the gene encoding the adaptor protein Salvador (Sav aka Shar-pie, Shrp after the dog species of the same name as the mutant flies showed a characteristic phenotype of folded dark cuticle on the overgrown heads) from complementation groups isolated from the big-head genetic screens (Kango-Singh et al. 2002; Tapon et al. 2002). Interestingly, both Wts and Sav regulated growth by suppressing proliferation and promoting apoptosis. Hippo (Hpo) was the name given to another complementation group from the "big-head" screens that showed a phenotype that was very similar to Wts and Sav mutants (Harvey et al. 2003; Jia et al. 2003; Pantalacci et al. 2003; Udan et al. 2003; Wu et al. 2003).

Molecular analysis of the three genes revealed that Wts and Hpo genes encode for serine-threonine (S-T) kinases whereas Sav is a WW domain containing adaptor protein. By this time it was clear that Warts, Salvador, and Hippo all show similar loss of function phenotypes and control organ size by a common signaling pathway that promotes apoptosis and restricts cell proliferation (Edgar 2006; O'Neill and Kolch 2005; Rothenberg and Jan 2002), and the pathway got its name from the last member of this trio of genes. A complete pathway that relays a growth regulatory signal from the plasma membrane to the nucleus has emerged over the last decade. Although genetic mutagenesis screens led to the initial discovery of this pathway, several components were identified by other genetic screening strategies and biochemical approaches (e.g., yeast-two hybrid screens, TAP-TAG based protein interaction assays; for review, see Halder and Johnson 2011; Kango-Singh and Singh 2009; Staley and Irvine 2012; Tumaneng et al. 2012a; Varelas and Wrana 2012). Today the Hippo pathway has grown to a large network of tumor suppressor genes that function upstream and downstream of the three initial members of the Hippo pathway (also known as the core kinase cascade) that control several aspects of tissue homeostasis. Overall, the Hippo signalling pathway is a key size regulatory pathway that controls organ size in flies and vertebrates, and misregulation of Hippo signalling is implicated in several diseases including cancer (for review, see Harvey and Hariharan 2012; Schroeder and Halder 2012; Staley and Irvine 2012; Zhao et al. 2011b; Fig. 2).

Regulation by Core Kinase Cascade of the Hippo Pathway

The molecular analysis of the three initial members of the Hippo pathway in *Drosophila* revealed that Hpo codes for an S-T kinase of the mammalian Sterile-20 family of kinases (Harvey et al. 2003; Jia et al. 2003; Pantalacci et al. 2003; Udan et al. 2003; Wu et al. 2003), and can physically associate with the WW-domain containing adaptor protein Sav (Harvey et al. 2003; Jia et al. 2003; Pantalacci et al. 2003;

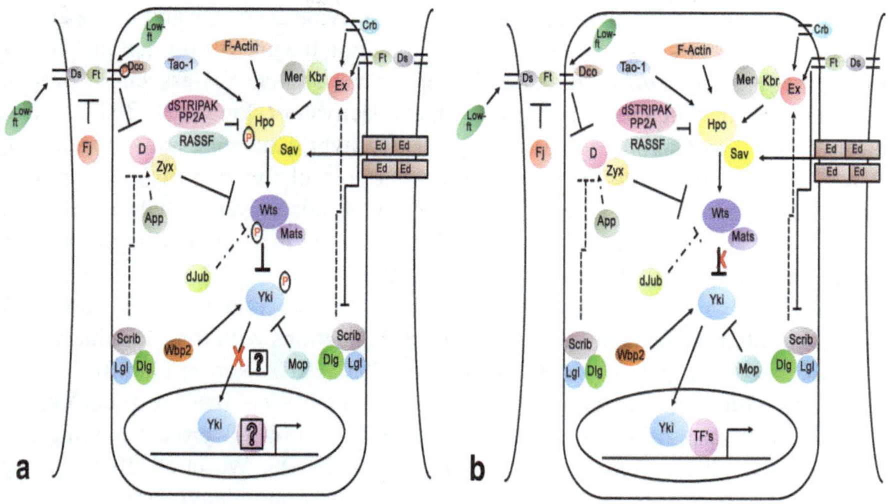

Fig. 2 Schematic representation of the Hippo pathway in *D. melanogaster*. **a** Signaling interactions when Hippo pathway is downregulated in response to extracellular signals. Hpo fails to phosphorylate Wts. Inactive Wts cannot phosphorylate Yki, and allows Yki to enter the nucleus to bind cognate transcription factors and induce expression of target genes. **b** Signaling interactions when Hippo pathway is activated by stress. Hpo is phosphorylated and in turn phosphorylates Wts with the help of adaptor proteins Sav and Mats. Activated Wts phosphorylates Yki and prevents it from entering the nucleus, thus preventing transcription of target genes. In addition, cell death is induced when the pathway is hyperactivated

Udan et al. 2003; Wu et al. 2003). Wts is an S-T kinase protein of the dystrophia myotonica protein kinase (DMPK) family that associates with another adaptor protein Mob as tumor suppressor (Mats; Justice et al. 1995; Lai et al. 2005; Shimizu et al. 2008; Wei et al. 2007; Xu et al. 1995). Loss of function of these genes in genetic mosaics revealed strong overgrowth phenotype caused by increased cell proliferation and diminished sensitivity to apoptosis. Hyperactivation of the pathway by over-expression of Hpo, Sav, Wts, or Mats leads to formation of smaller organs due to increased apoptosis (Harvey et al. 2003; Pantalacci et al. 2003; Udan et al. 2003; Wei et al. 2007; Wu et al. 2003). Biochemical analysis showed that the Hpo kinase phosphorylates and can physically associate with Sav, Wts, and Mats to form protein complexes in vitro (Wei et al. 2007). However, Hpo associates with its cognate adaptor protein Sav to form the Hpo-Sav complex for efficient activation of the downstream kinase Wts (Huang et al. 2005; Wu et al. 2003). Wts itself associates with Mats to form the downstream Wts-Mats complex of the core kinase cascade of the Hippo pathway (Wei et al. 2007). Association of these adaptor proteins is known to stimulate the catalytic activity of the Hpo and Wts kinases (Dong et al. 2007; Pan 2007; Wei et al. 2007). Moreover, phosphorylation of Mats by the Hpo kinase increases its affinity for the Wts kinase (Dong et al. 2007; Pan 2007; Pan 2010; Wei et al. 2007). Wts is activated by autophosphorylation and phosphorylation by Hpo-kinase. Activated Wts associates with Mats (thus Mats cannot simultaneously associate with Hpo and Wts), which acts as a coactivator for the kinase activity of

Wts (Dong et al. 2007; Huang et al. 2005; Oh and Irvine 2008; Oh and Irvine 2009). A major output of the core kinase cascade is to inhibit the growth-promoting activity of Yorkie (Yki), the *Drosophila* homolog of the mammalian Yes-associated protein (YAP) oncogene that acts as a transcriptional coactivator (Dong et al. 2007; Huang et al. 2005). Yki was identified via a yeast two-hybrid screen as an interactor of Warts. Over-expression of Yki phenocopies the loss of function of *hpo, sav, wts,* and *mats* (all genes of the core kinase cascade) and causes over-growth (Dong et al. 2007; Wei et al. 2007). Loss of function of *yki* results in formation of smaller organs due to induction of cell death (Huang et al. 2005).

Yki activity is regulated by controlling its subcellular localization via phosphorylation-dependent and -independent interactions with the core kinase cascade of the Hippo pathway (Oh and Irvine 2008, 2010; Ren et al. 2010b). Yki associates with Wts, and one mechanism by which the Wts-kinase restricts Yki activity is via phosphorylation at Ser168 that creates a 14-3-3 protein-binding site (Goulev et al. 2008; Peng et al. 2009; Ren et al. 2010b; Wu et al. 2008; Zhang et al. 2008b; Zhao et al. 2008b). Interestingly, only phosphorylated forms of Yki can associate with 14-3-3 proteins. Yki is phosphorylated at multiple sites (e.g., Ser 111 and S250), making it less sensitive to Hpo/Wts-mediated inhibition. These phosphorylation events act in parallel to phosphoYki/14-3-3 mediated mechanisms and inhibit Yki nuclear localization and activity. It is suggested that nuclear export is required for shuttling Yki to the nucleus in response to Hpo signaling, and binding of 14-3-3 proteins is thought to impede nuclear import and/or promote nuclear export thereby facilitating nucleocytoplasmic shuttling of target proteins (Brunet et al. 2002; Kumagai and Dunphy 1999). Nuclear transport of Yki depends on its binding with cognate transcription factors as Yki does not have an intrinsic nuclear localization signal (NLS) (Goulev et al. 2008; Zhang et al. 2008b). Currently, it is unclear if binding of 14-3-3 proteins to Yki prevents its binding with cognate transcription factors, or masks the NLSs or promotes export from the nucleus. Nevertheless, coactivator Yki/YAP is the critical downstream regulatory target of the Hpo kinase cascade, and regulation of its subcellular localization is the primary mechanism by which the Hpo pathway influences target gene expression (Goulev et al. 2008; Huang et al. 2005; Oh and Irvine 2008, 2009, 2010; Oh et al. 2009; Peng et al. 2009; Ren et al. 2010b).

Yki (like Sav) is a WW-domain-containing protein and interacts with the PPxY (where P = Proline; x = any amino acid; Y = Tyrosine) motifs in Wts (Huang et al. 2005). Besides Wts, the WW-domains of Yki interact with the PPxY motifs present in other components of Hippo signaling pathway like Expanded (Ex), Hpo, WW-domain-binding protein 2, and Myopic to regulate Hippo signaling via phosphorylation-independent mechanisms (Badouel et al. 2009; Gilbert et al. 2011; Oh et al. 2009; Zhang et al. 2011b). Another protein that acts via its WW-domains is Kibra which associates with the PPxY motifs in Ex (and binds Mer in a WW-domain independent manner; Baumgartner et al. 2010; Genevet et al. 2010). The identification of multiple proteins that act through the interaction between WW-domains and PPxY motifs in the Hippo pathway suggests that these motif-specific interactions are important for regulation of Hippo signaling (reviewed in Sudol (2010); Sudol and Harvey (2010)).

Yki Activity and Regulation of Expression of Target Genes

Hyperactivation of the pathway, for example, by overexpression of Hpo, leads to phosphorylation and activation of Hpo and Wts with the help of adaptor proteins Sav and Mats. Wts, in turn, phosphorylates the transcriptional coactivator Yki, which associates with 14-3-3 proteins and remains sequestered in the cytoplasm (Dong et al. 2007; Huang et al. 2005; Oh and Irvine 2008; Oh et al. 2009; Ren et al. 2010b). Analysis of adult and imaginal disc phenotypes reveals that over-expression of Hpo results in induction of ectopic apoptosis early in development in imaginal disc cells due to induction of caspase-dependent cell death (Hamaratoglu et al. 2006; Harvey et al. 2003; Udan et al. 2003; Verghese et al. 2012a). In mammalian cells, activation of MST (Mammalian Sterile-20 like kinase)1/2 and hyperphosphorylation of YAP2 by MST2 and LATS1 kinase leads to activation of cell death. Interestingly, MST1/2 are known targets of caspases and YAP1/2 are known to interact with p73 via a PDZ domain in YAP, and induce apoptotic target genes (Bertini et al. 2009; Sudol 2010; Sudol and Harvey 2010). However, these mechanisms of regulating apoptosis may not be conserved in flies because the site for caspase cleavage is not conserved in *Drosophila* Hpo (Wu et al. 2003), and *Drosophila* Yki does not have the conserved PDZ domain (Sudol and Harvey 2010). Nevertheless, Hpo overexpression in flies induces apoptosis through an alternate mechanism that does not involve caspase cleavage or p73. Recently, it was shown that the effector caspase Dronc (*Drosophila* homolog of mammalian Caspase 9) is induced in conditions when Hippo pathway is hyperactivated. Further, using reporter genes, it was shown that *dronc* transcription is induced during gain-of-function and downregulated during loss-of-function conditions of the Hippo pathway, suggesting that *dronc* is a transcriptional target of the Hippo pathway (Verghese et al. 2012a). However, the molecular mechanism by which Yki interacts with Dronc remains unclear. Both phosphorylation-dependent (e.g., with 14-3-3 by phosphorylation-dependent mechanisms) and phosphorylation-independent mechanisms (binding with Hpo, Wts, or Ex) result in cytoplasmic retention of Yki in multiple protein complexes. Thus, the possibility remains that hyperactivation of Hippo pathway, releases Yki from one or more cytoplasmic complexes to allow its binding to transcription factors and shuttle into the nucleus to induce *dronc* transcription. Alternatively, hyperactivation of the Hippo pathway involves a transcriptional repressor that acts together with or independent of Yki to control *dronc* expression. Thus, although it is clear that hyperactivation of the Hippo pathway leads to induction of apoptosis, the molecular mechanisms underlying this process are yet unidentified.

When the pathway is downregulated, the genes of the core kinase cascade act as tumor suppressors by suppressing the growth-promoting activity of Yki. Under these conditions, Yorkie can partner with transcription factors like the TEAD family protein, Scalloped (Sd) and enter the nucleus and cause transcription of target genes which regulate cell proliferation and apoptosis. Sd was identified as the transcriptional factor of the pathway via yeast two-hybrid screen, and in-vitro Yki activity assays (luciferase assay) (Goulev et al. 2008; Wu et al. 2008; Zhang et al. 2008b). Sd is required for wing development (Campbell et al. 1992;

Liu et al. 2000), whereas Yki is required for regulating growth of all imaginal disc cells. Other transcription factors that bind Yki to regulate growth via Hippo signaling have since been discovered. These include Mothers Against Dpp (Mad) (Alarcon et al. 2009; Oh and Irvine 2010; Peng et al. 2009), Homothorax (Hth), and Teashirt (Tsh) (Peng et al. 2009). Mad is a known transcription factor within the Dpp/tumor growth factor (TGFβ) signaling pathway, and Mad and Hth were shown to control the activity of the *bantam miRNA* (Alarcon et al. 2009; Peng et al. 2009). Mad, Hth, and Tsh are known transcription factors that respond to other signals and are required for patterning of imaginal discs during development.

Yki activity is controlled by the upstream signals (Grusche et al. 2010; Oh and Irvine, 2010). A large number of target genes have been identified over the past decade, which include the cell cycle regulators E2F1, and *cyclin E, A, B, D*; the growth promoter *Myc,* and cell survival-promoting miRNA *bantam*, genes regulating cell death like the *Drosophila inhibitor of apoptosis diap1, hid, dronc*; and cytoskeletal proteins like *f-actin*, which drive cell proliferation and cell survival (Fig. 3)(Goulev et al. 2008; Harvey et al. 2003; Huang et al. 2005; Jia et al. 2003; Kango-Singh et al. 2002; Neto-Silva et al. 2010; Nolo et al. 2006; Pantalacci et al. 2003; Peng et al. 2009; Tapon et al. 2002; Thompson and Cohen 2006; Udan et al. 2003; Wu et al. 2003; Wu et al. 2008; Zhang et al. 2008a; Ziosi et al. 2010). Yki also controls the expression of several upstream components of the Hpo pathway like Ex, Mer, Kibra, Crumbs (Crb) and Four-jointed (Fjose et al. 1984) by a negative feedback loop (Cho et al. 2006; Fjose et al. 1984; Genevet et al. 2009, 2010; Hamaratoglu et al. 2006). Recently, Yki was shown to affect the expression of components of other signaling pathways, such as ligands for the Notch, Wnt, EGFR, and Jak-Stat pathways (Cho et al. 2006; Karpowicz et al. 2010; Ren et al. 2010a; Shaw et al. 2010; Staley and Irvine 2010, 2012; Zhang et al. 2009a). These interactions suggest that Hippo pathway interacts with the major signal transduction pathways, and these points of contact between different pathways may play an important role in controlling correct tissue sizes and maintaining homeostasis (Fig. 4).

Genetic and biochemical studies thus provide a basic premise for how Yki activity is modulated when Hippo signaling is downregulated or upregulated (Halder and Johnson 2011; Harvey and Hariharan 2012; Schroeder and Halder 2012; Staley and Irvine 2012). Studies in imaginal discs and other cell types like intestinal stem cells and fat cells revealed that Hippo signaling is needed in all cell types to regulate growth, and that the activity of the pathway is modulated to achieve tissue homeostasis (Halder et al. 2012; Halder and Johnson 2011; Harvey and Hariharan 2012; Tumaneng et al. 2012a; Zhao et al. 2008a; Zhao et al. 2010a). Whether Hippo signaling pathway is regulated by other global instructive signals (e.g., morphogen gradients) or if the pathway is constitutively active remains unknown. However, several inputs that communicate a growth regulatory signal to the core kinase cascade have been identified. We will discuss the key inputs, and their connection to the core kinase cascade in the following sections.

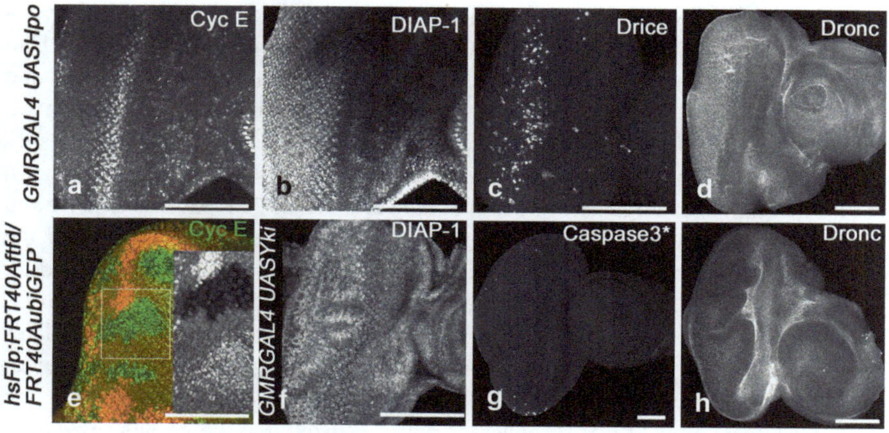

Fig. 3 Hippo pathway target genes regulate cell proliferation and apoptosis. (**a–d**) *GMRGAL4 UASHpo* third instar eye-antennal imaginal disc showing affect on target proteins upon pathway hyperactivation in the GMR domain. **a** Cyc E is downregulated, **b** DIAP-1 levels remain unaffected, and **c** Drice is activated (Drice is the homolog of *Drosophila* Caspase3* and is a read-out of active Dronc), **d** Dronc is upregulated in the GMR domain upon Hpo overexpression. **e** Loss of function clones of *ft* (GFP negative) made with yw hsFLP; UbiGFP [*hsFLP; FRT40A* ft^{fd}*/FRT40A ubiGFP*] show upregulation of Cyc E in the mutant cells. This effect is very strong in the region of the SMW. (**f–h**) *GMRGAL4 UASYki* third instar eye-antennal imaginal discs. **f** DIAP-1 is up-regulated, **g** Caspase3* staining is not observed, and **h** Dronc is down-regulated in the GMR domain consistent with overproliferation and no apoptosis

Upstream Regulators of the Hippo Pathway

Since the discovery of the core kinase cascade, several upstream regulators of the Hippo pathway were identified. These discoveries highlighted two remarkable properties of the Hippo pathway—one, that the Hippo pathway is a signaling network with multiple points of signal integration rather than a linear system of epistatic genes; and two, the interactions between various protein complexes (at the signal integration points) may play a decisive role in shaping the outcome, i.e., Yki activity levels. Although our understanding of the network is incomplete in both these areas, it is clear that signaling interactions within this pathway are shaped by several distinct inputs.

I. Fat signaling and the Hippo Pathway *Fat* (*ft*) alleles were spontaneous mutations first described by Mohr (1923, 1929). Subsequent analysis of mutations in the *ft* locus revealed both viable and lethal alleles, of which the null alleles are larval lethal and show hyperplastic overgrowth of imaginal discs thereby acting as tumor suppressor genes (Bryant et al. 1988). Molecular cloning of *ft* revealed that it codes for a transmembrane protein, which is an atypical Cadherin (Mahoney et al. 1991). Loss of *ft* affects two distinct aspects of imaginal disc growth and development, restriction of cell proliferation and generation of correctly oriented cells within the epithelial sheet, phenotypes that were mapped to two distinct signaling pathways—the Hippo and the Planar Cell Polarity (PCP) pathway (see (Brittle et al. 2010; Cho et al. 2006; Matakatsu and Blair 2006; Matakatsu and Blair 2008; Matakatsu and Blair 2012)). Ft is ubiquitously expressed, however its functions are regulated by two genes, Dachsous (Ds) and Fj, which are expressed in gradients in developing tissues (Matakatsu and Blair 2004; Reddy and Irvine 2008). Ds is another proto-cadherin in flies that acts as the ligand for Ft for both the Hippo and PCP pathways (reviewed in (Thomas and Strutt 2012)). Fj is a Golgi-localized kinase that phosphorylates the extracellular Cadherin domains of Ft and Ds to promote their binding (Ishikawa et al. 2008; Simon et al. 2010). Phosphorylation of Fat by Fj increases its affinity to Ds while phosphorylation of Ds reduces its affinity to Ft. One way in which Fat regulates growth and PCP is based on the slope and vector of the Ds and Fj gradients (Halder and Johnson 2011; Willecke et al. 2008; Zecca and Struhl 2010) .

Several years after Ft was discovered, it was realized that the growth regulatory functions of Fat were tied to the Hippo pathway (Bennett and Harvey 2006; Cho et al. 2006; Silva et al. 2006; Willecke et al. 2006). Loss of *ft* in mutant clones phenocopied the loss of function phenotypes of genes within the core kinase cascade of the Hippo pathway. Imaginal discs containing somatic clones of *ft* mutant cells continued to proliferate when normal cells had stopped, thereby forming large overgrown discs. Transcriptional targets of Hippo pathway are induced within the *ft* mutant cells, a phenotype similar to loss of function of positive regulators of Hippo pathway (e.g., *wts, hpo, sav, mats*). Ft affects the levels and localization of Hippo pathway components, including Wts, Ex, and Yki (Bennett and Harvey 2006; Cho et al. 2006; Oh and Irvine 2008; Silva et al. 2006; Tyler and Baker 2007; Willecke et al. 2006). Ft influences Hippo signaling independent of other upstream regulators like *expanded, merlin* (*mer*), and *kibra* which form a heteromeric complex (Ex-Mer-Kibra), and other genes like the Tao-1 kinase (Boggiano et al. 2011; Poon et al. 2011) that act upstream of Hpo (Boggiano and Fehon 2012). However, several other genes were recently identified, which specifically act downstream of Ft and integrate with the Hippo pathway by influencing the activity of the downstream kinase Wts. Thus, the Fat branch of the Hippo pathway has emerged, which independently influences Wts activity and tissue growth (Halder and Johnson 2011; Kango-Singh and Singh 2009; Reddy and Irvine 2008; Staley and Irvine 2012).

Several components of the Ft branch influence the intracellular domain of Ft—the region critical for transducing the signal within cells. These include the *Drosophila* Discs overgrown (Dco), a homolog of Casein Kinase I, that phosphorylates the Ft intracellular cytoplasmic domain in a Ds-dependent manner (Cho et al. 2006; Feng and Irvine 2009; Sopko et al. 2009); and the unconventional myosin Dachs (D) (Cho et al. 2006; Cho and Irvine 2004; Mao et al. 2006). Loss of function of dco^3, a hypomorphic allele, in homozygous discs and in somatic clones result in tissue overgrowth, and shows elevated levels of Fj and Diap-1 (Bryant and Schmidt 1990; Feng and Irvine 2009; Guan et al. 2007). Dco binds to the cytoplasmic domain of Fat, and in *dco* mutants, Fat intracellular domains fail to phosphorylate. Ds enriches availability of Fat at the point of cell contacts by forming *cis*-dimers with Fat. This promotes the transphosphorylation of Fat by Dco. Lowfat is a novel protein that interacts with the intracellular domains of Fat and Ds, and stabilizes the Fat-Ds interaction (Mao et al. 2009). Lowfat was identified in a genome-wide yeast two-hybrid screen as a Fat- and Ds-interacting protein (Mao et al. 2006, 2009). In addition, the palmitoyltraserase Approximated (App) acts downstream of Ft, and Ft regulates the localization of D to the membrane through APP (Matakatsu and Blair 2008). Recently, the apical-basal polarity gene *scribble* (*scrib*) (Verghese et al. 2012b) and the LIM (Lin-1; Isl-1; Mec-3)-domain protein *zyxin 102* (*zyx*) (Rauskolb et al. 2011) were shown to act in the Fat branch of Hippo signaling pathway (Bennett and Harvey 2006; Cho et al. 2006; Meignin et al. 2007; Polesello and Tapon 2007; Reddy et al. 2010; Silva et al. 2006; Willecke et al. 2006).

The differences in Ds and Fj expression between neighboring cells stimulate Yki activity, whereas the vector property of the gradients effects PCP signaling. Localization of D to the membrane is regulated by Fj, Ds, and Ft (Cho et al. 2006; Mao et al. 2006; Rogulja et al. 2008; Willecke et al. 2008). D controls Yki activity by two alternative mechanisms, one, involves post-translational effects of Ft on Wts, and the second involves the localization of Ex to the subapical membrane (Bennett and Harvey 2006). The apical basal polarity gene *scrib* and the atypical myosin D are responsible for partitioning the growth regulatory signal from Ft to downstream genes. Genetic epistasis experiments placed Ft upstream of D, and the apical regulator of the pathway—Ex (Cho et al. 2006; Mao et al. 2006; Silva et al. 2006; Willecke et al. 2008; Verghese et al. 2012b). D can reverse the effects of loss of *ft* on growth, and expression of Fat target genes like *wg*, *serrate*, and *fj* (Mao et al. 2006). Scrib was also placed upstream of D and Ex, and downstream of Ft based on genetic epistasis experiments (Verghese et al. 2012b). When Ft is inactive, D is regulated by Approximated (App) (Matakatsu and Blair 2008). App post-transcriptionally modifies D and affects its localization at the apical cell cortex. Hence, App functions in the Hippo pathway by affecting the availability of D at the apical cell cortex. When Ft is activated, D is released from App and binds to Zyxin (Zyx), which in turn interacts with Wts and stabilizes Wts activity (Rauskolb et al. 2011). Genetic epistasis experiments placed Zyx downstream of Ft and Dco, and upstream of Wts (Feng and Irvine 2007, 2009; Rauskolb et al. 2011). Thus, influencing Wts stability is a primary mechanism by which Ft controls growth via Hippo signaling. The other input via Ex remains less clear although there is clearly an input from Ft to Ex that also contributes to the Fat-branch-related phenotypes and regulation of the Hippo signaling pathway. Does Fat signaling simultaneously signal through Ex (and the core kinase cascade) and D; or the signals downstream of Ft are partitioned to allow maximum and more efficient signal transduction to the core kinase cascade remains unknown. Currently, the possibility that certain extracellular signals preferentially transmit the signal to Ex or D downstream of Ft has not been addressed.

II. Apical membrane proteins of the Hippo pathway Over the last 5 years, it has become clear that membrane-localized proteins are an intrinsic part of the Hippo signaling pathway (Genevet and Tapon 2011; Grusche et al. 2011; Halder et al. 2012; Schroeder and Halder 2012). Amongst these are the cell polarity proteins and proteins required for maintaining the cytoskeleton. The FERM (N-Terminal Globular domain (Band4.1, Ezrin, Radixin, Moesin)) domain-containing adaptor proteins Ex and Merlin (Mer) were amongst the earliest Hippo pathway components that were known to localize to the apical membrane (Hamaratoglu et al. 2006; McCartney et al. 2000). Ex and Mer act upstream of the Hpo kinase and regulate pathway activation (Hamaratoglu et al. 2006). Loss of *mer* and *ex* together in somatic clones caused dramatic overproliferation of cells leading to overgrowths. These effects were synergistic because loss of function of *ex* or *mer* alone does not cause similar defects. These genes function together to control proliferation by regulating expression of transcriptional targets of Hippo pathway (e.g., Cyclin E and DIAP1). Expanded can also regulate the pathway by independently interacting with Yki and sequestering it in the cytoplasm (Badouel et al. 2009; Oh et al. 2009).

Another protein that binds Ex and Mer, and acts upstream of Hpo is the WW- and C2-domain-containing adapter protein Kibra. Ex, Mer, and Kibra form a complex at the apical membrane in epithelial cells, which then activates the downstream core kinase cascade (Baumgartner et al. 2010; Cho et al. 2006; Genevet et al. 2010; Hamaratoglu et al. 2006; Pellock et al. 2007; Tyler and Baker 2007; Yu et al. 2010). Kibra was identified via a genome wide screen in *Drosophila* and in S2 cells for candidates that modified Yki activity (Baumgartner et al. 2010; Genevet et al. 2010; Yu et al. 2010). Genetic epistasis experiments placed Kibra upstream of Hpo and Yorkie. Kibra affects the phosphorylation of Hpo and Yorkie. Kibra acts synergistically with Ex and Mer to regulate Wts phosphorylation, and Kibra binds to Sav, and Hpo in a Sav-dependent manner (Baumgartner et al. 2010; Genevet et al. 2010; Yu et al. 2010).

Cell-polarity genes have been well characterized in flies and mammalian model systems, and recent studies reveals a role for cell polarity genes in the regulation of Hippo signaling (Genevet and Tapon 2011; Grusche et al. 2010; Grzeschik et al. 2007; Grzeschik et al. 2010a; Grzeschik et al. 2010b; Schroeder and Halder 2012). Crumbs (Crb), a trans-membrane protein is the upstream regulator that regulates Ex activity (Chen et al. 2010; Ling et al. 2010; Robinson et al. 2010). Crb is required for proper localization of Ex. Crb regulates Yki activity by interacting with Expanded (Chen et al. 2010; Grzeschik et al. 2010a; Robinson et al. 2010). Crb was found through a genetic screen, and loss and gain of function of Crb cause overgrowth of tissues and up-regulation of the Hippo pathway target genes. Echinoid (Ed) is another upstream regulator of the Hippo pathway, that like *kibra* interacts with both Ex and Yki (Baumgartner et al. 2010; Genevet et al. 2010; Yu et al. 2010; Yue et al. 2012). Cells mutant for *ed* cause mislocalization of Sav from the subapical membrane without affecting Ex or Mer localization. Ed also interacts physically with Hpo, Ex, Mer, and Kibra (Yue et al. 2012).

F-actin acts as an upstream regulator of the Hippo pathway. Increased levels of F-actin inhibit the pathway and activation of Hippo pathway inhibits F-actin accumulation (Fernandez et al. 2011; Richardson 2011; Sansores-Garcia et al. 2011). Tao-1 phosphorylates Hpo at T195 and acts upstream of Hpo (Boggiano and Fehon 2012; Boggiano et al. 2011; Poon et al. 2011). RNAi knockdown of Kibra, Ex, and Mer resulted in a significant decrease of endogenous Hpo protein in the membrane fraction (Boggiano and Fehon 2012; Boggiano et al. 2011; Poon et al. 2011). Thus, the apical proteins regulate Hpo at least in part by bringing the latter to the membrane, where Hpo may be activated via mechanisms yet to be determined.

Negative Regulators of the Hippo Pathway

Several members of the Hippo pathway were identified based on their effects on tissue growth, and the loss of function phenotypes of these components showed dramatic outgrowths and benign lesions in fly epithelia. It was clear that additional components that keep this pathway in check (for example, phosphatases or kinase inhibitors) must exist, as Hippo activity would need to be modulated both positively and negatively for maintaining tissue homeostasis. Thus, the search for negative regulators began, which yielded many important and critical regulators of the Hippo pathway. Amongst the first genes identified in this category, was the Ras Association Family (RASSF) gene, *dRASSF1* (Polesello et al. 2006). The dRASSF protein negatively regulates the pathway by inhibiting the phosphorylation of Hpo, thus interrupting the Hpo kinase from signaling to the downstream kinase Wts (Polesello et al. 2006; Scheel and Hofmann 2003). Other inhibitors that act by dephosphorylating Hpo are the phosphatases—Striatin-interacting phosphatase and protein phosphatase 2A (PP2A) (Ribeiro et al. 2010). A second mechanism of inhibition of Yki activity was identified by the *Drosophila* Ajuba family gene, *djub* (Das Thakur et al. 2010). Loss of *djub* in mutant clones in imaginal discs caused reduced proliferation and increased apoptosis, akin to *yki* mutant clones. Genetic interaction studies showed that *djub* acts downstream of Hpo but upstream of Yki and Wts (Das Thakur et al. 2010). Furthermore, Djub can physically associate with Wts and Sav and influence the signaling activity of Yki. Thus, *djub* negatively regulates Hippo signaling by interfering with Yki phosphorylation and its subcellular localization (Das Thakur et al. 2010). Recently, another negative regulator, *myopic* (*mop*) was identified in a genetic screen for conditional growth suppressors (Gilbert et al. 2011). *mop* encodes the *Drosophila* homolog of human *His-domain protein tyrosine phosphatase* gene (*HD-PTP* or *PTPN23*) (Toyooka et al. 2000). *mop* mutant cells show overgrowth phenotypes due to a block in cell death. This growth is accompanied by upregulation of a subset of Yki transcriptional targets but not the antiapoptotic gene *diap1*. *mop* interacts genetically with *yki* and acts downstream of *wts*, but at the level of *ex* and *yki*. Myopic PPxY motifs bind conserved residues in the WW domains of the transcriptional coactivator Yorkie, and Myopic colocalizes with Yorkie at endosomes (Gilbert et al. 2011). Thus, several negative regulators of the Hippo pathway are now known; however, much remains unknown about their mechanism of action and their influence on growth regulation during development (Table 1).

Table 1 Regulators of the Hippo pathway

	Gene name, *symbol*, [Chr]	Nature of protein	Role	References
Upstream regulators	Crumbs *Crb* [3]	Protein kinase C binding	Organization of adherens junction, establishment of cell polarity, photoreceptor & rhabdomere development	Fan et al. 2003; Pichaud and Desplan 2002; Fan et al. 2003; Tepass et al. 1990
	Expanded *ex* [2]	Protein binding	Compound eye, photoreceptor cell differentiation, negative regulation of hippo signaling cascade	Maitra et al. 2006; Pellock et al. 2007; Badouel et al. 2009; McCartney et al. 2000
	Merlin *Mer* [1]	Protein binding	Regulation of programmed cell death, negative regulator of hippo signaling	Pellock et al. 2007; Hamaratoglu et al. 2006
	Kibra *Kibra* [3]	Protein binding	Compound eye morphogenesis, regulation of hippo signaling cascade	Ling et al. 2010; Genevet et al. 2010; Yu et al. 2010; Baumgartner et al. 2010
Fat branch	Fat *ft* [2]	Cell adhesion molecule binding	Establishment of planar polarity, negative regulation of growth, imaginal disc growth	Yang et al. 2002; Mao et al. 2006; Torok et al. 1993; Garoia et al. 2000; Matakatsu and Blair 2006
	Lowfat *lft* [2]	Protein binding	Wing morphogenesis	Mao et al. 2009
	Dachs *D* [2]	ATPase activity (predicted nature)	Establishment of ommatidial planar polarity, positive regulation of growth	Mao et al. 2006
	Dachsous *Ds* [2]	Cell adhesion molecule binding	Eye morphogenesis, establishment of cell polarity, cell proliferation	Baena-Lopez et al. 2005; Clark et al. 1995
	Four-jointed *Fj* [2]	Wnt-protein binding; protein kinase activity	Imaginal disc growth, establishment of planar polarity	Villano and Katz 1995; Bosveld et al. 2012
	Scribbled *Scrib* [2]	Protein binding	Establishment of ommatidial planar polarity, negative regulation of imaginal disc growth	Courbard et al. 2009; Zeitler et al. 2004; Verghese et al. 2012a
	Zyxin *Zyx* [4]	Protein binding	Positive regulation of imaginal disc growth	Rauskolb et al. 2011
	Approximated *App* [3]	Protein-cysteine S-palmitoleyltransferase activity (predicted nature)	Establishment of body hair or bristle planar orientation	Matakatsu and Blair 2008
	Discs Overgrown *Dco* [3]	Kinase activity	Establishment of ommatidial planar polarity, positive regulation of cell growth	Strutt et al. 2006; Klein et al. 2006; Guan et al. 2007

Table 1 (continued)

	Gene name, *symbol,* [Chr]	Nature of protein	Role	References
Core kinase cascade	Warts *Wts* [3]	Protein binding, kinase activity	Negative regulation of cell proliferation, R8 cell fate specification	Justice et al. 1995; Mikeladze-Dvali et al. 2005
	Mob as tumor suppressor *Mats* [3]	Protein binding	Cell proliferation	Lai et al. 2005
	Hippo *Hpo* [2]	Protein binding; serine/threonine kinase activity	Negative regulation of cell proliferation, R8 cell fate specification	Udan et al. 2003; Mikeladze-Dvali et al. 2005
	Salvador *Sav* [3]	Protein binding	Negative regulation of cell proliferation, R8 cell fate specification	Kango-Singh et al. 2002; Mikeladze-Dvali et al. 2005
Other regulators	Ajuba *Jub* [1]	Ligand-dependent nuclear receptor binding	Positive regulation of organ growth	Das Thakur et al. 2010
	Tao *Tao* [1]	Serine/threonine kinase activity	Negative regulation of organ growth	Poon et al. 2011
	Echinoid *Ed* [2]	Protein binding	Negative regulation of hippo signaling cascade	Yue et al. 2012
	Pez *Pez* [2]	Protein tyrosine phosphatase activity	Negative regulation of hippo signaling cascade	Poernbacher et al. 2012
	d-STRIPAK PP2A *Pp2A-29B* [2]	Serine/threonine phosphatase activity	Centrosome organization	Dobbelaere et al. 2008
	Ras association family member *Rassf* [3]	Protein binding	Negative regulation of signal transduction	Polesello et al. 2006
	Par-6 *Par-6* [1]	Protein binding	Cell adhesion	Kiger et al. 2003
	Atypical protein kinase C *a-PKC* [2]	Protein binding; serine/threonine kinase activity	Compound eye retinal cell programmed cell death, establishment of epithelial cell planar polarity	Ogawa et al. 2009; Kaplan et al. 2011
	Stardust *Sdt* [1]	Protein binding	Zonula adherens assembly	Nam and Choi 2003; Bachmann et al. 2001

Table 1 (continued)

	Gene name, *symbol,* [Chr]	Nature of protein	Role	References
	Lethal 2 giant Larvae *L2gl* [2]	Myosin II binding; myosin binding	Cell competition in a multicellular organism, establishment of epithelial cell planar polarity	Tamori et al. 2010; Kaplan and Tolwinski 2010
	Myopic *Mop* [3]	Protein tyrosine phosphatase activity	Regulation of growth	Gilbert et al. 2011
	Patj *dPatj* [3]	Protein kinase C binding	Adherens junction organization	Nam and Choi 2006
Transcription factors / co-activators	Yorkie *Yki* [2]	Protein binding; transcription co-activator activity	Cell competition in a multicellular organism, cell proliferation	Ziosi et al. 2010; Huang et al. 2005; Thompson and Cohen 2006
	Scalloped *Sd* [1]	Transcription factor binding	Compound eye morphogenesis	Garg et al. 2007
	Homothorax *Hth* [3]	Protein binding; transcription factor	Compound eye photoreceptor fate determination	Wernet et al. 2003
	Teashirt *Tsh* [2]	Transcription factor activity	Eye-antennal disc development	Singh et al. 2004
	Wpb2 *Wbp2* [3]	Transcription factor binding	Positive regulation of imaginal disc growth	Zhang et al. 2011b
	Mothers against dpp *Mad* [2]	Transcription factor activity	Compound eye morphogenesis, negative regulation of gene expression	Cordero et al. 2007; Anderson et al. 2006

Hippo Pathway Cross-talks With Other Pathways

Hippo pathway is known to interact with other pathways to regulate growth. In mice it has been shown that Mst2 interacts with Raf-1 of the ERK/MAPK pathway (Graves et al. 1998). Raf-1 inhibits dimerization of Mst2 and recruits a phosphatase to dephosphorylate Mst2, thereby inactivating it, a function independent of the MAPK pathway (O'Neill and Kolch 2005). More recently, many points of intersection between Hippo and other signaling pathways have come to light. For example, in the last 5 years, Hippo pathway was shown to interact with JNK pathway to regulate compensatory proliferation, regeneration, and tumor progression (Chen et al. 2012; Doggett et al. 2011; Grzeschik et al. 2010a; Staley and Irvine 2010; Sun and Irvine 2010, 2011; Tyler et al. 2007; Varelas et al. 2010a). Furthermore, Hippo pathway interacts with Wingless/Wnt pathways in flies and mammals (Varelas et al. 2010a). Hippo pathway restricts Wnt/beta-Catenin signaling by promoting an interaction between TAZ and dishevelled (DVL) in the cytoplasm. TAZ inhibits the CK1delta/epsilon-mediated phosphorylation of DVL, thereby inhibiting Wnt/beta-Catenin signaling (Azzolin et al. 2012; Tsai et al. 2012; Varelas et al. 2010a). In *Drosophila*, Hippo signaling modulates Wg target gene expression (Varelas et al. 2010a). More connections of Hippo signaling with pathways that control morphogenetic patterning and growth have been uncovered which include the discovery of the regulation of TGF-beta Transforming Growth Factor/SMAD (refers to a family of transcription factors: Sma from *Caenorhabditis elegans*, Mad 1 from *Drosophila*, and SMAD1 from human) complexes by YAP/TAZ (transcriptional co-activator with PDZ) in mammalian models and Yki in flies (Chan et al. 2011; Meignin et al. 2007; Polesello and Tapon 2007; Rogulja et al. 2008; Sudol and Harvey 2010; Varelas et al. 2010b). Dpp (Decapentaplegic) signaling interacts with D to maintain Fj and Ds gradient in order to regulate proliferation in the wing (Rogulja et al. 2008). Hippo pathway also intersects the phosphoinositide 3-kinase (PI3K)/TOR pathway via multiple interactions (Bellosta and Gallant 2010; Collak et al. 2012; Karni et al. 2008; Kim et al. 2010; Mills et al. 2008; Sekido 2008; Strassburger et al. 2012; Tumaneng et al. 2012a; Tumaneng et al. 2012b; Wehr et al. 2012); with G-protein coupled receptor signaling (Yu et al. 2012) and Receptor Tyrosine Kinase signaling (Gadd et al. 2012; Garami et al. 2003). In fact, the web of interactions has grown exponentially over the last few years such that oftentimes the Hippo pathway is referred to as a network or superhighway (Barry and Camargo 2013; Table 2).

Mammalian Hippo Pathway

In vertebrate models, Hippo pathway is responsible for regulating organ size, and is involved in regeneration (Bertini et al. 2009; Hiemer and Varelas 2013; Hong and Guan 2012; Liu et al. 2012a). The core kinase pathway is highly conserved in mammals (Hong and Guan 2012; Liu et al. 2012a; Zhao et al. 2008a), and consists of Mst1/2 (Hpo homolog) and LATS1/2 (Wts homolog) along with their adaptor proteins WW45 (Sav) and MOB1 (Mats homolog), which control growth by regulating

Table 2 Pathways known to interact with the Hippo network

Pathway interactions	Responses	References
JNK pathway	Cell-competition, compensatory proliferation, regeneration, cytoskeletal integrity, tumorigenesis	Chen et al. 2012; Sun et al. 2011; Densham et al. 2009; Enomoto et al. 2011
Wingless pathway	Growth control	Varelas et al. 2010a
EGFR pathway	Growth control	Herranz et al. 2012
Decapentaplegic pathway	Growth control	Rogulja et al. 2008
Hedhehog pathway	Growth control, neuronal differentiation	Kagey et al. 2012; Lin et al. 2012
Notch pathway	Neural stem-cell maintenance, polar cell fate during oogenesis, cell-differentiation, proliferation	Li et al. 2012; Chen et al. 2011; Yu et al. 2008
TSC-TOR pathway		Wehr et al. 2013; Tumaneng et al., 2012a, b; Strassburger et al., 2012

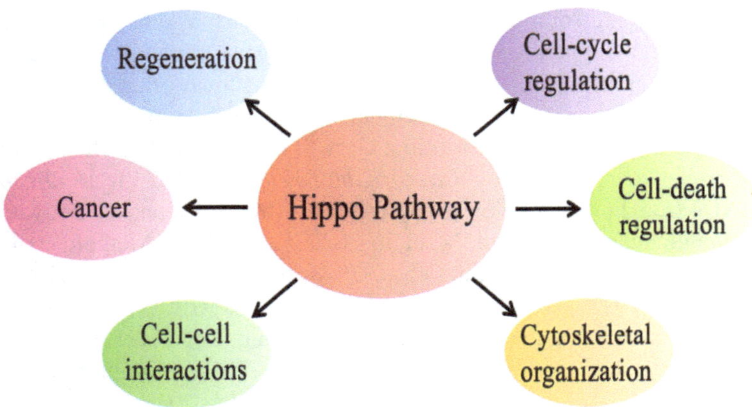

Fig. 4 Hippo pathway is linked to many biological and developmental processes. Hippo signaling has been shown to participate in generating myriad cellular responses that are aimed at attaining tissue homeostasis in addition to regulating organ size. Thus, the role of Hippo signaling is implicated not only during organ development but also in differentiated tissues. Further, tumorigenesis has also been attributed to dysregulation of Hippo signaling pathway placing it in the global network of regulatory mechanisms required for proper growth

phosphorylation of YAP (Yki homolog) and TAZ (transcriptional coactivator with PDZ-binding domain); Hong and Guan 2012; Liu et al. 2012a; Zhao et al. 2008a). Ft1-4 (Ft homolog), Dchs1-2 (Ds homolog), and Fjx1 (Fj homolog) are known to regulate PCP; however, their connection to other Hippo pathway components still needs to be explored (Brittle et al. 2010; Hiemer and Varelas 2013; Skouloudaki et al. 2009; Sopko et al. 2009; Zhao et al. 2007). The other downstream components like Dco and Lowfat homolog have not been shown yet to function within the Hippo pathway (Sopko et al. 2009; Zhang et al. 2008a, 2011a; Zhao et al. 2010a). However, Dco homolog CK1δ/ε has been shown to be involved in YAP/TAZ degradation (Zhao et al. 2010b).

Neurofibromatosis type II (NF2), the Mer homolog is the most extensively studied upstream regulator in mammals (Sekido 2011; Striedinger et al. 2008; Zhang et al. 2009b; Zhao et al. 2007). NF2 interacts with CD44 and adherens junction to relay the signal downstream to other Hippo pathway components during contact inhibition (Li et al. 2012; Morrison et al. 2001; Zhao et al. 2007). Kibra is known to interact with LATS2 to promote its phosphorylation (Zhang et al. 2012). It also protects LATS2 from proteosomal degradation by preventing its ubiquitination. Kibra is also the transcriptional target of Hippo pathway (Angus et al. 2012; Ishiuchi and Takeichi 2012; Visser-Grieve et al. 2012; Xiao et al. 2011). Angiomotin family interacts with its PPxY domain to YAP WW domain and TAZ PDZ domain independent of the upstream components. This interaction inhibits the activity of YAP/TAZ (Chan et al. 2011; Paramasivam et al. 2011; Skouloudaki and Walz 2012; Wang et al. 2012a; Wang et al. 2009; Zhao et al. 2011a). Ex1/FRMD6/Willin (Ex homolog) interacts with upstream Hippo pathway components like Mer (Angus et al. 2012; Ishiuchi and Takeichi 2012; Visser-Grieve et al. 2012). Crb interacts with YAP/TAZ and promotes its phosphorylation, which is dependent on cell density and at the same time inhibits TGF-beta SMAD pathway (Varelas et al. 2010b). Unlike *Drosophila* RASSF1, mammalian RASSF homologs activate MST1/2 (Avruch et al. 2012; Guo et al. 2007; Hergovich 2012; Hwang et al. 2007; Polesello et al. 2006; Ribeiro et al. 2010; Schagdarsurengin et al. 2010; Seidel et al. 2007).

Nephronophthisis4 (NPHP4), a known cilia-associated protein that is mutated in the severe degenerative renal disease nephronophthisis, acts as a potent negative regulator of mammalian Hippo signaling (Habbig et al. 2011, 2012). NPHP4 directly interacts with the kinase LATS1 and inhibits LATS1-mediated phosphorylation of the YAP and TAZ, leading to derepression of these proto-oncogenic transcriptional regulators. Moreover, NPHP4 induces release of YAP and TAZ from 14-3-3 binding and their nuclear translocation promoting TEA domain (TEAD)/TAZ/YAP-dependent transcriptional activity (Habbig et al. 2011). ITCH interacts with LATS to negatively regulate its stability (Ho et al. 2011; Salah et al. 2011; Wang et al. 2012a). α-catenin interacts with YAP and affects its stability by stabilizing the YAP/14-3-3 complex to restrict YAP activity, and by preventing PP2A to interact with YAP (Azzolin et al. 2012; Schlegelmilch et al. 2011; Silvis et al. 2011; Tsai et al. 2012; Varelas 2010a; Konsavage 2013; Mauviel 2012). Zona occludens-2 (ZO-2) promotes the pro-apoptotic function of YAP (Oka et al. 2010). The ASPP (apoptosis-stimulating protein of p53) family of proteins can function in the nucleus to modulate the transcriptional activity of p53, with ASPP1 and ASPP2 contributing to the expression of apoptotic target genes (Vigneron et al. 2010). ASPP increases YAP/TAZ nuclear availability by preventing LATS interaction with YAP/TAZ (Vigneron et al. 2010). Similarly, PP1A interacts with ASPP1 to dephosphorylate TAZ leading to increased TAZ nuclear availability (Liu et al. 2010, 2011).

In mammalian cell lines, E-cadherin acts as an upstream regulator of the pathway, which activates the pathway in response to contact inhibition. YAP and TAZ interact with several transcriptional factors. YAP/TAZ interacts with TEAD1/4 and Runx2. TAZ interacts with thyroid transcription factor-1, peroxisome proliferator-activated receptor gamma (PPARγ), Tbx5, Pax3, and Smad2/3/4. Yap interacts with p73 to mediate its pro-apoptotic functions. Various target genes are as follows: *CTGF*, *AREG*,

BIRC5–2, *GLI-2* (Liu et al. 2012b; Zhao et al. 2008a, 2010a). YAP1 interacts with Sonic Hedgehog pathway to promote the proliferation of cerebellar granule neuron precursors (CGNPs). TAZ inhibits Wnt signaling by inhibiting the phosphorylation of DVL by CKIδε. YAP/TAZ has also been shown to interact with SMAD to regulate tumorigenesis (Zhang et al. 2011a, b). Thus, our understanding of the mammalian Hippo pathway continues to grow with new insights on its molecular and signaling interactions with components from Hippo and other pathways.

The Insulin-Receptor Signaling Pathway: Regulation of Cell Size

The pin-head screens showed a large number of mutations that primarily caused decreased growth due to formation of smaller cells (Oldham et al. 2000a; Stocker and Hafen 2000). These mutants were subsequently categorized into two well-studied signaling pathways—the insulin/Phospho inositol 3 kinase (PI3K) pathway and the Target of Rapamycin (TOR) pathway. Using genetic and biochemical strategies, the epistatic and molecular interactions were elucidated for genes that comprise these pathways.

The Regulation of Cell Size and Not Cell Numbers

The PI3K Pathway *Drosophila* has one insulin/insulin-like growth factor (IGF) receptor homolog known as dINR (*Drosophila* insulin receptor) (Chen et al. 1996; Fernandez et al. 1995), and several insulin-like peptides (Brogiolo et al. 2001). These together control the carbohydrate metabolism and growth in flies (Ikeya et al. 2002; Rulifson et al. 2002). Through a mechanism that involves phosphorylation of its carboxy-terminal end, the dINR recruits downstream signaling molecules without the need for adaptor proteins. The signaling also involves the insulin receptor substrate protein Chico, which contains a phosphotyrosine binding domain, which facilitates its binding with activated dINR (Bohni et al. 1999; Poltilove et al. 2000). Subsequently, the pathway functions by activating the PI3K pathway, via activation of the Drosophila PI3K—Dp110 and its adapter subunit Dp60 (Leevers 2001; Leevers et al. 1996; Weinkove et al. 1999). Dp110/Dp60 heterodimers are recruited to the plasma membrane following the binding of p60 SH2 domain to phosphorylated dINR and Chico, which allows the PI3K access to the phosphinositide substrates in the plasma membrane. This sets up a signaling cascade in which PIP3 (phosphatidyl inositol (3, 4, 5 triphosphate)) transduces the signal to downstream effectors that contain the PIP3-binding PH (Pleckstrin homology) domains, and causes relocalization of these proteins to the plasma membrane.

In flies, two such effectors exist, which are the *Drosophila* homolog of phosphoinositide-dependent kinase 1 (PDK1) and its substrate AKT (AK: mouse strain that develops thymic lymphomas; T: thymoma) aka protein kinase B (PKB). PDK1 localizes to the membrane during low levels of PI3K activity via its affinity to PIP3, whereas AKT requires high levels of PI3K activity to become membrane localized, through a process involving binding of PIP3 to its PH-domain and phosphorylation by PDK1 (Vanhaesebroeck and Alessi 2000). In flies, the activity of dAKT is reduced in the absence of Dp110 and coexpression of dPDK1 and dAKT activates dAKT and induce growth (Cho et al. 2001; Radimerski et al. 2002b; Rintelen et al. 2001).

A negative regulator of the PI3K activity is the lipid phosphatase PTEN (phosphate and tensin homolog), which removes the 3' phosphate from three phosphoinositides generated by PI3K (Gao et al. 2000; Goberdhan et al. 1999; Huang et al. 1999). Genetic interaction studies support the model where PTEN directly antagonizes PI3K. Loss of PTEN leads to overgrowths due to increased levels of PIP3 (Oldham et al. 2002). Recently, the FOXO (Forkhead box)family of transcription factors was identified as the target that enabled AKT to regulate growth (Tran et al. 2003). AKT-mediated phosphorylation of FOXO antagonizes its transcriptional activity by creating a 14-3-3 binding site that leads to cytoplasmic sequestration of FOXO (Brunet et al. 1999, 2002; Burgering and Kops 2002). *Drosophila* has one FOXO family transcription factor (dFOXO)—which functions downstream of AKT. Interestingly, loss of function of dFOXO has no apparent effect on cell size or growth as flies homozygous mutant for dFOXO are viable and normal in size (Junger et al. 2003).

The loss of function of Dp110, p60, chico, dINR, dPDK1, and dAKT show similar effects on cell size and tissue growth. For example, twin-spot analysis revealed that loss of function clones of mutations in these genes are smaller than the corresponding wild-type twin clones that lead to formation of smaller structures (Bohni et al. 1999; Brogiolo et al. 2001; Rintelen et al. 2001; Verdu et al. 1999; Weinkove et al. 1999). Overexpression of PI3K pathway components like Dp110 leads to increased insulin/PI3K signaling and a corresponding increase in cell size, cell number, and tissue growth (Goberdhan et al. 1999; Huang et al. 1999; Leevers et al. 1996). Overall, changes in levels of insulin/PI3K signaling have profound effects on organ and organismal size due to effects on cell growth and cell division throughout development and affect the final body/organ size (Fig. 5).

The TSC-TOR Pathway

Two TOR genes, *TOR1* and *TOR2*, were initially identified in yeast, and were shown to be kinases that regulate growth in all organisms by acting as nutrient sensors that couple signaling to nutrient availability (Gingras et al. 2001). *Drosophila* TOR (dTOR) promotes growth by stimulating translation via promoting the activity of the *Drosophila* S6Kinase (Montagne et al. 1999), and inhibiting the *Drosophila* 4E-BP1 (a homolog of the Eukaryotic translation initiator 4E)—the translational inhibitor of eIF4E, which is a part of the translation initiation complex (Gingras et al. 2001; Lasko 2000). Hyperphosphorylation of d4E-BP1, which is in part controlled by the TOR kinase, relieves its interaction with eIF4E leading to translation initiation.

TOR signaling is negatively regulated by a complex formed by the tuberous sclerosis complex tumor suppressors, TSC1 and TSC2 (Marygold and Leevers 2002).

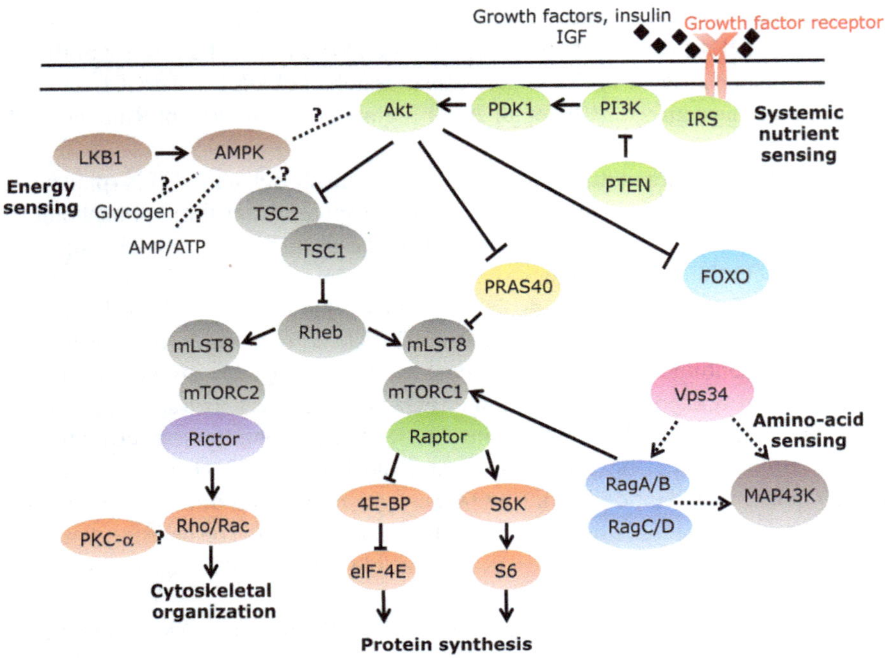

Fig. 5 Model depicting regulation of INR/TOR signaling pathway governed by nutritional status in *Drosophila*. Cellular growth in part is also dependent on the availability of nutrients. This aspect of growth regulation is mainly regulated by the insulin/TOR signaling pathway. Some of the well-studied players of the pathway include phosphatidylinositide 3-kinase and Akt that integrate upstream signaling from growth factor receptors and relay it to TSC1 and TSC2 to regulate ribosomal and protein biosynthesis in addition to actin organization. Other energy sensing and amino-acid sensing mechanisms are also thought to interact with the core TSC/TOR pathway. However, the exact role or the mechanism by which this takes place remains largely unknown

Mutations in *Tsc1/Tsc2* cause formation of large cells, and are implicated in the inherited benign hamartomas observed in the tuberous sclerosis patients (Kandt 2002; Montagne et al. 2001). The *Drosophila Tsc1/2* genes show similar effects on cell size, and were identified by several groups in the *eyFLP cell lethal* screens as mutants with overgrown heads (Gao and Pan 2001; Potter et al. 2001; Tapon et al. 2001). Loss of TSC1/2 causes increased growth whereas overexpression of TSC1/2 causes reduced growth due to slow cell cycle progression in the mutant cells. Growth regulation via TSC1/2 happens through preventing dS6K activation via dTOR (Gao et al. 2002; Radimerski et al. 2002a; Radimerski et al. 2002b). Another important component of this pathway is the GTPase (guanosine triphosphate hydrolase) *Rheb*, which is a target of TSC (Saucedo et al. 2003; Stocker et al. 2003; Zhang et al. 2003). The Rheb-GTP levels play a central role in regulating the activity of TOR pathway, and the TOR protein that exists in two large multimeric complexes in the cell, viz., the rapamycin-sensitive TORC1 complex, and the rapamycin-resistant TORC2 complex (Hara et al. 2002; Kim et al. 2002; Kim et al. 2003; Loewith et al. 2002; Sarbassov et al. 2004).

The TORC1 complex consists of TOR, Raptor, and GßL, and responds to the presence of growth factors and nutrients to control protein synthesis. The small GTPase protein Rheb (Ras homolog enriched in brain) is a direct activator of TORC1 (Long et al. 2004; Saucedo et al. 2003; Stocker et al. 2003), and the TSC complex (TSC1/TSC2) negatively regulates TORC1 by functioning as a GTPase-activating protein for Rheb (Potter and Xu 2001; Zhang et al. 2003). Growth factors such as insulin or IGFs activate TORC1 signaling upstream of the TSC1/TSC2 (TSC1/2) complex through the insulin receptor (InR)/PI3K/AKT signaling pathway (Inoki et al. 2002; Potter et al. 2002). TORC1 also senses nutrient availability. Amino acids regulate TORC1 through mechanisms independent or downstream of TSC complex, and recently the Rag small GTPases have been shown to interact with TOR and promote TORC1 activity by controlling its subcellular localization (Nellist et al. 2008; Sancak et al. 2010).

TORC2 complex consists of TOR, rictor, Sin1 (stress-activated map kinase-interacting protein 1) and GßL; and phosphorylates and activates several AGC family kinases, including AKT, serum and glucocorticoid-regulated kinase (SGK), and protein kinase C, and thereby regulates cell survival, cell cycle progression, and metabolism (Pearce et al. 2010; Li 2010; Gao 2010). In contrast to TORC1, little is known about the upstream activators of mTORC2. Although the general mechanisms have not been accepted, PI3K, TSC, and Rheb have been shown to regulate TORC2 activity, and Rictor has been identified as a substrate of S6 kinase (S6K), suggesting possible regulation of TORC2 through the TORC1 pathway (Dibble et al. 2009; Treins et al. 2010; Yang et al. 2006). Nevertheless, it is generally thought that growth factors may control TORC2, either directly or indirectly (Zinzalla et al. 2011). TORC2 has been proposed to function independent of amino acid availability (Jacinto et al. 2006); however, recent findings show that amino acids may also activate TORC2 (Tato et al. 2011).

The central role of TOR in cell growth has been largely attributed to TORC1, but mounting evidence points to a role for TORC2 as well in this basic cellular process. For instance, TORC2 localizes in polysomal fractions and associates with ribosomal proteins, indicating a potential role for TORC2 in protein synthesis and maturation (Cybulski and Hall 2009; Zinzalla et al. 2011). *lst8* knockout flies are viable but small, similar to *rictor* mutants but dissimilar to files with *tor* or *rheb* mutations, which are lethal (Avruch et al. 2009; Liao et al. 2008; Wang et al. 2012b). Neither loss nor overexpression of LST8 affected the kinase activity of TORC1 toward S6K or autophagy, whereas the kinase activity of TORC2 toward AKT was completely lost in the *lst8* mutants (Avruch et al. 2009; Liao et al. 2008; Wang et al. 2012b).

In terms of effects of TOR signaling on growth phenotypes in *Drosophila*, loss of dTOR leads to a decrease in larvae size; however, the larvae fail to mature and die before reaching adulthood. In mosaic *Drosophila*, loss of dTOR leads to a decrease in cell size while maintaining the general organization of the tissue (Oldham et al. 2000b; Zhang et al. 2000). However, it is less clear how cell size is regulated downstream of mTOR. One of the most potent candidates in this regulation is S6K. In *Drosophila*, knockout of *S6K* results in high rates of embryonic lethality. In the surviving adults, however, there is a decrease in body size. Knockdown of either dPTEN or dTSC1is sufficient to increase cell size; however, a double knockdown of dPTEN and dTSC1 has additive effects on cell size regulation. This suggests that in *Drosophila*, the pathways may have independent components in the regulation of cell size (Gao and Pan 2001). It may also highlight the differences in the regulation of TSC2 by AKT in *Drosophila* as seen by mutations of the AKT phosphorylation sites on TSC2 (Dong and Pan 2004; Pan et al. 2004). Loss of either dPTEN or dTSC1 can lead to increase in cell size; however, a report has suggested that only knockdown of dTSC1 leads to increase in dS6K (Radimerski et al. 2002a), whereas other reports have also seen increases in dS6K with the knockdown of dPTEN (Sarbassov et al. 2004; Yang et al. 2006). It is possible that dTSC1 regulates cell size in a dTOR-dependent manner, whereas dPTEN partially regulates cell size in a dTOR-independent manner (Radimerski et al. 2002b).

In conclusion, the TOR signaling pathway is a complex network of cell size regulators that is also implicated in tumorigenesis and cell survival. Several pathways interact and intersect with the TOR pathway at multiple points upstream and downstream of TOR.

Growth Regulation: A Network of Tumor Suppressors

Overall, growth control occurs through the Hippo and TSC-TOR pathways in conjunction with pathways regulating pattern formation during development. These pathways intersect in complicated signaling networks in all cell types, and coordinately regulate overall growth of an organism. Our progress in understanding of these pathways has lead the way to find molecules and interactions important for regenerative growth and wound healing—phenomena that have been well-documented but not well-understood at the molecular level for a long time. In addition, the establishment of these growth regulatory networks has generated many insights in the fields of cancer (e.g., the underlying genetics and biology link between hamartomas and TSC genes; Schwannomma's and NF2; YAP and Hepatocellular carcinoma; TAZ and Breast cancer etc.). In the future, it will be interesting to learn about the regulation of these pathways by extracellular and intracellular mechanisms, an area expected to expand rapidly with our increased understanding of the integration points in the circuitry of these networks.

References

Acquisti C, Kumar S, Elser JJ (2009) Signatures of nitrogen limitation in the elemental composition of the proteins involved in the metabolic apparatus. Proc Biol Sci 276:2605–2610

Alarcon C, Zaromytidou AI, Xi Q, Gao S, Yu J, Fujisawa S, Barlas A, Miller AN, Manova-Todorova K, Macias MJ et al (2009) Nuclear CDKs drive Smad transcriptional activation and turnover in BMP and TGF-beta pathways. Cell 139:757–769

Anderson J, Salzer CL, Kumar JP (2006) Regulation of the retinal determination gene dachshund in the embryonic head and developing eye of Drosophila. Dev Biol 297(2):536–549

Angus L, Moleirinho S, Herron L, Sinha A, Zhang X, Niestrata M, Dholakia K, Prystowsky MB, Harvey KF, Reynolds PA et al (2012) Willin/FRMD6 expression activates the Hippo signaling pathway kinases in mammals and antagonizes oncogenic YAP. Oncogene 31:238–250

Avruch J, Long X, Ortiz-Vega S, Rapley J, Papageorgiou A, Dai N (2009) Amino acid regulation of TOR complex 1. Am J Physiol Endocrinol Metab 296:E592–602

Avruch J, Zhou D, Fitamant J, Bardeesy N, Mou F, Barrufet LR (2012) Protein kinases of the Hippo pathway: regulation and substrates. Semin Cell Dev Biol 23:770–784

Azzolin L, Zanconato F, Bresolin S, Forcato M, Basso G, Bicciato S, Cordenonsi M, Piccolo S (2012) Role of TAZ as mediator of Wnt signaling. Cell 151:1443–1456

Bachmann A, Schneider M, Theilenberg E, Grawe F, Knust E (2001) Drosophila Stardust is a partner of Crumbs in the control of epithelial cell polarity. Nature 414(6864):638–643

Badouel C, Gardano L, Amin N, Garg A, Rosenfeld R, Le Bihan T, McNeill H (2009) The FERM-domain protein expanded regulates Hippo pathway activity via direct interactions with the transcriptional activator Yorkie. Dev Cell 16:411–420

Baena-Lopez LA, Baonza A, Garcia-Bellido A (2005) The orientation of cell divisions determines the shape of Drosophila organs. Curr Biol 15(18):1640–1644

Baker NE (2001) Cell proliferation, survival, and death in the Drosophila eye. Semin Cell Dev Biol 12:499–507

Baker NE, Yu S, Han D (1996) Evolution of proneural atonal expression during distinct regulatory phases in the developing Drosophila eye. Curr Biol 6:1290–1301

Bangs P, White K (2000) Regulation and execution of apoptosis during Drosophila development. Dev Dyn 218:68–79

Barry ER, Camargo FD (2013) The Hippo superhighway: signaling crossroads converging on the Hippo/Yap pathway in stem cells and development. Curr Opin Cell Biol 25(2):247–253

Baumgartner R, Poernbacher I, Buser N, Hafen E, Stocker H (2010) The WW domain protein Kibra acts upstream of Hippo in Drosophila. Dev Cell 18:309–316

Bellen HJ, O'Kane CJ, Wilson C, Grossniklaus U, Pearson RK, Gehring WJ (1989) P-element-mediated enhancer detection: a versatile method to study development in Drosophila. Genes Dev 3:1288–1300

Bellen HJ, Levis RW, He Y, Carlson JW, Evans-Holm M, Bae E, Kim J, Metaxakis A, Savakis C, Schulze KL et al (2011) The Drosophila gene disruption project: progress using transposons with distinctive site specificities. Genetics 188:731–743

Bellosta P, Gallant P (2010) Myc function in Drosophila. Genes Cancer 1:542–546

Bennett FC, Harvey KF (2006) Fat cadherin modulates organ size in Drosophila via the Salvador/Warts/Hippo signaling pathway. Curr Biol 16:2101–2110

Bergantinos C, Vilana X, Corominas M, Serras F (2010) Imaginal discs: renaissance of a model for regenerative biology. Bioessays 32:207–217

Bertini E, Oka T, Sudol M, Strano S, Blandino G (2009) YAP: at the crossroad between transformation and tumor suppression. Cell Cycle 8:49–57

Bier E (2005) Drosophila, the golden bug, emerges as a tool for human genetics. Nat Rev Genet 6:9–23

Blair SS (2003) Genetic mosaic techniques for studying Drosophila development. Development 130:5065–5072

Boggiano JC, Fehon RG (2012) Growth control by committee: intercellular junctions, cell polarity, and the cytoskeleton regulate Hippo signaling. Dev Cell 22:695–702

Boggiano JC, Vanderzalm PJ, Fehon RG (2011) Tao-1 phosphorylates Hippo/MST kinases to regulate the Hippo-Salvador-Warts tumor suppressor pathway. Dev Cell 21:888–895

Bohni R, Riesgo-Escovar J, Oldham S, Brogiolo W, Stocker H, Andruss BF, Beckingham K, Hafen E (1999) Autonomous control of cell and organ size by CHICO, a Drosophila homolog of vertebrate IRS1–4. Cell 97:865–875

Bonini NM, Fortini ME (1999) Surviving Drosophila eye development: integrating cell death with differentiation during formation of a neural structure. Bioessays 21:991–1003

Bosveld F, Bonnet I, Guirao B, Tlili S, Wang Z, Petitalot A et al (2012) Mechanical control of morphogenesis by Fat/Dachsous/Four-jointed planar cell polarity pathway. Science 336(6082):724–727

Boutros M, Ahringer J (2008) The art and design of genetic screens: RNA interference. Nat Rev Genet 9:554–566

Brittle AL, Repiso A, Casal J, Lawrence PA, Strutt D (2010) Four-jointed modulates growth and planar polarity by reducing the affinity of dachsous for fat. Curr Biol 20:803–810

Brogiolo W, Stocker H, Ikeya T, Rintelen F, Fernandez R, Hafen E (2001) An evolutionarily conserved function of the Drosophila insulin receptor and insulin-like peptides in growth control. Curr Biol 11:213–221

Brumby AM, Richardson HE (2003) Scribble mutants cooperate with oncogenic Ras or Notch to cause neoplastic overgrowth in Drosophila. EMBO J 22:5769–5779

Brunet A, Bonni A, Zigmond MJ, Lin MZ, Juo P, Hu LS, Anderson MJ, Arden KC, Blenis J, Greenberg ME (1999) Akt promotes cell survival by phosphorylating and inhibiting a Forkhead transcription factor. Cell 96:857–868

Brunet A, Kanai F, Stehn J, Xu J, Sarbassova D, Frangioni JV, Dalal SN, DeCaprio JA, Greenberg ME, Yaffe MB (2002) 14-3-3 transits to the nucleus and participates in dynamic nucleocytoplasmic transport. J Cell Biol 156:817–828

Bryant PJ (1978) Pattern formation in imaginal discs. In: Ashburner M, Wright TRF (eds) Genetics biology of Drosophila, vol 2C, pp 229–335

Bryant PJ (1987) Experimental and genetic analysis of growth and cell proliferation in Drosophila imaginal discs. In: Loomis WF (ed) Genetic regulation of development. AR Liss, New York, pp 339–372

Bryant PJ (2001) Growth factors controlling imaginal disc growth in Drosophila. Novartis Found Symp 237:182–194; discussion 194–202

Bryant PJ, Schmidt O (1990) The genetic control of cell proliferation in Drosophila imaginal discs. J Cell Sci Suppl 13:169–189

Bryant PJ, Huettner B, Held LI Jr, Ryerse J, Szidonya J (1988) Mutations at the fat locus interfere with cell proliferation control and epithelial morphogenesis in Drosophila. Dev Biol 129:541–554

Burgering BM, Kops GJ (2002) Cell cycle and death control: long live Forkheads. Trends Biochem Sci 27:352–360

Burke R, Basler K (1997) Hedgehog signaling in Drosophila eye and limb development—conserved machinery, divergent roles? Curr Opin Neurobiol 7:55–61

Cagan R (1993) Cell fate specification in the developing Drosophila retina. Dev Suppl 1993:19–28

Cagan R (2009) Principles of Drosophila eye differentiation. Curr Top Dev Biol 89:115–135

Campbell S, Inamdar M, Rodrigues V, Raghavan V, Palazzolo M, Chovnick A (1992) The scalloped gene encodes a novel, evolutionarily conserved transcription factor required for sensory organ differentiation in Drosophila. Genes Dev 6:367–379

Chan SW, Lim CJ, Chen L, Chong YF, Huang C, Song H, Hong W (2011) The Hippo pathway in biological control and cancer development. J Cell Physiol 226:928–939

Chen CK, Chien CT (1999) Negative regulation of atonal in proneural cluster formation of Drosophila R8 photoreceptors. Proc Natl Acad Sci USA 96:5055–5060

Chen C, Jack J, Garofalo RS (1996) The Drosophila insulin receptor is required for normal growth. Endocrinology 137:846–856

Chen CL, Gajewski KM, Hamaratoglu F, Bossuyt W, Sansores-Garcia L, Tao C, Halder G (2010) The apical-basal cell polarity determinant Crumbs regulates Hippo signaling in Drosophila. Proc Natl Acad Sci USA 107:15810–15815

Chen CL, Schroeder MC, Kango-Singh M, Tao C, Halder G (2012) Tumor suppression by cell competition through regulation of the Hippo pathway. Proc Natl Acad Sci USA 109:484–489

Chen HJ, Wang CM, Wang TW, Liaw GJ, Hsu TH, Lin TH et al (2011) The Hippo pathway controls polar cell fate through Notch signaling during Drosophila oogenesis. Dev Biol 357(2):370–379

Chen L, Qin F, Deng X, Avruch J, Zhou D (2012) Hippo pathway in intestinal homeostasis and tumorigenesis. Protein Cell 3(4):305–310

Cho E, Irvine KD (2004) Action of fat, four-jointed, dachsous and dachs in distal-to-proximal wing signaling. Development 131:4489–4500

Cho KS, Lee JH, Kim S, Kim D, Koh H, Lee J, Kim C, Kim J, Chung J (2001) Drosophila phosphoinositide-dependent kinase-1 regulates apoptosis and growth via the phosphoinositide 3-kinase-dependent signaling pathway. Proc Natl Acad Sci USA 98:6144–6149

Cho E, Feng Y, Rauskolb C, Maitra S, Fehon R, Irvine KD (2006) Delineation of a fat tumor suppressor pathway. Nat Genet 38:1142–1150

Clark HF, Brentrup D, Schneitz K, Bieber A, Goodman C, Noll M (1995) Dachsous encodes a member of the cadherin superfamily that controls imaginal disc morphogenesis in Drosophila. Genes Dev 9(12):1530–1542

Collak FK, Yagiz K, Luthringer DJ, Erkaya B, Cinar B (2012) Threonine-120 phosphorylation regulated by phosphoinositide-3-kinase/Akt and mammalian target of rapamycin pathway signaling limits the antitumor activity of mammalian sterile 20-like kinase 1. J Biol Chem 287:23698–23709

Conlon I, Raff M (1999) Size control in animal development. Cell 96:235–244

Cook M, Tyers M (2007) Size control goes global. Curr Opin Biotechnol 18:341–350

Cooper S (2004) Control and maintenance of mammalian cell size. BMC Cell Biol 5:35

Cordero JB, Larson DE, Craig CR, Hays R, Cagan R (2007) Dynamic decapentaplegic signaling regulates patterning and adhesion in the Drosophila pupal retina. Development 134(10):1861–1871

Courbard JR, Djiane A, Wu J, Mlodzik M (2009) The apical/basal-polarity determinant Scribble cooperates with the PCP core factor Stbm/Vang and functions as one of its effectors. Dev Biol 333(1):67–77

Crickmore MA, Mann RS (2008) The control of size in animals: insights from selector genes. Bioessays 30:843–853

Cybulski N, Hall MN (2009) TOR complex 2: a signaling pathway of its own. Trends Biochem Sci 34:620–627

Daniel A, Dumstrei K, Lengyel JA, Hartenstein V (1999) The control of cell fate in the embryonic visual system by atonal, tailless and EGFR signaling. Development 126:2945–2954

Das TM, Feng Y, Jagannathan R, Seppa MJ, Skeath JB, Longmore GD (2010) Ajuba LIM proteins are negative regulators of the Hippo signaling pathway. Curr Biol 20:657–662

de Nooij JC, Hariharan IK (1995) Uncoupling cell fate determination from patterned cell division in the Drosophila eye. Science 270:983–985

Densham RM, O'Neill E, Munro J, Konig I, Anderson K, Kolch W et al (2009) MST kinases monitor actin cytoskeletal integrity and signal via c-Jun N-terminal kinase stress-activated kinase to regulate p21Waf1/Cip1 stability. Mol Cell Biol 29(24):6380–6390

Dibble CC, Asara JM, Manning BD (2009) Characterization of rictor phosphorylation sites reveals direct regulation of mTOR complex 2 by S6K1. Mol Cell Biol 29:5657–5670

Dickson B, Hafen E (1993) Genetic dissection of eye development in Drosophila. In: Bate M, Martinez Arias A (eds) The development of *Drosophila melanogaster*. Cold Spring Harbor Laboratory Press, Cold Spring Harbor, pp 1327–1362

Dobbelaere J, Josue F, Suijkerbuijk S, Baum B, Tapon N, Raff J (2008) A genome-wide RNAi screen to dissect centriole duplication and centrosome maturation in Drosophila. PLoS Biol 6(9):e224

Doggett K, Grusche FA, Richardson HE, Brumby AM (2011) Loss of the Drosophila cell polarity regulator scribbled promotes epithelial tissue overgrowth and cooperation with oncogenic Ras-Raf through impaired Hippo pathway signaling. BMC Dev Biol 11:57

Dominguez M (1999) Dual role for Hedgehog in the regulation of the proneural gene atonal during ommatidia development. Development 126:2345–2353

Dominguez M, Casares F (2005) Organ specification-growth control connection: new in-sights from the Drosophila eye-antennal disc. Dev Dyn 232:673–684

Dong J, Pan D (2004) Tsc2 is not a critical target of Akt during normal Drosophila development. Genes Dev 18:2479–2484

Dong J, Feldmann G, Huang J, Wu S, Zhang N, Comerford SA, Gayyed MF, Anders RA, Maitra A, Pan D (2007) Elucidation of a universal size-control mechanism in Drosophila and mammals. Cell 130:1120–1133

Edgar BA (1999) From small flies come big discoveries about size control. Nat Cell Biol 1:E191–E193

Edgar BA (2006) From cell structure to transcription: Hippo forges a new path. Cell 124:267–273

Enomoto M, Igaki T (2011) Deciphering tumor-suppressor signaling in flies: genetic link between Scribble/Dlg/Lgl and the Hippo pathways. J Genet Genomics 38(10):461–470

Fan SS, Chen MS, Lin JF, Chao WT, Yang VC (2003) Use of gain-of-function study to delineate the roles of crumbs in Drosophila eye development. J Biomed Sci 10(6 Pt 2):766–773

Feng Y, Irvine KD (2007) Fat and expanded act in parallel to regulate growth through warts. Proc Natl Acad Sci USA 104:20362–20367

Feng Y, Irvine KD (2009) Processing and phosphorylation of the fat receptor. Proc Natl Acad Sci USA 106:11989–11994

Fernandez R, Tabarini D, Azpiazu N, Frasch M, Schlessinger J (1995) The Drosophila insulin receptor homolog: a gene essential for embryonic development encodes two receptor isoforms with different signaling potential. EMBO J 14:3373–3384

Fernandez BG, Gaspar P, Bras-Pereira C, Jezowska B, Rebelo SR, Janody F (2011) Actin-Capping protein and the Hippo pathway regulate F-actin and tissue growth in Drosophila. Development 138:2337–2346

Firth LC, Bhattacharya A, Baker NE (2010) Cell cycle arrest by a gradient of Dpp signaling during Drosophila eye development. BMC Dev Biol 10:28

Fjose A, Polito LC, Weber U, Gehring WJ (1984) Developmental expression of the white locus of *Drosophila melanogaster*. EMBO J 3:2087–2094

Gadd S, Beezhold P, Jennings L, George D, Leuer K, Huang CC, Huff V, Tognon C, Sorensen PH, Triche T et al (2012) Mediators of receptor tyrosine kinase activation in infantile fibrosarcoma: a Children's Oncology Group study. J Pathol 228:119–130

Gao D, Wan L, Inuzuka H, Berg AH, Tseng A, Zhai B et al (2010) Rictor forms a complex with Cullin-1 to promote SGK1 ubiquitination and destruction. Mol Cell 39(5):797–808

Gao X, Pan D (2001) TSC1 and TSC2 tumor suppressors antagonize insulin signaling in cell growth. Genes Dev 15:1383–1392

Gao X, Neufeld TP, Pan D (2000) Drosophila PTEN regulates cell growth and proliferation through PI3K- dependent and -independent pathways. Dev Biol 221:404–418

Gao X, Zhang Y, Arrazola P, Hino O, Kobayashi T, Yeung RS, Ru B, Pan D (2002) Tsc tumour suppressor proteins antagonize amino-acid-TOR signalling. Nat Cell Biol 4:699–704

Garami A, Zwartkruis FJ, Nobukuni T, Joaquin M, Roccio M, Stocker H, Kozma SC, Hafen E, Bos JL, Thomas G (2003) Insulin activation of Rheb, a mediator of mTOR/S6K/4E-BP signaling, is inhibited by TSC1 and 2. Mol Cell 11:1457–1466

Garg A, Srivastava A, Davis MM, O'Keefe SL, Chow L, Bell JB (2007) Antagonizing scalloped with a novel vestigial construct reveals an important role for scalloped in Drosophila melanogaster leg, eye and optic lobe development. Genetics 175(2):659–669

Garoia F, Guerra D, Pezzoli MC, Lopez-Varea A, Cavicchi S, Garcia-Bellido A (2000) Cell behaviour of Drosophila fat cadherin mutations in wing development. Mech Develop 94(1-2):95–109

Genevet A, Tapon N (2011) The Hippo pathway and apico-basal cell polarity. Biochem J 436:213–224

Genevet A, Polesello C, Blight K, Robertson F, Collinson LM, Pichaud F, Tapon N (2009) The Hippo pathway regulates apical-domain size independently of its growth-control function. J Cell Sci 122:2360–2370

Genevet A, Wehr MC, Brain R, Thompson BJ, Tapon N (2010) Kibra is a regulator of the Salvador/Warts/Hippo signaling network. Dev Cell 18:300–308

Gilbert MM, Tipping M, Veraksa A, Moberg KH (2011) A screen for conditional growth suppressor genes identifies the Drosophila homolog of HD-PTP as a regulator of the oncoprotein Yorkie. Dev Cell 20:700–712

Gingras AC, Raught B, Sonenberg N (2001) Regulation of translation initiation by FRAP/mTOR. Genes Dev 15:807–826

Goberdhan DC, Paricio N, Goodman EC, Mlodzik M, Wilson C (1999) Drosophila tumor suppressor PTEN controls cell size and number by antagonizing the Chico/PI3-kinase signaling pathway. Genes Dev 13:3244–3258

Goulev Y, Fauny JD, Gonzalez-Marti B, Flagiello D, Silber J, Zider A (2008) SCALLOPED interacts with YORKIE, the nuclear effector of the hippo tumor-suppressor pathway in Drosophila. Curr Biol 18:435–441

Graves JD, Gotoh Y, Draves KE, Ambrose D, Han DK, Wright M, Chernoff J, Clark EA, Krebs EG (1998) Caspase-mediated activation and induction of apoptosis by the mammalian Ste20-like kinase Mst1. EMBO J 17:2224–2234

Grebien F, Dolznig H, Beug H, Mullner EW (2005) Cell size control: new evidence for a general mechanism. Cell Cycle 4:418–421

Greenwood S, Struhl G (1999) Progression of the morphogenetic furrow in the Drosophila eye: the roles of Hedgehog, Decapentaplegic and the Raf pathway. Development 126:5795–5808

Grusche FA, Richardson HE, Harvey KF (2010) Upstream regulation of the hippo size control pathway. Curr Biol 20:R574–582

Grusche FA, Degoutin JL, Richardson HE, Harvey KF (2011) The Salvador/Warts/Hippo pathway controls regenerative tissue growth in *Drosophila melanogaster*. Dev Biol 350:255–266

Grzeschik NA, Amin N, Secombe J, Brumby AM, Richardson HE (2007) Abnormalities in cell proliferation and apico-basal cell polarity are separable in Drosophila lgl mutant clones in the developing eye. Dev Biol 311:106–123

Grzeschik NA, Parsons LM, Allott ML, Harvey KF, Richardson HE (2010a) Lgl, aPKC, and Crumbs regulate the Salvador/Warts/Hippo pathway through two distinct mechanisms. Curr Biol 20:573–581

Grzeschik NA, Parsons LM, Richardson HE (2010b) Lgl, the SWH pathway and tumorigenesis: it's a matter of context & competition!. Cell Cycle 9:3202–3212

Guan J, Li H, Rogulja A, Axelrod JD, Cadigan KM (2007) The Drosophila casein kinase Iepsilon/delta Discs overgrown promotes cell survival via activation of DIAP1 expression. Dev Biol 303:16–28

Guo C, Tommasi S, Liu L, Yee JK, Dammann R, Pfeifer GP (2007) RASSF1A is part of a complex similar to the Drosophila Hippo/Salvador/Lats tumor-suppressor network. Curr Biol 17:700–705

Habbig S, Bartram MP, Muller RU, Schwarz R, Andriopoulos N, Chen S, Sagmuller JG, Hoehne M, Burst V, Liebau MC et al (2011) NPHP4, a cilia-associated protein, negatively regulates the Hippo pathway. J Cell Biol 193:633–642

Habbig S, Bartram MP, Sagmuller JG, Griessmann A, Franke M, Muller RU, Schwarz R, Hoehne M, Bergmann C, Tessmer C et al (2012) The ciliopathy disease protein NPHP9 promotes nuclear delivery and activation of the oncogenic transcriptional regulator TAZ. Hum Mol Genet 21:5528–5538

Hafen E (1991) Patterning by cell recruitment in the Drosophila eye. Curr Opin Genet Dev 1:268–274

Hafen E (2004) Interplay between growth factor and nutrient signaling: lessons from Drosophila TOR. Curr Top Microbiol Immunol 279:153–167

Halder G, Johnson RL (2011) Hippo signaling: growth control and beyond. Development 138:9–22

Halder G, Dupont S, Piccolo S (2012) Transduction of mechanical and cytoskeletal cues by YAP and TAZ. Nat Rev Mol Cell Biol 13:591–600

Hamaratoglu F, Willecke M, Kango-Singh M, Nolo R, Hyun E, Tao C, Jafar-Nejad H, Halder G (2006) The tumour-suppressor genes NF2/Merlin and Expanded act through Hippo signalling to regulate cell proliferation and apoptosis. Nat Cell Biol 8:27–36

Hara K, Maruki Y, Long X, Yoshino K, Oshiro N, Hidayat S, Tokunaga C, Avruch J, Yonezawa K (2002) Raptor, a binding partner of target of rapamycin (TOR), mediates TOR action. Cell 110:177–189

Harvey KF, Hariharan IK (2012) The hippo pathway. Cold Spring Harb Perspect Biol 4:a011288

Harvey NL, Daish T, Mills K, Dorstyn L, Quinn LM, Read SH, Richardson H, Kumar S (2001) Characterization of the Drosophila caspase, DAMM. J Biol Chem 276:25342–25350

Harvey KF, Pfleger CM, Hariharan IK (2003) The Drosophila Mst ortholog, hippo, restricts growth and cell proliferation and promotes apoptosis. Cell 114:457–467

Hergovich A (2012) Mammalian Hippo signalling: a kinase network regulated by protein-protein interactions. Biochem Soc Trans 40:124–128
Herranz H, Hong X, Cohen SM (2012) Mutual repression by bantam miRNA and Capicua links the EGFR/MAPK and Hippo pathways in growth control. Curr Biol 22(8):651–657
Hiemer SE, Varelas X (2013) Stem cell regulation by the Hippo pathway. Biochim Biophys Acta 1830:2323–2334
Ho KC, Zhou Z, She YM, Chun A, Cyr TD, Yang X (2011) Itch E3 ubiquitin ligase regulates large tumor suppressor 1 stability [corrected]. Proc Natl Acad Sci USA 108:4870–4875
Hong W, Guan KL (2012) The YAP and TAZ transcription co-activators: key downstream effectors of the mammalian Hippo pathway. Semin Cell Dev Biol 23:785–793
Huang H, Potter CJ, Tao W, Li DM, Brogiolo W, Hafen E, Sun H, Xu T (1999) PTEN affects cell size, cell proliferation and apoptosis during Drosophila eye development. Development 126:5365–5372
Huang J, Wu S, Barrera J, Matthews K, Pan D (2005) The Hippo signaling pathway coordinately regulates cell proliferation and apoptosis by inactivating Yorkie, the Drosophila Homolog of YAP. Cell 122:421–434
Hwang E, Ryu KS, Paakkonen K, Guntert P, Cheong HK, Lim DS, Lee JO, Jeon YH, Cheong C (2007) Structural insight into dimeric interaction of the SARAH domains from Mst1 and RASSF family proteins in the apoptosis pathway. Proc Natl Acad Sci USA 104:9236–9241
Ikeya T, Galic M, Belawat P, Nairz K, Hafen E (2002) Nutrient-dependent expression of insulin-like peptides from neuroendocrine cells in the CNS contributes to growth regulation in Drosophila. Curr Biol 12:1293–1300
Inoki K, Li Y, Zhu T, Wu J, Guan KL (2002) TSC2 is phosphorylated and inhibited by Akt and suppresses mTOR signalling. Nat Cell Biol 4:648–657
Ishikawa HO, Takeuchi H, Haltiwanger RS, Irvine KD (2008) Four-jointed is a Golgi kinase that phosphorylates a subset of cadherin domains. Science 321:401–404
Ishiuchi T, Takeichi M (2012) Nectins localize Willin to cell-cell junctions. Genes Cells 17:387–397
Jacinto E, Facchinetti V, Liu D, Soto N, Wei S, Jung SY, Huang Q, Qin J, Su B (2006) SIN1/MIP1 maintains rictor-mTOR complex integrity and regulates Akt phosphorylation and substrate specificity. Cell 127:125–137
Jacobson MD, Weil M, Raff MC (1997) Programmed cell death in animal development. Cell 88:347–354
Jarman AP, Grell EH, Ackerman L, Jan LY, Jan YN (1994) Atonal is the proneural gene for Drosophila photoreceptors. Nature 369:398–400
Jia J, Zhang W, Wang B, Trinko R, Jiang J (2003) The Drosophila Ste20 family kinase dMST functions as a tumor suppressor by restricting cell proliferation and promoting apoptosis. Genes Dev 17:2514–2519
Johnston LA, Gallant P (2002) Control of growth and organ size in Drosophila. Bioessays 24:54–64
Junger MA, Rintelen F, Stocker H, Wasserman JD, Vegh M, Radimerski T, Greenberg ME, Hafen E (2003) The Drosophila forkhead transcription factor FOXO mediates the reduction in cell number associated with reduced insulin signaling. J Biol 2:20
Justice RW, Zilian O, Woods DF, Noll M, Bryant PJ (1995) The Drosophila tumor suppressor gene warts encodes a homolog of human myotonic dystrophy kinase and is required for the control of cell shape and proliferation. Genes Dev 9:534–546
Kagey JD, Brown JA, Moberg KH (2012) Regulation of Yorkie activity in Drosophila imaginal discs by the Hedgehog receptor gene patched. Mech Dev 129(9-12):339–349
Kandt RS (2002) Tuberous sclerosis complex and neurofibromatosis type 1: the two most common neurocutaneous diseases. Neurol Clin 20:941–964
Kango-Singh M, Singh A (2009) Regulation of organ size: insights from the Drosophila Hippo signaling pathway. Dev Dyn 238:1627–1637
Kango-Singh M, Nolo R, Tao C, Verstreken P, Hiesinger PR, Bellen HJ, Halder G (2002) Shar-pei mediates cell proliferation arrest during imaginal disc growth in Drosophila. Development 129:5719–5730

Kango-Singh M, Singh A, Henry Sun Y (2003) Eyeless collaborates with Hedgehog and Decapentaplegic signaling in Drosophila eye induction. Dev Biol 256:49–60

Kaplan NA, Colosimo PF, Liu X, Tolwinski NS (2011) Complex interactions between GSK3 and aPKC in Drosophila embryonic epithelial morphogenesis. PLoS One 6(4):e18616

Kaplan NA, Tolwinski NS (2010) Spatially defined Dsh-Lgl interaction contributes to directional tissue morphogenesis. J Cell Sci 123(Pt 18):3157–3165

Karni R, Hippo Y, Lowe SW, Krainer AR (2008) The splicing-factor oncoprotein SF2/ASF activates mTORC1. Proc Natl Acad Sci USA 105:15323–15327

Karpowicz P, Perez J, Perrimon N (2010) The Hippo tumor suppressor pathway regulates intestinal stem cell regeneration. Development 137:4135–4145

Kiger AA, Baum B, Jones S, Jones MR, Coulson A, Echeverri C et al (2003) A functional genomic analysis of cell morphology using RNA interference. J Biol 2(4):27

Kim DH, Sarbassov DD, Ali SM, King JE, Latek RR, Erdjument-Bromage H, Tempst P, Sabatini DM (2002) mTOR interacts with raptor to form a nutrient-sensitive complex that signals to the cell growth machinery. Cell 110:163–175

Kim DH, Sarbassov DD, Ali SM, Latek RR, Guntur KV, Erdjument-Bromage H, Tempst P, Sabatini DM (2003) GbetaL, a positive regulator of the rapamycin-sensitive pathway required for the nutrient-sensitive interaction between raptor and mTOR. Mol Cell 11:895–904

Kim D, Shu S, Coppola MD, Kaneko S, Yuan ZQ, Cheng JQ (2010) Regulation of proapoptotic mammalian ste20-like kinase MST2 by the IGF1-Akt pathway. PLoS ONE 5:e9616

Kim J, Guan KL (2011) Amino acid signaling in TOR activation. Annu Rev Biochem 80:1001–1032

Klein TJ, Jenny A, Djiane A, Mlodzik M (2006) CKIepsilon/discs overgrown promotes both Wnt-Fz/beta-catenin and Fz/PCP signaling in Drosophila. Curr Biol 16(13):1337–43

Konsavage WM, Jr., Yochum GS (2013) Intersection of Hippo/YAP and Wnt/beta-catenin signaling pathways. Acta Biochim Biophys Sin (Shanghai) 45(2):71–79

Kramer H, Cagan RL (1994) Determination of photoreceptor cell fate in the Drosophila retina. Curr Opin Neurobiol 4:14–20

Kumagai A, Dunphy WG (1999) Binding of 14-3-3 proteins and nuclear export control the intracellular localization of the mitotic inducer Cdc25. Genes Dev 13:1067–1072

Kumar JP (2001) Signalling pathways in Drosophila and vertebrate retinal development. Nat Rev Genet 2:846–857

Kumar JP (2009) The molecular circuitry governing retinal determination. Biochim Biophys Acta 1789:306–314

Kumar JP, Moses K (2000) Cell fate specification in the Drosophila retina. Results Probl Cell Differ 31:93–114

Kumar JP, Moses K (2001) Eye specification in Drosophila: perspectives and implications. Semin Cell Dev Biol 12:469–474

Lai ZC, Wei X, Shimizu T, Ramos E, Rohrbaugh M, Nikolaidis N, Ho LL, Li Y (2005) Control of cell proliferation and apoptosis by mob as tumor suppressor, mats. Cell 120:675–685

Lasko P (2000) The Drosophila melanogaster genome: translation factors and RNA binding proteins. J Cell Biol 150:F51–56

Leevers SJ (2001) Growth control: invertebrate insulin surprises! Curr Biol 11:R209–212

Leevers SJ, Weinkove D, MacDougall LK, Hafen E, Waterfield MD (1996) The Drosophila phosphoinositide 3-kinase Dp110 promotes cell growth. EMBO J 15:6584–6594

Li W, Cooper J, Karajannis MA, Giancotti FG (2012) Merlin: a tumour suppressor with functions at the cell cortex and in the nucleus. EMBO Rep 13:204–215

Li X, Gianoulis TA, Yip KY, Gerstein M, Snyder M (2010) Extensive in vivo metabolite-protein interactions revealed by large-scale systematic analyses. Cell 143(4):639–650

Liao XH, Majithia A, Huang X, Kimmel AR (2008) Growth control via TOR kinase signaling, an intracellular sensor of amino acid and energy availability, with crosstalk potential to proline metabolism. Amino Acids 35:761–770

Lin YT, Ding JY, Li MY, Yeh TS, Wang TW, Yu JY (2012) YAP regulates neuronal differentiation through Sonic hedgehog signaling pathway. Exp Cell Res 318(15):1877–1888

Ling C, Zheng Y, Yin F, Yu J, Huang J, Hong Y, Wu S, Pan D (2010) The apical transmembrane protein Crumbs functions as a tumor suppressor that regulates Hippo signaling by binding to expanded. Proc Natl Acad Sci USA 107:10532–10537

Liu X, Grammont M, Irvine KD (2000) Roles for scalloped and vestigial in regulating cell affinity and interactions between the wing blade and the wing hinge. Dev Biol 228:287–303

Liu CY, Zha ZY, Zhou X, Zhang H, Huang W, Zhao D, Li T, Chan SW, Lim CJ, Hong W et al (2010) The hippo tumor pathway promotes TAZ degradation by phosphorylating a phosphodegron and recruiting the SCF{beta}-TrCP E3 ligase. J Biol Chem 285:37159–37169

Liu C, Huang W, Lei Q (2011) Regulation and function of the TAZ transcription co-activator. Int J Biochem Mol Biol 2:247–256

Liu AM, Wong KF, Jiang X, Qiao Y, Luk JM (2012a) Regulators of mammalian Hippo pathway in cancer. Biochim Biophys Acta 1826:357–364

Liu H, Jiang D, Chi F, Zhao B (2012b) The Hippo pathway regulates stem cell proliferation, self-renewal, and differentiation. Protein Cell 3:291–304

Loewith R (2011) A brief history of TOR. Biochem Soc Trans 39:437–442

Loewith R, Jacinto E, Wullschleger S, Lorberg A, Crespo JL, Bonenfant D, Oppliger W, Jenoe P, Hall MN (2002) Two TOR complexes, only one of which is rapamycin sensitive, have distinct roles in cell growth control. Mol Cell 10:457–468

Long X, Muller F, Avruch J (2004) TOR action in mammalian cells and in Caenorhabditis elegans. Curr Top Microbiol Immunol 279:115–138

Mahoney PA, Weber U, Onofrechuk P, Biessmann H, Bryant PJ, Goodman CS (1991) The fat tumor suppressor gene in Drosophila encodes a novel member of the cadherin gene superfamily. Cell 67:853–868

Maitra S, Kulikauskas RM, Gavilan H, Fehon RG (2006) The tumor suppressors Merlin and Expanded function cooperatively to modulate receptor endocytosis and signaling. Curr Biol 16(7):702–709

Mao Y, Rauskolb C, Cho E, Hu WL, Hayter H, Minihan G, Katz FN, Irvine KD (2006) Dachs: an unconventional myosin that functions downstream of Fat to regulate growth, affinity and gene expression in Drosophila. Development 133:2539–2551

Mao Y, Kucuk B, Irvine KD (2009) Drosophila lowfat, a novel modulator of Fat signaling. Development 136:3223–3233

Martin FA, Perez-Garijo A, Morata G (2009) Apoptosis in Drosophila: compensatory proliferation and undead cells. Int J Dev Biol 53:1341–1347

Marygold SJ, Leevers SJ (2002) Growth signaling: TSC takes its place. Curr Biol 12:R785–787

Matakatsu H, Blair SS (2004) Interactions between Fat and Dachsous and the regulation of planar cell polarity in the Drosophila wing. Development 131:3785–3794

Matakatsu H, Blair SS (2006) Separating the adhesive and signaling functions of the Fat and Dachsous protocadherins. Development 133:2315–2324

Matakatsu H, Blair SS (2008) The DHHC palmitoyltransferase approximated regulates Fat signaling and Dachs localization and activity. Curr Biol 18:1390–1395

Matakatsu H, Blair SS (2012) Separating planar cell polarity and Hippo pathway activities of the protocadherins Fat and Dachsous. Development 139:1498–1508

Mauviel A, Nallet-Staub F, Varelas X (2012) Integrating developmental signals: a Hippo in the (path)way. Oncogene 31(14):1743–1756

McCartney BM, Kulikauskas RM, LaJeunesse DR, Fehon RG (2000) The Neurofibromatosis-2 homologue, Merlin, and the tumor suppressor expanded function together in Drosophila to regulate cell proliferation and differentiation. Development 127:1315–1324

Meignin C, Alvarez-Garcia I, Davis I, Palacios IM (2007) The salvador-warts-hippo pathway is required for epithelial proliferation and axis specification in Drosophila. Curr Biol 17:1871–1878

Mikeladze-Dvali T, Wernet MF, Pistillo D, Mazzoni EO, Teleman AA, Chen YW et al (2005) The growth regulators warts/lats and melted interact in a bistable loop to specify opposite fates in Drosophila R8 photoreceptors. Cell 122(5):775–787

Mills JR, Hippo Y, Robert F, Chen SM, Malina A, Lin CJ, Trojahn U, Wendel HG, Charest A, Bronson RT et al (2008) mTORC1 promotes survival through translational control of Mcl-1. Proc Natl Acad Sci USA 105:10853–10858

Mitchison JM, Novak B, Sveiczer A (1997) Size control in the cell cycle. Cell Biol Int 21:461–463

Mohr OL (1923) Modifications of the sex-ratio through a sex-linked semi-lethal in Drosophila melanogaster. (Besides notes on an autosomal section deficiency). Studia Mendeliana, ad centesimum diem natalem Gregorii Mendelii a grata patria celebrandum, adiuvante ministerio Pragensi edita. 266–287

Mohr OL (1929) Exaggeration and inhibition phenomena encountered in the analysis of an autosomal dominant. Zeitschrift fur induktive Abstammungs- und Vererbungslehre 50:113–200

Montagne J (2000) Genetic and molecular mechanisms of cell size control. Mol Cell Biol Res Commun 4:195–202

Montagne J, Stewart MJ, Stocker H, Hafen E, Kozma SC, Thomas G (1999) Drosophila S6 kinase: a regulator of cell size. Science 285:2126–2129

Montagne J, Radimerski T, Thomas G (2001) Insulin signaling: lessons from the Drosophila tuberous sclerosis complex, a tumor suppressor. Sci STKE 2001:pe36.

Morrison H, Sherman LS, Legg J, Banine F, Isacke C, Haipek CA, Gutmann DH, Ponta H, Herrlich P (2001) The NF2 tumor suppressor gene product, merlin, mediates contact inhibition of growth through interactions with CD44. Genes Dev 15:968–980

Nam SC, Choi KW (2003) Interaction of Par-6 and Crumbs complexes is essential for photoreceptor morphogenesis in Drosophila. Development 130(18):4363–4372

Nam SC, Choi KW (2006) Domain-specific early and late function of Dpatj in Drosophila photoreceptor cells. Dev Dyn 235(6):1501–1507

Nellist M, Sancak O, Goedbloed M, Adriaans A, Wessels M, Maat-Kievit A, Baars M, Dommering C, den Ouweland A van, Halley D (2008) Functional characterisation of the TSC1-TSC2 complex to assess multiple TSC2 variants identified in single families affected by tuberous sclerosis complex. BMC Med Genet 9:10

Neto-Silva RM, Beco S de, Johnston LA (2010) Evidence for a growth-stabilizing regulatory feedback mechanism between Myc and Yorkie, the Drosophila homolog of Yap. Dev Cell 19:507–520

Newsome TP, Asling B, Dickson BJ (2000) Analysis of Drosophila photoreceptor axon guidance in eye-specific mosaics. Development 127:851–860

Nolo R, Morrison CM, Tao C, Zhang X, Halder G (2006) The bantam microRNA is a target of the hippo tumor-suppressor pathway. Curr Biol 16:1895–1904

Ogawa H, Ohta N, Moon W, Matsuzaki F (2009) Protein phosphatase 2A negatively regulates aPKC signaling by modulating phosphorylation of Par-6 in Drosophila neuroblast asymmetric divisions. J Cell Sci 122(Pt 18):3242–3249

Oh h, Irvine KD (2008) In vivo regulation of Yorkie phosphorylation and localization. Development 135:1081–1088

Oh H, Irvine KD (2009) In vivo analysis of Yorkie phosphorylation sites. Oncogene 28:1916–1927

Oh H, Irvine KD (2010) Yorkie: the final destination of Hippo signaling. Trends Cell Biol 20:410–417

Oh H, Reddy BV, Irvine KD (2009) Phosphorylation-independent repression of Yorkie in Fat-Hippo signaling. Dev Biol 335(1):188–197

Oka T, Remue E, Meerschaert K, Vanloo B, Boucherie C, Gfeller D, Bader G, Sidhu S, Vandekerckhove J, Gettemans J et al (2010) Functional complex between YAP2 and ZO-2 is PDZ domain dependent, regulates YAP2 nuclear localization and signaling. Biochem J 432(3):461–472

Oldham S, Hafen E (2003) Insulin/IGF and target of rapamycin signaling: a TOR de force in growth control. Trends Cell Biol 13:79–85

Oldham S, Bohni R, Stocker H, Brogiolo W, Hafen E (2000a) Genetic control of size in Drosophila. Philos Trans R Soc Lond B Biol Sci 355:945–952

Oldham S, Montagne J, Radimerski T, Thomas G, Hafen E (2000b) Genetic and biochemical characterization of dTOR, the Drosophila homolog of the target of rapamycin. Genes Dev 14:2689–2694

Oldham S, Stocker H, Laffargue M, Wittwer F, Wymann M, Hafen E (2002) The Drosophila insulin/IGF receptor controls growth and size by modulating PtdInsP(3) levels. Development 129:4103–4109
O'Neill E, Kolch W (2005) Taming the Hippo: Raf-1 controls apoptosis by suppressing MST2/Hippo. Cell Cycle 4:365–367
Pagliarini RA, Quinones AT, Xu T (2003) Analyzing the function of tumor suppressor genes using a Drosophila model. Methods Mol Biol 223:349–382
Pan D (2007) Hippo signaling in organ size control. Genes Dev 21:886–897
Pan D (2010) The hippo signaling pathway in development and cancer. Dev Cell 19:491–505
Pan D, Dong J, Zhang Y, Gao X (2004) Tuberous sclerosis complex: from Drosophila to human disease. Trends Cell Biol 14:78–85
Pantalacci S, Tapon N, Leopold P (2003) The Salvador partner Hippo promotes apoptosis and cell-cycle exit in Drosophila. Nat Cell Biol 5:921–927
Paramasivam M, Sarkeshik A, Yates JR 3rd, Fernandes MJ, McCollum D (2011) Angiomotin family proteins are novel activators of the LATS2 kinase tumor suppressor. Mol Biol Cell 22:3725–3733
Pearce LR, Komander D, Alessi DR (2010) The nuts and bolts of AGC protein kinases. Nat Rev Mol Cell Biol 11:9–22
Pellock BJ, Buff E, White K, Hariharan IK (2007) The Drosophila tumor suppressors Expanded and Merlin differentially regulate cell cycle exit, apoptosis, and Wingless signaling. Dev Biol 304:102–115
Peng HW, Slattery M, Mann RS (2009) Transcription factor choice in the Hippo signaling pathway: homothorax and yorkie regulation of the microRNA bantam in the progenitor domain of the Drosophila eye imaginal disc. Genes Dev 23:2307–3219
Penton A, Selleck SB, Hoffmann FM (1997) Regulation of cell cycle synchronization by decapentaplegic during Drosophila eye development. Science 275:203–206
Pfeiffer BD, Ngo TT, Hibbard KL, Murphy C, Jenett A, Truman JW, Rubin GM (2010) Refinement of tools for targeted gene expression in Drosophila. Genetics 186:735–755
Pichaud F, Desplan C (2002) Cell biology: a new view of photoreceptors. Nature 416(6877):139–140
Poernbacher I, Baumgartner R, Marada SK, Edwards K, Stocker H (2012) Drosophila Pez acts in Hippo signaling to restrict intestinal stem cell proliferation. Curr Biol 22(5):389–396
Polesello C, Tapon N (2007) salvador-warts-hippo signaling promotes Drosophila posterior follicle cell maturation downstream of notch. Curr Biol 17:1864–1870
Polesello C, Huelsmann S, Brown NH, Tapon N (2006) The Drosophila RASSF homolog antagonizes the hippo pathway. Curr Biol 16:2459–2465
Poltilove RM, Jacobs AR, Haft CR, Xu P, Taylor SI (2000) Characterization of Drosophila insulin receptor substrate. J Biol Chem 275:23346–23354
Poon CL, Lin JI, Zhang X, Harvey KF (2011) The sterile 20-like kinase Tao-1 controls tissue growth by regulating the Salvador-Warts-Hippo pathway. Dev Cell 21:896–906
Potter CJ, Xu T (2001) Mechanisms of size control. Curr Opin Genet Dev 11:279–286
Potter CJ, Huang H, Xu T (2001) Drosophila Tsc1 functions with Tsc2 to antagonize insulin signaling in regulating cell growth, cell proliferation, and organ size. Cell 105:357–368
Potter CJ, Pedraza LG, Xu T (2002) Akt regulates growth by directly phosphorylating Tsc2. Nat Cell Biol 4:658–665
Potter CJ, Pedraza LG, Huang H, Xu T (2003) The tuberous sclerosis complex (TSC) pathway and mechanism of size control. Biochem Soc Trans 31:584–586
Price DM, Jin Z, Rabinovitch S, Campbell SD (2002) Ectopic expression of the Drosophila Cdk1 inhibitory kinases, Wee1 and Myt1, interferes with the second mitotic wave and disrupts pattern formation during eye development. Genetics 161:721–731
Radimerski T, Montagne J, Hemmings-Mieszczak M, Thomas G (2002a) Lethality of Drosophila lacking TSC tumor suppressor function rescued by reducing dS6K signaling. Genes Dev 16:2627–2632

Radimerski T, Montagne J, Rintelen F, Stocker H, Kaay J van der, Downes CP, Hafen E, Thomas G (2002b) dS6K-regulated cell growth is dPKB/dPI(3)K-independent, but requires dPDK1. Nat Cell Biol 4:251–255
Raff MC (1996) Size control: the regulation of cell numbers in animal development. Cell 86:173–175
Rauskolb C, Pan G, Reddy BV, H Oh, Irvine KD (2011) Zyxin links fat signaling to the hippo pathway. PLoS Biol 9:e1000624
Reddy BV, Irvine KD (2008) The Fat and Warts signaling pathways: new insights into their regulation, mechanism and conservation. Development 135:2827–2838
Reddy BV, Rauskolb C, Irvine KD (2010) Influence of fat-hippo and notch signaling on the proliferation and differentiation of Drosophila optic neuroepithelia. Development 137:2397–2408
Ren F, Wang B, Yue T, Yun EY, Ip YT, Jiang J (2010a) Hippo signaling regulates Drosophila intestine stem cell proliferation through multiple pathways. Proc Natl Acad Sci USA 107:21064–21069
Ren F, Zhang L, Jiang J (2010b) Hippo signaling regulates Yorkie nuclear localization and activity through 14-3-3 dependent and independent mechanisms. Dev Biol 337:303–312
Ribeiro PS, Josue F, Wepf A, Wehr MC, Rinner O, Kelly G, Tapon N, Gstaiger M (2010) Combined functional genomic and proteomic approaches identify a PP2A complex as a negative regulator of Hippo signaling. Mol Cell 39:521–534
Richardson HE (2011) Actin up for Hippo. EMBO J 30:2307–2309
Richardson H, Kumar S (2002) Death to flies: Drosophila as a model system to study programmed cell death. J Immunol Methods 265:21–38
Rintelen F, Stocker H, Thomas G, Hafen E (2001) PDK1 regulates growth through Akt and S6K in Drosophila. Proc Natl Acad Sci USA 98:15020–15025
Robinson BS, Huang J, Hong Y, Moberg KH (2010) Crumbs regulates Salvador/Warts/Hippo signaling in Drosophila via the FERM-domain protein Expanded. Curr Biol 20:582–590
Rogulja D, Rauskolb C, Irvine KD (2008) Morphogen control of wing growth through the Fat signaling pathway. Dev Cell 15:309–321
Rothenberg ME, Jan YN (2002) Salvador—the persistence of proliferation. Cancer Cell 2:171–173
Rubin GM (1989) Development of the Drosophila retina: inductive events studied at single cell resolution. Cell 57:519–520
Rulifson EJ, Kim SK, Nusse R (2002) Ablation of insulin-producing neurons in flies: growth and diabetic phenotypes. Science 296:1118–1120
Rusconi JC, Hays R, Cagan RL (2000) Programmed cell death and patterning in Drosophila. Cell Death Differ 7:1063–1070
Salah Z, Melino G, Aqeilan RI (2011) Negative regulation of the Hippo pathway by E3 ubiquitin ligase ITCH is sufficient to promote tumorigenicity. Cancer Res 71:2010–2020
Sancak Y, Bar-Peled L, Zoncu R, Markhard AL, Nada S, Sabatini DM (2010) Ragulator-Rag complex targets mTORC1 to the lysosomal surface and is necessary for its activation by amino acids. Cell 141:290–303
Sansores-Garcia L, Bossuyt W, Wada K, Yonemura S, Tao C, Sasaki H, Halder G (2011) Modulating F-actin organization induces organ growth by affecting the Hippo pathway. EMBO J 30:2325–2335
Sarbassov DD, Ali SM, Kim DH, Guertin DA, Latek RR, Erdjument-Bromage H, Tempst P, Sabatini DM (2004) Rictor, a novel binding partner of mTOR, defines a rapamycin-insensitive and raptor-independent pathway that regulates the cytoskeleton. Curr Biol 14:1296–1302
Saucedo LJ, Edgar BA (2007) Filling out the Hippo pathway. Nat Rev Mol Cell Biol 8:613–621
Saucedo LJ, Gao X, Chiarelli DA, Li L, Pan D, Edgar BA (2003) Rheb promotes cell growth as a component of the insulin/TOR signalling network. Nat Cell Biol 5:566–571
Sawamoto K, Okano H (1996) Cell-cell interactions during neural development: multiple types of lateral inhibitions involved in Drosophila eye development. Neurosci Res 26:205–214
Schagdarsurengin U, Richter AM, Hornung J, Lange C, Steinmann K, Dammann RH (2010) Frequent epigenetic inactivation of RASSF2 in thyroid cancer and functional consequences. Mol Cancer 9:264

Scheel H, Hofmann K (2003) A novel interaction motif, SARAH, connects three classes of tumor suppressor. Curr Biol 13:R899–900

Schlegelmilch K, Mohseni M, Kirak O, Pruszak J, Rodriguez JR, Zhou D, Kreger BT, Vasioukhin V, Avruch J, Brummelkamp TR et al (2011) Yap1 acts downstream of alpha-catenin to control epidermal proliferation. Cell 144:782–795

Schroeder MC, Halder G (2012) Regulation of the Hippo pathway by cell architecture and mechanical signals. Semin Cell Dev Biol 23(7):803–811

Seidel C, Schagdarsurengin U, Blumke K, Wurl P, Pfeifer GP, Hauptmann S, Taubert H, Dammann R (2007) Frequent hypermethylation of MST1 and MST2 in soft tissue sarcoma. Mol Carcinog 46:865–71

Sekido Y (2008) Molecular biology of malignant mesothelioma. Environ Health Prev Med 13:65–70

Sekido Y (2011) Inactivation of Merlin in malignant mesothelioma cells and the Hippo signaling cascade dysregulation. Pathol Int 61:331–344

Shaw RL, Kohlmaier A, Polesello C, Veelken C, Edgar BA, Tapon N (2010) The Hippo pathway regulates intestinal stem cell proliferation during Drosophila adult midgut regeneration. Development 137:4147–4158

Shimizu T, Ho LL, Lai ZC (2008) The mob as tumor suppressor gene is essential for early development and regulates tissue growth in Drosophila. Genetics 178:957–965

Silva E, Tsatskis Y, Gardano L, Tapon N, McNeill H (2006) The tumor-suppressor gene fat controls tissue growth upstream of expanded in the hippo signaling pathway. Curr Biol 16:2081–2089

Silvis MR, Kreger BT, Lien WH, Klezovitch O, Rudakova GM, Camargo FD, Lantz DM, Seykora JT, Vasioukhin V (2011) alpha-catenin is a tumor suppressor that controls cell accumulation by regulating the localization and activity of the transcriptional coactivator Yap1. Sci Signal 4:ra33

Simon MA, Xu A, Ishikawa HO, Irvine KD (2010) Modulation of fat: dachsous binding by the cadherin domain kinase four-jointed. Curr Biol 20:811–817

Singh A, Kango-Singh M, Choi KW, Sun YH (2004) Dorso-ventral asymmetric functions of teashirt in Drosophila eye development depend on spatial cues provided by early DV patterning genes. Mech Dev 121(4):365–370

Singh A, Tare M, Puli OR, Kango-Singh M (2012) A glimpse into dorso-ventral patterning of the Drosophila eye. Dev Dyn 241:69–84

Skouloudaki K, Walz G (2012) YAP1 recruits c-Abl to protect angiomotin-like 1 from Nedd4-mediated degradation. PLoS ONE 7:e35735

Skouloudaki K, Puetz M, Simons M, Courbard JR, Boehlke C, Hartleben B, Engel C, Moeller MJ, Englert C, Bollig F et al (2009) Scribble participates in Hippo signaling and is required for normal zebrafish pronephros development. Proc Natl Acad Sci USA 106:8579–8584

Sopko R, Silva E, Clayton L, Gardano L, Barrios-Rodiles M, Wrana J, Varelas X, Arbouzova NI, Shaw S, Saburi S et al (2009) Phosphorylation of the tumor suppressor fat is regulated by its ligand Dachsous and the kinase discs overgrown. Curr Biol 19:1112–1117

Soulard A, Cohen A, Hall MN (2009) TOR signaling in invertebrates. Curr Opin Cell Biol 21:825–836

St. Johnston D (2002) The art and design of genetic screens: Drosophila melanogaster. Nat Rev Genet 3:176–188

Staley BK, Irvine KD (2010) Warts and Yorkie mediate intestinal regeneration by influencing stem cell proliferation. Curr Biol 20:1580–1587

Staley BK, Irvine KD (2012) Hippo signaling in Drosophila: recent advances and insights. Dev Dyn 241:3–15

Stocker H, Hafen E (2000) Genetic control of cell size. Curr Opin Genet Dev 10:529–535

Stocker H, Radimerski T, Schindelholz B, Wittwer F, Belawat P, Daram P, Breuer S, Thomas G, Hafen E (2003) Rheb is an essential regulator of S6K in controlling cell growth in Drosophila. Nat Cell Biol 5:559–565

Strassburger K, Tiebe M, Pinna F, Breuhahn K, Teleman AA (2012) Insulin/IGF signaling drives cell proliferation in part via Yorkie/YAP. Dev Biol 367:187–196

Striedinger K, VandenBerg SR, Baia GS, McDermott MW, Gutmann DH, Lal A (2008) The neurofibromatosis 2 tumor suppressor gene product, merlin, regulates human meningioma cell growth by signaling through YAP. Neoplasia 10:1204–1212

Strutt H, Price MA, Strutt D (2006) Planar polarity is positively regulated by casein kinase Iepsilon in Drosophila. Curr Biol 16(13):1329–1336

Su TT, O'Farrell PH (1998) Size control: cell proliferation does not equal growth. Curr Biol 8:R687–689

Sudol M (2010) Newcomers to the WW domain-mediated network of the Hippo tumor suppressor pathway. Genes Cancer 1:1115–1118

Sudol M, Harvey KF (2010) Modularity in the Hippo signaling pathway. Trends Biochem Sci 35(11):627–633

Sun G, Irvine KD (2010) Regulation of Hippo signaling by Jun kinase signaling during compensatory cell proliferation and regeneration, and in neoplastic tumors. Dev Biol 350(1):139–151

Sun G, Irvine KD (2011) Regulation of Hippo signaling by Jun kinase signaling during compensatory cell proliferation and regeneration, and in neoplastic tumors. Dev Biol 350:139–151

Sun Q (2007) The mechanism of pattern formation in the developing Drosophila retina. Sci China C Life Sci 50:120–4

Tamori Y, Bialucha CU, Tian AG, Kajita M, Huang YC, Norman M et al (2010) Involvement of Lgl and Mahjong/VprBP in cell competition. PLoS Biol 8(7):e1000422

Tapon N, Ito N, Dickson BJ, Treisman JE, Hariharan IK (2001) The Drosophila tuberous sclerosis complex gene homologs restrict cell growth and cell proliferation. Cell 105:345–355

Tapon N, Harvey K, Bell D, Wahrer D, Schiripo T, Haber D, Hariharan I (2002) salvador promotes both cell cycle exit and apoptosis in Drosophila and is mutated in human cancer cell lines. Cell 110:467

Tato I, Bartrons R, Ventura F, Rosa JL (2011) Amino acids activate mammalian target of rapamycin complex 2 (mTORC2) via PI3K/Akt signaling. J Biol Chem 286:6128–6142

Tepass U, Theres C, Knust E (1990) crumbs encodes an EGF-like protein expressed on apical membranes of Drosophila epithelial cells and required for organization of epithelia. Cell 61(5):787–799

Thomas C, Strutt D (2012) The roles of the cadherins Fat and Dachsous in planar polarity specification in Drosophila. Dev Dyn 241:27–39

Thompson BJ, Cohen SM (2006) The Hippo pathway regulates the bantam microRNA to control cell proliferation and apoptosis in Drosophila. Cell 126:767–774

Torok T, Tick G, Alvarado M, Kiss I (1993) P-lacW insertional mutagenesis on the second chromosome of Drosophila melanogaster: isolation of lethals with different overgrowth phenotypes. Genetics 135(1):71–80

Toyooka S, Ouchida M, Jitsumori Y, Tsukuda K, Sakai A, Nakamura A, Shimizu N, Shimizu K (2000) HD-PTP: a novel protein tyrosine phosphatase gene on human chromosome 3p21.3. Biochem Biophys Res Commun 278:671–678

Tran H, Brunet A, Griffith EC, Greenberg ME (2003) The many forks in FOXO's road. Sci STKE 2003:RE5

Treins C, Warne PH, Magnuson MA, Pende M, Downward J (2010) Rictor is a novel target of p70 S6 kinase-1. Oncogene 29:1003–1016

Treisman JE, Heberlein U (1998) Eye development in Drosophila: formation of the eye field and control of differentiation. Curr Top Dev Biol 39:119–158

Tsachaki M, Sprecher SG (2012) Genetic and developmental mechanisms underlying the formation of the Drosophila compound eye. Dev Dyn 241:40–56

Tsai BP, Hoverter NP, Waterman ML (2012) Blending hippo and WNT: sharing messengers and regulation. Cell 151:1401–1403

Tumaneng K, Russell RC, Guan KL (2012a) Organ size control by Hippo and TOR pathways. Curr Biol 22:R368–379

Tumaneng K, Schlegelmilch K, Russell RC, Yimlamai D, Basnet H, Mahadevan N, Fitamant J, Bardeesy N, Camargo FD, Guan KL (2012b) YAP mediates crosstalk between the Hippo and PI(3)K-TOR pathways by suppressing PTEN via miR-29. Nat Cell Biol 14:1322–1329

Tyler DM, Baker NE (2007) Expanded and fat regulate growth and differentiation in the Drosophila eye through multiple signaling pathways. Dev Biol 305:187–201

Tyler DM, Li W, Zhuo N, Pellock B, Baker NE (2007) Genes affecting cell competition in Drosophila. Genetics 175:643–657

Udan RS, Kango-Singh M, Nolo R, Tao C, Halder G (2003) Hippo promotes proliferation arrest and apoptosis in the Salvador/Warts pathway. Nat Cell Biol 5(10):914–20

Vanhaesebroeck B, Alessi DR (2000) The PI3K-PDK1 connection: more than just a road to PKB. Biochem J 346 Pt 3:561–576

Varelas X, Wrana JL (2012) Coordinating developmental signaling: novel roles for the Hippo pathway. Trends Cell Biol 22:88–96

Varelas X, Miller BW, Sopko R, Song S, Gregorieff A, Fellouse FA, Sakuma R, Pawson T, Hunziker W, McNeill H et al (2010a) The Hippo pathway regulates Wnt/beta-catenin signaling. Dev Cell 18:579–591

Varelas X, Samavarchi-Tehrani P, Narimatsu M, Weiss A, Cockburn K, Larsen BG, Rossant J, Wrana JL (2010b) The crumbs complex couples cell density sensing to Hippo-dependent control of the TGF-beta-SMAD pathway. Dev Cell 19:831–844

Venken KJ, Bellen HJ (2012) Genome-wide manipulations of *Drosophila melanogaster* with transposons, Flp recombinase, and PhiC31 integrase. Methods Mol Biol 859:203–228

Verdu J, Buratovich MA, Wilder EL, Birnbaum MJ (1999) Cell-autonomous regulation of cell and organ growth in Drosophila by Akt/PKB. Nat Cell Biol 1:500–506

Verghese S, Bedi S, Kango-Singh M (2012a) Hippo signalling controls Dronc activity to regulate organ size in Drosophila. Cell Death Differ 19(10):1664–1676

Verghese S, Waghmare I, Kwon H, Hanes K, Kango-Singh M (2012b) Scribble acts in the Drosophila fat-hippo pathway to regulate warts activity. PLoS ONE 7:e47173

Vidal M, Cagan RL (2006) Drosophila models for cancer research. Curr Opin Genet Dev 16L:10–6

Vigneron AM, Ludwig RL, Vousden KH (2010) Cytoplasmic ASPP1 inhibits apoptosis through the control of YAP. Genes Dev 24:2430–2439

Villano JL, Katz FN (1995) four-jointed is required for intermediate growth in the proximal-distal axis in Drosophila. Development 121(9):2767–2777

Visser-Grieve S, Hao Y, Yang X (2012) Human homolog of Drosophila expanded, hEx, functions as a putative tumor suppressor in human cancer cell lines independently of the Hippo pathway. Oncogene 31:1189–1195

Wang C, An J, Zhang P, Xu C, Gao K, Wu D, Wang D, Yu H, Liu JO, Yu L (2012a) The Nedd4-like ubiquitin E3 ligases target angiomotin/p130 to ubiquitin-dependent degradation. Biochem J 444:279–289

Wang K, Degerny C, Xu M, Yang XJ (2009) YAP, TAZ, and Yorkie: a conserved family of signal-responsive transcriptional coregulators in animal development and human disease. Biochem Cell Biol 87:77–91

Wang T, Blumhagen R, Lao U, Kuo Y, Edgar BA (2012b) LST8 regulates cell growth via target-of-rapamycin complex 2 (TORC2). Mol Cell Biol 32:2203–2213

Wehr MC, Holder MV, Gailite I, Saunders RE, Maile TM, Ciirdaeva E, Instrell R, Jiang M, Howell M, Rossner MJ et al (2012) Salt-inducible kinases regulate growth through the Hippo signalling pathway in Drosophila. Nat Cell Biol 15(1):61–71

Wei X, Shimizu T, Lai ZC (2007) Mob as tumor suppressor is activated by Hippo kinase for growth inhibition in Drosophila. EMBO J 26:1772–1781

Weinkove D, Neufeld TP, Twardzik T, Waterfield MD, Leevers SJ (1999) Regulation of imaginal disc cell size, cell number and organ size by Drosophila class I(A) phosphoinositide 3-kinase and its adaptor. Curr Biol 9:1019–1029

Wernet MF, Labhart T, Baumann F, Mazzoni EO, Pichaud F, Desplan C (2003) Homothorax switches function of Drosophila photoreceptors from color to polarized light sensors. Cell 115(3):267–279

Willecke M, Hamaratoglu F, Kango-Singh M, Udan R, Chen CL, Tao C, Zhang X, Halder G (2006) The fat cadherin acts through the Hippo tumorsuppressor pathway to regulate tissue size. Curr Biol 16(21):2090–2100

Willecke M, Hamaratoglu F, Sansores-Garcia L, Tao C, Halder G (2008) Boundaries of Dachsous Cadherin activity modulate the Hippo signaling pathway to induce cell proliferation. Proc Natl Acad Sci USA 105:14897–14902

Wolff T, Ready DF (1991) The beginning of pattern formation in the Drosophila compound eye: the morphogenetic furrow and the second mitotic wave. Development 113:841–850

Wolff T, Ready DF (1993) Pattern formation in the Drosophila retina. In: Bate, Martinez Arias (eds), pp 1277–1325

Wu S, Huang J, Dong J, Pan D (2003) Hippo encodes a Ste-20 family protein kinase that restricts cell proliferation and promotes apoptosis in conjunction with salvador and warts. Cell 114:445–456

Wu S, Liu Y, Zheng Y, Dong J, Pan D (2008) The TEAD/TEF family protein Scalloped mediates transcriptional output of the Hippo growth-regulatory pathway. Dev Cell 14:388–398

Xiao L, Chen Y, Ji M, Dong J (2011) KIBRA regulates Hippo signaling activity via interactions with large tumor suppressor kinases. J Biol Chem 286:7788–7796

Xu T, Rubin GM (1993) Analysis of genetic mosaics in developing and adult Drosophila tissues. Development 117:1223–1237

Xu T, Wang W, Zhang S, Stewart RA, Yu W (1995) Identifying tumor suppressors in genetic mosaics: the Drosophila lats gene encodes a putative protein kinase. Development 121:1053–1063

Yamamoto D (1993) Positive and negative signaling mechanisms in the regulation of photoreceptor induction in the developing Drosophila retina. Review. Genetica 88:153–164

Yang Q, Inoki K, Kim E, Guan KL (2006) TSC1/TSC2 and Rheb have different effects on TORC1 and TORC2 activity. Proc Natl Acad Sci USA 103:6811–6816

Yu FX, Zhao B, Panupinthu N, Jewell JL, Lian I, Wang LH, Zhao J, Yuan H, Tumaneng K, Li H et al (2012) Regulation of the Hippo-YAP pathway by G-protein-coupled receptor signaling. Cell 150:780–791

Yu J, Poulton J, Huang YC, Deng WM (2008) The hippo pathway promotes Notch signaling in regulation of cell differentiation, proliferation, and oocyte polarity. PLoS One 3(3):e1761

Yu J, Zheng Y, Dong J, Klusza S, Deng WM, Pan D (2010) Kibra functions as a tumor suppressor protein that regulates Hippo signaling in conjunction with Merlin and Expanded. Dev Cell 18:288–299

Yue T, Tian A, Jiang J (2012) The cell adhesion molecule echinoid functions as a tumor suppressor and upstream regulator of the Hippo signaling pathway. Dev Cell 22:255–267

Zecca M, Struhl G (2010) A feed-forward circuit linking wingless, fat-dachsous signaling, and the warts-hippo pathway to Drosophila wing growth. PLoS Biol 8:e1000386

Zeitler J, Hsu CP, Dionne H, Bilder D (2004) Domains controlling cell polarity and proliferation in the Drosophila tumor suppressor Scribble. J Cell Biol 167(6):1137–1146

Zhang H, Stallock JP, Ng JC, Reinhard C, Neufeld TP (2000) Regulation of cellular growth by the Drosophila target of rapamycin dTOR. Genes Dev 14:2712–2724

Zhang Y, Gao X, Saucedo LJ, Ru B, Edgar BA, Pan D (2003) Rheb is a direct target of the tuberous sclerosis tumour suppressor proteins. Nat Cell Biol 5:578–581

Zhang J, Smolen GA, Haber DA (2008a) Negative regulation of YAP by LATS1 underscores evolutionary conservation of the Drosophila Hippo pathway. Cancer Res 68:2789–2794

Zhang L, Ren F, Zhang Q, Chen Y, Wang B, Jiang J (2008b) The TEAD/TEF family of transcription factor Scalloped mediates Hippo signaling in organ size control. Dev Cell 14:377–387

Zhang L, Yue T, Jiang J (2009a) Hippo signaling pathway and organ size control. Fly (Austin) 3:68–73

Zhang X, Milton CC, Humbert PO, Harvey KF (2009b) Transcriptional output of the Salvador/warts/hippo pathway is controlled in distinct fashions in Drosophila melanogaster and mammalian cell lines. Cancer Res 69:6033–6041

Zhang X, George J, Deb S, Degoutin JL, Takano EA, Fox SB, Bowtell DD, Harvey KF (2011a) The Hippo pathway transcriptional co-activator, YAP, is an ovarian cancer oncogene. Oncogene 30:2810–2822

Zhang X, Milton CC, Poon CL, Hong W, Harvey KF (2011b) Wbp2 cooperates with Yorkie to drive tissue growth downstream of the Salvador-Warts-Hippo pathway. Cell Death Differ 18:1346–1355

Zhang L, Iyer J, Chowdhury A, Ji M, Xiao L, Yang S, Chen Y, Tsai MY, Dong J (2012) KIBRA regulates aurora kinase activity and is required for precise chromosome alignment during mitosis. J Biol Chem 287:34069–34077

Zhao B, Wei X, Li W, Udan RS, Yang Q, Kim J, Xie J, Ikenoue T, Yu J, Li L et al (2007) Inactivation of YAP oncoprotein by the Hippo pathway is involved in cell contact inhibition and tissue growth control. Genes Dev 21:2747–2761

Zhao B, Lei QY, Guan KL (2008a) The Hippo-YAP pathway: new connections between regulation of organ size and cancer. Curr Opin Cell Biol 20(6):638–646

Zhao B, Ye X, Yu J, Li L, Li W, Li S, Lin JD, Wang CY, Chinnaiyan AM, Lai ZC et al (2008b) TEAD mediates YAP-dependent gene induction and growth control. Genes Dev 22:1962–1971

Zhao B, Li L, Lei Q, Guan KL (2010a) The Hippo-YAP pathway in organ size control and tumorigenesis: an updated version. Genes Dev 24:862–874

Zhao B, Li L, Tumaneng K, Wang CY, Guan KL (2010b) A coordinated phosphorylation by Lats and CK1 regulates YAP stability through SCF(beta-TRCP). Genes Dev 24:72–85

Zhao B, Li L, Lu Q, Wang LH, Liu CY, Lei Q, Guan KL (2011a) Angiomotin is a novel Hippo pathway component that inhibits YAP oncoprotein. Genes Dev 25:51–63

Zhao B, Tumaneng K, Guan KL (2011b) The Hippo pathway in organ size control, tissue regeneration and stem cell self-renewal. Nat Cell Biol 13:877–883

Zinzalla V, Stracka D, Oppliger W, Hall MN (2011) Activation of mTORC2 by association with the ribosome. Cell 144:757–768

Ziosi M, Baena-Lopez LA, Grifoni D, Froldi F, Pession A, Garoia F, Trotta V, Bellosta P, Cavicchi S (2010) dMyc functions downstream of Yorkie to promote the supercompetitive behavior of hippo pathway mutant cells. PLoS Genet 6

Zipursky SL (1989) Molecular and genetic analysis of Drosophila eye development: sevenless, bride of sevenless and rough. Trends Neurosci 12:183–189

Unraveling the Basis of Neurodegeneration using the *Drosophila* Eye

Pedro Fernandez-Funez, Jonatan Sanchez-Garcia and Diego E. Rincon-Limas

Abbreviations

AD	Alzheimer's disease
ALS	Amyotrophic lateral sclerosis
Aβ	Amyloid-β peptide
Atx	Ataxin
FMR1	Fragile X mental retardation 1
FTD	Frontotemporal dementia
GFP	Green fluorescent protein
GOF	Gain-of-function
Hsp70	Heat shock protein 70
HD	Huntington's disease
LOF	Loss-of-function
MD	Myotonic dystrophy
PD	Parkinson's disease
PrP	Prion protein
SCA	Spinocerebellar ataxia
TDP-43	TAR DNA-binding protein 43
WT	Wild type

P. Fernandez-Funez · D. E. Rincon-Limas (✉) · J. Sanchez-Garcia
Department of Neurology, McKnight Brain Institute, 1149 Newell Dr.,
University of Florida, 32611 Gainesville, FL, USA
e-mail: pedro.fernandez@neurology.ufl.edu
e-mail: diego.rincon@neurology.ufl.edu

P. Fernandez-Funez · D. E. Rincon-Limas
Department of Neuroscience, Genetics Institute and Center for Translational Research on Neurodegenerative Diseases,
University of Florida, 32611 Gainesville, FL, USA

P. Fernandez-Funez
Center for Movement Disorders and Neurorestoration,
University of Florida, 32611 Gainesville, FL, USA

A. Singh, M. Kango-Singh (eds.), *Molecular Genetics of Axial Patterning, Growth and Disease in the Drosophila Eye,* DOI 10.1007/978-1-4614-8232-1_10,

Human Neurodegenerative Diseases

A complex class of debilitating brain diseases afflicting humans is characterized by mid- to-late onset and progressive neuronal loss. Among these, Alzheimer's disease (AD) and Parkinson's disease (PD) are the most prevalent and the source of serious economic and societal burdens in advanced societies. Like AD and PD, the most prevalent proteinopathies in humans are sporadic in which the neurotoxic agents are misfolded, aggregated conformers of wild-type (WT), proteins in different parts of the brain. These so-called protein misfolding disorders or proteinopathies involve structurally and functionally diverse proteins that contain aggregation-prone, amyloidogenic domains (Aguzzi and O'Connor 2010). Some widely recognized WT amyloids include the amyloid-β peptide (Aβ) in AD, α-Synuclein (α-Syn) in PD, superoxide dismutase 1 (SOD1) in amyotrophic lateral sclerosis (ALS) as well as Tau and transactivation response (TAR) DNA-binding protein 43 (TDP-43) in several memory and movement disorders. In addition, a large number of brain disorders present with dominant familial inheritance. The most representative examples are Huntington's disease (HD), at least 15 spinocerebellar ataxias (SCA), and several frontotemporal dementias (FTD) as well as familial forms of AD, PD, and ALS. In addition to these diverse proteinopathies, non-coding RNAs bearing repeat expansions are responsible for several neurodegenerative conditions, including Fragile X syndrome, myotonic dystrophy (MD), SCA8, and SCA10. Finally, a small class of disorders such as Friedrich's ataxia and several storage disorders are linked to loss-of-function (LOF) mutations. Although this is a diverse group of diseases, the common denominator is the activation of quality control and stress pathways and the induction of progressive neuronal loss.

Most protein and RNA pathologies are dominant and implicate gain-of-function (GOF) mechanisms. Thus, these pathogenic agents easily replicate their deleterious effects upon misexpression, not only in traditional mammalian models, but also in alternative models such as the round worm *Caenorhabditis elegans*, the fruit fly *Drosophila*, and the zebrafish *Danio rerio*. In those cases where the pathogenic genes are conserved, alternative laboratory animals can model the LOF disorders, thanks to the existence of large collections of mutant strains (*Drosophila*) or the easy generation of mutants (all three). Here, we review the use of the *Drosophila* eye to better understand how these disease-related genes cause progressive neuronal loss.

The *Drosophila* Eye as a Gateway to Discoveries

Here, we briefly describe the features that make the eye so powerful to model neurotoxicity. More detailed descriptions of the development of the eye are available in a recent review (Kumar 2012) and in other chapters of this book. The first important feature of the *Drosophila* eye is that it develops from a simple epithelial monolayer in the eye-antenna imaginal disc that grows during the larval stages. Specification of photoreceptor (R1–8) cells follows a wave of coordinated cell divisions through the combined activity of the Notch and Epidermal growth factor receptor (EGFR) pathways. When all the cone and pigment cells are recruited to complete the ommatidia, the leftover cells undergo apoptosis. Final differentiation of the eye involves complex changes in cell morphology that create a precisely arranged three-dimensional eye (Fig. 1a-b) from an essentially uniform two-dimensional imaginal disc.

The features described earlier (simple origin, complex structure) make the eye a very attractive organ to study both developmental and neurodegenerative processes. First, the precise lattice of the 800 ommatidia makes the adult eye very sensitive to small developmental disruptions visible under a dissecting stereoscope, saving time-consuming manipulations. Second, the adult eye has very robust external (lenses and bristles) and internal (retina) structures that undergo little change during normal aging, a precious quality for detecting aberrant degenerative changes. Third, the detailed knowledge of the genetic and cellular events regulating eye development enables the easy interpretation of any perturbation. Fourth, the eye is dispensable for fertility and survival, thus strong perturbations of its structure or physiology can be documented in adult flies.

Modeling Neurodegenerative Diseases in *Drosophila*

Early Models of Neurodegeneration in the Eye

The potential for modeling monogenic, dominant diseases in flies was finally realized in 1998, when two independent groups expressed polyglutamine (polyQ) expansions linked to Machado–Joseph disease (MJD or spinocerebellar ataxia type 3 (SCA3)) and HD (Jackson et al. 1998; Warrick et al. 1998). Quite appropriately, both groups expressed the disease genes in the eye, hoping to exploit this easily accessible organ. In fact, Jackson expressed mutant Huntingtin (Htt) directly under the control of eye regulatory sequences, betting that those flies would display abnormal eyes. In contrast, Bonini expressed mutant Ataxin (Atx)3 under the control of upstream activation sequence (UAS), which provides the flexibility of expressing the construct under different neuronal patterns, including the eye (Brand and Perrimon 1993). Interestingly, expression of these two genes bearing polyQ expansions induced distinct eye phenotypes. Flies expressing Htt-120Q eclosed with normal eyes, but the retina underwent aggressive loss of photoreceptors over the next 20 days (Jackson et al. 1998; Fig. 1j). In contrast, Atx3–78Q led to highly disorganized and depigmented eyes (Warrick et al. 1998; Fig. 1l) that experienced little change over time. These results confirmed the toxicity of the polyQ-bearing proteins in flies, supported the specific deleterious consequences of each mutant protein, and validated the use of flies for modeling neurodegenerative diseases. In all, these high profile papers launched the interest for disease-oriented research in an animal model that, until then, had produced substantial contributions to basic biological questions, particularly genetics, development, cell biology, and neurobiology (Bellen 2010).

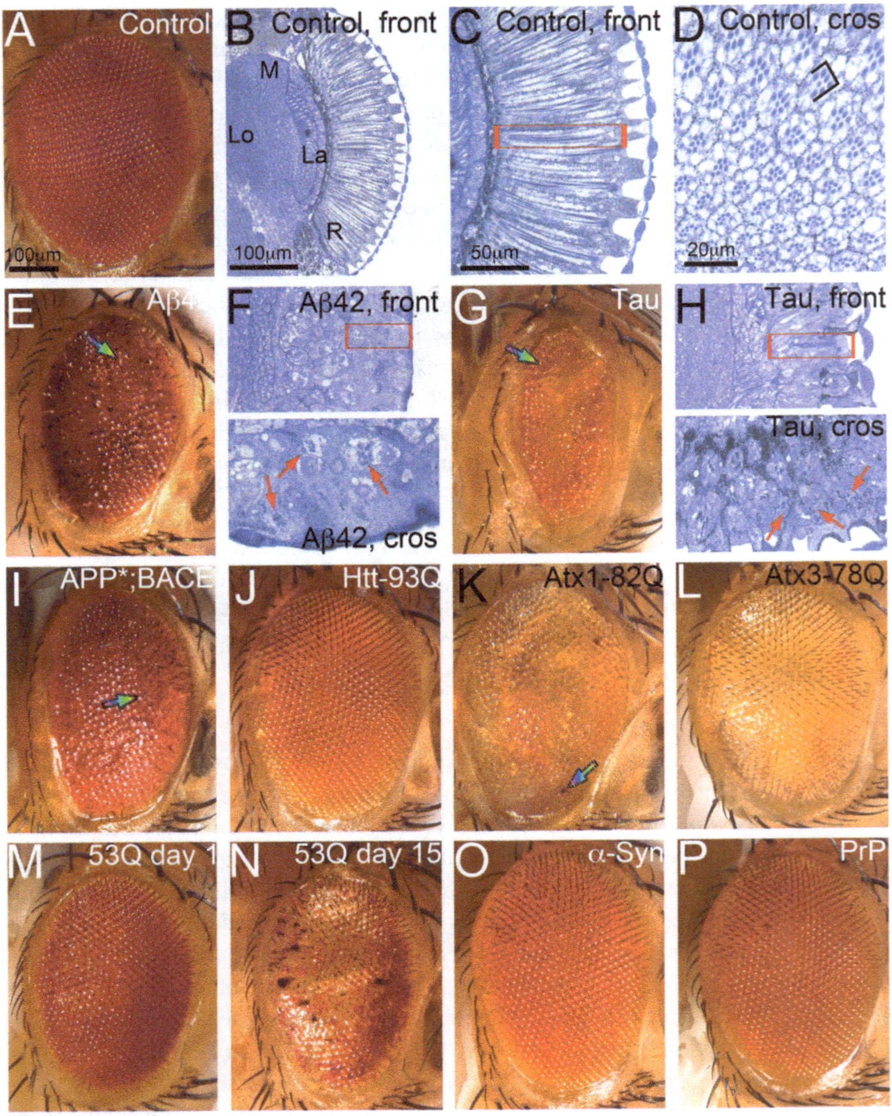

Fig. 1 Representative eye phenotypes induced by neurotoxic proteins involved in neurodegenerative diseases. **a** Micrograph of fresh eyes from wild type (WT) flies display the neat arrangement of ~800 ommatidia. **b–d** Sections of resin-embedded eyes. **b** and **c** are frontal sections showing the vertical position of rhabdomeres in each ommatidium (*red box* in **c**). The retina (ret) contains photoreceptors, pigment cells, and lens-secreting cone cells (**b**). The lamina (lam) consists of the axonal projections of the photoreceptors and glial projections. The medulla and lobula receive the synaptic terminals of the photoreceptors and process the visual information for the brain. **d** A tangential (cross) section of the retina provides a view of the highly organized ommatidia and the stereotyped arrangement of the seven visible rhabdomeres. **e** and **g** Micrographs of fresh eyes from flies expressing (Aβ42) (**e**) and Tau (**g**) under the control of gmr-Gal4. Both Aβ42 and Tau induce small, highly disorganized, and depigmented eyes and are characterized by

Other Drosophila *Models of Proteinopathies in the Eye*

Over the ensuing decade, multiple *Drosophila* models focused on inherited diseases in which the neurotoxic agents were proteins with expanded polyQ stretches (Rincon-Limas et al. 2012). Expression of full-length Atx1–82Q induced dramatically disorganized and depigmented eyes, a phenotype reminiscent of the Atx3–78Q phenotype, but slightly stronger due to the lack of interommatidial bristles in Atx1 flies (Fernandez-Funez et al. 2000; Fig. 1k). Marsh and Thompson later expressed Htt exon 1 with 93Q under the control of UAS, resulting in phenotypes similar to those described by Jackson, but with the flexibility of inducing expression in neuron-specific patterns using *elav-Gal4* (Steffan 2001). This N-terminal fragment of Htt contains the polyQ expansion and is known to accumulate in nuclear inclusions in HD patients. But, Htt is a large protein (3,144 amino acids) involved in vesicular trafficking and different cleavage products are present in HD patients (Ross and Tabrizi 2011). Additional models of HD containing longer N-terminal fragments of Htt (90, 171, 336 amino acids)—as well as full-length Htt—also showed eye phenotypes, with the longer constructs showing weaker phenotypes (Romero et al. 2008; Mugat et al. 2008; Branco et al. 2008). These models are highly relevant to understand the physiological function of Htt and the mechanisms leading to Htt cleavage and neurotoxicity. A later model of SCA3 expressing full length Atx3 also demonstrated the importance of caspase-dependent cleavage for neurotoxicity (Jung et al. 2009). In a model of spinal and bulbar muscular atrophy (SBMA), expression of the human X-linked androgen receptor (AR) with 52Q or 112Q resulted in small, disorganized, and depigmented eyes, but only in the presence of its male-specific ligand (Chan et al. 2002; Takeyama et al. 2002), which supports the male-linked nature of SBMA. These results also indicate that mutant AR is toxic only after ligand-dependent nuclear translocation. In the case of SCA17, which is linked to the TATA box-binding protein (TBP), expression of human TBP with 34Q, 54Q, or 80Q induced disorganized and depigmented eyes that progressively worsened over 20 days,

fusion of ommatidia (*arrows*). **f** and **h** The retinas of these flies are thin and lack proper organization of the different cell types of each ommatidium. The rhabdomeres are poorly organized and differentiated. **i–p** Micrographs of fresh eyes from flies expressing human amyloids under the control of gmr-Gal4. **i** Expression of APPswe; (BACE) induces disorganization of the eye with fused ommatidia (*arrow*) similar to but weaker than Aβ42. Expression of Htt-93Q results in well-organized eyes with slight depigmentation than lose pigmentation over time (not shown). **k** and **l** Expression of (Atx)1–82Q and Atx3–78Q produce depigmented eyes with thin retinas. In addition, Atx1–82Q has small interommatidial bristles and necrotic spots (*arrow*), whereas Atx3–78Q shows a better external structure. **m** and **n** A 53Q construct results in weak depigmentation in young flies, but progresses to strong depigmentation and necrotic spots over 15 days. **o** and **p** WT and mutant (α-Syn) and prion protein (PrP) (only WT shown) have no effect on the eye in young and older flies. Methods: Fresh eye images were collected as Z-stacks with a Leica Z16 APO using a 2 × Plan Apo objective and single projections were produced with Leica Application Software Montage Multifocus. Eye sections were collected with AxioVision (Zeiss) in an Axio-Observer Z1 microscope (Zeiss) with 20 × air numerical aperture (NA): 0.8, 40 × air NA: 0.75, and 63 × oil NA: 1.4 objectives with the longer repeats showing the stronger phenotypes (Ren et al. 2011). PolyQ expansions in Atrophin-1 are linked to dentatorubral-pallidoluysian atrophy (DRPLA). Expression of Atrophin-75Q induced slightly disorganized eyes that showed progressive loss of photoreceptors and increased autophagy (Nisoli et al. 2010). Finally, the Marsh and Kazemi-Esfarjani laboratories demonstrated early on that polyQ-only peptides were highly toxic in flies, resulting in small, glassy, and depigmented eyes (Kazemi-Esfarjani and Benzer 2000; Marsh et al. 2000; Fig. 1m, n), although these models lost their appeal after disease-specific constructs showed distinct phenotypes. Overall, these results supported the idea that, although the polyQ expansion is responsible for pathogenesis in these diseases, the protein context determines the cellular phenotypes.

In addition to these models of polyQ toxicity, Feany and Bender created a fly model of α-Syn toxicity based on the expression of WT and mutant α-Syn (A30P, A53T) (Feany and Bender 2000). All three forms of α-Syn showed subtle phenotypes in the central nervous system (CNS) and in the eye: flies eclosed with normal eyes, but the retina showed a slight increase in vacuolation after 30 days. This weak-eye phenotype (Fig. 1o) has precluded the screening for suppressors of α-Syn neurotoxicity in high-throughput conditions. In contrast, expression of WT and mutant leucine-rich repeat kinase 2 (LRRK2), another gene linked to familial PD, induced severe disruption of ommatidia, depigmentation, and poor differentiation of photoreceptors (Venderova 2009; Liu et al. 2008).

Tau is a protein with key roles in several neurodegenerative diseases, including AD, FTD, and several movement disorders. Two independent models of tauopathies appeared in 2001 and 2002, but only one was expressed in the eye (Jackson et al. 2002; Wittmann et al. 2001). Jackson expressed WT and FTDP-17-linked mutant human Tau with four repeats in the eye with the *gmr* regulatory region. Both WT and mutant Tau induced smaller, disorganized eyes with fused ommatidia (Fig. 1g, h) that accumulated insoluble, hyperphosphorylated Tau (Jackson et al. 2002). In addition to Tau, the other main pathological hallmark of AD is the Aβ peptide, which is the main component of amyloid plaques. One model of Aβ neurotoxicity successfully replicated the production of Aβ42 from human amyloid precursor protein (APP) bearing the Swedish mutation (K595N/M596L) (Greeve et al. 2004). Since APP does not undergo amyloid processing in flies, the investigators introduced human beta-site amyloid precursor protein cleaving enzyme (BACE) and mutant Presenilin (Psen). These flies accumulated Aβ42 two days after eclosion, which induced age-dependent degeneration of the retina (Fig. 1i) and the axonal projections of the photoreceptors. Although this model reproduced closely the physiology of Aβ42 in the human brain, the weak eye phenotype and the dual construct made it inadequate for genetic screens. Consequently, several groups followed an alternative strategy: express Aβ fused to an efficient signal peptide. The first model expressing Aβ40 and Aβ42 induced weak eye phenotypes that required multiple copies to produce robust phenotypes (Finelli et al. 2004). Later models expressing higher levels or pathogenic mutants of Aβ42 induced small, disorganized, and depigmented eyes (Fig. 1e, f), phenotypes compatible with further genetic and pharmacologic manipulations (Crowther et al. 2005; Iijima et al. 2008; Casas-Tinto et al. 2011).

Mutations in SOD1 were the first clue to familial ALS, while WT SOD1 accumulates in aggregates in sporadic forms of ALS. Surprisingly, misexpression of human WT or mutant SOD1 in the eye did not lead to retinal degeneration (Watson et al. 2008). But, misexpression of TDP-43 or FUS (Fused in sarcoma/translated in liposarcoma), two RNA-binding proteins recently linked to ALS, resulted in aberrant phenotypes in the fly eye (Lanson et al. 2011; Li et al. 2010). While WT FUS induced weak depigmentation, mutant FUS alleles induced a range of defects including small, disorganized, and completely depigmented eyes (Lanson et al. 2011). WT TDP-43 induced retinal degeneration in a dose-dependent fashion and pathogenic mutants resulted in slightly stronger depigmentation (Li et al. 2010). Interestingly, combinations of WT and mutant TDP-43 and FUS constructs resulted in stronger eye phenotypes, arguing for deleterious effects on the same pathway (Lanson et al. 2011). Finally, expression of WT Ewing sarcoma breakpoint region 1 (EWSR1), a new gene recently associated to ALS and structurally related to TDP-43 and FUS, also resulted in eye depigmentation and a thinner retina (Couthouis et al. 2012). Missense mutations in EWSR1 found in ALS patients enhanced the eye phenotype, as was observed for FUS mutants. These studies suggest that these three related proteins cause pathogenesis by perturbing the same cellular mechanisms.

Most SCAs are linked to repeat expansions, but SCA5 is a rare example associated with point mutations in β-III-spectrin. Expression of either human or fly WT β-III-spectrin did not affect the eye, but β-III-spectrin carrying the American or German mutations resulted in severely disorganized and depigmented eyes reminiscent of SCA1 (Lorenzo et al. 2010).

Overall, the intrinsic susceptibility of these amyloidogenic proteins to form toxic assemblies allowed the generation of a wide array of disease models with highly unique perturbations of the eye (Fig. 1). Thus, amyloid proteins do not simply kill photoreceptors as toxins, they actually interfere with specific aspects of eye development and physiology. Thus, the fruit fly may be an excellent model to identify the cellular mechanisms responsible for these specific perturbations. Intriguingly, several amyloids induce more potent neuronal loss in flies than in rodent models (e.g., Aβ42, APP), making *Drosophila* a unique tool for AD research. However, not all amyloidogenic proteins induce eye phenotypes. The WT Prion protein (PrP) is a well-known amyloid involved in prion diseases that induces neuronal dysfunction and cell loss in the *Drosophila* brain (Fernandez-Funez et al. 2009), but neither WT nor pathogenic mutants induce eye degeneration (Fig. 1p). The lack of eye phenotype in the PrP model has considerably limited our ability to validate candidate genes or screen for new modifiers. These exceptions, although rare, support the specific phenotypes of other amyloids and further prove the advantages of the eye.

Modeling RNA Toxicity in the Eye

In addition to the highly prevalent and well-known protein misfolding diseases, non-coding RNA repeat expansions also play a significant role in several brain and neuromuscular disorders in humans. These mRNAs bearing repeat expansions are retained in the nucleus, producing the so-called nuclear foci, and lead to aberrant interaction with the splicing machinery. Fragile X is a common form of mental retardation caused by long (200+) CGG repeats in the 5′-UTR of the *Fragile X mental retardation 1* (*FMR1*) gene associated with LOF on the FMR1 protein. However, smaller expansions (premutations) are linked to progressive neurodegeneration with ataxia and dementia. Expression of a CGG repeat in an intronic sequence preceding the green fluorescent protein (GFP) induced disorganized and depigmented eyes in flies (Jin et al. 2003). This was the first in vivo demonstration that non-coding RNA repeats are neurotoxic. Non-coding expansions also play a key role in neuromuscular diseases. Myotonic dystrophy 1 (MD1) is a relatively common (1:8,000) neuromuscular disorder linked to a non-coding CTG repeat in the 3′-UTR of *DMPK* (dystrophiamyotonica protein kinase). Expression of 480 uninterrupted CTG repeats induced smaller and slightly disorganized eyes with nuclear foci pathology (de Haro et al. 2006). Another example of the complexity of the cerebellar ataxias is SCA8, which is mediated by noncoding CTG repeats. Expression of the human *SCA8* gene with 9 and 112 interrupted CTG repeats induced similar eye disorganization with underdeveloped retinas (Mutsuddi et al. 2004). The toxicity of these non-coding triplet repeats suggested that polyQ-encoding CAG repeats might also exert RNA toxicity. Bonini demonstrated that CAACAG repeats also encoding polyQ showed weaker toxicity in the eye than pure CAG repeats (Li et al. 2008). Moreover, expression of non-coding CAG repeats in the 3′-UTR of DsRed induced retinal and brain degeneration, supporting a role for RNA toxicity in polyQ diseases.

Modeling Loss-of-Function Diseases

Neurodegenerative diseases are enriched in dominant and sporadic etiologies in which the toxic agents are inherently toxic proteins and mRNAs. However, LOF mutations can also cause progressive neuronal loss, although these disorders are less common because they require two mutant alleles to induce disease. Interestingly, LOF mutations lead essentially to the same neurotoxic phenotypes as dominant misfolded proteins and expanded RNAs, suggesting that they perturb the same mechanism of neuronal protection and survival. In some cases, LOF mutations may cause pathologies within the spectrum of complex disorders, such as PD and the broader parkinsonian syndrome. In fact, *Parkin*, *Pink1*, and *DJ-1* LOF cause early onset familial PD, as do dominant mutations in α-Syn and LRKK2 (Hardy et al. 2009), which illustrate the complexity of PD. So far, 44 genes controlling neural integrity have been identified in *Drosophila* as (i) regulators of lifespan, locomotion, or behavior, (ii) candidate genes with a known role in human disease, or (iii) serendipitous findings (Lessing and Bonini 2008). These diverse genes seem to play a major role in five cellular processes: protein degradation, lipid homeostasis, mitochondria, signaling, and cytoskeleton. Approximately half of these genes have human orthologues involved in disease or are functionally related to processes important for human disease. The other half with no direct human disease connection may still contribute to the identification of novel mechanisms important for neuronal integrity in humans.

Thus, these LOF diseases play a critical role in uncovering key physiological functions that prevent neuronal loss, an activity comparable to that of tumor-suppressor genes in cancer.

Among the LOF genes, the contribution of the eye to the study of genes associated with autosomal recessive PD deserves special consideration. For instance, inactivation of *dj-1* and *pink1* using specific RNA interference (RNAi) constructs led to abnormal eyes and retinal degeneration, although *dj-1* LOF was more critical for eye development, resulting in very small and disorganized eyes (Wang et al. 2006; Yang et al. 2005). Interestingly, both *dj-1* and *pink1* increased the accumulation of reactive oxygen species, supporting their antioxidative function linked to the mitochondria. Overexpression of *pink1* also induced aberrant eye morphology, a phenotype that was suppressed by *parkin* LOF alleles and enhanced by coexpression of *parkin* (Whitworth et al. 2008). This genetic interaction supported observations in *Drosophila* muscle and sperm cells that Parkin functions downstream of Pink1 to regulate mitochondria morphology (Clark et al. 2006; Park et al. 2006; Yang et al. 2006). Moreover, mutations in mitochondria fission genes also rescued the *pink1* misexpression phenotype in the eye, indicating the role of Pink1/Parkin in mitochondria fission (Poole et al. 2008). Similar genetic epistasis experiments indicated that LOF alleles of *high temperature requirement A2* (*htrA2/omi*), another gene linked to PD, rescued the *pink1* misexpression phenotype in the eye, but functions independently of *parkin*. In addition, Rhomboid-7, a mitochondrial kinase not previously implicated in PD, showed strong genetic interactions with *pink1*, *parkin*, and *omi* in the eye, and demonstrated to be an upstream regulator of the Pink1 pathway in the mitochondria (Whitworth et al. 2008). Finally, *pink1*, *parkin*, and *dj-1* misexpression partially suppressed the abnormal eye induced by *LRKK2* misexpression, supporting the involvement of all these PD genes in a common pathway (Venderova et al. 2009).

LOF approaches have contributed to model other recessive neurodegenerative conditions in the eye. Disruption of endogenous *Ataxia telangiectasia mutated* (*ATM*), a gene implicated in ataxia telangiectasia, resulted in disorganized eyes with fused ommatidia and progressive degeneration of photoreceptors that was suppressed by LOF mutations in cell cycle regulators (Rimkus et al. 2008). These results suggested that loss of *ATM* induces cell cycle reentry (Khurana et al. 2006). Similarly, LOF mutations in *pantothenate kinase* (*Pank*)/*fumble* induced retinal and brain vacuolation, phenotypes that were rescued only by expression of human or fly *PanK* in the mitochondria, suggesting that Pantothenate kinase-associated neurodegeneration is a mitochondrial disease (Wu et al. 2009). Another LOF mutation in a mitochondrial protein, Methionyl-tRNA synthetase (Aats-met), was identified in a screen for genes that induce progressive photoreceptor dysfunction (Bayat et al. 2012). Interestingly, the human homologue turned out to be affected in autosomal recessive spastic ataxia with leukoencephalopathy, indicating the prowess of *Drosophila* in identifying candidate genes in human diseases. Deficiency in succinate dehydrogenase (SdhA) is linked to Leigh Syndrome, an early-onset disease characterized by mitochondrial encephalopathy. Flies mutant for *SdhA* eclose with normal eyes, but the photoreceptors degenerate over the next few days associated to the loss of mitochondria, while antioxidant therapy prevented degeneration of the retina (Mast et al. 2008). Another large category of LOF mutants characterized in the eye involves aberrant lysosomal function. Niemann–Pick disease type C is a lipid storage disease associated with mutations in the *NPC1* gene. LOF in the fly *dnpc1a* gene induced potent retinal degeneration that was rescued by both neuronal and glial expression of *dnpc1* (Phillips et al. 2008). Mutations in the Transient receptor potential mucolipin 1 (TRPML1) channel cause another lipid storage disorder, mucolipidosis type IV. LOF mutations in the fly *trpml 1* gene induced progressive photoreceptor loss, defective autophagy, and was suppressed by heat shock protein 70 (Hsp70) misexpression (Venkatachalam et al. 2008). Infantile neuronal ceroid lipofuscinoses (INCL) are caused by LOF mutations in *Palmitoyl-protein thioesterase 1* (*Ppt1*). Interestingly, misexpression of *Ppt1* in flies led to small and disorganized eyes, suggesting that both up- and down-regulation of Ppt1 can induce neurodegeneration (Korey and MacDonald 2003). Finally, LOF in *benchwarmer*, which encodes a multipass membrane transporter, caused retinal degeneration associated to lysosomal storage deficits (Dermaut et al. 2005).

Elucidating the Mechanisms of Neurodegeneration

Assays to Analyze Eye Degeneration

The main advantage of the eye is the direct visualization of the cellular and developmental perturbations induced by these neurotoxic agents. Gross analysis under the stereoscope allows for fast inspection of changes in the eye, which is very useful in large modifier screens (see below). However, this rapid visualization offers low resolution, which hinders the ability to interpret the molecular mechanisms underlying the structural damage. A variety of structural, functional, and molecular assays provide additional resolution toward the study of eye degeneration. Lessing and Bonini 2009), including molecular markers of cell fate and differentiation by immunofluorescence, semithin sectioning for gross morphology, and ultrathin sectioning for subcellular detail. However, the adoption of rapid assays that indirectly reflect eye structure and/or function allowed carrying out high-throughput screening for novel genes involved in progressive neuronal degeneration. For instance, a fast technique for evaluating the integrity of photoreceptors is optical neutralization (also known as pseudopupil assay), which allows the visualization of photoreceptors in live specimens by transmission of bright light (Franceschini 1972). Complex, but highly specific functional assays are also available, including electroretinogram (Bayat et al. 2012) and vision-dependent behavioral assays (phototaxis) (Benzer 1967).

Identification of Intrinsic Mechanisms Mediating Neurotoxicity

Fly models of proteinopathies have been instrumental in the discovery of mechanisms regulating the toxicity of several pathogenic proteins, particularly those associated with posttranslational modifications reviewed in (Rincon-Limas et al. 2012). For instance, phosphorylation plays a key role in regulating the misfolding, aggregation, and toxicity of several amyloid proteins, including Tau, α-Syn, and Atx1. Tau is a large protein with 79 potential Ser/Thr phosphorylation sites, making its in vivo analysis very complex. Manipulation of the activity of GSK-3 and Cdk5 in the eye demonstrated that these kinases phosphorylate Tau and regulate its aggregation (Jackson et al. 2002; Shulman and Feany 2003; Steinhilb et al. 2007a). Also, mutant analysis indicated that the 14 Ser-Pro/Thr-Pro sites seemed to work coordinately (mutating a few had no effect); in contrast, the non-Ser-Pro sites S262 and S356 played key roles by promoting the phosphorylation of Ser-Pro/Thr-Pro sites (Nishimura et al. 2004; Steinhilb 2007b; Chatterjee et al. 2009). Along the same line, α-Syn phosphorylation at T125 and S129 had opposite effects, the first being neuroprotective and the second neurotoxic, thus providing reversible mechanisms to modulate α-Syn toxicity (Chen et al. 2009). Finally, Atx1 phosphorylation at S776 by Akt1 promotes binding to 14-3-3, aggregation, and neurotoxicity (Chen et al. 2003). As a consequence, *Akt* LOF mutations suppress the toxicity of Atx1–82Q in flies, offering clear therapeutic opportunities. Other relevant posttranslational modifications have been also uncovered using the fly eye. For instance, elimination of three SUMOylation (small ubiquitin-related modifier) sites completely abrogated the toxicity of Htt-93Q in the eye and CNS (Steffan et al. 2004).

Since polyQ expansions induce different eye phenotypes alone or in the context of a disease-related protein, the protein context was found to be a key determinant of polyQ toxicity (Jackson et al. 1998; Warrick et al. 1998; Fernandez-Funez et al. 2000; Kazemi-Esfarjani and Benzer 2000). In addition, CAG expansions in SCA1 promote the formation of complexes containing RNA-binding motif protein 17 (RBM17), which contributes to Atx1–82Q toxicity, whereas they attenuate the formation of complexes with Capicua, which causes a LOF mechanism (Lim et al. 2008; Lam et al. 2006). Interestingly, genetic interactions revealed that the activity of Atx2 (SCA2) enhances the toxicity of both expanded Atx1 and Atx3, suggesting common mechanisms of neurodegeneration relevant for therapeutics (Lessing and Bonini 2008; Al-Ramahi et al. 2007). To illustrate the complexity of the repeat expansions, a non-coding CTG implicated in MD1 was found to transcribe in both directions, resulting in a CUG/CAG RNA duplex that is cleaved into small RNAi fragments that target endogenous genes carrying CAG repeats, including Atx2 and TBP (Yu et al. 2011; Lawlor et al. 2011).

Finally, the eye was crucial to shed light on the pathogenesis of TDP-43 and FUS, two related proteins that contain similar domains, including RNA-binding domain, nuclear export signal (NES), and nuclear localization signal (NLS). The analysis of these motifs led to the following key results in the eye: (i) deletion of the FUS NES suppressed degeneration, (ii) only cytoplasmic accumulation of TDP-43 was neurotoxic in the eye, while nuclear TDP-43 was also toxic in the CNS, and (iii) FUS genetically interacts with TDP-43 in a mutation-dependent fashion (Lanson et al. 2011; Miguel et al. 2011).

Understanding the Function of Disease-Related Genes

Since most disease genes induce pathology through dominant mutations or spontaneous protein misfolding, the mechanisms mediating this GOF toxicity receive the most attention. However, it is also important to understand the physiological function of these proteins to determine how they contribute to diseases. An example of this approach is the generation of an ingenious sensor of APP proteolysis to identify factors that regulate γ-secretase activity (Guo et al. 2003). The fusion of Gal4 to the intracellular domain of APP resulted in the regulation of a proapoptotic protein (Grim) under the control of γ-secretase activity. As expected, the size of the eye was highly sensitive to the components of the γ-secretase complex, including Psen, Aph1, Pen2, and Nicastrin (Guo et al. 2003). Moreover, this assay was utilized to identify other factors regulating the release of the APP intracellular domain (AICD): Ubiquilin (Ubqln, see later) and X11-L (Gross et al. 2008). Interestingly, Ubqln had been identified previously as a risk locus for AD in genome-wide association studies (GWAS) (Bertram et al. 2005) and its function had been linked to Psen function and APP processing. Two independent studies demonstrated that Ubqln binds *Drosophila* Psen and antagonizes its activity, although both studies failed to show that Ubqln promotes proteasome-dependent degradation of Psen (Ganguly et al. 2008; Li et al. 2007). Based on the progressive eye degeneration induced by Ubqln misexpression, Guo proposed the existence of amyloid-independent neurodegenerative pathways mediated by Psen LOF or Ubqln GOF variants (Ganguly et al. 2008).

As described earlier, polyQ expansions in Atrophin-1 (Atro-1) lead to DRPLA. In addition to its role in DRPLA, *Atro*-1 seems to exert a complex activity as a nuclear hormone receptor, including several developmental processes. Since *Atro* LOF caused embryonic lethality, researchers generated somatic mosaics of $Atro^-$ that reduced the size of the eye and induced abnormal specification and differentiation of photoreceptors (Zhang et al. 2002). Further studies identified the cadherin Fat as a strong interactor of Atro, a partner mediating planar polarity and specification of R3 in the eye through repression of *four-jointed* expression (Fanto et al. 2003). In addition, Atro was identified as a negative regulator of EGFR signaling in the wing and eye, phenotypes possibly mediated through the activity of the transcriptional repressor Yan (Charroux et al. 2006). Overall, these studies on Atro in the eye provided new insight into its function as a transcriptional corepressor in multiple developmental processes, including early embryogenesis and the development of adult epithelial structures (wing and eye) through inhibition of EGFR activity.

Using the Eye to Validate Candidate Genes

One of the main contributions of *Drosophila* to the field is the efficient validation of genes suspected to be involved in neurodegeneration. Early on, *Drosophila* played a big role in examining the protective activity of molecular chaperones. Mounting evidence suggested the protective activity of the (Hsp70) against misfolded proteins in purified folding assays and in cell culture. Then, Bonini demonstrated that misexpression of human Hsp70 (HSP1AL) suppressed the eye phenotype of Atx3–78Q and AR-112Q (Chan et al. 2002; Warrick et al. 1999). In contrast, Hsp70 LOF mutations enhanced the eye phenotype of Atx1–82Q and AR-112Q (Fernandez-Funez et al. 2000; Chan et al. 2002). Moreover, the Hsp70 co-chaperone Hsp40 demonstrated its protective activity against the highly toxic polyQ-only (Kazemi-Esfarjani and Benzer 2000), which we also identified as a genetic suppressor of Atx1–82Q (Fernandez-Funez et al. 2000). NMNAT (NAD synthase nicotinamide mononucleotide adenyltransferase) is a chaperone with known neuroprotective activity that is upregulated following Atx1–82Q expression, and binds and suppresses Atx1–82Q neurotoxicity in the eye independently of its catalytic activity (Zhai et al. 2008). Recognition of these toxic proteins by chaperones is typically followed by degradation through the ubiquitin-proteasome system. Overexpression of the E3 ligase CHIP (C-terminus of Hsc-70 interacting protein) interacted with and rescued Atx1–82Q pathology in the eye (Al-Ramahi et al. 2006). CHIP also rescued the eye phenotype of Htt-127Q, but not of 127Q-only, further arguing for the importance of the protein context in polyQ toxicity.

Another significant contribution was the identification of several acetyltransferases as targets of polyQ toxicity (Steffan et al. 2001). To compensate for this LOF, several authors demonstrated the protective activity of overexpressing Creb-binding protein, a key histone acetyltransferase in neurons, or mutating histone deacetylases (HDAC) (Steffan et al. 2001; Taylor et al. 2003). Further studies showed that inhibition of the classic HDAC Rpd3 as well as Sir2 and Sirt2 by RNAi exert the best neuroprotective effects against Httt-93Q (Pallos et al. 2008). However, overexpression of HDAC6, a microtubule deacetylase, prevented the toxicity of AR-52Q by activating autophagy (Pandey et al. 2007).

Another likely pathway involved in neurodegeneration is cell death/apoptosis. It is obvious that the end stage of these diseases is neuronal cell death, so the relevant question is whether inhibiting apoptosis can exert neuroprotection despite being a distal pathway. For instance, Apaf-1 is a key regulator of apoptosis activated by Cytochrome c released from mitochondria. LOF mutations in Dark1/Apaf1, the fly homologue, rescued the toxicity of 108Q-only, Htt-93Q, and Atx1–82Q in the eye (Sang et al. 2005). Expression of Aβ42 also induced prominent cell death during eye development that was rescued by the baculovirus Caspase inhibitor p35 (Tare et al. 2011). Interestingly, Aβ42 also induced c-Jun N-terminal kinase (JNK) signaling, another pathway involved in apoptosis. Inhibiting JNK signaling in combination with p35 led to normal eyes, arguing for the protective activity of antiapoptotic interventions.

The expanded RNA disorders are characterized by the aberrant interaction with RNA-binding proteins, which are proposed to cause LOF in key RNA processing functions. Overexpression of Muscleblind 1 (MBNL1) and CUG binding protein 1 (CUGBP1) rescued the eye phenotype of CUG repeats in a model of MD1, which confirmed their LOF as a disease mechanism (de Haro et al. 2006). Similarly, overexpression of CUGBP1 and hnRNP rescued the eye phenotype of CGG repeats in a model of fragile X-associated tremor/ataxia syndrome (FXTAS) (Sofola et al. 2007a).

In addition, RNAi directed against the CGG repeat was also protective, opening an interesting therapeutic avenue (Sofola et al. 2007b).

Unleashing the Beast: Insight from Genetic Screens

Without a doubt, unbiased genetic screens have played a significant role in expanding our understanding of the molecular mechanisms that regulate protein misfolding and progressive neuronal loss. These modifier screens are the main distinguishing feature of *Drosophila*, providing this little fly a special place in the search for disease mechanisms. The fruit fly offers many options for large-scale screens for genes involved in neuronal development, differentiation, and physiology. Among the most successful are screens that assess the role of genes in viability, longevity, locomotion, vision, and other complex behaviors (Bellen et al. 2010). The eye has been particularly useful because it combines a complex structure sensitive to small perturbations with easy detection, allowing the efficient screening of thousands of strains. The first large-scale screen of modifiers of neurotoxic genes was performed in flies expressing Atx1–82Q. Since these flies showed a strong eye perturbation (see Fig. 1k), they were combined with 4,500 insertions that cause LOF or misexpression of nearby genes. This screen identified a handful of suppressors of the Atx1–82Q phenotype, including multiple components of the protein quality control pathways (chaperones, ubiquitin-dependent protein degradation, and cellular detoxification) (Fernandez-Funez et al. 2000). Along with those expected modifiers, we also found several RNA-binding proteins and transcription factors as specific modifiers of Atx1–82Q, suggesting new roles for Atx1 in RNA metabolism and transcriptional regulation (Fernandez-Funez et al. 2000). Interestingly, these new mechanisms implicated in Atx1 toxicity also showed potent interaction with Htt-128Q in the eye, suggesting that both polyQ diseases share common mechanisms (Branco et al. 2008). However, other modifiers had distinct effects on the two models, supporting the implication of disease-specific pathways. Few of these modifiers affected the Tau phenotype in the eye, but Atx2 was a potent enhancer of Tau neurotoxicity (Shulman and Feany 2011). Interestingly, a genetic screen for modifiers of Atx3 neurotoxicity yielded the same classes of suppressors as the Atx1 screen, underscoring the key role of RNA metabolism, cellular detoxification, and transcriptional regulation in several polyQ diseases (Bilen and Bonini 2007). However, one suppressor of Atx3–78Q (*bantam*) revealed a new role for miRNA pathways in the prevention of cell death (Bilen et al. 2006).

A search for genetic modifiers of Tau toxicity in the eye uncovered the involvement of various kinases and phosphatases, which did not modify Atx1 toxicity (Shulman and Feany 2003), and several cytoskeleton proteins including filamin, myosin VI, and paxillin (Blard et al. 2007). A genetic screen for modifiers of Aβ42 identified components of the secretory pathway, cholesterol homeostasis, chromatin regulation, and copper transport as mediators of neurotoxicity (Cao et al. 2008). A few of these genes also interacted with Tau and Htt-93Q, suggesting that alterations in vesicular trafficking are common to various neurodegenerative conditions. We also characterized X-box binding protein 1 (XBP1), a key ER stress response factor, as a suppressor of Aβ42 neurotoxicity in the eye (Casas-Tinto et al. 2011). Since XBP1 misexpression reduced calcium release from the ER and transcriptional downregulation of the ER-associated ryanodine calcium channel, we showed that LOF of *Rya-r* was also protective.

Fewer modifier screens have been performed in models of LOF diseases, but a search for modifiers of Ppt1-induced degeneration implicated endo-lysosomal trafficking, synaptic vesicle cycling, and synaptic development in the pathogenesis of INCL (Buff et al. 2007).

An alternative to the classic screens based on genetic interactions is the application of genomic technologies to uncover perturbations in gene expression. Microarray on mice expressing FTD-linked Tau^{P301L} led to the identification of Puromycin-sensitive aminopeptidase (PSA) as a candidate suppressor of Tau pathology (Karsten et al. 2006). Coexpression of Tau^{P301L} and PSA in the fly eye demonstrated the protective activity of PSA, thus serving as a fast method for validating the microarray data. Microarray performed with flies overexpressing noncoding CAG repeats identified several common pathways between RNA and polyQ toxicity, including chaperones, transcriptional regulation, and RNA metabolism (Shieh and Bonini 2011). Similarly, microarray analysis in flies expressing non-coding CAG, CUG, and AUUCU found common alterations in GSK3 signaling (Eyk et al. 2011). A targeted expression profiling of miRNAs detected elevated levels of *mir-277* in flies expressing non-coding CGG repeats, while *mir-277* LOF suppressed the eye phenotype of the CGG repeats (Tan et al. 2012). Furthermore, 15 risk loci for AD identified by GWAS were assayed for functional interaction with Tau in the *Drosophila* eye, resulting in six genes that modified Tau neurotoxicity, including a glucose transporter (*glut1*) (Shulman et al. 2011).

More recently, we have witnessed the original and powerful combination of in vitro screening platforms with genetic manipulations in the fly eye. For instance, parallel identification of Htt-binding proteins in yeast two-hybrid and affinity pull-down assays led to a comprehensive Htt interactome (Kaltenbach et al. 2007). Then, the functional relevance of these interactors was confirmed in a *Drosophila* model of HD, which identified 17 highly relevant targets for therapeutics. A different approach consisted in identifying genes that regulate Htt aggregation in a genome-wide cell-based RNAi screen, which yielded new and known genes related to nuclear transport, nucleotide processing, and signaling (Zhang et al. 2010). Then, the authors validated the new modifiers, including chaperone Hsp110 and the transcriptional regulator Tra1, in the eye. Altogether, these studies highlight the value of integrating complex platforms of gene discovery with genetic manipulation in the fly eye.

Role of the Fly Eye in Drug Discovery

Testing Candidate Drugs

The generation of disease models with neurotoxic phenotypes sensitive to genetic modifiers raised the next challenge: can we exploit these fly models to identify new, more effective drugs with disease-modifying activity? Initial evidence for the effectiveness of this approach came from feeding flies with the histone deacetylase (HDAC) inhibitors suberoyl anilide hydroxamic acid (SAHA) and butyrate, which protected against Htt neurotoxicity (Steffan et al 2001). Other pharmacologic suppressors of Htt toxicity in the eye include: the mTOR inhibitor rapamycin, which promotes autophagy, a key clearance pathway (Ravikumar et al. 2004); the transglutaminase 2 inhibitor ZDON, which reverts the aberrant transcriptional repression induced by Htt-dependent up-regulation of transglutaminase 2 (McConoughey et al. 2010); inhibitors of the protein deacetylase Sirtuin 2 (Pallos et al. 2010; Luthi-Carter et al. 2010); activation of ERK signaling by the polyphenols fisetin and resveratrol (Maher et al. 2011); reduction of oxidative stress by meclizine, a Food and Drug Administration (FDA)-approved drug (Gohil et al. 2011), and expression of intracellular single chain antibodies (intrabodies) against Htt (Wolfgang et al. 2005).

Other contributions of the eye system in different models include the suppression of Atx3 and Htt neurotoxicity by stimulating heat shock factor 1 (HSF1) with the Hsp90 inhibitor 17-AAG (Fujikake et al. 2008), the suppression of CGG repeat toxicity with histone deacetylase inhibitors in a fly model of FXTAS (Todd et al. 2010), the inhibition of Pink1-dependent retinal degeneration with the antioxidant vitamin E (Wang et al. 2006), and the modulation of Aβ42-induced toxicity with zinc/copper chelators (Hua et al. 2011) or with a mixed extract from 15 crude herbs known as KSOP1009 (Hong et al. 2011).

Combinatorial Therapies and Secondary Drug Screens

Neurodegenerative diseases cause multiple cellular perturbations, an indication of their mechanistic complexity. Therefore, the development of disease-modifying therapies may require the elimination of the toxic agents by targeting disease-associated alleles or conformations. However, these ideal therapies have proven hard to achieve. In the meantime, the next best option is to target several independent targets simultaneously. The flexible manipulation and rapid turnaround of flies allow for testing drug combinations with the objective of inhibiting multiple targets with synergistic effects. Proof of principle for this approach came from testing combinations of compounds with unique mechanism of action: SAHA, the amyloid dye Congo red, the transglutaminase inhibitor cystamine, the Hsp90 inhibitor geldanamycin, and the rho-associated coil kinase (ROCK) inhibitor Y-27632. This seminal work identified combinatorial regimens of SAHA and either Congo red, cystamine, Y-27632, or geldanamycin at low concentrations that alleviated the degeneration of photoreceptor neurons in a fly model of HD (Agrawal et al. 2005). Also, the combined inhibition of HDAC3 and Sirtuins (Rpd3 and Sir2) with low doses of butyrate and nicotinamide enhanced neuroprotection against Htt-induced neurotoxicity (Pallos et al. 2008). Moreover, the combination of rapamycin and lithium-enhanced neuroprotection in HD flies through the stimulation of mTOR-dependent and -independent autophagy (Sarkar et al. 2008). More recently, feeding cystamine to adult HD flies improved the protection exerted by anti-Htt intrabodies (Bortvedt et al. 2010).

In addition to validating known targets, the fly eye can also play a significant role in assessing the in vivo function of compounds identified in high-throughput, in vitro platforms. For instance, a fluorescence resonance energy transfer-based cellular aggregation assay allowed screening over 2,800 small molecules that inhibit polyQ aggregation (Pollitt et al. 2003). Y-27632, one of the strongest hits, reduced protein aggregation and photoreceptor loss in a fly model of HD and improved rotarod performance in a mouse model of HD (Li et al. 2009). In another study, researchers used a yeast-based Htt aggregation assay to screen 16,000 compounds and later tested the positive hits in multiple secondary assays that included analysis of photoreceptor neurons in HD flies (Zhang et al. 2005). This led to the identification of C2–8, whose therapeutic potential was later confirmed in a mouse model of HD (Chopra et al. 2007), although the mechanism of action is unknown. These studies demonstrate that the use of *Drosophila* in secondary screens not only saves time and funds, but also helps identify the most promising candidates with good pharmacokinetic properties for clinical trials.

Concluding Remarks

In this chapter, we illustrated the vast contributions and potential of the *Drosophila* eye as a model for understanding the cellular and molecular mechanisms of neurodegeneration. The eye has been a productive tool for candidate validation, gene discovery efforts, and drug testing in vivo. Remarkably, the inclusion of the eye as validation tool in the drug discovery pipeline has gained momentum in recent years, contributing to speedup the identification of compounds likely to be effective in mammalian models. Therefore, genetic and pharmacological screens in the fly eye are likely to be an instrumental tool in this field for many years, particularly if researchers continue their inventive approaches to exploit this little fly.

Acknowledgments We thank the Bloomington *Drosophila* Stock Center for strains. We want to apologize for those whose work we could not cite due to space constraints. This work was supported by the NIH grants DP2OD002721 to PF-F and R21NS081356 to DER-L, and star-up funding from the Department of Neurology (UF) to PF-F and DER-L.

References

Agrawal N, Pallos J, Slepko N, Apostol BL, Bodai L et al (2005) Identification of combinatorial drug regimens for treatment of Huntington's disease using *Drosophila*. Proc Natl Acad Sci U S A 102:3777–3781

Aguzzi A, O'Connor T (2010) Protein aggregation diseases: pathogenicity and therapeutic perspectives. Nat Rev Drug Discov 9:237–248

Al-Ramahi I, Lam YC, Chen HK, de Gouyon B, Zhang M et al (2006) CHIP protects from the neurotoxicity of expanded and wild-type ataxin-1 and promotes their ubiquitination and degradation. J Biol Chem 281:26714–26724

Al-Ramahi I, Perez AM, Lim J, Zhang M, Sorensen R et al (2007) dAtaxin-2 mediates expanded Ataxin-1-induced neurodegeneration in a *Drosophila* model of SCA1. PLoS Genet 3:e234

Bayat V, Thiffault I, Jaiswal M, Tetreault M, Donti T et al (2012) Mutations in the mitochondrial methionyl-tRNA synthetase cause a neurodegenerative phenotype in flies and a recessive ataxia (ARSAL) in humans. PLoS Biol 10:e1001288

Bellen HJ, Tong C, Tsuda H (2010) 100 years of *Drosophila* research and its impact on vertebrate neuroscience: a history lesson for the future. Nat Rev Neurosci 11:514–522

Benzer S (1967) Behavioral mutants of *Drosophila* isolated by countercurrent distribution. Proc Natl Acad Sci U S A 58:1112–1119

Bertram L, Hiltunen M, Parkinson M, Ingelsson M, Lange C et al (2005) Family-based association between Alzheimer's disease and variants in UBQLN1. N Engl J Med 352:884–894

Bilen J, Liu N, Burnett BG, Pittman RN, Bonini NM (2006) MicroRNA pathways modulate polyglutamine-induced neurodegeneration. Mol Cell 24:157–163

Bilen J, Bonini NM (2007) Genome-wide screen for modifiers of ataxin-3 neurodegeneration in *Drosophila*. PLoS Genet 3:1950–1964

Blard O, Feuillette S, Bou J, Chaumette B, Frebourg T et al (2007) Cytoskeleton proteins are modulators of mutant tau-induced neurodegeneration in *Drosophila*. Hum Mol Genet 16:555–566

Bortvedt SF, McLear JA, Messer A, Ahern-Rindell AJ, Wolfgang WJ (2010) Cystamine and intrabody co-treatment confers additional benefits in a fly model of Huntington's disease. Neurobiol Dis 40:130–134

Branco J, Al-Ramahi I, Ukani L, Perez AM, Fernandez-Funez P et al (2008) Comparative analysis of genetic modifiers in *Drosophila* points to common and distinct mechanisms of pathogenesis among polyglutamine diseases. Hum Mol Genet 17:376–390

Brand AH, Perrimon N (1993) Targeted gene expression as a means of altering cell fates and generating dominant phenotypes. Development 118:401–415

Buff H, Smith AC, Korey CA (2007) Genetic modifiers of *Drosophila* palmitoyl-protein thioesterase 1-induced degeneration. Genetics 176:209–220

Cao W, Song HJ, Gangi T, Kelkar A, Antani I et al (2008) Identification of novel genes that modify phenotypes induced by Alzheimer's beta-amyloid overexpression in *Drosophila*. Genetics 178:1457–1471

Casas-Tinto S, Zhang Y, Sanchez-Garcia J, Gomez-Velazquez M, Rincon-Limas DE et al (2011) The ER stress factor XBP1s prevents amyloid-beta neurotoxicity. Hum Mol Genet 20:2144–2160

Chan HY, Warrick JM, Andriola I, Merry D, Bonini NM (2002) Genetic modulation of polyglutamine toxicity by protein conjugation pathways in *Drosophila*. Hum Mol Genet 11:2895–2904

Charroux B, Freeman M, Kerridge S, Baonza A (2006) Atrophin contributes to the negative regulation of epidermal growth factor receptor signaling in *Drosophila*. Dev Biol 291:278–290

Chatterjee S, Sang TK, Lawless GM, Jackson GR (2009) Dissociation of tau toxicity and phosphorylation: role of GSK-3beta, MARK and Cdk5 in a *Drosophila* model. Hum Mol Genet 18:164–177

Chen HK, Fernandez-Funez P, Acevedo SF, Lam YC, Kaytor MD et al (2003) Interaction of Akt-phosphorylated ataxin-1 with 14–3-3 mediates neurodegeneration in spinocerebellar ataxia type 1. Cell 113:457–468

Chen L, Periquet M, Wang X, Negro A, McLean PJ et al (2009) Tyrosine and serine phosphorylation of alpha-synuclein have opposing effects on neurotoxicity and soluble oligomer formation. J Clin Invest 119:3257–3265

Chopra V, Fox JH, Lieberman G, Dorsey K, Matson W et al (2007) A small-molecule therapeutic lead for Huntington's disease: preclinical pharmacology and efficacy of C2–8 in the R6/2 transgenic mouse. Proc Natl Acad Sci U S A 104:16685–16689

Clark IE, Dodson MW, Jiang C, Cao JH, Huh JR et al (2006) *Drosophila* pink1 is required for mitochondrial function and interacts genetically with parkin. Nature 441:1162–1166

Couthouis J, Hart MP, Erion R, King OD, Diaz Z et al (2012) Evaluating the role of the FUS/TLS-related gene EWSR1 in amyotrophic lateral sclerosis. Hum Mol Genet 21:2899–2911

Crowther DC, Kinghorn KJ, Miranda E, Page R, Curry JA et al (2005) Intraneuronal Abeta, non-amyloid aggregates and neurodegeneration in a *Drosophila* model of Alzheimer's disease. Neuroscience 132:123–135

Haro M de, Al-Ramahi I, De Gouyon B, Ukani L, Rosa A et al (2006) MBNL1 and CUGBP1 modify expanded CUG-induced toxicity in a *Drosophila* model of myotonic dystrophy type 1. Hum Mol Genet 15:2138–2145

Dermaut B, Norga KK, Kania A, Verstreken P, Pan H et al (2005) Aberrant lysosomal carbohydrate storage accompanies endocytic defects and neurodegeneration in *Drosophila* benchwarmer. J Cell Biol 170:127–139

Fanto M, Clayton L, Meredith J, Hardiman K, Charroux B et al (2003) The tumor-suppressor and cell adhesion molecule Fat controls planar polarity via physical interactions with Atrophin, a transcriptional co-repressor. Development 130:763–774

Feany MB, Bender WW (2000) A *Drosophila* model of Parkinson's disease. Nature 404:394–398

Fernandez-Funez P, Nino-Rosales ML, de Gouyon B, She WC, Luchak JM et al (2000) Identification of genes that modify ataxin-1-induced neurodegeneration. Nature 408:101–106

Fernandez-Funez P, Casas-Tinto S, Zhang Y, Gomez-Velazquez M, Morales-Garza MA et al (2009) In vivo generation of neurotoxic prion protein: role for hsp70 in accumulation of misfolded isoforms. PLoS Genet 5:e1000507

Finelli A, Kelkar A, Song HJ, Yang H, Konsolaki M (2004) A model for studying Alzheimer's Abeta42-induced toxicity in *Drosophila melanogaster*. Mol Cell Neurosci 26:365–375

Franceschini N (1972) Pupil and pseudopupil in the compound eye of *Drosophila*. In: Wehner R (ed) Information processing in the visual systems of arthropods. Springer, Berlin, pp 75–82

Fujikake N, Nagai Y, Popiel HA, Okamoto Y, Yamaguchi M et al (2008) Heat shock transcription factor 1-activating compounds suppress polyglutamine-induced neurodegeneration through induction of multiple molecular chaperones. J Biol Chem 283:26188–26197

Ganguly A, Feldman RM, Guo M (2008) ubiquilin antagonizes presenilin and promotes neurodegeneration in *Drosophila*. Hum Mol Genet 17:293–302

Greeve I, Kretzschmar D, Tschape JA, Beyn A, Brellinger C et al (2004) Age-dependent neurodegeneration and Alzheimer-amyloid plaque formation in transgenic *Drosophila*. J Neurosci 24:3899–3906

Gross GG, Feldman RM, Ganguly A, Wang J, Yu H et al (2008) Role of X11 and ubiquilin as in vivo regulators of the amyloid precursor protein in *Drosophila*. PLoS ONE 3:e2495

Gohil VM, Offner N, Walker JA, Sheth SA, Fossale E et al (2011) Meclizine is neuroprotective in models of Huntington's disease. Hum Mol Genet 20:294–300

Guo M, Hong EJ, Fernandes J, Zipursky SL, Hay BA (2003) A reporter for amyloid precursor protein gamma-secretase activity in *Drosophila*. Hum Mol Genet 12:2669–2678

Hardy J, Lewis P, Revesz T, Lees A, Paisan-Ruiz C (2009) The genetics of Parkinson's syndromes: a critical review. Curr Opin Genet Dev 19:254–265

Hong YK, Park SH, Lee S, Hwang S, Lee MJ et al (2011) Neuroprotective effect of SuHeXiang Wan in *Drosophila* models of Alzheimer's disease. J Ethnopharmacol 134:1028–1032

Hua H, Munter L, Harmeier A, Georgiev O, Multhaup G et al (2011) Toxicity of Alzheimer's disease-associated Abeta peptide is ameliorated in a *Drosophila* model by tight control of zinc and copper availability. Biol Chem 392:919–926

Iijima K, Chiang HC, Hearn SA, Hakker I, Gatt A et al (2008) Abeta42 mutants with different aggregation profiles induce distinct pathologies in *Drosophila*. PLoS ONE 3:e1703

Jackson GR, Salecker I, Dong X, Yao X, Arnheim N et al (1998) Polyglutamine-expanded human huntingtin transgenes induce degeneration of *Drosophila* photoreceptor neurons. Neuron 21:633–642

Jackson GR, Wiedau-Pazos M, Sang TK, Wagle N, Brown CA et al (2002) Human wild-type tau interacts with wingless pathway components and produces neurofibrillary pathology in *Drosophila*. Neuron 34:509–519

Jin P, Zarnescu DC, Zhang F, Pearson CE, Lucchesi JC et al (2003) RNA-mediated neurodegeneration caused by the fragile X premutation rCGG repeats in *Drosophila*. Neuron 39:739–747

Jung J, Xu K, Lessing D, Bonini NM (2009) Preventing Ataxin-3 protein cleavage mitigates degeneration in a *Drosophila* model of SCA3. Hum Mol Genet 18:4843–4852

Kaltenbach LS, Romero E, Becklin RR, Chettier R, Bell R et al (2007) Huntingtin interacting proteins are genetic modifiers of neurodegeneration. PLoS Genet 3:e82

Karsten SL, Sang TK, Gehman LT, Chatterjee S, Liu J et al (2006) A genomic screen for modifiers of tauopathy identifies puromycin-sensitive aminopeptidase as an inhibitor of tau-induced neurodegeneration. Neuron 51:549–560

Kazemi-Esfarjani P, Benzer S (2000) Genetic suppression of polyglutamine toxicity in *Drosophila*. Science 287:1837–1840

Khurana V, Lu Y, Steinhilb ML, Oldham S, Shulman JM et al (2006) TOR-mediated cell-cycle activation causes neurodegeneration in a *Drosophila* tauopathy model. Curr Biol 16:230–241

Korey CA, MacDonald ME (2003) An over-expression system for characterizing Ppt1 function in *Drosophila*. BMC Neurosci 4:30

Kumar JP (2012) Building an ommatidium one cell at a time. Dev Dyn 241:136–149

Lam YC, Bowman AB, Jafar-Nejad P, Lim J, Richman R et al (2006) ATAXIN-1 interacts with the repressor Capicua in its native complex to cause SCA1 neuropathology. Cell 127:1335–1347

Lawlor KT, O'Keefe LV, Samaraweera SE, van Eyk CL, McLeod CJ et al (2011) Double-stranded RNA is pathogenic in *Drosophila* models of expanded repeat neurodegenerative diseases. Hum Mol Genet 20:3757–3768

Lanson NA Jr, Maltare A, King H, Smith R, Kim JH et al (2011) A *Drosophila* model of FUS-related neurodegeneration reveals genetic interaction between FUS and TDP-43. Hum Mol Genet 20:2510–2523

Lessing D, Bonini NM (2008) Polyglutamine genes interact to modulate the severity and progression of neurodegeneration in *Drosophila*. PLoS Biol 6:e29

Lessing D, Bonini NM (2009) Maintaining the brain: insight into human neurodegeneration from *Drosophila melanogaster* mutants. Nat Rev Genet 10:359–370

Li A, Xie Z, Dong Y, McKay KM, McKee ML et al (2007) Isolation and characterization of the *Drosophila* ubiquilin ortholog dUbqln: in vivo interaction with early-onset Alzheimer disease genes. Hum Mol Genet 16:2626–2639

Li LB, Yu Z, Teng X, Bonini NM (2008) RNA toxicity is a component of ataxin-3 degeneration in *Drosophila*. Nature 453:1107–1111

Li M, Huang Y, Ma AA, Lin E, Diamond MI (2009) Y-27632 improves rotarod performance and reduces huntingtin levels in R6/2 mice. Neurobiol Dis 36:413–420

Li Y, Ray P, Rao EJ, Shi C, Guo W et al (2010) A *Drosophila* model for TDP-43 proteinopathy. Proc Natl Acad Sci U S A 107:3169–3174

Lim J, Crespo-Barreto J, Jafar-Nejad P, Bowman AB, Richman R et al (2008) Opposing effects of polyglutamine expansion on native protein complexes contribute to SCA1. Nature 452:713–718

Liu Z, Wang X, Yu Y, Li X, Wang T et al (2008) A *Drosophila* model for LRRK2-linked parkinsonism. Proc Natl Acad Sci U S A 105:2693–2698

Lorenzo DN, Li MG, Mische SE, Armbrust KR, Ranum LP et al (2010) Spectrin mutations that cause spinocerebellar ataxia type 5 impair axonal transport and induce neurodegeneration in *Drosophila*. J Cell Biol 189:143–158

Luthi-Carter R, Taylor DM, Pallos J, Lambert E, Amore A et al (2010) SIRT2 inhibition achieves neuroprotection by decreasing sterol biosynthesis. Proc Natl Acad Sci U S A 107:7927–7932

Maher P, Dargusch R, Bodai L, Gerard PE, Purcell JM et al (2011) ERK activation by the polyphenols fisetin and resveratrol provides neuroprotection in multiple models of Huntington's disease. Hum Mol Genet 20:261–270

Marsh JL, Walker H, Theisen H, Zhu YZ, Fielder T et al (2000) Expanded polyglutamine peptides alone are intrinsically cytotoxic and cause neurodegeneration in *Drosophila*. Hum Mol Genet 9:13–25

Mast JD, Tomalty KM, Vogel H, Clandinin TR (2008) Reactive oxygen species act remotely to cause synapse loss in a *Drosophila* model of developmental mitochondrial encephalopathy. Development 135:2669–2679

McConoughey SJ, Basso M, Niatsetskaya ZV, Sleiman SF, Smirnova NA et al (2010) Inhibition of transglutaminase 2 mitigates transcriptional dysregulation in models of Huntington disease. EMBO Mol Med 2:349–370

Miguel L, Frebourg T, Campion D, Lecourtois M (2011) Both cytoplasmic and nuclear accumulations of the protein are neurotoxic in *Drosophila* models of TDP-43 proteinopathies. Neurobiol Dis 41:398–406

Mugat B, Parmentier ML, Bonneaud N, Chan HY, Maschat F (2008) Protective role of Engrailed in a *Drosophila* model of Huntington's disease. Hum Mol Genet 17:3601–3616

Mutsuddi M, Marshall CM, Benzow KA, Koob MD, Rebay I (2004) The spinocerebellar ataxia 8 noncoding RNA causes neurodegeneration and associates with staufen in *Drosophila*. Curr Biol 14:302–308

Nishimura I, Yang Y, Lu B (2004) PAR-1 kinase plays an initiator role in a temporally ordered phosphorylation process that confers tau toxicity in *Drosophila*. Cell 116:671–682

Nisoli I, Chauvin JP, Napoletano F, Calamita P, Zanin V et al (2010) Neurodegeneration by polyglutamine Atrophin is not rescued by induction of autophagy. Cell Death Differ 17:1577–1587

Pallos J, Bodai L, Lukacsovich T, Purcell JM, Steffan JS et al (2008) Inhibition of specific HDACs and sirtuins suppresses pathogenesis in a *Drosophila* model of Huntington's disease. Hum Mol Genet 17:3767–3775

Pandey UB, Nie Z, Batlevi Y, McCray BA, Ritson GP et al (2007) HDAC6 rescues neurodegeneration and provides an essential link between autophagy and the UPS. Nature 447:859–863

Park J, Lee SB, Lee S, Kim Y, Song S et al (2006) Mitochondrial dysfunction in *Drosophila* PINK1 mutants is complemented by parkin. Nature 441:1157–1161

Phillips SE, Woodruff EA 3rd, Liang P, Patten M, Broadie K (2008) Neuronal loss of *Drosophila* NPC1a causes cholesterol aggregation and age-progressive neurodegeneration. J Neurosci 28:6569–6582

Pollitt SK, Pallos J, Shao J, Desai UA, Ma AA et al (2003) A rapid cellular FRET assay of polyglutamine aggregation identifies a novel inhibitor. Neuron 40:685–694

Poole AC, Thomas RE, Andrews LA, McBride HM, Whitworth AJ et al (2008) The PINK1/Parkin pathway regulates mitochondrial morphology. Proc Natl Acad Sci U S A 105:1638–1643

Ravikumar B, Vacher C, Berger Z, Davies JE, Luo S et al (2004) Inhibition of mTOR induces autophagy and reduces toxicity of polyglutamine expansions in fly and mouse models of Huntington disease. Nat Genet 36:585–595

Ren J, Jegga AG, Zhang M, Deng J, Liu J et al (2011) A *Drosophila* model of the neurodegenerative disease SCA17 reveals a role of RBP-J/Su(H) in modulating the pathological outcome. Hum Mol Genet 20:3424–3436

Rimkus SA, Katzenberger RJ, Trinh AT, Dodson GE, Tibbetts RS et al (2008) Mutations in String/CDC25 inhibit cell cycle re-entry and neurodegeneration in a *Drosophila* model of Ataxia telangiectasia. Genes Dev 22:1205–1220

Rincon-Limas D, Jensen K, Fernandez Funez A (2012) *Drosophila* models of proteinopathies: the little fly that could. Curr Pharm Des 18:1108–1122

Ross CA, Tabrizi SJ (2011) Huntington's disease: from molecular pathogenesis to clinical treatment. Lancet Neurol 10:83–98

Romero E, Cha GH, Verstreken P, Ly CV, Hughes RE et al (2008) Suppression of neurodegeneration and increased neurotransmission caused by expanded full-length huntingtin accumulating in the cytoplasm. Neuron 57:27–40

Sang TK, Li C, Liu W, Rodriguez A, Abrams JM et al (2005) Inactivation of *Drosophila* Apaf-1 related killer suppresses formation of polyglutamine aggregates and blocks polyglutamine pathogenesis. Hum Mol Genet 14:357–372

Sarkar S, Krishna G, Imarisio S, Saiki S, O'Kane CJ et al (2008) A rational mechanism for combination treatment of Huntington's disease using lithium and rapamycin. Hum Mol Genet 17:170–178

Shieh SY, Bonini NM (2011) Genes and pathways affected by CAG-repeat RNA-based toxicity in *Drosophila*. Hum Mol Genet 20:4810–4821

Shulman JM, Feany MB (2003) Genetic modifiers of tauopathy in *Drosophila*. Genetics 165:1233–1242
Shulman JM, Chipendo P, Chibnik LB, Aubin C, Tran D et al (2011) Functional screening of Alzheimer pathology genome-wide association signals in *Drosophila*. Am J Hum Genet 88:232–238
Sofola OA, Jin P, Qin Y, Duan R, Liu H et al (2007a) RNA-binding proteins hnRNP A2/B1 and CUGBP1 suppress fragile X CGG premutation repeat-induced neurodegeneration in a *Drosophila* model of FXTAS. Neuron 55:565–571
Sofola OA, Jin P, Botas J, Nelson DL (2007b) Argonaute-2-dependent rescue of a *Drosophila* model of FXTAS by FRAXE premutation repeat. Hum Mol Genet 16:2326–2332
Steffan JS, Bodai L, Pallos J, Poelman M, McCampbell A et al (2001) Histone deacetylase inhibitors arrest polyglutamine-dependent neurodegeneration in *Drosophila*. Nature 413:739–743
Steffan JS, Agrawal N, Pallos J, Rockabrand E, Trotman LC et al (2004) SUMO modification of Huntingtin and Huntington's disease pathology. Science 304:100–104
Steinhilb ML, Dias-Santagata D, Mulkearns EE, Shulman JM, Biernat J et al (2007a) S/P and T/P phosphorylation is critical for tau neurotoxicity in *Drosophila*. J Neurosci Res 85:1271–1278
Steinhilb ML, Dias-Santagata D, Fulga TA, Felch DL, Feany MB (2007b) Tau phosphorylation sites work in concert to promote neurotoxicity in vivo. Mol Biol Cell 18:5060–5068
Takeyama K, Ito S, Yamamoto A, Tanimoto H, Furutani T et al (2002) Androgen-dependent neurodegeneration by polyglutamine-expanded human androgen receptor in *Drosophila*. Neuron 35:855–864
Tan H, Poidevin M, Li H, Chen D, Jin P (2012) MicroRNA-277 modulates the neurodegeneration caused by Fragile X premutation rCGG repeats. PLoS Genet 8:e1002681
Tare M, Modi RM, Nainaparampil JJ, Puli OR, Bedi S et al (2011) Activation of JNK signaling mediates amyloid-ss-dependent cell death. PLoS One 6:e24361
Taylor JP, Taye AA, Campbell C, Kazemi-Esfarjani P, Fischbeck KH et al (2003) Aberrant histone acetylation, altered transcription, and retinal degeneration in a *Drosophila* model of polyglutamine disease are rescued by CREB-binding protein. Genes Dev 17:1463–1468
Todd PK, SY Oh, Krans A, Pandey UB, Di Prospero NA et al (2010) Histone deacetylases suppress CGG repeat-induced neurodegeneration via transcriptional silencing in models of fragile X tremor ataxia syndrome. PLoS Genet 6:e1001240
Eyk CL van, O'Keefe LV, Lawlor KT, Samaraweera SE, McLeod CJ et al (2011) Perturbation of the Akt/Gsk3-beta signalling pathway is common to *Drosophila* expressing expanded untranslated CAG, CUG and AUUCU repeat RNAs. Hum Mol Genet 20:2783–2794
Venderova K, Kabbach G, Abdel-Messih E, Zhang Y, Parks RJ et al (2009) Leucine-Rich Repeat Kinase 2 interacts with Parkin, DJ-1 and PINK-1 in a *Drosophila melanogaster* model of Parkinson's disease. Hum Mol Genet 18:4390–4404
Venkatachalam K, Long AA, Elsaesser R, Nikolaeva D, Broadie K et al (2008) Motor deficit in a *Drosophila* model of mucolipidosis type IV due to defective clearance of apoptotic cells. Cell 135:838–851
Wang D, Qian L, Xiong H, Liu J, Neckameyer WS et al (2006) Antioxidants protect PINK1-dependent dopaminergic neurons in *Drosophila*. Proc Natl Acad Sci U S A 103:13520–13525
Warrick JM, Paulson HL, Gray-Board GL, Bui QT, Fischbeck KH et al (1998) Expanded polyglutamine protein forms nuclear inclusions and causes neural degeneration in *Drosophila*. Cell 93:939–949
Warrick JM, Chan HY, Gray-Board GL, Chai Y, Paulson HL et al (1999) Suppression of polyglutamine-mediated neurodegeneration in *Drosophila* by the molecular chaperone HSP70. Nat Genet 23:425–428
Watson MR, Lagow RD, Xu K, Zhang B, Bonini NM (2008) A drosophila model for amyotrophic lateral sclerosis reveals motor neuron damage by human SOD1. J Biol Chem 283:24972–24981
Whitworth AJ, Lee JR, Ho VM, Flick R, Chowdhury R et al (2008) Rhomboid-7 and HtrA2/Omi act in a common pathway with the Parkinson's disease factors Pink1 and Parkin. Dis Model Mech 1: 168–174; discussion 173.

Wittmann CW, Wszolek MF, Shulman JM, Salvaterra PM, Lewis J et al (2001) Tauopathy in *Drosophila*: neurodegeneration without neurofibrillary tangles. Science 293:711–714

Wolfgang WJ, Miller TW, Webster JM, Huston JS, Thompson LM et al (2005) Suppression of Huntington's disease pathology in *Drosophila* by human single-chain Fv antibodies. Proc Natl Acad Sci U S A 102:11563–11568

Wu Z, Li C, S Lv, Zhou B (2009) Pantothenate kinase-associated neurodegeneration: insights from a *Drosophila* model. Hum Mol Genet 18:3659–3672

Yang Y, Gehrke S, Haque ME, Imai Y, Kosek J et al (2005) Inactivation of *Drosophila* DJ-1 leads to impairments of oxidative stress response and phosphatidylinositol 3-kinase/Akt signaling. Proc Natl Acad Sci U S A 102:13670–13675

Yang Y, Gehrke S, Imai Y, Huang Z, Ouyang Y et al (2006) Mitochondrial pathology and muscle and dopaminergic neuron degeneration caused by inactivation of *Drosophila* Pink1 is rescued by Parkin. Proc Natl Acad Sci U S A 103:10793–10798

Yu Z, Teng X, Bonini NM (2011) Triplet repeat-derived siRNAs enhance RNA-mediated toxicity in a *Drosophila* model for myotonic dystrophy. PLoS Genet 7:e1001340

Zhai RG, Zhang F, Hiesinger PR, Cao Y, Haueter CM et al (2008) NAD synthase NMNAT acts as a chaperone to protect against neurodegeneration. Nature 452:887–891

Zhang S, Xu L, Lee J, Xu T (2002) *Drosophila* atrophin homolog functions as a transcriptional corepressor in multiple developmental processes. Cell 108:45–56

Zhang X, Smith DL, Meriin AB, Engemann S, Russel DE et al (2005) A potent small molecule inhibits polyglutamine aggregation in Huntington's disease neurons and suppresses neurodegeneration in vivo. Proc Natl Acad Sci U S A 102:892–897

Zhang S, Binari R, Zhou R, Perrimon N (2010) A genomewide RNA interference screen for modifiers of aggregates formation by mutant Huntingtin in *Drosophila*. Genetics 184:1165–1179

Genetic Regulation of Early Eye Development in Non-dipteran Insects

Markus Friedrich, Ying Dong, Zhenyi Liu and Iris Yang

Comparative analyses of eye development in *Drosophila* and distantly related phyla have fundamentally changed the way we think about the evolution of animal eyes today. On the one hand, it is clear that select eye-patterning mechanisms have deep evolutionary roots, such as the involvement of Pax6 and an ever-extending catalogue of additional transcription factors with selector gene-like functions in development (Donner and Maas 2004; Gehring 2002; Kozmik 2008; Pichaud and Desplan 2002). On the other hand, the diversity of distinct eye types in extant animals implies the evolution of lineage-specific patterning processes, superimposed onto the ancient gene interactions inherited from the prototype eye at the dawn of animal evolution (Lamb 2011; Nilsson 1996; Salvini-Plawen and Mayr 1977; Zuker 1994). Therefore, an important question to consider is how far back the regulatory program organizing the development of the compound eye in *Drosophila* can be traced to arthropod evolution.

Elaborate compound eyes are found in living representatives of all arthropod phyla, namely Crustacea, Chelicerates, and Myriapods, in addition to the insects (Buschbeck and Friedrich 2008; Fahrenbach 1969; Müller et al. 2003). The earliest fossils of advanced compound eye design have been discovered in deposits of the early Cambrian, which dates 515 million years before present (Lee et al. 2011;

M. Friedrich (✉) · I. Yang
Department of Biological Sciences, Wayne State University,
5047 Gullen Mall, Detroit, MI 48202, USA
e-mail: mf@biology.biosci.wayne.edu

M. Friedrich
Department of Anatomy and Cell Biology, School of Medicine, Wayne State University,
540 East Canfield Avenue, Detroit, MI 48201, USA

Y. Dong
Harold C. Simmons Cancer Center,
UT Southwestern Medical Center at Dallas, Dallas, USA

Z. Liu
Washington University,
3600 Cancer Research Building, St. Louis, USA

A. Singh, M. Kango-Singh (eds.), *Molecular Genetics of Axial Patterning, Growth and Disease in the Drosophila Eye,* DOI 10.1007/978-1-4614-8232-1_11,

Paterson et al. 2011). This implies that the regulatory program patterning the *Drosophila* compound eye retina is hundreds of millions of years of age. Comparative analysis in arthropods, therefore, offers unique opportunities to dissect the conserved and evolutionary younger components in the genetic control networks which pattern the *Drosophila* eye. To this end, a number of gene-specific studies have been carried out in representatives of other arthropod phyla, such as crustaceans and the horseshoe crab *Limulus polyphemus*, the only extant chelicerate with compound eyes (Blackburn et al. 2008; Duman-Scheel et al. 2002; Smith et al. 1993). Also, the cellular organization of growth and differentiation of the visual system has been studied in non-insect arthropods (Hafner and Tokarski 1998, 2001; Harzsch and Walossek 2001; Melzer et al. 2000). However, the most comprehensive comparative molecular studies of compound eye development have focused on non-dipteran insect species up to this point.

Here, I introduce the satellite model organisms in current comparative genetic studies of insect compound eye development and their phylogenetic relationships. This is followed by a systematic review of the molecular findings that concern the patterning of the retinal precursor tissues in these organisms, which, at this point, are based on gene expression pattern analysis and lack-of-function analyses by RNA interference (RNAi)-mediated gene knockdown. The cellular assembly of retinal precursor cells in the differentiating retina is strongly conserved in arthropods and has been previously reviewed in depth (Buschbeck and Friedrich 2008; Friedrich et al. 2006). It will not be further explored here. I will conclude pointing out broader insights and the most important pending questions regarding the developmental evolution of the *Drosophila* compound eye, a story of profound sensory organ primordium reorganization.

The Phylogenetic Framework

Against the backdrop of insect diversity, the number of non-dipteran species that have been studied with comparative questions regarding the developing eye is dwindlingly small (Fig. 1). Besides studies looking at the morphogenesis of very unusual visual systems, such as stalk-eyed flies or the enigmatic Strepsiptera (Buschbeck 2005; Buschbeck et al. 2001), molecular work boils down to five species. Two of these belong to the same basal order of hemimetabolous insects. This refers to the bispotted cricket *Gryllus bimaculatus* and the American desert locust *Schistocerca americana*, both of which are members of the Orthoptera, although of distantly related subgroups. *G. bimaculatus* belongs to the suborder Ensifera while *S. americana* is part of the second orthopteran suborder, the Caelifera.

The insect order Orthoptera is one of the 22 currently recognized direct-developing insect orders. The latter refers to the direct development of most adult body structures in the embryo, which continue to gain size during the postembryonic growth stages of the nymphs. Except for wing and genital appendages, the nymph disposes over all essential body structures of the future adult form (Truman and Riddiford 2002). The ancestral lack of wings distinguishes ametabolous direct-developers from

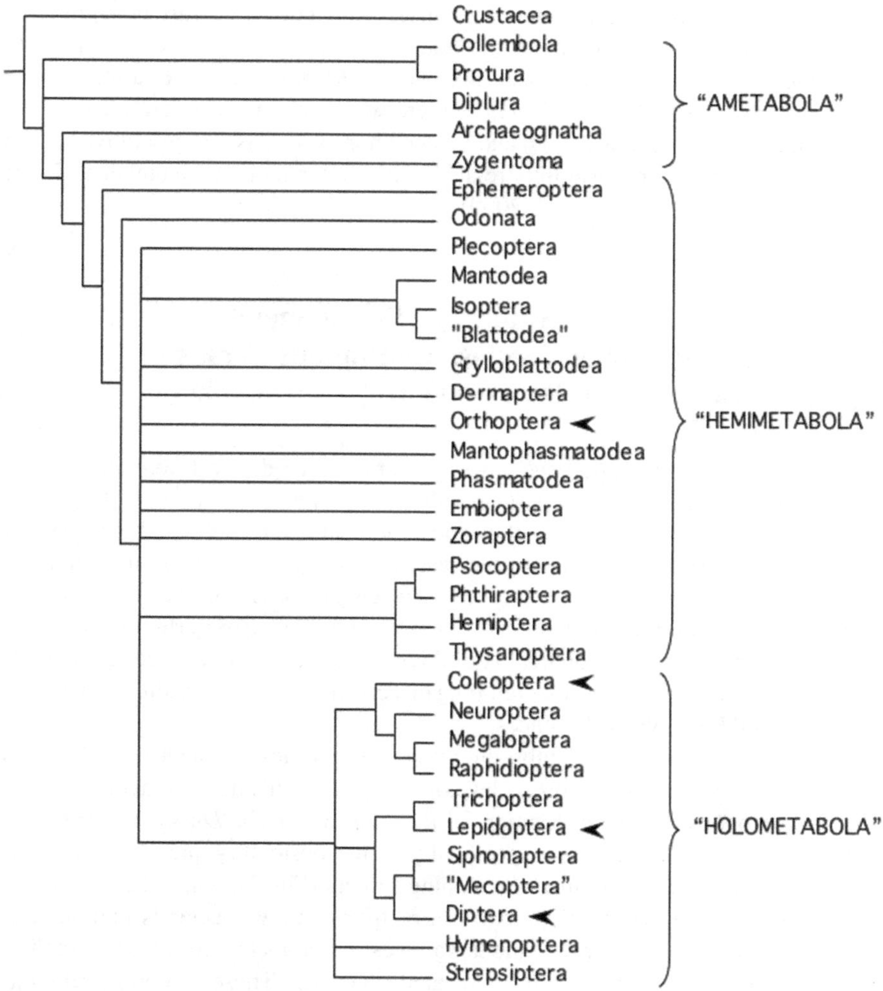

Fig. 1 Phylogenetic framework. *Arrowheads* indicate groups that include model system used in studies of insect eye development. *Quotation marks* indicate paraphyletic groups. Ametabolous insects are primitively wingless and undergo less postembryonic changes than hemi- and holometabolous forms. (Adapted from Friedrich et al. 2006)

hemimetabolous direct developers like orthopterans due to the final differentiation of the wings in the transition from the last nymphal growth instar to the adult. The Orthoptera are considered to have split at least 350 million years ago from the lineage that eventually gave rise to the ancestor of the large superclade of endopterygote or holometabolous insects, which transition through a larval growth stage and the pupal-resting stage before acquiring adult morphology (Beutel et al. 2011; Kristensen 1999; Figs. 1 and 2).

Besides *Drosophila*, holometabolous insects include three further significant models of insect eye development: the flour beetle *Tribolium castaneum*, the silk moth *Bombyx mori*, and the tobacco hornworm *Manduca sexta*. As a representative of the Coleoptera (beetles), *Tribolium* represents one of the oldest orders in the Holometabola, while the silk moth and tobacco hornworm, as representatives of the order Lepidoptera, are more closely related to the dipteran order (Beutel et al. 2011; Kristensen 1999; Wiegmann et al. 2009).

Comparing Drosophila Adult Eye Development with Direct-Developing Species: Continuous Versus Biphasic Visual System Development

The comparison of compound eye development between direct-developing species and the holometabolous *Drosophila* requires the pointing out of homology relationships between specific phases of eye development, which are not obvious at first glance (Fig. 2). In direct-developing species, a significant part of the adult compound eye differentiates already in the embryo. As a result, about 20 % of the posterior adult compound eye is of embryonic origin. The remaining anterior portion is added on during postembryonic development (Friedrich 2006). This mode of compound eye development is typical of direct-developing insects where larval and adult form shows relatively mild body plan differences.

Importantly, although the embryonic phase of eye development contributes to structures of the adult eye in direct-developing species, this developmental process is not homologous to the development of the adult eye in the *Drosophila* eye disc. The latter corresponds, instead, specifically to the postembryonic phase of compound eye development in direct-developing insects (Fig. 2), while the embryonic phase of compound eye development in direct-developing species is homologous to the embryonic development of the larval eyes of holometabolous insects such as the *Drosophila* Bolwig organs (see associated Chap. 12). These homology relationships follow from comparative morphogenetic and molecular evidence (Friedrich 2006, 2008) and, as will emerge later, have important consequences regarding the comparison of retinal primordium-patterning mechanisms.

The postembryonic phase of eye development in direct-developing insects is, thus, the closest evolutionary reference point for comparisons with the development of the *Drosophila* compound eye. Notwithstanding this, it remains a meaningful and evolutionarily significant question to ask whether and to which extent mechanisms regulating the commitment and differentiation of retinal precursor cells during the embryonic phase of eye development in direct-developing insects are recapitulated in the de novo development of the retinal primordium of *Drosophila* eye disc.

Direct-developing insects also differ from holometabolous insects with respect to the transition from embryonic to postembryonic visual development. In direct-developing insects, this transition proceeds with continued retinal differentiation.

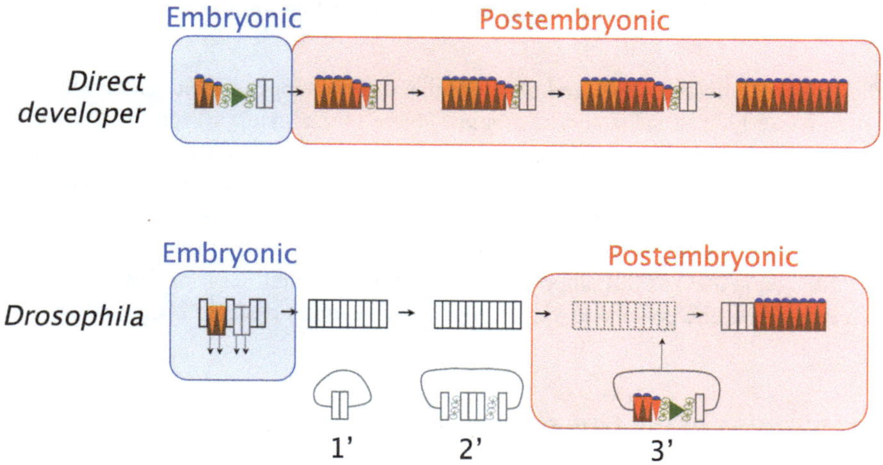

Fig. 2 Homology of embryonic and postembryonic visual system development between direct-developing species and *Drosophila*. Conceptual alignment of homologous phases of visual system development in the direct-developing species and the holometabolous *Drosophila*. In direct-developing species, ommatidia develop during both embryogenesis (*blue* backdrop shade) and postembryogenesis (*red* backdrop shade). Ommatidia of both embryonic (*orange* cell bodies) and postembryonic (*red* cell bodies) origin become part of the adult eye. In *Drosophila*, the development of the visual system is split in two discrete phases. The embryonic phase produces larval eyes, which are not integrated into the adult eye. The postembryonic phase begins with the initiation of retinal determination and differentiation in the eye–antennal imaginal disc of the third (3') larval instar. As a result, the adult *Drosophila* eye consists entirely of postembryonic ommatidia. The eye–antennal disc precursor disc separates from the larval epidermis during embryogenesis and experiences continued growth during the first (1') and second (2') larval instar. During metamorphosis, the eye–antennal imaginal disc derivatives completely replace the larval epidermis during pupation. Apoptosis of larval epidermis is indicated by *dotted outlines*. Color code of cellular components: *gray* = epithelial cells which persist from the embryo into adult, *black* = epithelial cells which are disposed during postembryogenesis, *dark blue* = cone cells, *brown* = pigment cells, *orange cones* = embryonic photoreceptor cells, *red cones* = postembryonic photoreceptor cells, *green* = mitotic cells. Progressing front of retinal differentiation is represented by forward pointing *green arrowhead*

In holometabolous insects, however, larval and adult eye development are temporally and spatially separate processes (Fig. 2). It has been hypothesized that the developmental evolution of this separation began with the transient arrest of retinal differentiation (Dong and Friedrich 2010). In support of this, a transient arrest of retinal differentiation can be enforced by the specific manipulation of eye developmental regulators in direct-developing insects like grasshopper (Dong and Friedrich 2010). Of note, the transient arrest model of biphasic eye development evolution is also consistent with the intermittent developmental arrest of other organs such as the leg appendages in the larval stage of holometabolous insects (Singh et al. 2007; Suzuki et al. 2009).

The American Desert Locust *Schistocerca americana*

The American desert locust and closely related grasshopper species, including the African desert locust *Schistocerca gregaria*, have a long history of serving as experimental models in developmental and neurobiological research due to the accessibility of neural elements in both the embryo and the adult form (Moreaux and Laurent 2007; Rogers et al. 2010; Sanchez et al. 1995). More recently, the grasshopper system has been adopted for the comparative developmental analysis of insect segmentation (Dearden and Akam 2000), appendage development (Mahfooz et al. 2004), and the development of the peripheral visual system (Dong and Friedrich 2005, 2010).

Organization of the Grasshopper Retina

Desert locusts are famous for their voracious food consumption, large body size, and coordinated long distance flights, translating into their economic importance as major pest species (Lomer et al. 2001). These features are supported by an enormous visual system. First instar grasshopper nymphs hatch with compound eyes of close to 2,500 ommatidia (Anderson 1978). This number increases to approximately 9,400 in the adult eye by the addition of new ommatidia at the anterior margin of the eye during the total of 5–6 nymphal intermolt stages (Dong and Friedrich 2010). Grasshopper ommatidia contain a conserved set of 8 photoreceptor cells, 4 cone cells, and 2 primary pigment cells, surrounded by 16 secondary pigment cells (Wilson et al. 1978). The photoreceptor cells exhibit three morphological subtypes. There are two photoreceptors with proximally restricted rhabdomeres, five photoreceptors with rhabdomeres extending along the entire proximodistal axis of the ommatidium, and a single photoreceptor with a distally restricted rhabdomere that corresponds to the *Drosophila* R7 cell (Wilson et al. 1978). Electrophysiological data suggest the presence of green-sensitive, blue-sensitive, and UV-sensitive photoreceptors (Bennet et al. 1967; Vishnevskaya et al. 1985). However, the spatial patterns of opsin gene expression have not yet been investigated, despite the isolation of green-sensitive and UV-sensitive opsin gene family paralogs (Towner et al. 1997). So, it is not yet known whether the grasshopper retina is subdivided into specialized subcompartments. There is, however, a detailed analysis of the retinal organization of the distinct dorsal rim area (DRA) at the dorsal margin of the eye that is populated with anatomically specialized photoreceptor cells (Homberg and Paech 2002). The DRA is a polarized light-sensitive compartment of the insect eye, which is found with varying outlines including the DRA in *Drosophila* (Labhart and Meyer 1999).

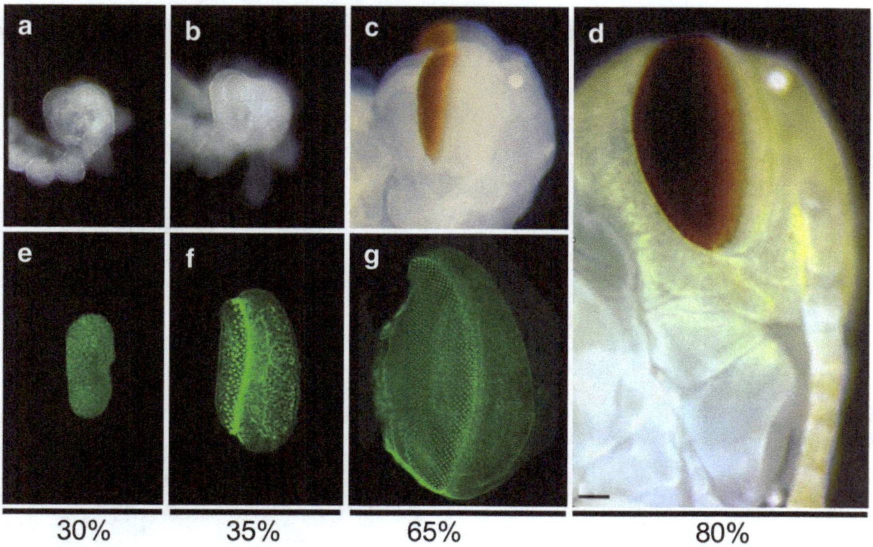

Fig. 3 Embryonic eye development in the grasshopper *S. americana*. **a–d** Lateral stereomicroscopy view of embryonic head at 30 % (**a**), 35 % (**b**), 65 % (**c**), and 80 % (**d**) of embryonic development. **e–g** Laser-scanning confocal images of differentiating embryonic retina labeled with phalloidin, which highlights cell morphogenesis by binding to f-actin, at respective stages of development. A morphogenetic furrow-like differentiation front can be seen starting from 35 % of development (**f**)

Embryonic Phase of Grasshopper Eye Development

The embryonic development of grasshopper species like *S. gregaria* takes about 20 days, which means that development advances by approximately 5 % per day (Bentley et al. 1979). At about 20 % embryogenesis, the grasshopper embryo has formed a distinct head region with two prominent lateral extensions, i.e., the head lobes. The posterior region of the head lobes will then transform to produce a secondary set of lobe-like compartments that are exclusively occupied by precursor tissue of the visual system. These compartments are the eye lobes (Fig. 3a; Dong et al. 2003; Roonwal 1936). The outermost epithelial layer of the eye lobes represents the precursor tissue, i.e., primordium of the retina. In addition, the optic lobes house the developing outer and inner optic neuropiles: lamina, medulla, and lobula (Dong et al. 2003).

Retinal differentiation initiates between 30 and 35 % of development, leading to the formation of a morphogenetic furrow-like front of differentiation, which travels across the eye lobe ectoderm from posterior to anterior (Fig. 3b, f). Of note, the nonhomology of embryonic eye development in direct-developing insects and the *Drosophila* eye–antennal imaginal disc implies that the similarity of the *Drosophila* morphogenetic furrow and the differentiation front in the grasshopper embryonic eye lobe ectoderm reflects generic cell morphological consequences of neurogenesis in cellular epithelia.

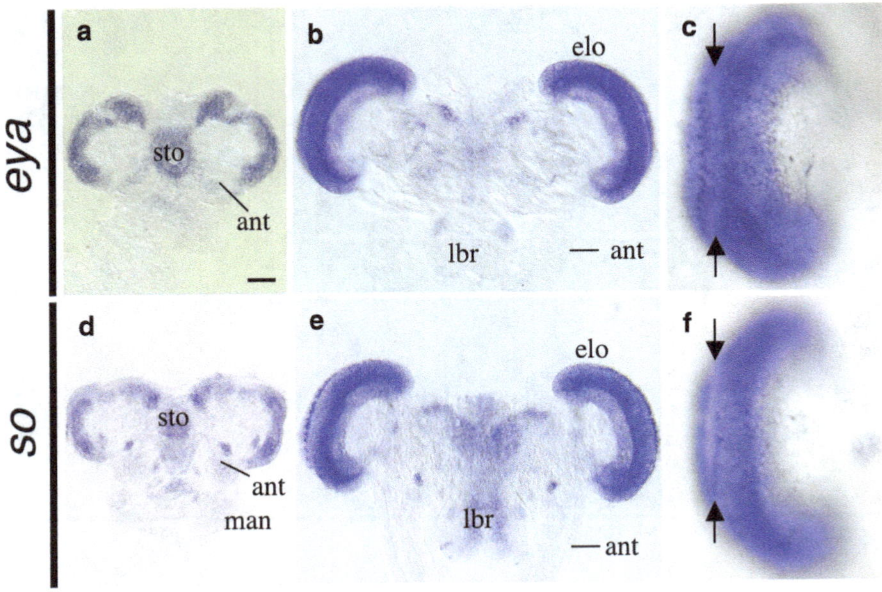

Fig. 4 Expression of *eya* and *so* in the grasshopper eye lobes. **a**, **b**, **d**, **e** Frontal view of grasshopper embryonic head. Dorsal *up*. **c**, **f** Optical section of eye lobe from a lateral perspective at the level of the peripheral ectoderm. Specimens labeled by whole mount in situ hybridization for transcript detection of *eya* (**a–c**) and *so* (**d–f**). *Black arrows* indicate retinal front of differentiation. Dorsal *up* and anterior to the *right*. *ant* antenna, *elo* eye lobe, *lbr* labrum, *man* mandible, *sto* stomodeum

Coexpression of so *and* eya *in the Grasshopper Embryonic Eye Lobes*

The transcription factor genes *eyes absent* (*eya*) and *sine oculis* (*so*) represent the earliest markers of the visual anlage in the *Drosophila* embryo, a neuroectermal field in the median head that contains the precursor cells of the entire visual system (Chang et al. 2001). Consistent with a conserved function of *eya* and *so* in the specification of the embryonic visual anlagen, the grasshopper orthologs of *so* and *eya* are coexpressed in the periphery of the head lobes and, thus, soon after grastrulation (Dong and Friedrich 2005; Fig. 4a, d). As the optic lobes emerge, *eya* and *so* continue to be strongly coexpressed in the retina, lamina, and medulla tissue layers (Fig. 4b, c, e, f).

After the initiation of retinal differentiation, *eya* and *so* are detected throughout the differentiating retina and the morphogenetic furrow as well as extending into a wide area of the undifferentiated neuroectoderm ahead of the morphogenetic furrow (Fig. 4d, f). The *eya* and *so* expressing field ahead of the furrow is limited to a range defined by its distance to the morphogenetic furrow. This observation, and the gradient-like decrease of the *eya* and *so* expression levels toward the anterior margin of their coexpression domain, have been taken as circumstantial evidence that the expression of *eya* and *so* may be primarily transcriptionally activated by signals emanating from the morphogenetic furrow in a manner comparable to the induction of the preproneural (PPN) field in the *Drosophila* eye disc (Bessa et al. 2002; Dong and Friedrich 2005; Greenwood and Struhl 1999).

In *Drosophila*, the PPN field is activated through the long-distance signaling impact by the Transforming Growth Factor β homolog *decapentaplegic* (*dpp*; Heberlein et al. 1993), which is associated with the strong and specific expression of *dpp* in the morphogenetic furrow. In the grasshopper, however, *dpp* is not expressed in the morphogenetic furrow (Friedrich and Benzer 2000). Instead, a low transcript level of *dpp* is detected throughout the anterior eye lobe ectoderm ahead of the morphogenetic furrow (Fig. 8). While *dpp* may function in this domain as a growth activating factor, this pattern rules out a similar furrow movement organizing function as in the *Drosophila* eye–antennal disc. That leaves the signaling factor *hedgehog* (*hh*) as a candidate inducer of the PPN expression domain in the grasshopper based on the *Drosophila* paradigm (Heberlein et al. 1993; Ma et al. 1993). The expression of *hh* in the grasshopper eye lobe remains to be explored, but this scenario is supported by the reported expression of *hh* in crickets (see further text; Niwa et al. 2000).

Expression and Function of wg

The investigation of the complex expression patterns of the signaling factor *wingless* (*wg*) in the grasshopper has produced evidence that *wg* functions as an antagonist of *eya* and *so* transcription at the anterior poles of the embryonic eye lobes, very similar to the situation in the anterior eye–antennal disc of *Drosophila* (Dong and Friedrich 2005; Pichaud and Casares 2000). In the embryonic eye lobe, *wg* is expressed in two prominent polar domains (Friedrich and Benzer 2000; Liu et al. 2006). In these areas, *eya* as well as *so* expression seems to be nonoverlapping with *wg* (Fig. 5).

The suggested repressive effect of *wg* in retinal specification and differentiation was tested by LiCl incubation experiments with cultured embryonic eye discs (Dong and Friedrich 2005). Through its inhibition of glycogen synthase kinase 3β, LiCl application is known to stimulate Wg signaling (Stambolic et al. 1996). In cultured eye lobes, the addition of LiCl caused a stalling of retinal differentiation. This was associated with a strong increase of cell division anterior to the morphogenetic furrow and strong increase of cell death, specifically posterior to the morphogenetic furrow (Dong and Friedrich 2005). These findings are consistent with the role of *wg* as a growth activator in the anterior *Drosophila* eye disc and its impact on differentiation in the posterior *Drosophila* eye disc (Baonza and Freeman 2002; Lee and Treisman 2001; Treisman and Rubin 1995), suggesting deeply conserved functions of *wg* in the control of retinal patterning.

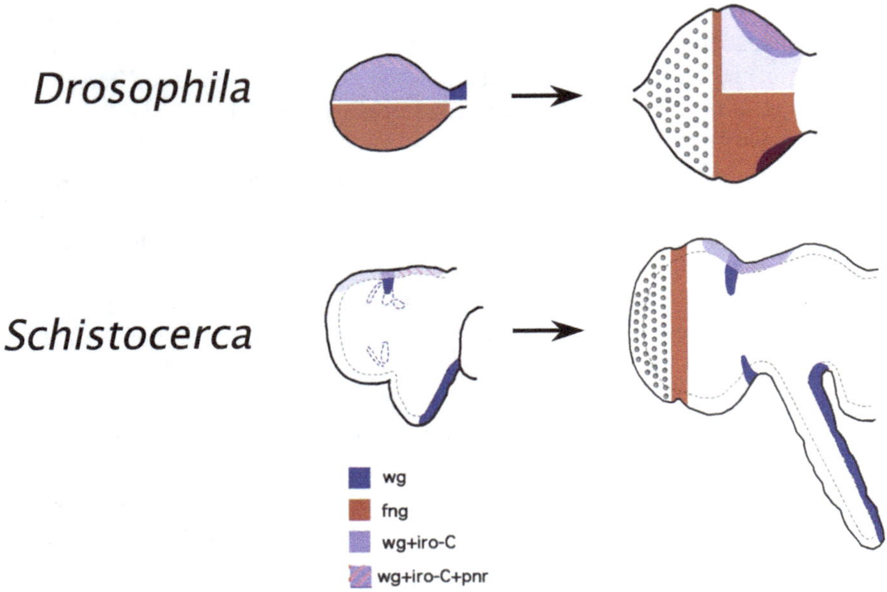

Fig. 5 Dorsoventral patterning gene expression in *Drosophila* and grasshopper. Schematic comparison of the expression domains of *wg* and *fng* as well as areas with overlapping expression of *wg* with *Iro-C* or *wg* with *Iro-C* and *pnr*. *Left column* shows the *Drosophila* eye disc and the grasshopper head hemisphere at an early developmental stage that precedes the onset of retinal differentiation (2nd larval instar eye–antennal imaginal disc in *Drosophila* and 30 % stage of *Schistocerca*). The *right column* compares the late 3rd larval instar eye imaginal disc of *Drosophila* with the *left* grasshopper head hemisphere at about 45 % stage of *Schistocerca* embryo. Dorsal *up* and anterior to the *right*. (Adapted from Dong and Friedrich 2005)

Dorsoventral Patterning

In *Drosophila*, the activation of focal Notch (N) signaling along the midline of the early eye disc is essential for stimulating the rapid expansion of the eye primordium by cell proliferation (Cho and Choi 1998; Dominguez and de Celis 1998; Dominguez et al. 2004; Kenyon et al. 2003; Papayannopoulos et al. 1998). In addition, the differential expression of N-signaling components in, precisely, the dorsal or ventral half of the eye disc anticipates the compartmentalization of the adult eye into dorsoventral compartments (Reifegerste and Moses 1999). Together with *wg*, the analysis of the expression of the grasshopper homologs of the N-signaling modifier glycosyltransferase *fringe* (*fng*), and the transcription factor genes *Delta* (*Dl*), *pannier* (*pnr*), and *Iroquois-C* (*Iro-C*) provided insights into the dorsoventral patterning organization of the grasshopper eye (Dong and Friedrich 2005).

Similar to the *Drosophila* situation (Cavodeassi et al. 1999, 2000; Maurel-Zaffran and Treisman 2000), *pnr* and *Iro-C* are expressed in dorsal cell populations of the embryonic head. However, in contrast to *Drosophila*, the expression of *pnr* remains outside the eye lobes, representing an extension of the dorsal margin cells. Further, the expression of *Iro-C* extended only 10 % into the dorsal of the anterior embryonic eye lobe, consistent with a role in patterning the grasshopper DRA ommatidia but incompatible with a role in subdividing the retina field into a dorsal and ventral half. In combination, the data indicate conserved genetic mechanisms in DRA specification but divergence with regards to the dorsoventral patterning in the retina of grasshopper and *Drosophila* (Fig. 5). Also, in further support of the latter notion as well as the lack of a N-induced growth-promoting organizer in the embryonic grasshopper eye, the expression of *Dl* and *fng* shows no evidence of dorsoventral compartmentalization ahead of the morphogenetic furrow or prior to its initiation (Fig. 5; Dong and Friedrich 2005). Instead, the expression of these genes is associated with the initiation and progression of the morphogenetic furrow itself indicating roles in regulating the progress of neural differentiation.

Postembryonic Phase of Grasshopper Eye Development

During the transition from embryonic to postembryonic development, the retinal precursor cell population of the anterior eye lobe neuroectoderm transforms into a growth zone margin, outlining the anterior edge of the nymphal eye in direct-developing insects like *S. americana* (Figs. 2 and 6a, b; Dong et al. 2003; Friedrich 2006). The cellular organization of the growth zone, which is heavily enriched with mitotic cells, has been described in early histological and experimental papers (Anderson 1978; Bodenstein 1953). Today, it is interesting to note its organizational similarity to the ciliary margin region of the fish or amphibian eye (Perron et al. 1998; Raymond et al. 2006). Posterior to the proliferation zone, the transition into the fully differentiated retina is filled with intermediate stages of ommatidial development defining the differentiation zone (Fig. 6b; Anderson 1978; Dong and Friedrich 2010).

Unfortunately, the molecular organization of the grasshopper eye proliferation zone is still little investigated. Yet, RNAi-mediated gene knockdown experiments targeting *eya* and *so* produced first insights into the function of eye selector genes during postembryonic eye development in the grasshopper (Dong and Friedrich 2010). For both genes, a transient arrest of postembryonic retina differentiation was observed in nymphs which completed development into adult form, generating adult eyes with a pronounced vertical scar area (Fig. 6). These findings were interpreted as suggesting that the downregulation of *so* and *eya* does not irreversibly affect the organization of the mitotic activity in the growth zone (Dong and Friedrich 2010). Thus, *eya* and *so* have been proposed to act in a similar manner in the postembryonic grasshopper eye, as in the PPN zone of the *Drosophila* eye disc, by making cells responsive and competent to undergo retinal differentiation.

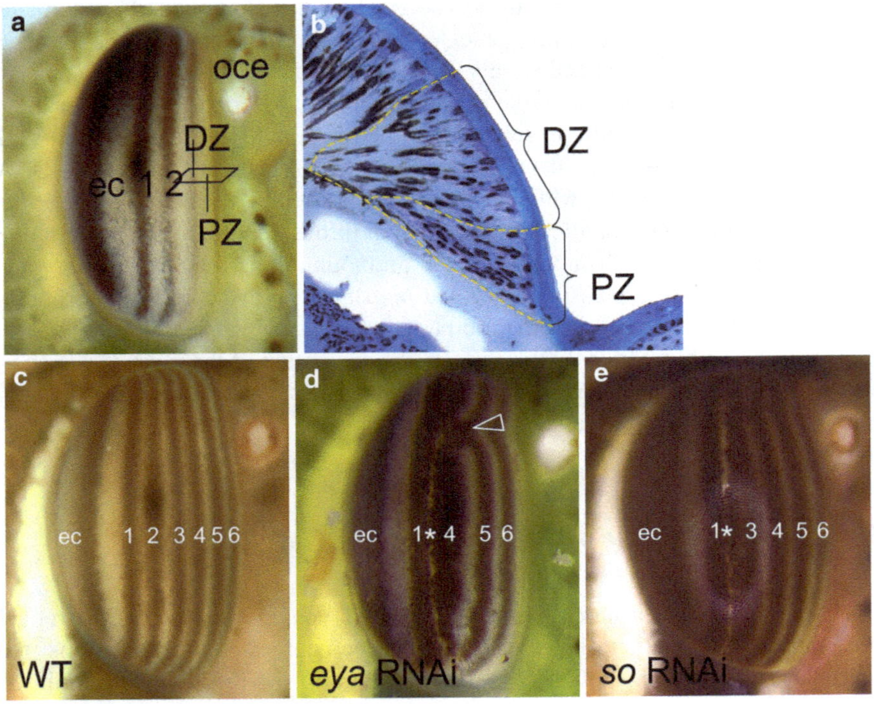

Fig. 6 Effect of *eya* and *so* knockdown on the postembryonic development of the grasshopper compound eye. **a** Frontolateral view of fourth instar grasshopper nymphal eye. Relative position of differentiation zone (*DZ*) and proliferation zone (*PZ*) are indicated and related to section plane of panel **b**. The posterior *dark* pigmented region of the eye that is generated in the embryo is labeled as the embryonic cap (*ec*). Numbers label pigment stripe areas formed during postembryonic retina differentiation in the first two nymphal instars. **b** Toluidine blue stained sagittal semithin section through the anterior compound eye of a first instar grasshopper nymph. Cells in the DZ elongate and accumulate pigment. Cells in the PZ are densely packed and indifferentiated. **c–e** Lateral view of the adult compound eye. **c** Untreated wild type animal. **d** Strongly affected *eya* knockdown animal. *Asterisk* in panel **d** indicates position of scar between stripes *1* and *4*. *Arrowhead* in **d** points at disrupted anterior stripe pattern. **e** Phenotypic *so* knockdown animal. *Asterisk* indicates position of scar between stripes *1* and *3*. In all panels anterior is to the *left* and dorsal *up*. Numbers identify specific lateral pigment stripes. *ec* embryonic cap, *gen* gena, *oce* ocellus. (Adapted from Dong and Friedrich 2010)

The Bispotted Cricket *Gryllus bimaculatus*

Driven by a major effort in developing tools for molecular analysis, including whole mount in situ hybridization, RNAi-mediated gene knockdown, and germline transformation, the cricket *G. bimaculatus* has evolved into a versatile and efficient model system for comparative development (Fig. 7; Mito and Noji 2008). With regards to vision-mediated behaviors, it is noteworthy that crickets are generally crepuscular and less prominent in the aerial insect fauna. Despite the fact that crickets do not

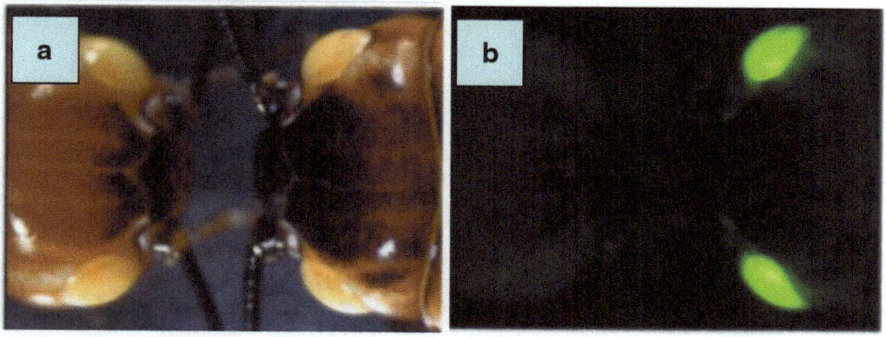

Fig. 7 Eye morphology of the cricket *Gryllus bimaculatus*. **a** Stereomicroscope view of dorsal head of white-eyed wild type (*left*) and transgenic (*right*) animal. **b** Epifluorescence image of the same, note strong EGFP expression in the compound eye of the transgenic animal. (Kindly provided by Dr. Sumihare Noji)

exhibit flight behavior under laboratory conditions unless artificially stimulated, female crickets are known for their extensive prereproductive flight dispersal, mostly at evening hours (Lorenz 2007).

Organization of the Cricket Retina

The eyes of adult *G. bimaculatus* consist of approximately 4,600 ommatidia (Labhart and Keller 1992). Like in the grasshopper, the *G. bimaculatus* eye includes a structurally and functionally distinct DRA, which is populated by blue-opsin and UV-opsin expressing photoreceptors (Blum and Labhart 2000; Henze et al. 2012). The recent analysis of opsin gene expression patterns in the cricket uncovered further compartmentalization in the retina (Henze et al. 2012). Accordingly, the *G. bimaculatus* main retina encompasses a blue-opsin and green-opsin expressing ventral area while the remainder of the retina expresses UV-opsin and green-opsin. The photoreceptor-specificity, as well as the ecological significance of these differential opsin expression patterns, awaits future study.

Patterning Gene Expression and Function During the Embryonic Phase of Cricket Eye Development

The early developing cricket visual system is organized in the same way as the eye lobe compartments in grasshoppers (Inoue et al. 2004). Likewise, in correspondence to the organization in the grasshopper, retinal differentiation is initiated in the posterior margin of the eye lobe ectoderm and a morphogenetic furrow-like front of

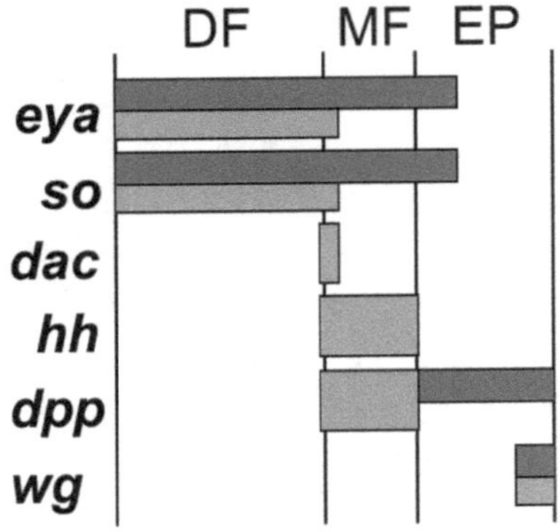

Fig. 8 Summary of eye developmental expression patterns in orthopteran species. *Gray* expression domain in cricket, *black* expression domain in grasshopper. *DF* differentiating retina, *EP* eye primordium, *MF* morphogenetic furrow

differentiation travels the cricket eye lobe neuroectoderm in posterior to anterior direction (Inoue et al. 2004; Takagi et al. 2012).

The available expression data on the cricket homologs of *wg*, *hh*, and *dpp* suggest that *wg* is expressed in the anterior margins of the eye lobe, while *hh* and *dpp* are expressed in different dorsoventral domains across the eye (Fig. 8; Niwa et al. 2000). *hh*, in particular, appears to be strongly expressed in the differentiating retina (Niwa et al. 2000). These data are prima facie consistent with conserved roles of *dpp* and *hh* in promoting eye development, and the grasshopper supported conserved role of *wg* as tissue growth-stimulating antagonist of retinal differentiation (Friedrich 2006; Liu et al. 2006).

At the transcription factor gene level, the expression of *so* and *eya* as well as *dachshund* (*dac*) has been studied in detail (Fig. 8; Inoue et al. 2004; Takagi et al. 2012). The expression of *dac* is detected in the eye lobe neuroectoderm prior to morphogenetic furrow initiation (Inoue et al. 2004). In the differentiating eye, *dac* transcript levels are concentrated in the morphogenetic furrow yet below detection level both anterior and posterior to the morphogenetic furrow (Inoue et al. 2004).

The *so* and *eya* orthologs of the cricket are strongly expressed in the nondifferentiated area of the eye lobes prior to the initiation of eye differentiation (Takagi et al. 2012). Thereafter, *so* and *eya* expression extends from the morphogenetic furrow uniformly across the differentiating retina in the posterior head lobe, much the same as in grasshopper. However, the expression of *so* and *eya* seems more confined anterior to the morphogenetic furrow raising the possibility of differences in the transcriptional organization of retinal induction between the two species (Fig. 8). Consistent with the predicted important function of *eya* in specification and differentiation of the eye during embryonic development, parental RNAi-mediated knockdown resulted in strong eye depletion phenotypes, including complete loss (Takagi et al. 2012).

Expression and Function of eya *and* so *During the Postembryonic Phase of Cricket Eye Development*

The role of *eya* and *so* has also been studied in the nymphal eye of *G. bimaculatus* (Takagi et al. 2012). This analysis revealed the presence of defined anterior proliferation and differentiation zones as in the nymphal eye of grasshopper. In situ hybridization analysis of the expression of *eya* revealed the differential accumulation of transcripts in the proliferation zone and posterior to it, in both differentiating and differentiated pigment cells (Takagi et al. 2012). The RNAi-mediated knockdown of *eya* or *so* by dsRNA injection into third instar nymphs resulted in highly informative phenotypes. In the strongest *eya* knockdown animals, the proliferation zone appeared completely missing in contrast to the preservation of the growth zone in the corresponding *eya* knockdown experiments with grasshopper. Moreover, the posterior retina region of the cricket, which had differentiated prior to injection, reorganized into a nonsensory head cuticle (Takagi et al. 2012).

While these data are consistent with the expected role of *eya* in specification and differentiation of the postembryonic cricket eye, the mechanism explaining its role in the maintenance of the differentiated state will require further investigation. In contrast to grasshopper, the data suggest that *eya* and *so* are not only essential for the differentiation of the nymphal retina but also for the maintenance of the proliferation zone. Before mechanistic conclusions can be drawn with confidence, it will be important to address whether these differences reflect differences in gene knockdown efficiencies, stage of the injected nymphs, or lineage-specific differences in regulatory mechanisms.

Comparing Drosophila Adult Eye Development with Other Holometabolous Species: Early Versus Late Eye Discs

The physical separation of the products of embryonic and postembryonic eye development in holometabolous species dominates the comparison of *Drosophila* to direct-developing species (Fig. 2). The comparison of eye development within holometabolous species attracts interest because of the dramatic differences in the morphogenetic organization of postembryonic eye primordium formation (Fig. 9). In the most ancestrally organized Holometabola, the retina differentiates in the lateral head epidermis of the adult-like head capsule of the eucephalic larva. Pending the size of the prospective adult eye, this can be associated with the formation of an eye disc during metamorphosis, i.e., the last larval instar and the pupa. This contrasts with the early formation of the *Drosophila* eye–antennal imaginal disc during embryogenesis.

Correlated with this, there is a second fundamental morphogenetic difference between the ancestral late eye disc formation and the early eye disc development in *Drosophila*. In the first case, the eye disc is the growth-accommodating intermediate

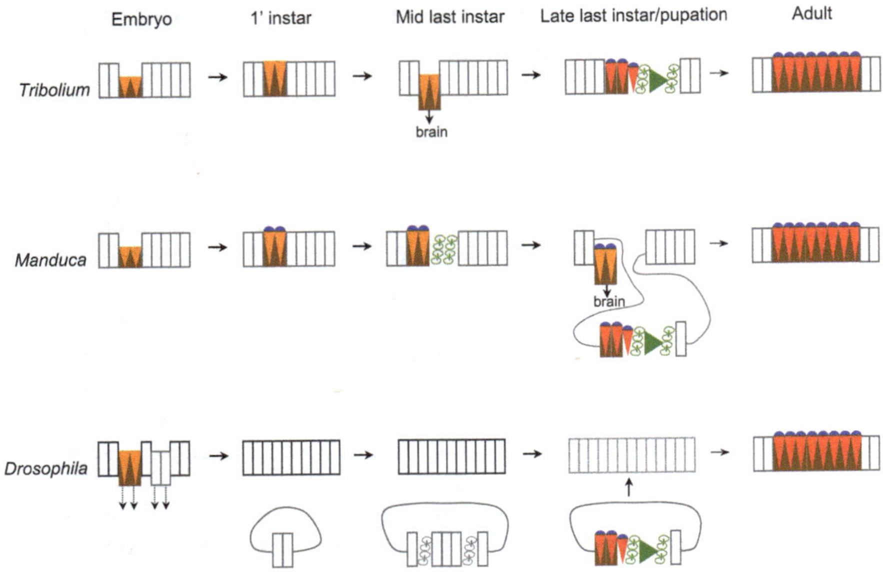

Fig. 9 Early and late eye disc formation in holometabolous insects. Cell body color-coding as in Fig. 2. Note the differentiation of photoreceptors with cone cells in *M. sexta*. In *Tribolium*, the adult retina differentiates in the lateral head epidermis without eye disc formation. In *Manduca*, a later eye disc is formed in the last larval instar and the pupa. The *Drosophila* eye–antennal imaginal disc is an example of early imaginal disc formation in the embryo

structure of single organ. In the second case, the eye–antennal imaginal disc functions as the precursor structures of many head cuticle structures and sensory organs (see also Fig. 15). This has the effect that organ-specific primordium have to be patterned via postembryonic regional specification in addition to their coordinated growth (for review, see Dominguez and Casares 2005). This compaction of head patterning processes into a single composite imaginal disc represents a derived state that emerged during the evolution of the acephalic morphology of the maggot-type larva (Melzer and Paulus 1989). The latter characterizes not only *Drosophila* and closely related flies but also one of the larger groups of the Diptera: the Cyclorrhapha. The early eye disc of *Drosophila* and other cyclorrhaphan flies, thus, represents an evolutionary novelty at the level of developmental precursor tissue organization.

The Red Flour Beetle *Tribolium castaneum*

The publication of the genome sequence in 2008 cemented the pivotal position of *Tribolium* in comparative evolutionary developmental biology (Klingler 2004; Richards et al. 2008). The recent surge in *Tribolium* research benefited profoundly from earlier genetic and population genetic studies exploring the biology of this major economic pest (Sokoloff 1972). The taxonomic significance of *Tribolium*

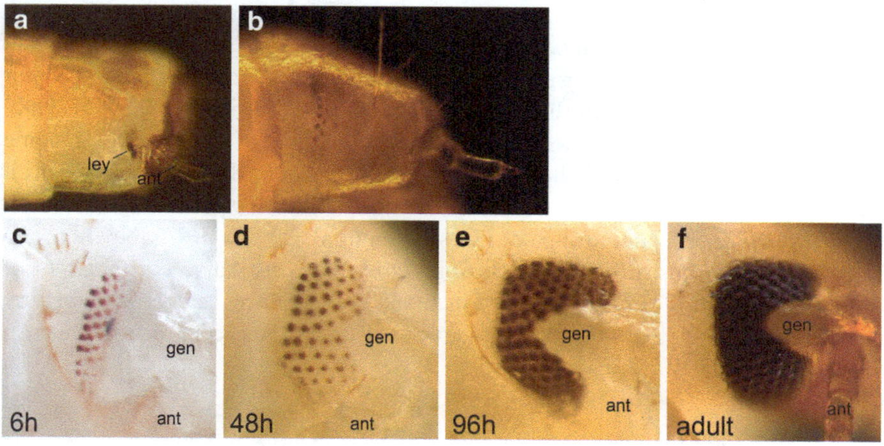

Fig. 10 Adult eye development in *Tribolium*. **a** Lateral view of last instar larval head before entering the resting stage. Note position of larval eyes (*ley*) posterior to the antenna (*ant*) and the gena (*gen*). **b** Lateral view of resting stage larva. The larval eyes have relocated from their antenna-associated position toward the brain (not shown). The first two rows of photoreceptors, visible by virtue of their pigment accumulation, have become visible in the posterior half of the lateral head capsule. **c–f** Lateral view of pupal (**c–e**) and freshly hatched adult (**f**) *Tribolium* head. (Adapted from Liu and Friedrich 2004; Yang et al. 2009b)

arises from representing the largest order of insects (Coleoptera) and the intermediate phylogenetic position between *Drosophila* and hemimetabolous insects (Fig. 1; Kristensen 1999; Savard et al. 2006; Wiegmann et al. 2009). These aspects and the short germband type of embryonic development have attracted considerable interest by comparative developmental biologists, leading to the development of refined and effective protocols for in situ hybridization, RNAi-mediated gene knockdown, transgenesis (Brown et al. 2009), and most recently, ectopic gene expression (Schinko et al. 2012). *Tribolium* has been used to gain insights into early embryonic patterning (Schroder 2003), segmentation (Maderspacher et al. 1998), appendage (Prpic et al. 2001), and head development (Posnien et al. 2010), including the visual system (Liu and Friedrich 2004).

Organization of the Tribolium *Compound Eye*

A first notable difference of the *Tribolium* eye to *Drosophila* is its smaller size: an average of 95 ommatidia in the *Tribolium* eye compared to the 800 ommatidia in the *Drosophila* eye (Fig. 10f; Friedrich et al. 1996). This size difference can be attributed to the crepuscular biology of *Tribolium*, which tends to spend much of its life span burrowed in nutritional substrate (Park 1934). However, recent studies document a previously underestimated frequency of flight-facilitated adult dispersal (Perez-Mendoza et al. 2011; Ridley et al. 2011). A second eye-catching difference between the *Tribolium* and *Drosophila* eye is the midline notch at the anterior margin of the *Tribolium* eye, accommodating a posteriorly extended gena (Fig. 10e, f).

At the cellular level, the fused rhabdom formed by the *Tribolium* photoreceptor cells contrasts with the open rhabdom in *Drosophila* (Friedrich et al. 1996). Only two compound eye vision-related opsin genes are conserved in the *Tribolium* genome (Richards et al. 2008). This includes a green-sensitive opsin, which is expressed in all retinal photoreceptor cells, and a UV-sensitive opsin, which is specifically conserved in the *Tribolium* R7 photoreceptors (Jackowska et al. 2007). In combination, the *Tribolium* retina thus differs from *Drosophila* by the constitutive coexpression of opsin paralogs in all ommatidia. The functional consequences and gene regulatory mechanisms associated with this unique retinal opsin mosaic have not yet been investigated in detail.

Morphogenesis of the Tribolium *Compound Eye*

Like *Drosophila*, *Tribolium* develops a separate pair of lateral larval eyes in the embryo that are structurally very distinct from the adult compound eye. The larval eyes are situated close to the larval antenna from where they withdraw into the brain during metamorphosis (Fig. 10a, b; see Chap. 12 for further details; Liu and Friedrich 2004). The relative small size of the adult *Tribolium* eye allows for the differentiation of the retina in the lateral head epithelium without the detachment of the latter from the head cuticle (Figs. 9 and 10). Due to the early accumulation of retinal pigment granules in differentiating photoreceptor cells, the morphogenesis of the *Tribolium* compound eye can be conveniently followed by external observation (Fig. 9; Friedrich et al. 1996; Liu and Friedrich 2004). The first row of photoreceptors are recognizable at the end of the last larval instar (Fig. 10b), in preparation of pupation. At this point, the larvae enter a similar premetamorphic resting stage that is equivalent to the wandering stage of the *Drosophila* larva. In the case of *Tribolium*, however, the larvae simply remain motionless without food uptake (Parthasarathy et al. 2008).

In the freshly hatched pupa, the number of photoreceptor columns extends in the anterior direction along the longitudinal body axis over the first 48 h after pupa formation (Fig. 10c, d; Liu and Friedrich 2004; Yang et al. 2009b). In the midline area, the progression of photoreceptor differentiation stalls earlier than in the dorsal and ventral halves (Fig. 10d, e). Investigations of cellular morphogenesis revealed that this process is associated with the split of the contiguous morphogenetic furrow in the midline region (Friedrich and Benzer 2000). About 96 h after pupa formation, the retinal field becomes homogeneously filled with dark color following the specification and differentiation of the pigment cells (Yang et al. 2009b).

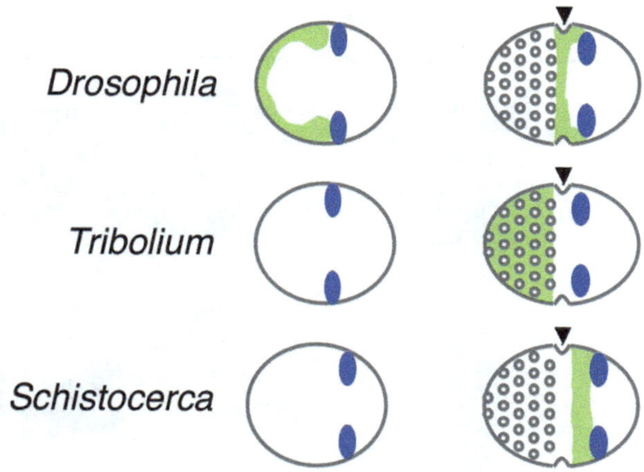

Fig. 11 Comparison of *wg* and *dpp* expression domains in *Drosophila*, *Tribolium*, and *Schistocerca*. *Left column* represents the eye field before the onset of retinal differentiation. The *right column* represents the eye field after the onset of retinal differentiation. *Arrowheads* point at the front of retina differentiation. Posterior to the right. Color code of gene expression domains: *green* = dpp, *blue* = wg. (Modified from Friedrich and Benzer 2000)

Signaling Factor Expression Patterns in the Developing **Tribolium** *Adult Eye*

The first molecular study of *Tribolium* eye development explored the expression patterns of *wg* and *dpp* (Fig. 11; Friedrich and Benzer 2000). Similar to the situation in grasshopper and *Drosophila*, *wg* is expressed in separate dorsal and ventral domains, consistent with evolutionary conservation of the repressive effect of Wg signaling on retinal differentiation in *Drosophila* and the grasshopper (Dong and Friedrich 2005).

The dorsoventral *wg* domains transform into a circumferential domain along the entire retinal field margin at about 36 h after pupal formation, thereby resembling the late expression of *wg* around the *Drosophila* eye (Friedrich and Benzer 2000). These data suggest that *wg* is also involved in eye margin patterning of the *Tribolium* eye, although this has not yet been functionally tested.

The expression of *dpp* in *Tribolium* is different from both grasshopper and *Drosophila* (Friedrich and Benzer 2000). At the onset of retinal differentiation, *dpp* is weakly expressed in the presumptive eye primordium (Fig. 11). After the initiation of retinal differentiation, *dpp* was detected through the entire differentiating retina in a pattern, which suggested the repression of *dpp* specifically in the differentiating photoreceptor cells.

Eye Selector Gene Expression in the Developing **Tribolium** *Adult Eye*

Following the candidate gene approach, the expression and function of *eya*, *so*, *dac*, and the Pax6 transcription factor genes *eyeless* (*ey*) and *twin of eyeless* (*toy*) have been studied in detail with respect to their role in *Tribolium* eye development (Figs. 12 and 13; Yang et al. 2009a, b). All of these genes are expressed in the

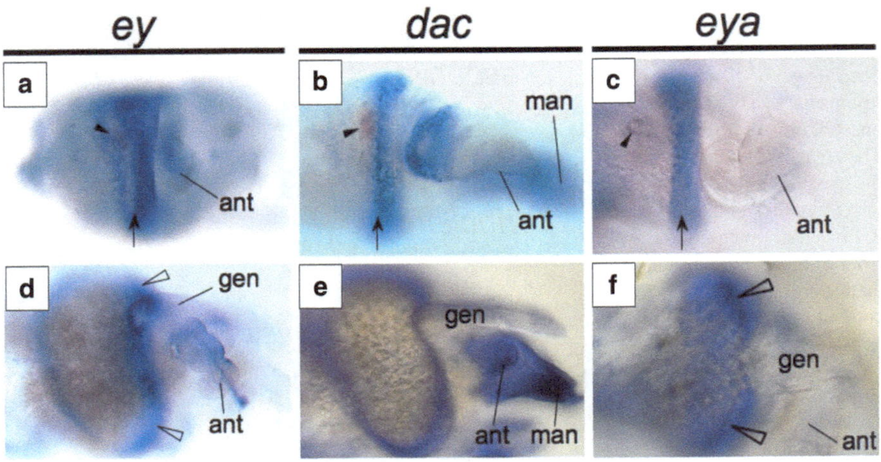

Fig. 12 Developmental transcription factor gene expression in the developing *Tribolium* compound eye. **a–c** Lateral view of dissected last instar larval head. **d–f** Lateral view of pupal head at approximately 48 h after pupal formation. Dorsal *up* and anterior to the *right*. *ant* antenna, *gen* gena, *man* mandible

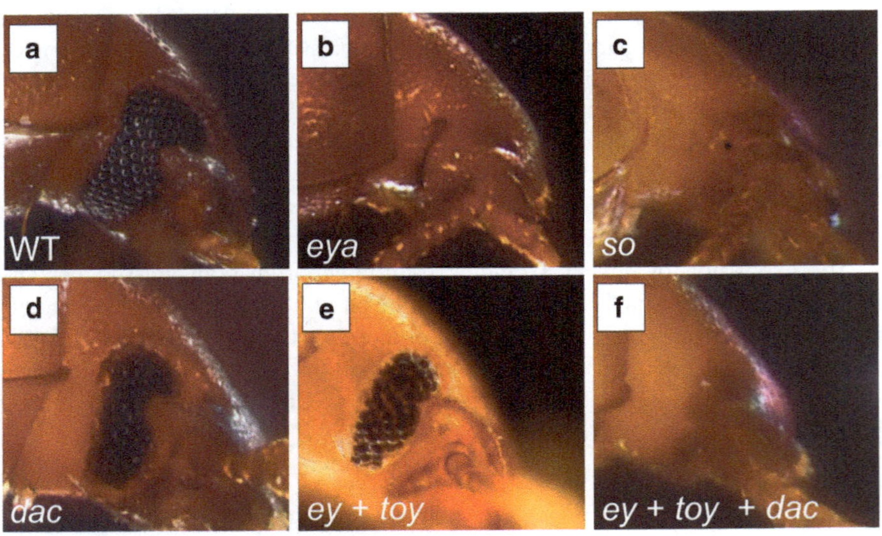

Fig. 13 Eye selector gene expression and function in *Tribolium* compound eye development. **a–f** Lateral view of adult head of wild type (**a**) and strongly phenotypic knockdown animals (**b–f**). See text for details. Dorsal *up* and anterior to the *right* (Adapted from Yang et al. 2009a, b)

undifferentiated eye primordium prior to retinal differentiation and subsequent to the initiation of differentiation ahead of the morphogenetic furrow, suggesting their coexpression in the early eye primordium (Fig. 12a–c). The extent of these expression domains, however, differs. The most restricted expression domain was detected for *eyg* (ZarinKamar et al. 2011). *eya* and *so* appear to be more specifically expressed in the retinal precursor tissue of the lateral head (Fig. 12c). *ey*, *toy*, and *dac*, by contrast, are characterized by wider expression domains, exceeding that of *so* and *eya*, suggesting broader roles in the patterning of the lateral head (Fig. 12a, b; Yang et al. 2009a).

Informative expression pattern differences were also observed in the differentiating retina. While *eya* and *so* continue to be expressed in the developing photoreceptor cells, *ey*, *toy*, and *dac* are downregulated as cells pass through the morphogenetic furrow. These expression dynamics are largely consistent with the expression and function of *eya* and *so* as early retina determination genes versus *toy* and *ey* as upstream specification genes in the *Drosophila* eye–antennal disc (Kumar 2009). Most noteworthy, perhaps, is the higher coordination of *dac* expression with *ey* and *toy* in *Tribolium* (Fig. 12d, e), considering the downstream position of *dac* in the *Drosophila* retina determination gene network.

These three genes are also coexpressed in a domain surrounding the late differentiating *Tribolium* retina, suggesting roles in eye margin patterning (Fig. 12d, e; Yang et al. 2009a, b).

Knockdown Analysis of Tribolium *Eye Development*

Lack-of-function analyses by RNAi have been very informative regarding the roles of *eya*, *so*, *ey*, *toy*, and *dac* in *Tribolium*. The strongest impact of larval RNAi-mediated gene knockdown was observed in the case of *eya* and *so*, which ranged from partial to complete depletion of the compound eye (Fig. 13b, c; Yang et al. 2009b). The analysis of *ey* and *toy*, however, revealed a first major difference of *Tribolium* from *Drosophila*. Knockdown of *ey* or *toy* individually or in combination leads to only a subtle, although significant, decrease in eye size as measured by number of ommatidia (Fig. 13e; Yang et al. 2009a). This result contrasts strongly with the sensitivity of adult head and eye development to the reduction of these genes in *Drosophila* (Kronhamn et al. 2002). However, the combinatorial knockdown of *ey* and *toy* in the developing embryonic head results in a high penetrance larval eye deletion phenotype (Yang et al. 2009a), suggesting similarly important functions of *ey* and *toy* in the developing visual system of *Tribolium* as in *Drosophila*.

In the adult eye, the knockdown of *dac* also yielded only partial reduction of the eye, although more dramatic in comparison to the average of 10 % eye reduction in *ey* and *toy* knockdown animals (Yang et al. 2009a). Most important, the combinatorial knockdown of *ey* and *toy* with *dac* leads to complete eye depletion phenotypes (Fig. 13f; Yang et al. 2009a). The model inferred from these data poses that the Pax6 genes *ey* and *toy* play roles in visual system specification during embryogenesis and remain essential for eye primordium maintenance throughout the postembryonic phase of development in functional redundance with *dac* (Yang et al. 2009a).

An Unexpected Role of eyg *in the* Tribolium *Eye*

The second major deviation in gene function between *Tribolium* and *Drosophila* concerns the role of the Pax gene *eyegone* (*eyg*) (ZarinKamar et al. 2011). Reducing *eyg* levels in the *Drosophila* eye–antennal disc has strong eye depletion effects (Dominguez et al. 2004; Jun et al. 1998). In *Tribolium*, the knockdown of *eyg* leads to the opposite: a 5 % increase in eye size (ZarinKamar et al. 2011). Analysis of the morphogenetic origin of the *eyg* phenotype in *Tribolium* revealed that the morphogenetic furrow is not suppressed in the midline when approaching the introducing gena tissue. In this case, retinal differentiation in the median head appears to gain dominance over the developmental program involved in gena formation. The result is the differentiation of six surplus ommatidia on an average, in the median anterior *Tribolium* eye (ZarinKamar et al. 2011).

Given that *eyg* is not expressed in the gena, it is currently assumed that *eyg* functions as a competence factor that renders the anterior eye field sensitive to retina suppressing factors released by the developing gena (ZarinKamar et al. 2011). Such eye-antagonistic role of *eyg* is striking given the contrast to its facilitating role in the *Drosophila* eye, which leads to the idea that *eyg* may represent a functional homolog of the primordium growth-activating Pax6(a) isoform (Moses and Rodrigues 2004). A parallel investigation into the evolutionary origin of *eyg*, however, showed that *eyg* represents a deeply conserved Pax gene subfamily of its own (Friedrich and Caravas 2011).

The Tobacco Hornworm *Manduca sexta*

Compared to *Tribolium*, the tobacco hornworm *M.sexta* has thus far played a lesser role in the comparative analysis of visual system development. Early work described basic aspects of the differentiation of the retina, which align well with the events in the wake of the morphogenetic furrow in *Drosophila* and other species (Champlin and Truman 1998; Egelhaaf 1988; Friedrich et al. 1996). Even more significant is the body of work, which elucidated the mechanisms that regulate the postembryonic activation of the adult eye primordium (Champlin and Truman 1998; Truman et al. 2006), thereby coordinating eye disc development with other metamorphic events. In vivo and in vitro experiments revealed that the early initiation of the adult eye primordium occurs because nutritional signals mediated through the insulin signal pathway begin to overrule the differentiation-suppressing effect of juvenile hormone (Koyama et al. 2008; Truman et al. 2006).

As mentioned earlier (Fig. 9), *Manduca* is a significant point of comparison in insect eye development because of the late formation of an eye-specific imaginal disc

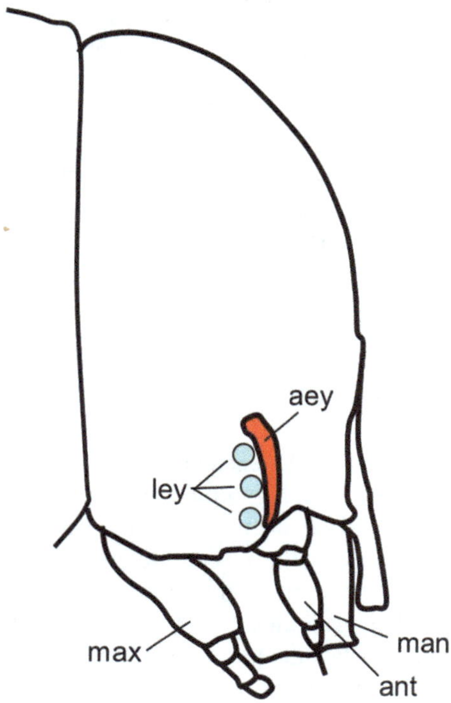

Fig. 14 Spatial organization of adult eye primordium initiation in relation to the larval eyes in *Manduca*. Drawing of lateral view on *Manduca* final instar larval head based on Allee et al. (2006). The adult eye primordium is initiated as a wedge of proliferating tissue anterior to the three ommatidia-like larval eyes (*turquoise*). Dorsal is *up* and anterior to the *right*. *aey* adult eye primordium, *ant* antenna, *ley* larval eye, *man* mandible, *max* maxilla

(Allee et al. 2006; Friedrich 2006; Truman and Riddiford 2002). It is reasonable to assume that the late-forming disc type of *Manduca* resembles an ancestral precursor stage toward the evolution of the *Drosophila* eye–antennal imaginal disc.

Early Development of the Manduca *Compound Eye Primordium*

The adult eye primordium of *Manduca* becomes detectable in the late final instar larva. Morphologically, it has been described as a half moon crest-shaped rim of compacted, proliferating tissue that begins to delaminate from the larval head capsule cuticle, thus forming the eye disc (Fig. 14; Allee et al. 2006; MacWhinnie et al. 2005; Monsma and Booker 1996). This position of the emerging eye disc is notable because it is consistent with the transient arrest model of the larval eyes in holometabolous insects. The latter predicts that the larval eye primordium is initiated as a continuation of larval eye development in the anterior direction (Fig. 2).

Unfortunately, no data are as yet available regarding the expression of head and eye determination genes during eye disc activation in *Manduca*. However, the expression and function of specific isoforms of the zinc finger transcription factor *broad* (*br*), which is a molecular signature of primordium commitment to the pupal state in holometabolous insects, have been studied in detail (Konopova and Jindra 2008; Parthasarathy et al. 2008; Suzuki et al. 2008; Uhlirova et al. 2003). The expression of *br* is specifically activated in the early *Manduca* eye primordium (Allee et al.

2006). Functional data regarding the role of *br* are not yet available in *Manduca*. However, *br* knockdown in *B. mori* and in *Tribolium* leads to an attenuation of eye development, demonstrating the importance of *br* for eye primordium commitment (Parthasarathy et al. 2008; Uhlirova et al. 2003).

Of note, in direct-developing insects *br* is expressed throughout the nymphal stages (Erezyilmaz et al. 2006), lending further molecular support to the homology of postembryonic eye development in the pupae of holometabolous species and the nymph of direct developers (Fig. 2; Erezyilmaz et al. 2006; Suzuki et al. 2008).

Eye Specification Across Insect Species: Summary and Perspectives

From both phylogenetic and developmental perspectives, the diversity of adult eye morphogenesis is enormous in insects, posing challenges to the experienced comparative biologist and the weathered *Drosophila* geneticist alike. Fortunately, some of the available molecular data allow for identifying shared ancestral themes in the early molecular development of the compound eye in both direct-developing and indirect-developing species. Arguably, the clearest example of this is the involvement of *eya* and *so* as facilitators of retinal precursor tissue determination and subsequent retinal differentiation (Figs. 4 and 12). A similar point may be made regarding *dac*, *ey*, and *toy*. These genes share broad expression patterns that include the retinal precursor tissue and are downregulated in the differentiating retina, pointing at a conserved role in implementing competence for retinal determination (Fig. 12). Taken together, these data are consistent with the roles experimentally ascribed to *eya*, *so*, *dac*, *ey*, and *toy* in *Drosophila* (Kumar 2009), which in this regard serves as a confirmed general model. The conserved expression of *eya* and *so* is further suggestive of a broad conservation of the PPN state of retinal commitment, at least at the transcription factor landscape level (Bessa et al. 2002; Dong and Friedrich 2005; Greenwood and Struhl 1999).

At the signaling gene level, the repressive effect of *wg* in the anterior developing eye field is a highly conserved aspect of compound eye patterning. It is reflected in the conservation of the polar domains in the anterior eye precursor field of all insect species so far examined (Fig. 11) and has even been reported for crustacean species (Duman-Scheel et al. 2002). Although the spatial expression patterns of *dpp* are quite diversified in the developing eyes of different species (Fig. 11), the eye development-promoting role of *dpp* can likewise be presumed to be conserved but awaits functional test. The same applies to the retinal differentiation-promoting role of *hh*.

Breakdown of Genetic Redundancy of ey *and* toy *During Dipteran Evolution*

Some of the *dac*-, *ey*-, and *toy*-related data in *Tribolium* suggest substantial rewiring of the regulatory interactions among these conserved players in eye development. The prime example is the redundant interaction of *ey* and *toy* during adult eye development in *Tribolium*, in conjunction with *dac* (Yang et al. 2009a). These relationships contrast with the upstream roles of *ey* and *toy* in the *Drosophila* retinal gene network (Gehring 2002). The *Tribolium* findings are not surprising given that functional redundancy is one of the proximate and ultimate causes for the conservation of duplicated genes (Force et al. 1999). The fact that the level of redundancy is lower in the developing *Drosophila* system may be tied to the more dramatic reorganization of genetic interactions during the evolution of the eye–antennal disc-patterning mechanisms. This may have led to a stronger degree of functional differentiation between *ey* and *toy* due to novel subfunctionalization opportunities. Along these lines, Lynch and Wagner (2011) have initiated a debate regarding the ancestral regulatory status of *ey* in comparison to *toy* in *Drosophila*.

At this point, the lack of data on how *ey* and *toy* act in direct-developing species like the grasshopper and cricket represents one of the most glaring gaps in the comparative study of insect eye development. There is little doubt that these highly awaited data will yield further important insights regarding the developmental organization of the early embryonic head as well as the gene regulatory organization of cells in the postembryonic growth zone of the eye.

Divergence of Eye Primordium Growth Activation

The comparative analysis of *eyg* in eye development also points toward profound differences between *Drosophila* and more ancestrally organized insects. At the surface, the opposite effects of downregulating *eyg* in *Drosophila* and *Tribolium* could be considered to reflect changes in the architecture of the eye specification gene network. However, there are arguments to conclude that these differences are more likely to reflect fundamental differences specifically in primordium growth activation. In *Drosophila, eyg* is part of the N-signaling-induced growth-promoting genet network that is pivotal for triggering the rapid tissue growth in the developing eye disc (for review, see Dominguez and Casares 2005). The discrepancy of *eyg* function in *Tribolium* and *Drosophila* may thus be explained by the smaller size of the eye in *Tribolium*, requiring less tissue proliferation. A second possibility is that the N-signaling-mediated organizer originated more recently in conjunction with the evolution of the *Drosophila* eye disc during dipteran evolution (Melzer and Paulus 1989). Consistent with this, an evolutionarily derived status of the N-initiated growth activation mechanism would explain the noncompartmentalized expression patterns of *fng* and *Dl* in the grasshopper (Dong and Friedrich 2005). A new data point in support of this model has come from the silk moth. Similar to *Manduca*, this lepidopteran develops its 3,000-ommatidia large compound eye from a late-forming eye disc (Yu et al. 2012). The silk moth mutant *flügellos* has been found to represent a null allele of Bombyx *fng* (Sato et al. 2008). Importantly, while *fng* mutant animals are characterized by wing defects, the development of the compound eye is not affected in dramatic ways. This suggests that the dramatic growth of the lepidopteran eye does not depend on *fng* as in *Drosophila*. In conclusion, these data demonstrate that the N- and *eyg*-involving activation of growth in the *Drosophila* eye disc is not a conserved component of eye disc development in holometabolous insects. This compelling evidence notwithstanding, additional genes will need to be examined in the lepidopteran models before definitive conclusions can be drawn regarding the derived state of N-initiated growth activation module in the *Drosophila* eye disc.

Embryonic Versus Postembryonic Adult Eye Primordium Determination

Another fundamental question waiting to be addressed concerns the specification of the adult retina primordium in ancestrally organized holometabolous species like *Tribolium* and *Manduca*. To get a taste of the foundational nature of this issue, one has to remember that the late postembryonic specification of the adult eye primordium in *Drosophila*, based on molecular genetic analysis, came as a surprise to the *Drosophila* field (Baker 2001; Kumar and Moses 2001). The preceding consensus was that this step takes place in the embryo, during the subdivision of the embryonic visual anlage into its major constituents (Postlethwait and Schneiderman 1971; Wieschaus and Gehring 1976). Assuming that the late specification of the eye primordium is the consequence of the evolution of the highly derived integrated eye–antennal imaginal disc of *Drosophila* (Fig. 13), it is reasonable to hypothesize that the specification of the adult eye primordium in the lateral larval head capsule takes place during embryogenesis in species with late eye discs like *Manduca* or no disc formation like *Tribolium* (Fig. 9). Otherwise, one has to postulate a postembryonic patterning mechanism, which drives the specification and activation of the adult eye primordium in the static head epithelium of the last instar larva.

Also the comparative framework of the transient arrest model of holometabolous visual system development predicts that both larval eye and adult eye precursor cell populations are committed in the embryonic visual anlage (Fig. 2). In the embryo, differentiation is initiated in the larval eye precursor but suppressed in the adult eye precursor cells. The latter, embedded in the lateral head epidermis, are maintained as a quiescent primordium until activation at the beginning of metamorphosis. This scenario is consistent with the positioning of the adult eye primordium in front of the larval eye in *Manduca* (Allee et al. 2006).

Of note, this anteroposterior alignment of larval and adult eye primordium seems not conserved in *Tribolium*. This may be due to the more extreme modification of the

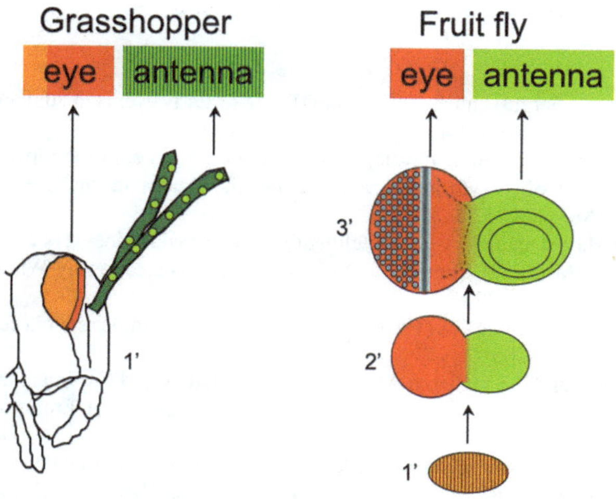

Fig. 15 Somatic stem cell reservoirs versus imaginal discs in insect eye development. In direct-developing insects like the grasshopper, the adult antenna and compound eye derive from organ-specific stem cell reservoirs (eye: *red*; antenna: *light green*) and differentiated cells of the nymph (eye: *orange*; antenna: *dark green*), which have been generated during embryogenesis. This mode of organ precursor tissue organization contrasts with the development of adult antenna and compound eye from the joint eye–antennal imaginal disc of *Drosophila*, which undergoes dramatic morphogenetic change through all three larval instars (1'–3')

Tribolium larval eyes in terms of accessory cell reduction and anatomical positioning in the larval head (Liu and Friedrich 2004). In *Manduca*, the larval eyes still form ommatidia-like subunits with lenses and pigment cells (Fig. 9; Allee et al. 2006).

Important work remains to be done to probe the previously discussed model by elucidating whether and how the precursor cells of the adult eye are set aside in more ancestrally organized systems like *Tribolium* and *Manduca* (Fig. 9). While interesting in its own right, answers to these questions will yield insights of broader significance. For one, they will add to our understanding of the molecular developmental evolution of holometabolous development, which after all was co-responsible for the unparalleled radiation of holometabolous insects (Kristensen 1999). Furthermore, the comparative evidence implies that the *Drosophila* eye–antennal imaginal disc is a derivative of the retinal growth zone in direct-developing insects, which most likely represents a tissue-specific stem cell population (Dong and Friedrich 2010; Fig. 15). If confirmed, the evolutionary transformation of the retinal growth zone in directly developing species to the *Drosophila* eye–antennal imaginal disc would be an example of how evolution reprogrammed stem cell populations to invent novel ways of body plan development.

Acknowledgments We are grateful to Amit Singh for the invitation to provide this book chapter, Rewaa Yas for meticulous proofreading of the manuscript, and Dr. Sumihare Noji for providing Fig. 7. Research in the Friedrich lab has been supported by NSF awards IOS 0951886, IOS 0091926, and EF-0334948.

References

Allee JP, Pelletier CL, Fergusson EK, Champlin DT (2006) Early events in adult eye development of the moth, *Manduca sexta*. J Insect Physiol 52:450–460

Anderson H (1978) Postembryonic development of the visual system of the locust, *Schistocerca gregaria*. I. Pattern of growth and developmental interactions in the retina and optic lobe. J Embryol Exp Morphol 45:55–83

Baker NE (2001) Master regulatory genes; telling them what to do. Bioessays 23:763–766

Baonza A, Freeman M (2002) Control of *Drosophila* eye specification by Wingless signalling. Development 129:5313–5322

Bennet RR, Tunstall J, Horridge GA (1967) Spectral sensitivity of single retinula cells in the locust. Z Vgl Physiol 55:195–206

Bentley D, Keshishian H, Shankland M, Toroian-Raymond A (1979) Quantitative staging of embryonic development of the grasshopper, *Schistocerca nitens*. J Embryol Exp Morphol 54:47–74

Bessa J, Gebelein B, Pichaud F, Casares F, Mann RS (2002) Combinatorial control of *Drosophila* eye development by *eyeless, homothorax,* and *teashirt*. Genes Dev 16:2415–2427

Beutel RG, Friedrich F, Hörnschemeyer T, Pohl H, Hünefeld F, Beckmann F, Meier R, Misof B, Whiting MF, Vilhelmsen L (2011) Morphological and molecular evidence converge upon a robust phylogeny of the megadiverse Holometabola. Cladistics 27:341–355

Blackburn DC, Conley KW, Plachetzki DC, Kempler K, Battelle BA, Brown NL (2008) Isolation and expression of *Pax6* and *atonal* homologues in the American horseshoe crab, *Limulus polyphemus*. Dev Dyn 237:2209–2219

Blum M, Labhart T (2000) Photoreceptor visual fields, ommatidial array, and receptor axon projections in the polarisation-sensitive dorsal rim area of the cricket compound eye. J Comp Physiol A Neuroethol Sens Neural Behav Physiol 186:119–128

Bodenstein D (1953) Postembryonic development. In: Roeder KD (ed) Insect Physiology. Wiley, New York, pp 275–367

Brown SJ, Shippy TD, Miller S, Bolognesi R, Beeman RW, Lorenzen MD, Bucher G, Wimmer EA, Klingler M (2009) The red flour beetle, *Tribolium castaneum* (Coleoptera): a model for studies of development and pest biology. Cold Spring Harb Protoc pdb.emo126

Buschbeck E, Friedrich M (2008) Evolution of insect eyes: tales of ancient heritage, deconstruction, reconstruction, remodeling and recycling. Evolution Education and Outreach 1:448–462

Buschbeck EK (2005) The compound lens eye of Strepsiptera: morphological development of larvae and pupae. Arthropod Struct Dev 34:315–326

Buschbeck EK, Roosevelt JL, Hoy RR (2001) Eye stalks or no eye stalks: a structural comparison of pupal development in the stalk-eyed fly *Cyrtodiopsis* and in *Drosophila*. J Comp Neurol 433:486–498

Cavodeassi F, Diez Del Corral R, Campuzano S, Dominguez M (1999) Compartments and organising boundaries in the *Drosophila* eye: the role of the homeodomain Iroquois proteins. Development 126:4933–4942

Cavodeassi F, Modolell J, Campuzano S (2000) The Iroquois homeobox genes function as dorsal selectors in the *Drosophila* head. Development 127:1921–1929

Champlin DT, Truman JW (1998) Ecdysteroids govern two phases of eye development during metamorphosis of the moth, *Manduca sexta*. Development 125:2009–2018

Chang T, Mazotta J, Dumstrei K, Dumitrescu A, Hartenstein V (2001) Dpp and Hh signaling in the *Drosophila* embryonic eye field. Development 128:4691–4704

Cho KO, Choi KW (1998) Fringe is essential for mirror symmetry and morphogenesis in the *Drosophila* eye. Nature 396:272–276

Dearden P, Akam M (2000) A role for *Fringe* in segment morphogenesis but not segment formation in the grasshopper, *Schistocerca gregaria*. Development Genes Evolution 210:329–336

Dominguez M, Casares F (2005) Organ specification-growth control connection: new in-sights from the *Drosophila* eye-antennal disc. Dev Dyn 232:673–684

Dominguez M, de Celis JF (1998) A dorsal/ventral boundary established by Notch controls growth and polarity in the *Drosophila* eye. Nature 396:276–278

Dominguez M, Ferres-Marco D, Gutierrez-Avino FJ, Speicher SA, Beneyto M (2004) Growth and specification of the eye are controlled independently by Eyegone and Eyeless in *Drosophila melanogaster*. Nat Genet 36:31–39

Dong Y, Friedrich M (2005) Comparative analysis of Wg patterning in the embryonic grasshopper eye. Dev Genes Evol 215:177–197

Dong Y, Friedrich M (2010) Enforcing biphasic eye development in a directly developing insect by transient knockdown of single eye selector genes. J Exp Zool B Mol Dev Evol 314B:104–114

Dong Y, Dinan L, Friedrich M (2003) The effect of manipulating ecdysteroid signaling on embryonic eye development in the locust *Schistocerca americana*. Dev Genes Evol 213:587–600

Donner AL, Maas RL (2004) Conservation and non-conservation of genetic pathways in eye specification. Int J Dev Biol 48:743–753

Duman-Scheel M, Pirkl N, Patel NH (2002) Analysis of the expression pattern of *Mysidium columbiae wingless* provides evidence for conserved mesodermal and retinal patterning processes among insects and crustaceans. Dev Genes Evol 212:114–123

Egelhaaf A (1988) Evidence for the priming role of the central retinula cell in ommatidium differentiation of *Ephestia kuehniella*. Rouxs Arch Dev Biol 197:184–189

Erezyilmaz DF, Riddiford LM, Truman JW (2006) The pupal specifier *broad* directs progressive morphogenesis in a direct-developing insect. Proc Natl Acad Sci 103:6925–6930

Fahrenbach WH (1969) The morphology of the eyes of *Limulus*. II. Ommatidia of the compound eye. Z Zellforsch Mikrosk Anat 93:451–483

Force A, Lynch M, Pickett FB, Amores A, Yan YL, Postlethwait J (1999) Preservation of duplicate genes by complementary, degenerative mutations. Genetics 151:1531–1545

Friedrich M (2006) Continuity versus split and reconstitution: exploring the molecular developmental corollaries of insect eye primordium evolution. Dev Biol 299:310–329

Friedrich M (2008) Opsins and cell fate in the *Drosophila* Bolwig organ: tricky lessons in homology inference. Bioessays 30:980–993

Friedrich M, Benzer S (2000) Divergent *decapentaplegic* expression patterns in compound eye development and the evolution of insect metamorphosis. J Exp Zool B Mol Dev E 288:39–55

Friedrich M, Caravas J (2011) New insights from hemichordate genomes: prebilaterian origin and parallel modifications in the paired domain of the Pax gene *eyegone*. J Exp Zool B Mol Dev E 316:387–392

Friedrich M, Rambold I, Melzer RR (1996) The early stages of ommatidial development in the flour beetle *Tribolium castaneum* (Coleoptera, Tenebrionidae). Dev Genes Evol 206:136–146

Friedrich M, Dong Y, Jackowska M (2006) Insect interordinal relationships: insights from the visual system. Arthropod Syst Phylogeny 64:133–148

Gehring WJ (2002) The genetic control of eye development and its implications for the evolution of the various eye-types. Int. J. Dev Biol 46:65–73

Greenwood S, Struhl G (1999) Progression of the morphogenetic furrow in the *Drosophila* eye: the roles of Hedgehog, Decapentaplegic and the Raf pathway. Development 126:5795–5808

Hafner GS, Tokarski TR (1998) Morphogenesis and pattern formation in the retina of the crayfish *Procambarus clarkii*. Cell Tissue Res 293:535–550

Hafner GS, Tokarski TR (2001) Retinal development in the lobster *Homarus americanus*. Comparison with compound eyes of insects and other crustaceans. Cell Tissue Res 305:147–158

Harzsch S, Walossek D (2001) Neurogenesis in the developing visual system of the branchiopod crustacean Triops longicaudatus (LeConte, 1846): corresponding patterns of compound-eye formation in Crustacea and Insecta? Dev Genes Evol 211:37–43

Heberlein U, Wolff T, Rubin GM (1993) The TGF beta homolog *dpp* and the segment polarity gene *hedgehog* are required for propagation of a morphogenetic wave in the *Drosophila* retina. Cell 75:913–926

Henze MJ, Dannenhauer K, Kohler M, Labhart T, Gesemann M (2012) Opsin evolution and expression in arthropod compound eyes and ocelli: Insights from the cricket *Gryllus bimaculatus*. BMC Evol Biol 12:163

Homberg U, Paech A (2002) Ultrastructure and orientation of ommatidia in the dorsal rim area of the locust compound eye. Arthropod Struct Dev 30:271–280

Inoue Y, Miyawaki K, Terasawa T, Matsushima K, Shinmyo Y, Niwa N, Mito T, Ohuchi H, Noji S (2004) Expression patterns of *dachshund* during head development of *Gryllus bimaculatus* (cricket). Gene Expr Patterns 4:725–731

Jackowska M, Bao R, Liu Z, McDonald EC, Cook TA, Friedrich M (2007) Genomic and gene regulatory signatures of cryptozoic adaptation: loss of blue sensitive photoreceptors through expansion of long wavelength-opsin expression in the red flour beetle *Tribolium castaneum*. Front Zool 4:24

Jun S, Wallen RV, Goriely A, Kalionis B, Desplan C (1998) Lune/eye gone, a Pax-like protein, uses a partial paired domain and a homeodomain for DNA recognition. Proc Natl Acad Sci U S A 95:13720–13725

Kenyon KL, Ranade SS, Curtiss J, Mlodzik M, Pignoni F (2003) Coordinating proliferation and tissue specification to promote regional identity in the *Drosophila* head. Dev Cell 5:403–414

Klingler M (2004) *Tribolium*. Curr Biol 14:R639–R640

Konopova B, Jindra M (2008) Broad-Complex acts downstream of Met in juvenile hormone signaling to coordinate primitive holometabolan metamorphosis. Development 135:559–568

Koyama T, Syropyatova MO, Riddiford LM (2008) Insulin/IGF signaling regulates the change in commitment in imaginal discs and primordia by overriding the effect of juvenile hormone. Dev Biol 324:258–265

Kozmik Z (2008) The role of Pax genes in eye evolution. Brain Res Bull 75:335–339

Kristensen N (1999) Phylogeny of edopterygote insects, the most successful lineage of living organisms. Eur J Entomol 96:237–253

Kronhamn J, Frei E, Daube M, Jiao R, Shi Y, Noll M, Rasmuson-Lestander A (2002) Headless flies produced by mutations in the paralogous Pax6 genes *eyeless* and *twin of eyeless*. Development 129:1015–1026

Kumar JP (2009) The molecular circuitry governing retinal determination. Biochim Biophys Acta 1789:306–314

Kumar JP, Moses K (2001) The EGF receptor and Notch signaling pathways control the initiation of the morphogenetic furrow during *Drosophila* eye development. Development 128:2689–2697

Labhart T, Keller K (1992) Fine structure and growth of the polarization-sensitive dorsal rim area in the compound eye of larval crickets. Naturwissenschaften 79:527–529

Labhart T, Meyer EP (1999) Detectors for polarized skylight in insects: a survey of ommatidial specializations in the dorsal rim area of the compound eye. Microsc Res Tech 47:368–379

Lamb T (2011) Evolution's witness: how eyes evolved. Oxford University Press, USA

Lee JD, Treisman JE (2001) The role of Wingless signaling in establishing the anteroposterior and dorsoventral axes of the eye disc. Development 128:1519–1529

Lee MSY, Jago JB, Garcia-Bellido DC, Edgecombe GD, Gehling JG, Paterson JR (2011) Modern optics in exceptionally preserved eyes of Early Cambrian arthropods from Australia. Nature 474:631–634

Liu Z, Friedrich M (2004) The *Tribolium* homologue of *glass* and the evolution of insect larval eyes. Dev Biol 269:36–54

Liu Z, Yang X, Dong Y, Friedrich M (2006) Tracking down the "head blob": comparative analysis of *wingless* expression in the embryonic insect procephalon reveals progressive reduction of ocular segment patterning in higher insects. Arthropod Struct Dev 35:341–356

Lomer CJ, Bateman RP, Johnson DL, Langewald J, Thomas M (2001) Biological control of locusts and grasshoppers. Annu Rev Entomol 46:667–702

Lorenz MW (2007) Oogenesis-flight syndrome in crickets: age-dependent egg production, flight performance, and biochemical composition of the flight muscles in adult female *Gryllus bimaculatus*. J Insect Physiol 53:819–832

Lynch V, Wagner G (2011) Revisiting a classic example of transcription factor functional equivalence: are Eyeless and Pax6 functionally equivalent or divergent? J Exp Zool B Mol Dev Evol 316B:93–98

Ma CY, Zhou Y, Beachy PA, Moses K (1993) The segment polarity gene *hedgehog* is required for progression of the morphogenetic furrow in the developing *Drosophila* eye. Cell 75:927–938

MacWhinnie SGB, Allee JP, Nelson CA, Riddiford LM, Truman JW, Champlin DT (2005) The role of nutrition in creation of the eye imaginal disc and initiation of metamorphosis in *Manduca sexta*. Dev Biol 285:285–297

Maderspacher F, Bucher G, Klingler M (1998) Pair-rule and gap gene mutants in the flour beetle *Tribolium castaneum*. Dev Genes Evol 208:558–568

Mahfooz NS, Li H, Popadić A (2004) Differential expression patterns of the hox gene are associated with differential growth of insect hind legs. Proc Natl Acad Sci U S A 101:4877–4882

Maurel-Zaffran C, Treisman JE (2000) *pannier* acts upstream of *wingless* to direct dorsal eye disc development in *Drosophila*. Development 127:1007–1016

Melzer RR, Paulus HF (1989) Evolutionswege zum Larvalauge der Insekten -Die Stemmata der höheren Dipteren und ihre Abwandlung zum Bolwig-Organ. Z Zool Syst Evolutionsforsch 27:200–245

Melzer RR, Michalke C, Smola U (2000) Walking on insect paths? Early ommatidial development in the compound eye of the ancestral crustacean, *Triops cancriformis*. Naturwissenschaften 87:308–311

Mito T, Noji S (2008) The two-spotted cricket *Gryllus bimaculatus*: an emerging model for developmental and regeneration studies. Cold Spring Harb Protoc: pdb.emo110

Monsma SA, Booker R (1996) Genesis of the adult retina and outer optic lobes of the moth, *Manduca sexta*. I. patterns of proliferation and cell death. J Comp Neurol 367:10–20

Moreaux L, Laurent G (2007) Estimating firing rates from calcium signals in locust projection neurons in vivo. Front Neural Circuits 1:2

Moses K, Rodrigues AB (2004) Growth and specification: fly Pax6 homologs *eyegone* and *eyeless* have distinct functions. Bioessays 26:600–603

Müller C, Rosenberg J, Richter S, Meyer-Rochow V (2003) The compound eye of *Scutigera coleoptrata* (Linnaeus, 1758)(Chilopoda: Notostigmophora): an ultrastructural reinvestigation that adds support to the Mandibulata concept. Zoomorphology 122:191–209

Nilsson D-E (1996) Eye ancestry: old genes for new eyes. Curr Biol 6:39–42

Niwa N, Inoue Y, Nozawa A, Saito M, Misumi Y, Ohuchi H, Yoshioka H, Noji S (2000) Correlation of diversity of leg morphology in *Gryllus bimaculatus* (cricket) with divergence in *dpp* expression pattern during leg development. Development 127:4373–4381

Papayannopoulos V, Tomlinson A, Panin VM, Rauskolb C, Irvine KD (1998) Dorsal-ventral signaling in the *Drosophila* eye. Science 281:2031–2034

Park T (1934) Observations on the general biology of the flour beetle, *Tribolium confusum*. Quart Rev Biol 9:36–64

Parthasarathy R, Tan A, Bai H, Palli SR (2008) Transcription factor *broad* suppresses precocious development of adult structures during larval–pupal metamorphosis in the red flour beetle, *Tribolium castaneum*. Mech Dev 125:299–313

Paterson JR, Garcia-Bellido DC, Lee MSY, Brock GA, Jago JB, Edgecombe GD (2011) Acute vision in the giant Cambrian predator *Anomalocaris* and the origin of compound eyes. Nature 480:237–240

Perez-Mendoza J, Campbell JF, Throne JE (2011) Effects of rearing density, age, sex, and food deprivation on flight initiation of the red flour beetle (Coleoptera: Tenebrionidae). J Econ Entomol 104:443–451

Perron M, Kanekar S, Vetter ML, Harris WA (1998) The genetic sequence of retinal development in the ciliary margin of the *Xenopus* eye. Dev Biol 199:185–200

Pichaud F, Casares F (2000) *homothorax* and *iroquois-C* genes are required for the establishment of territories within the developing eye disc. Mech Dev 96:15–25

Pichaud F, Desplan C (2002) Pax genes and eye organogenesis. Current Opionion in Genetics and. Development 12:430–434

Posnien N, Schinko JB, Kittelmann S, Bucher G (2010) Genetics, development and composition of the insect head–a beetle's view. Arthropod Struct Dev 39:399–410

Postlethwait JH, Schneiderman HA (1971) A clonal analysis of development in *Drosophila melanogaster*: morphogenesis, determination and and growth in the wild type antenna. Dev Biol 24:477–519

Prpic NM, Wigand B, Damen WG, Klingler M (2001) Expression of *dachshund* in wild-type and *Distal-less* mutant *Tribolium* corroborates serial homologies in insect appendages. Dev Genes Evol 211:467–477

Raymond PA, Barthel LK, Bernardos RL, Perkowski JJ (2006) Molecular characterization of retinal stem cells and their niches in adult zebrafish. BMC Dev Biol 6:36

Reifegerste R, Moses K (1999) Genetics of epithelial polarity and pattern in the *Drosophila* retina. Bioessays 21:275–285

Richards S, Gibbs RA, Weinstock GM, Brown SJ, Denell R, Beeman RW, Gibbs R, Bucher G, Friedrich M, Grimmelikhuijzen CJ, Klingler M, Lorenzen M, Roth S, Schroder R, Tautz D, Zdobnov EM, Muzny D, Attaway T, Bell S, Buhay CJ, Chandrabose MN, Chavez D, Clerk-Blankenburg KP, Cree A, Dao M, Davis C, Chacko J, Dinh H, Dugan-Rocha S, Fowler G, Garner TT, Garnes J, Gnirke A, Hawes A, Hernandez J, Hines S, Holder M, Hume J, Jhangiani SN, Joshi V, Khan ZM, Jackson L, Kovar C, Kowis A, Lee S, Lewis LR, Margolis J, Morgan M, Nazareth LV, Nguyen N, Okwuonu G, Parker D, Ruiz SJ, Santibanez J, Savard J, Scherer SE, Schneider B, Sodergren E, Vattahil S, Villasana D, White CS, Wright R, Park Y, Lord J, Oppert B, Brown S, Wang L, Weinstock G, Liu Y, Worley K, Elsik CG, Reese JT, Elhaik E, Landan G, Graur D, Arensburger P, Atkinson P, Beidler J, Demuth JP, Drury DW, Du YZ, Fujiwara H, Maselli V, Osanai M, Robertson HM, Tu Z, Wang JJ, Wang S, Song H, Zhang L, Werner D, Stanke M, Morgenstern B, Solovyev V, Kosarev P, Brown G, Chen HC, Ermolaeva O, Hlavina W, Kapustin Y, Kiryutin B, Kitts P, Maglott D, Pruitt K, Sapojnikov V, Souvorov A, Mackey AJ, Waterhouse RM, Wyder S, Kriventseva EV, Kadowaki T, Bork P, Aranda M, Bao R, Beermann A, Berns N, Bolognesi R, Bonneton F, Bopp D, Butts T, Chaumot A, Denell RE, Ferrier DE, Gordon CM, Jindra M, Lan Q, Lattorff HM, Laudet V, von Levetsow C, Liu Z, Lutz R, Lynch JA, da Fonseca RN, Posnien N, Reuter R, Schinko JB, Schmitt C, Schoppmeier M, Shippy TD, Simonnet F, Marques-Souza H, Tomoyasu Y, Trauner J, Van der Zee M, Vervoort M, Wittkopp N, Wimmer EA, Yang X, Jones AK, Sattelle DB, Ebert PR, Nelson D, Scott JG, Muthukrishnan S, Kramer KJ, Arakane Y, Zhu Q, Hogenkamp D, Dixit R, Jiang H, Zou Z, Marshall J, Elpidina E, Vinokurov K, Oppert C, Evans J, Lu Z, Zhao P, Sumathipala N, Altincicek B, Vilcinskas A, Williams M, Hultmark D, Hetru C, Hauser F, Cazzamali G, Williamson M, Li B, Tanaka Y, Predel R, Neupert S, Schachtner J, Verleyen P, Raible F, Walden KK, Angeli S, Foret S, Schuetz S, Maleszka R, Miller SC, Grossmann D (2008) The genome of the model beetle and pest *Tribolium castaneum*. Nature 452:949–955

Ridley AW, Hereward JP, Daglish GJ, Raghu S, Collins PJ, Walter GH (2011) The spatiotemporal dynamics of *Tribolium castaneum* (Herbst): adult flight and gene flow. Mol Ecol 20:1635–1646

Rogers SM, Harston GW, Kilburn-Toppin F, Matheson T, Burrows M, Gabbiani F, Krapp HG (2010) Spatiotemporal receptive field properties of a looming-sensitive neuron in solitarious and gregarious phases of the desert locust. J Neurophysiol 103:779–792

Roonwal ML (1936) Studies on the embryology of the African migratory locust, *Locusta migratoria migratorioides* Reiche and Frm. (Orthoptera, Acrididae) II-Organogeny. Philos Trans R Soc Lond B 227:175–244

Salvini-Plawen L, Mayr E (1977) On the evolution of photoreceptors and eyes. Evol Biol 10:207–263

Sanchez D, Ganfornina MD, Bastiani M (1995) Contributions of an orthopteran to the understanding of neuronal pathfinding. Immunol Cell Biol 73:565–574

Sato K, Matsunaga TM, Futahashi R, Kojima T, Mita K, Banno Y, Fujiwara H (2008) Positional cloning of a *Bombyx wingless* locus *flugellos* (*fl*) reveals a crucial role for *fringe* that is specific for wing morphogenesis. Genetics 179:875–885

Savard J, Tautz D, Richards S, Weinstock GM, Gibbs RA, Werren JH, Tettelin H, Lercher MJ (2006) Phylogenomic analysis reveals bees and wasps (Hymenoptera) at the base of the radiation of holometabolous insects. Genome Res 16:1334–1338

Schinko J, Hillebrand K, Bucher G (2012) Heat shock-mediated misexpression of genes in the beetle *Tribolium castaneum*. Dev Genes Evol 222:287–298

Schroder R (2003) The genes *orthodenticle* and *hunchback* substitute for *bicoid* in the beetle *Tribolium*. Nature 422:621–625

Singh A, Kango-Singh M, Parthasarathy R, Gopinathan KP (2007) Larval legs of mulberry silkworm *Bombyx mori* are prototypes for the adult legs. Genesis 45:169–176

Smith WC, Price DA, Greenberg RM, Battelle BA (1993) Opsins from the lateral eyes and ocelli of the horseshoe-crab, *Limulus polyphemus*. Proc Natl Acad Sci U S A 90:6150–6154.

Sokoloff A (1972) The biology of *Tribolium*. Clarendon Press, Oxford

Stambolic V, Ruel L, Woodgett JR (1996) Lithium inhibits *glycogen synthase kinase*-3 activity and mimics *wingless* signalling in intact cells. Curr Biol 6:1664–1668

Suzuki Y, Truman JW, Riddiford LM (2008) The role of *Broad* in the development of *Tribolium castaneum*: implications for the evolution of the holometabolous insect pupa. Development 135:569–577

Suzuki Y, Squires DC, Riddiford LM (2009) Larval leg integrity is maintained by Distal-less and is required for proper timing of metamorphosis in the flour beetle, *Tribolium castaneum*. Dev Biol 326:60–67

Takagi A, Kurita K, Terasawa T, Nakamura T, Bando T, Moriyama Y, Mito T, Noji S, Ohuchi H (2012) Functional analysis of the role of *eyes absent* and *sine oculis* in the developing eye of the cricket *Gryllus bimaculatus*. Dev Growth Differ 54:227–240

Towner P, Harris P, Wolstenholme AJ, Hill C, Worm K, Gartner W (1997) Primary structure of locust opsins: a speculative model which may account for ultraviolet wavelength light detection. Vision Res 37:495–503

Treisman JE, Rubin GM (1995) *wingless* inhibits morphogenetic furrow movement in the *Drosophila* eye disc. Development 121:3519–3527

Truman JW, Riddiford LM (2002) Endocrine insights into the evolution of metamorphosis in insects. Annu Rev Entomol 47:467–500

Truman JW, Hiruma K, Allee JP, MacWhinnie SGB, Champlin DT, Riddiford LM (2006) Juvenile hormone is required to couple imaginal disc formation with nutrition in insects. Science 312:1385–1388

Uhlirova M, Foy BD, Beaty BJ, Olson KE, Riddiford LM, Jindra M (2003) Use of Sindbis virus-mediated RNA interference to demonstrate a conserved role of Broad-Complex in insect metamorphosis. Proc Natl Acad Sci U S A 100:15607–15612

Vishnevskaya TM, Cherkasov AD, Shura-Bura TM (1985) Spectral sensitivity of photoreceptors in the compound eye of the locust. Neurophysiology 18:69–76

Wiegmann BM, Trautwein MD, Kim JW, Cassel BK, Bertone MA, Winterton SL, Yeates DK (2009) Single-copy nuclear genes resolve the phylogeny of the holometabolous insects. BMC Biol 7:34

Wieschaus E, Gehring W (1976) Clonal analysis of primordial disc cells in the early embryo of *Drosophila melanogaster*. Dev Biol 50:249–263

Wilson M, Garrard P, McGinness S (1978) The unit structure of the locust compound eye. Cell Tiss Res 195:205–226

Yang X, Weber M, ZarinKamar N, Wigand B, Posnien G, Friedrich R, Beutel R, Damen W, Bucher G, Klingler M, Friedrich M (2009a) Probing the *Drosophila* retinal determination gene network in *Tribolium* (II): the Pax6 genes *eyeless* and *twin of eyeless*. Dev Biol 333:215–227

Yang X, ZarinKamar N, Bao R, Friedrich M (2009b) Probing the *Drosophila* retinal determination gene network in *Tribolium* (I): the early retinal genes *dachshund*, *eyes absent* and *sine oculis*. Dev Biol 333:202–214

Yu L, Zhou Q, Zhang C, Pignoni F (2012) Identification of *Bombyx atonal* and functional comparison with the *Drosophila atonal* proneural factor in the developing fly eye. Genesis 50:393–403

ZarinKamar N, Yang X, Bao R, Friedrich F, Beutel R, Friedrich M, 2011. The Pax gene eyegone facilitates repression of eye development in *Tribolium*. Evodevo 2:8

Zuker CS (1994) On the evolution of eyes: would you like it simple or compound? Science 265:742–743

Development and Evolution of the *Drosophila* Bolwig's Organ: A Compound Eye Relict

Markus Friedrich

The development of the *Drosophila* adult compound eyes is preceded by comparatively minimalist larval eyes, which differentiate in the embryo and provide the larva with visual input, which facilitates phototactic and circadian behavior (Helfrich-Förster et al. 2002; Keene et al. 2011; Lilly and Carlson 1990; Malpel et al. 2002; Mazzoni et al. 2005; Sawin et al. 1995; Sprecher et al. 2011). The adjective "minimalist" is anything but understated, considering the fact that it took expert dissecting skills to locate these visual organs in a considerably larger dipteran species: the house fly *Musca domestica*. This feat was accomplished by Niels Bolwig, credited by the use of his name as the label for the larval eyes of higher (schizophoran) dipterans such as *Drosophila* and *Musca*, now known as Bolwig's organs (BOs; Bolwig 1946). Given the complete lack of accessory cells such as lens and pigment cells, the BOs can be assumed to function as simple directional light detectors as opposed to the landscape vision mediating adult eye. The BOs, thus, do not meet the definition of eye in the sense of an image forming peripheral sense organ (Land and Nilsson 2002).

Despite its small size and in some cases because of it, many research interests have been compacted into the BOs. Besides serving as an efficient system for the genetic analysis of visual and circadian behavior (Keene and Sprecher 2011; Keene et al. 2011; Mazzoni et al. 2005; Sprecher et al. 2011), the BO has also been used to study axonal targeting mechanisms (Holmes et al. 1998; Schmucker et al. 1997, 1994, 1992, 2000). The development of the BO is of particular interest because of its simplicity, offering comprehensive analysis at an affordable price, and the fact that the BO, first pointed out by Paulus based on structural comparative data (Melzer and Paulus 1989), is an evolutionarily related sibling of the adult eye. The first investigation of the morphogenesis of the BO arrived at similar preliminary conclusions but asked for further study (Green et al. 1993; Sprecher et al. 2007). Subsequent

M. Friedrich (✉)
Department of Biological Sciences, Wayne State University, 5047 Gullen Mall, Detroit, MI 48202, USA
e-mail:mf@biology.biosci.wayne.edu

Department of Anatomy and Cell Biology, School of Medicine, Wayne State University, 540 East Canfield Avenue, Detroit, MI 48201, USA

A. Singh, M. Kango-Singh (eds.), *Molecular Genetics of Axial Patterning, Growth and Disease in the Drosophila Eye*, DOI 10.1007/978-1-4614-8232-1_12,

comparisons at the developmental and gene regulatory level have nothing but corroborated that the *Drosophila* BO and adult eye have become separated during evolution like Siamese twins through surgery. That is, the BO corresponds to the embryonic compound eye portion of direct-developing insects, while the adult *Drosophila* eye corresponds to the portion of the eye, which is formed during postembryonic development in direct-developing insects (Friedrich 2008). The resulting deep-shared roots are reflected in the high degree of similarity of the genetic mechanisms that regulate the development of the BO and adult eye (Friedrich 2006a). Viewing the BO as a relict compound eye is pivotal for a deeper understanding of its structural, developmental, and functional organization in the spirit of the fact that "nothing makes sense in biology except in the light of evolution" (Dobzhansky 1973).

Morphology

Many noteworthy features of the *Drosophila* larva are characterized by reduction or absence. In contrast to the adult fly, numerous hallmark features of the insect body plan are lacking in the larva. This includes the subdivision of the longitudinal body axis into head, thorax, and abdominal tagmata, and the absence of appendages. Most importantly, this list further includes the complete reduction of the head capsule, a condition referred to as acephaly, which implies the absence of fully formed peripheral sensory organs such as antennae or lateral eyes (Fig. 1; see Liu et al. 2006 for detailed discussion regarding the evolutionary developmental implications). The evolutionary transformation from eu- to acephalic body plan organization in dipteran larvae took place approximately 150 million years ago in the lineage leading to the subgroup Cyclorrhapha (Wiegmann et al. 2011). Specific steps of this transition are still documented by intermediate morphologies in extant dipteran species (Melzer and Paulus 1989; see Fig. 2 for further details).

The acephalic condition of cyclorrhaphan larvae resulted from the shift of the head cuticle morphology components into the anterior thoracic segments, producing the prominently black cephalopharyngeal head skeleton (Fig. 1a, b; Jurgens 1987; Jurgens and Hartenstein 1993; Jurgens et al. 1986). This reorganization also affected the position of the sensory organs like larval eyes, which likewise shifted interiorly. As a result of this, the basic morphology of the BO can be described as a pair of small photoreceptor bundles, each attached to the lateral plate (tentorial phragma) of the cephalopharyngeal head skeleton (Fig. 1a, b).

The light-blocking tentorial phragma is presumed to serve as a light barrier at the median side of the BO (Melzer and Paulus 1989), to the effect that the excitatory difference between the two laterally corresponding organs conveys information on the direction of light influx. This situation seems analogous to the similarly simple visual system that provides the trochophora larvae of the marine annelid *Platynereis dumerelii* with the capacity for phototactic behavior (Jekely et al. 2008). Consistent with this, behavioral and detailed electrophysiological analyses indicate that the phototactic behavior mediated by the BOs is largely driven by spatial comparison of light intensity (Hinnemann et al. 2010).

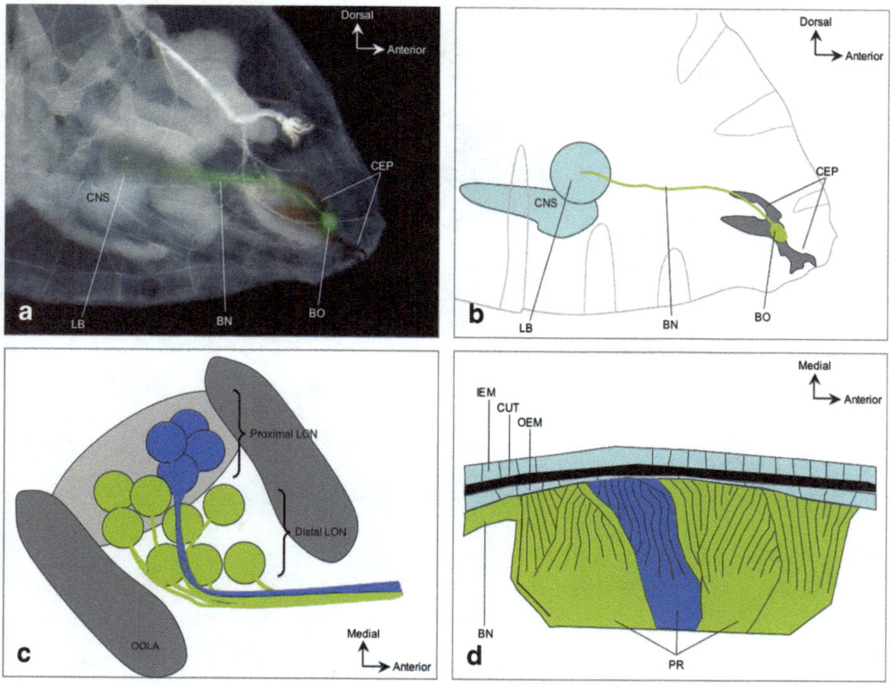

Fig. 1 Morphology and neuroanatomy of the *Drosophila* larval eye. **a** Bolwig's organ (BO) and Bolwig nerve (BN) visualized by Green Fluorescent Protein marker gene expression via activation by the Rh6 gene promoter. The anterior region of the third instar *Drosophila* larva is shown from lateral perspective. (Adapted from Friedrich 2011). **b** Schematic representation of panel **a**. Head cuticle outline, dorsal and ventral denticle belts, larval brain (LB) central nervous system (CNS), Bolwig organ and cephalopharyngeal head skeleton (CPN) outlined in *gray*. (Adapted from Friedrich 2011). **c** Schematic representation of Bolwig's organ photoreceptor projections in the larval optic neuropile (LON). (Based on Sprecher et al. 2011). **d** Morphology of the *Drosophila* Bolwig's organ. (Based on Melzer and Paulus 1989). Only a subset of the Bolwig's organ photoreceptors is depicted. See text for details. **c, d** Rh5- and Rh6-expressing photoreceptors are indicated in *blue* and *green* color, respectively. *CUT* cuticle, *IEM* inner epithelial membrane, *LB* larval brain, *OEM* outer epithelial membrane, *OOLA* outer optic lobe anlage, *PR* photoreceptor

The number of photoreceptors per BO bundle is usually described as 12 (Steller et al. 1987), although there seems to be some degree of plasticity with reported photoreceptor numbers ranging from 8 to 16 (Green et al. 1993; Sprecher et al. 2007). There is, however, consensus that the BO photoreceptors fall in two subtypes: 4–5 Rh5-positive or founder photoreceptors and 6–7 Rh6-positive or peripheral photoreceptors (Fig. 1d and 3). These terminologies reflect developmental and visual gene expression characteristics described further below in detail.

In their ultrastructural study of the visual system in higher dipterans, Melzer and Paulus (1989) confirmed the high structural correspondence between the *Drosophila* larval eye and the housefly BO described by Bolwig (1946). At the cell morphological level, it is most notable that the BO photoreceptors do not form rhabdomeres but slightly disorganized stacks of lamella-like sheets of a thickness similar to that of the rhabdomeric microvilli in the adult eye (50–90 nm; Fig. 1d). It is commonly assumed

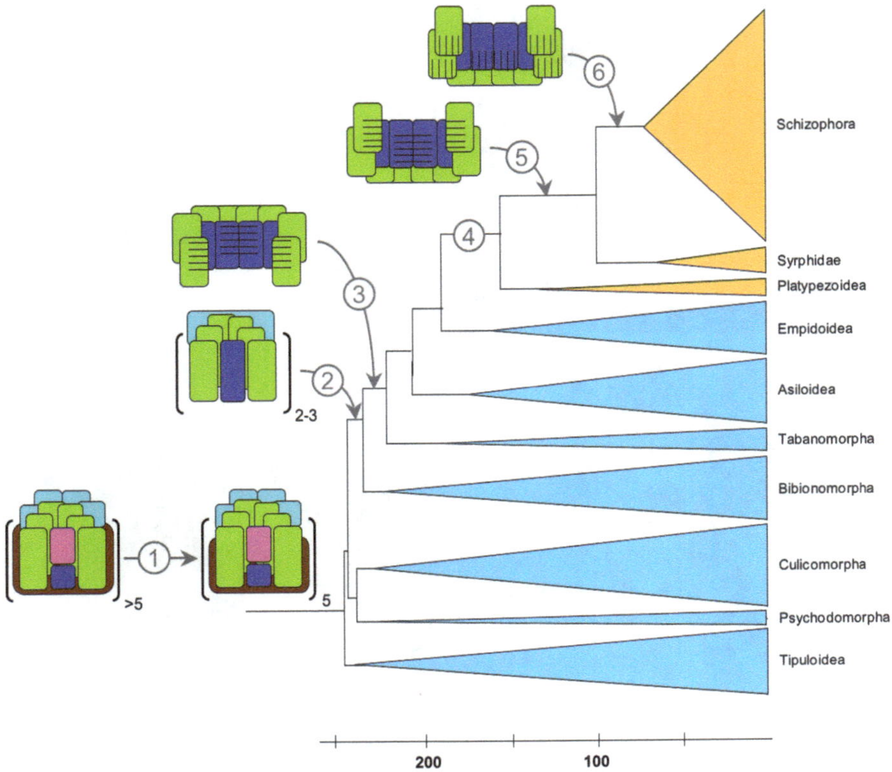

Fig. 2 Key steps in the regressive evolution of the *Drosophila* larval eye. Numbers in *gray circles* indicate structural transformations of the larval eye, which have been mapped to specific branches of dipteran evolution (Melzer and Paulus 1989). (Based on Caravas and Friedrich 2013; Wiegmann et al. 2011). The process is tracked starting from the ancestral organization of the larval eye in holometabolous insects that consists of usually five discrete ommatidia. Multiplicity of the larval ommatidia is indicated by subscribed *number* and *parentheses*. Transformations: 1 = Reduction of larval eye with compound eye organization comprising many ommatidia (scorpion flies) to five larval ommatidia with partially reduced pigment cells; 2 = Reduction of larval ommatidia number to two or three, complete reduction of pigment cells, and partial reduction of the lens; 3 = Fusion of individual larval ommatidia into a single visual organ. The concomitant reduction of R7 photoreceptors is hypothesized but has not been specifically investigated; 4 = Transition from eucephalic to acephalic larval body plan; 5 = Inversion of the larval eye; 6 = Restructuring of the rhabdomere membranes into lamellae-like protrusions in the ancestor of schizophoran Diptera which includes *Drosophila* and *Musca*. Photoreceptor now pointing medially as a result of the relocalization during head involution (see also Fig. 4f-i). Clade expansions drawn to relative scale with Schizophora representing an estimated 40,000 species (Wiegmann et al. 2011). *Yellow* coloring highlights the cyclorrhaphan clades characterized by acephalic larval morphology. Cell types color-coded as in Fig. 3

that the phototransduction protein machinery is localized in the BO photoreceptor lamellae compartment. However, this has not been demonstrated by protein localization studies yet, nor has the functional significance of the lamellar organization of the BO been investigated. A defining feature of the BO organization of the larval eyes in schizophoran Diptera like *Drosophila* and *Musca* is that the lamellae are directed against the head cuticle and thus pointing opposite to the direction of the incoming light, which represents a derived state (Melzer and Paulus 1989; Fig. 1d and 2).

Neural Anatomy

The BO photoreceptors extend their axons into a joint nerve fascicle: the Bolwig nerve (BN) (Steller et al. 1987; Fig. 1a, b). The BN projects into what is now recognized as the larval optic neuropile (LON) (Sprecher et al. 2011; Fig. 1a, c). The small LON is located in the protocerebrum and to a large extent surrounded by the outer optic lobe anlage (OOLA), which will give rise to the outer optic neuropiles of the adult visual system (Schmucker et al. 1997; Sprecher et al. 2011): the lamina and the medulla. In contrast to the massive expansion of the OOLA during postembryonic development due to the proliferation of visual interneurons, there is no evidence of proliferative growth of the LON (for detailed discussion, see Friedrich 2011). Moreover, in contrast to the adult photoreceptors, the BO photoreceptors are characterized by directly contacting a diverse set of circadian neurons and interneurons. Comparative evidence suggests that the LON represents an extremely reduced and compacted evolutionary derivative of a part of an ancestral OOLA (Friedrich 2011).

The neuroanatomical organization of the LON is complex and only partially understood at this point. Most relevant for understanding the organization of the BO is the fact that two photoreceptor projection target areas have been described in the LON (Sprecher et al. 2011). The majority, if not all, of BO photoreceptors have synaptic contacts in the distal layer, while the proximal layer is only reached by the Rh5 subset of BO neurons (Fig. 1c).

A number of nonvisual interneurons that innervate the LON have been identified. This includes: (1) The functionally still enigmatic three glutamatergic optic lobe pioneer (OLP) neurons (Sprecher et al. 2011; Tix et al. 1989a). (2) A subset of five circadian system controlling lateral neurons (LNs; Iyengar et al. 2006; Kaneko and Hall 2000; Malpel et al. 2002; Mazzoni et al. 2005). (3) Putatively, two cholinergic interneurons with cell bodies in the dorsal protocerebrum (Iyengar et al. 2006). (4) A subset of serotonergic neurons in the central brain, whose dendritic arborizations in the LON increase throughout larval development (Mukhopadhyay and Campos 1995; Rodriguez Moncalvo and Campos 2005). (5) Octopaminergic and tyraminergic neurons which have thus far not been characterized in detail (Sprecher et al. 2011). (6) Preliminary evidence has also been reported that the Rh5 photoreceptors make contacts outside the LON (Keene et al. 2011).

At this point, only the above-mentioned LNs and OLP neurons have been identified with confidence as postsynaptic targets of the BN (for detailed discussion, see Sprecher et al. 2011). Remarkably, molecular genetic analysis revealed that the negative phototactic behavior of the *Drosophila* larva is mediated through a specific set of pacemaker neurons in the larval brain, which are contacted by the BO through interneurons that are yet unknown (Keene et al. 2011). The integration of the larval visual system with the biological clock system thus extends to both the perception of light and the regulation of circadian behavior (Keene and Sprecher 2011).

BO Photoreceptor Subtypes and their Homologs in the Adult Eye

Except for the differential projection of photoreceptor axons into the larval optic neuropile (Sprecher et al. 2011), the structural organization of the BO gives little hint regarding the cellular diversity among the BO photoreceptors. However, the two photoreceptor subtypes in the BO are also differentiated by their sensory and functional characteristics. At the sensory level, they differ by the differential expression of the opsin paralogs Rh5 and Rh6 (Fig. 3a; Helfrich-Förster et al. 2002; Malpel et al. 2002; Sprecher et al. 2007). The four to five of the founder photoreceptors express Rh5, which is associated with maximal sensitivity to blue light. The remaining peripheral photoreceptors express the green-sensitive paralog Rh6. Consistent with the expression of green- and blue-sensitive opsins in the BO, the larvae of the schizophoran fly species *Calliphora vicina* show maximal physiological stimulation of the BN by these wavelengths compared to violet, red and yellow (Hinnemann et al. 2010). While only the Rh5-positive photoreceptors are required for the avoidance of blue light, both the founder and peripheral photoreceptors have been found to be involved in circadian clock regulation (Keene et al. 2011).

The differential expression of Rh5 and Rh6 is not only important for discriminating BO photoreceptor subpopulations but also reveals homology relationships in the adult eye in conjunction with additional structural and developmental genetic data (for detailed discussion see Friedrich 2008). The situation is most straightforward in the case of the Rh5-positive BO photoreceptors. In the adult eye, the Rh5 opsin is expressed in the adult R8 photoreceptors of the pale-type ommatidia (Fig. 3b). The homology between the Rh5-positive BO photoreceptors to the adult R8 photoreceptors is also supported by the correspondence of the deeper axonal projection of the Rh5-positive BO photoreceptors in the LON and of the adult R8 cells into the medulla (Fig. 3b). Finally, the axons of the Rh5-positive BO photoreceptors form the median core of the BN, similar to the central position of the R8 axon in the axonal fascicles that exit a single ommatidium in the adult eye (Sprecher et al. 2011).

The shorter projection of the Rh6-positive BO photoreceptors presents evidence that these correspond to the peripheral photoreceptors in the adult eye, which project into the peripheral-most layer of the adult optic brain: the lamina (Fig. 3b). The recognition of the homology of the peripheral photoreceptors in the larval and adult eye is obstructed by the fact that the adult peripheral photoreceptors express a different opsin paralog: the Rh1 gene. To make matters worse, Rh6 is expressed in the R8 cells of yellow ommatidium subtype photoreceptors of the adult eye. The resolution to this conundrum is that the Rh1 and Rh6 opsins are phylogenetically related. Both are long wavelength sensitive-opsins which originated through a gene duplication during dipteran evolution (Bao and Friedrich 2009). In the eyes of most other insects, the same long wavelength opsin is expressed in both R8 and the peripheral

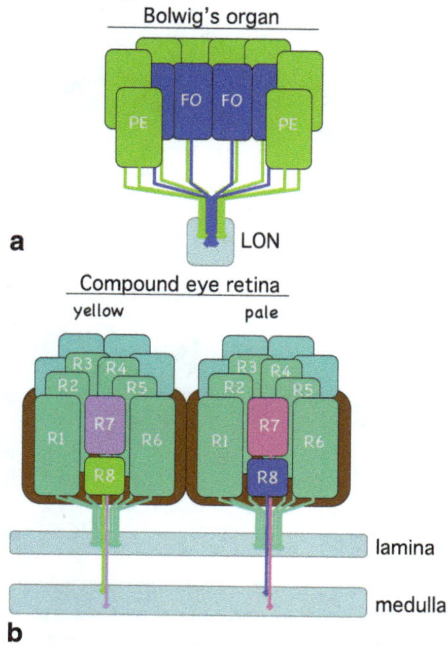

Fig. 3 Homologies of larval and adult photoreceptor subtypes. **a** Opsin expression and neuropile connectivity of the *Drosophila* Bolwig's organ. Opsin paralog Rh5 is expressed in the founder photoreceptors (*FO blue* filling). Opsin paralog Rh6 is expressed in the peripheral photoreceptors (*PE green* filling). *LON* larval optic neuropile. **b** Opsin expression and neuropile connectivity of the yellow- and pale-type ommatidia in the *Drosophila* main retina. Opsin paralog Rh1 is expressed in the six peripheral photoreceptors (R1–R6, *turquoise* filling) per ommatidium. Opsin paralogs Rh5 and Rh6 are expressed in the R8 photoreceptors of pale- and yellow-type ommatidia, respectively, as indicated by corresponding *blue* and *green* filling. Other cell type color codes: *Purple* R7 UV-sensitive photoreceptors; *brown* pigment cells, *light blue* cone cells

photoreceptors (Friedrich 2008; Friedrich et al. 2011). The expression of Rh6 in the peripheral BO photoreceptors is therefore consistent with their ancestral status as peripheral photoreceptors.

In summary, the 12 photoreceptors of the BO correspond to the fusion of 4–6 compound eye ommatidia after evolutionary reduction of all accessory cells as well as of the UV-sensitive R7 photoreceptor (compare Fig. 3a, b; see also Fig. 2).

Lineage and Morphogenesis

The BO photoreceptors derive from defined lateral fields of neuroectodermal cells in the developing head region of the *Drosophila* embryo: the optic placodes (Daniel et al. 1999; Green et al. 1993; Schmucker et al. 1997). These anlagen fields trace back to a cell population in the median dorsal head area of the embryo: the visual field or anlage (Fig. 4a; Chang et al. 2001). Establishing the lineage connection

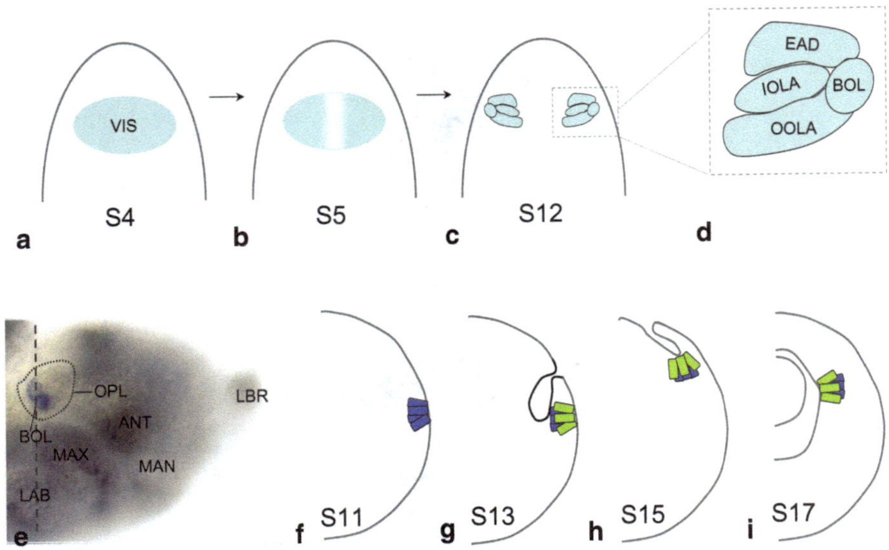

Fig. 4 Major steps of embryonic Bolwig's organ development and morphogenesis. **a–d** Schematics describing the development of the visual anlage. (Based on Chang et al. 2001). The *Drosophila* embryonic head region is shown from dorsal perspective with the anterior end pointing up. The visual anlage originates as dorsomedial field of neuroectodermal cells at developmental stage 4 (S4) and separates into lateral halves during developmental stage 5 (S5). In stage 12 embryos, the visual anlage divides into precursor tissues of different components of the *Drosophila* visual system including the outer optic lobe anlagen (OOLA) that will form the lamina and medulla, the inner optic lobe anlage (IOLA), which forms the lobula neuropile, the anlage of the larval eye or Bolwig's organ (BOL), and part of the eye-antennal imaginal disc (EAD). **d** Higher magnification view of relative positions of visual system primordia **e** Lateral view of *Drosophila* embryonic head labeled by in situ hybridization for the segmentation marker *wingless* (*brown*) and the photoreceptor marker *gl* (*blue*). Larval eye differentiation has initiated at the ventral tip of the optic placode (OPL). The dorsal sector of the optic placode develops in the outer and inner optic lobe anlagen. Segmental abbreviations: LAB labial segment, LBR labrum, MAX maxillary segment. *Hatched line* indicates section level of schematics in **f–i**. Anterior to the right and dorsal up. **f–i** Schematic cross-sectional view of the right side of the *Drosophila* embryo at different stages of development. **f** The Bolwig's organ founder cell cluster has formed in the lateral ectoderm. **g** The Bolwig's organ has acquired rosette morphology and the outer optic lobe anlagen tissue (*black outline*) has started to invaginate. **h** The outer optic lobe anlagen tissue has detached from the ectoderm. The Bolwig's organ has started to move medially, attached to the forming pharyngeal pocket during head involution. **i** The Bolwig's organ has reached its medial inverted position in the inverted embryonic head, attached to the forming pharyngeal pocket during head involution. (Adapted from Green et al. 1993)

between the optic placodes and the visual anlage has been challenging due to the complex tissue movements during embryonic head involution. Based on cell tracing studies, the visual anlage has been mapped to one of three mitotic domains in the early blastoderm head: mitotic domain 20 (Namba and Minden 1999). This area is characterized by the coexpression of the developmental transcription factor genes *sine oculis* (*so*) and *eyes absent* (*eya*; Fig. 4a; Bonini et al. 1997; 1993; Chang et al. 2001; Cheyette et al. 1994; Serikaku and O'Tousa 1994). Chang et al. (2001)

consider the expression domain of *so* as defining the boundaries of the visual anlage. Consistent with this, the *so* expressing cells give rise to all derivatives of the visual system, as well as a number of central brain neurons which develop connections between different visual processing centers (Chang et al. 2003).

By stage 8 of *Drosophila* embryonic development (Hartenstein 1993), the visual anlage is broken up into a pair of lateral optic placodes (Fig. 4). The BO photoreceptors develop from the ventral margin area of the optic placode (Fig. 4c, e). Morphologically visible BO photoreceptor differentiation and bundle formation initiates in stage 13 embryos (Daniel et al. 1999; Green et al. 1993; Schmucker et al. 1997). During this process, the early developing BO is layered into the four to five founder cells, which take a basal position to the surrounding peripheral cells during stage 13 and 14. In stage 15 to 17 embryos, the developing BO delaminates from the peripheral neuroectoderm and initiates its relocation towards the cephalopharyngeal head skeleton, coming first to rest in the dorsal side of an internal lumen: the dorsal pouch (Daniel et al. 1999; Green et al. 1993; Schmucker et al. 1997).

The BO develops in close association with delaminating precursor cells of the eye-antennal imaginal disc as well as the cell population of the OOLA neurons (Fig. 4a; Younossi-Hartenstein et al. 1993). This has been established by examining the BO-specific expression of *Kruppel* (*Kr*) and the OOLA-specific expression of *disconnected* (*disco*) (Schmucker et al. 1997). While the eye-antennal imaginal disc and BO primordial separate during subsequent development, the OOLA and BO remain connected by a bridge of glia-like cells, which play a role as guidepost cells during the projection of the BN (Schmucker et al. 1997).

Another important cell derivative from the optic placode are the three OLP neurons (Chang et al. 2003), which become situated in proximity of the LON and maintain synaptic contact with the BO axon endings in the LON (Sprecher et al. 2011).

Genetic Specification

The visual anlage of the *Drosophila* blastoderm embryo is characterized by the co-expression of *eya* and *so*, which is regulated by Decapentaplegic (Dpp) signaling in a dosage-responsive manner (Chang et al. 2001). Dpp signaling from the dorsal midline is essential for the induction of *eya* and *so*. Eventually, however, high Dpp signaling levels in the midline induces the expression of *zerknullt* (*zen*), which represses *eya* and *so*, leading to the split of the visual anlage into the optic placode domains (Fig. 4a-d and 4a).

Consistent with the expression of *eya* and *so* in the visual anlage, all components of the visual anlage, including the BO, are missing in *so*-mutant *Drosophila* (Cheyette et al. 1994). The BO primordium stands out as being derived from cells of the visual anlage which maintain *so* expression, while the expression of *so* ceases in the precursor cells of other derivatives of the visual system (Chang et al. 2003; Cheyette et al. 1994). It is not clear whether this latter event is due to repression or discontinued activation, which would shed light on the default state of visual anlagen cells.

Published expression data suggest that the visual anlage expression of *eya* and *so* may be partially overlapping with the Pax6 transcription factor *twin of eyeless* (*toy*; Czerny et al. 1999). Remarkably, however, the BOs form normally in embryos null mutant for *toy* and its sister paralog *eyeless* (*ey*; Suzuki and Saigo 2000). This is consistent with the expression of *ey* in the segregating eye-antennal disc in contrast to the lack of *ey* expression in the developing BO (Daniel et al. 1999). Thus, the Pax6 genes *toy* and *ey* are dispensable for the development of the BO in stark contrast to the sensitivity of the adult eye to Pax6 reduction (Quiring et al. 1994). These results further imply that the activation of *eya* and *so* is independent of *ey* and *toy*, marking a second difference to the gene regulatory interactions in the developing adult eye (Kumar and Moses 2001; Suzuki and Saigo 2000).

The transition from specification to differentiation in the prospective BO photoreceptors is marked by the onset of the expression of the proneural transcription factor *atonal* (*ato*) as early as stage 10 (Fig. 5b, c; Daniel et al. 1999; Jarman et al. 1993, 1994; Suzuki and Saigo 2000). The activation of *ato* depends on Hedgehog (Hh) signaling activity (Borod and Heberlein 1998; Curtiss and Mlodzik 2000; Daniel et al. 1999; Dominguez 1999; Dominguez and Hafen 1997; Greenwood and Struhl 1999; Heberlein et al. 1995; Jarman et al. 1993, 1994; Suzuki and Saigo 2000), drawing another parallel to the developmental regulation in the adult eye. Hh signaling has been identified as an essential agonist of both BO anlage and eye-antennal disc development (Chang et al. 2003; Schmucker et al. 1994; Suzuki and Saigo 2000).

There may be two phases of *hh* involvement during the early development of the BO. In the first, well-documented phase, *hh* expression in a preantennal domain anterior of the BO primordium promotes the specification of the BO at a distance (Chang et al. 2001; Suzuki and Saigo 2000). Subsequently, the preantennal *hh* domain relinquishes but *hh* becomes expressed in a subset of the early BO photoreceptor cells (Chang et al. 2001). Given the early role of *hh* in the differentiating adult R8 cells (Heberlein et al. 1993; Ma et al. 1993), it is tempting to speculate that *hh* in the founder cells continues to contribute to peripheral BO photoreceptor specification.

Ato expression is initially detected in a larger number of cells (14) but decreases by stage 12 until *ato* expression eventually disappears (Fig. 5c; Suzuki and Saigo 2000; Chang et al. 2001). This dynamic expression is consistent with the initial activation of *ato* in a larger number of precursor cells, as in the adult eye (Dokucu et al. 1996), and the subsequent restriction to the founder photoreceptors. Further consistent with this hypothesis, the specification of the peripheral BO photoreceptors is independent of *ato* (Daniel et al. 1999; Sprecher et al. 2007), relying instead on the epithelial growth factor receptor (EGFR) ligand source Spitz (Spi) expressed from the founder cells (Fig. 5c; Daniel et al. 1999; Suzuki and Saigo 2000; Daniel et al. 1999). In addition, *pointed* (*pnt*) and *argos* (*aos*) are expressed in the larval eye primordium (Daniel et al. 1999), consistent with the activation of EGFR signaling (Daniel et al. 1999; Suzuki and Saigo 2000). Interestingly, *aos* has also been connected to BO development based on mutagenesis data (Daniel et al. 1999; Schmucker et al. 1997).

In the absence of EGFR signaling, the prospective peripheral BO cells die (Daniel et al. 1999; Sprecher et al. 2007). Cell death rescue experiments suggest that EGFR is primarily a survival signal in the peripheral cells but not an instructive differentiation

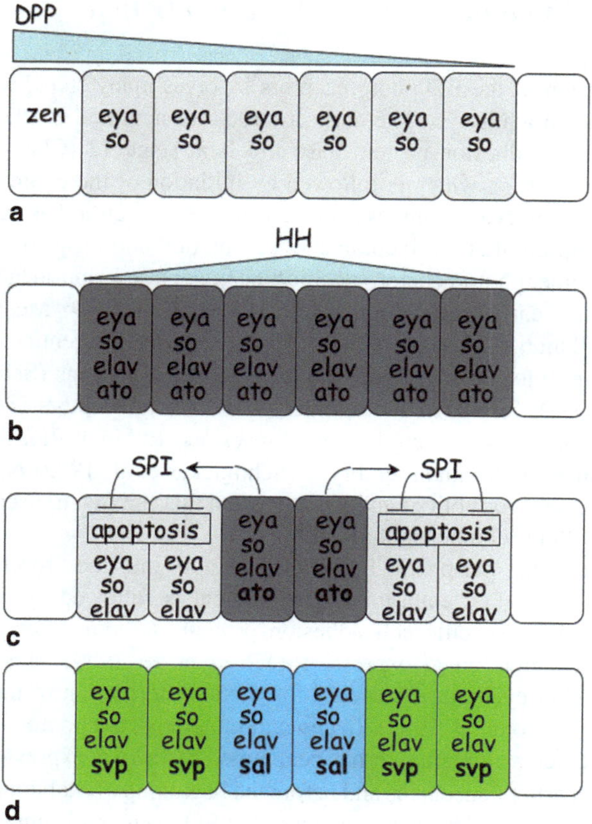

Fig. 5 Extracellular signals and differential gene expression during Bolwig's organ specification. *Bold font*: differentially expressed genes in Bolwig's organ precursor cells. **a** Activation of *so* and *eya* expression by Dpp signaling during visual anlagen and optic placode specification (see also Fig. 3a-d). **b** Positive regulation of Bolwig's organ precursor cell status and widespread activation of *ato* expression by Hh signaling. *Dark gray* indicates Bolwig's organ precursor cell population. **c** Restriction of *ato* expression to the Bolwig's organ founder cell subpopulation, which activates EGFR-signaling via Spi in adjacent Bolwig's organ precursor cells, repressing cell death. **d** Implementation of founder (*light blue*) versus peripheral Bolwig's organ photoreceptor cell state (*light green*) through the differential expression of *svp* and *sal*.

signal (Daniel et al. 1999; Sprecher et al. 2007). The precise mechanisms leading to peripheral photoreceptor specification are still unknown.

While *ato* appears to play a more sustained role in the founder cells given its longer expression compared to the peripheral cell precursors (Fig. 5c), *ato* may also be involved in or associated with proneural specification of the peripheral cells. This is tentatively suggested by the onset of the neural marker *embryonic lethal abnormal vision* (*elav*) in all BO precursor cells following the initiation of *ato* in the BO cell precursor population (Daniel et al. 1999; Suzuki and Saigo 2000). In contrast to *ato*, *elav* expression is sustained in both the peripheral and founder cell population, suggesting that the earliest expression *ato* is associated with the neural priming of all BO precursors.

Regulation of Photoreceptor Cell Differentiation

The differentiation of the BO photoreceptors involves many steps, which are also well known in the adult eye. The earliest structural neuronal marker detected in the BO is the cell adhesion protein Fasciclin II at stage 12 (Chang et al. 2003; Schmucker et al. 1997), which is followed by initiation of the expression of *Kr* in the late stage 12 embryo (Schmucker et al. 1997). *Kr* function has been suggested to control BO differentiation, including BN formation and projection (Schmucker et al. 1992). Another photoreceptor-specific zinc finger transcription factor expressed starting from this early stage is *glass* (*gl*; Moses et al. 1989; Moses and Rubin 1991). In *gl* null mutant *Drosophila*, adult photoreceptors differentiate but die or fail to enter photoreceptor quality, remaining unspecialized neurons (Schmucker et al. 1997, 1994, 1992). Similarly, the number of BO photoreceptors is reduced in *gl* null mutant embryos (Moses et al. 1989; Moses and Rubin 1991). Similar effects are observed in *Kr* null mutant embryos (Schmucker et al. 1992) but the analysis of embryos deficient for both *gl* and *Kr* suggested that the two transcription factors function in parallel (Schmucker et al. 1992).

The joining of the peripheral BO photoreceptors into the "rosette" formation starting from stage 13 of embryonic development is followed by the expression of the photoreceptor-specific cell adhesion protein *chaoptic* (*chp*; Reinke et al. 1988), which depends on *gl* expression (Moses et al. 1989; Moses and Rubin 1991). In the adult eye, *chp* is required for accurate formation of the rhabdomere microvilli (Van Vactor et al. 1988). In *chp* null mutations, the photoreceptors can be missing rhabdomeres altogether. It has been noted that *chp* is expressed in the adult photoreceptors earlier than opsins and before the beginning of rhabdomere formation (Van Vactor et al. 1988). The development of the BO in *chp* null-mutants has not yet been investigated. In the adult eye, Chp is required for the proper arrangement of the microvilli, the rhabdomeres, and the separation of the rhabdomeres in combination with the membrane proteins Prominin (Prom) and Eyes shut (Eys) (Zelhof et al. 2006). Interestingly, the Prom and Eys proteins are not expressed in the BO (Andrew Zehlof, personal communication). It is therefore tempting to hypothesize that the difference in rhabdomere organization between adult and larval photoreceptors is mediated through the differential regulation of Prom and Eys. It is further interesting to note that the elaboration of the BO photoreceptor lamellae occurs during postembryogenesis, in the first instar larva, consistent with the relatively late initiation of opsin expression in the late embryo (Sprecher et al. 2007).

Understanding the regulation and function of differential Prom and Eys expression in the BO might yield insights into the roles of additional photoreceptor-generic transcription factors in the BO photoreceptors. This includes the homeodomain transcription factors *orthodenticle* (*otd*) and *PvuII-PstI homology 13* (Pph13; Goriely et al. 1999; Vandendries et al. 1996), known to cooperate in the activation of opsin genes and other structural photoreceptor genes (Fichelson et al. 2012; Mishra et al. 2010; Ranade et al. 2008; Tahayato et al. 2003; Vandendries et al. 1996). As elaborated further below, *otd* also contributes to the regulation of differential opsin expression in the BO.

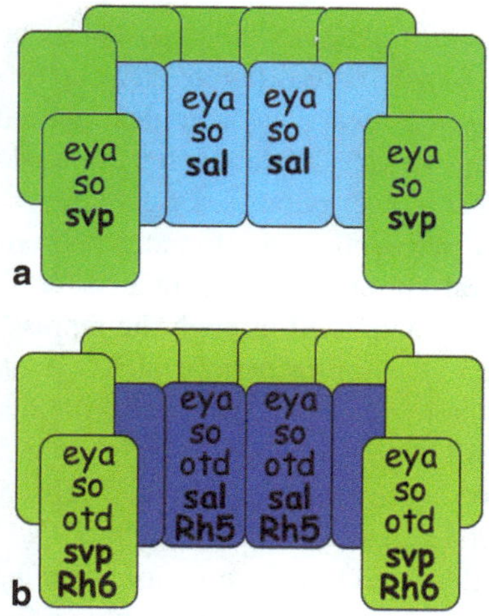

Fig. 6 Differential gene expression during Bolwig's organ photoreceptor subtype specification. *Bold font*: differentially expressed genes in Bolwig's organ precursor cells. **a** Bolwig's organ rosette cluster prior to opsin expression at stage 13 when *svp* and *sal* have become differentially expressed in peripheral and founder photoreceptors respectively, (Sprecher et al. 2007). **b** Differential expression of opsin genes Rh5 and Rh6 in founder (*dark blue*) and peripheral (*dark green*) Bolwig's organ photoreceptors respectively, starting from embryonic stage 16.

Genetic Regulation of Photoreceptor Subtype Development

Concomitant with the initiation of photoreceptor-generic gene products, the BO photoreceptors diversify genetically and structurally into the founder and peripheral photoreceptor subtypes (Fig. 5 and 6). This distinction is first implemented through the differential activation of *ato* in the course of photoreceptor specification (Fig. 5b, c). How this initial distinction translates into the subsequent steps of differential photoreceptor differentiation has not yet been fully elucidated. However, as stated previously, there is insight into the mechanisms that lead to the differential expression of the Rh5 and Rh6 opsin paralogs.

One important player in this process is *otd* (Sprecher et al. 2007), which is uniformly expressed in all BO photoreceptors starting from embryonic stage 12 (Fig. 6a). If *otd* is depleted from the differentiating BO, the founder photoreceptors fail to express Rh5 despite being correctly specified. Instead, the *otd*-founder cells express Rh6, reflecting a repressive role for *otd* on Rh6 during normal BO development. The differential action of *otd* in the founder and peripheral cells is thus identical to the effect of *otd* on opsin expression in the adult eye (Tahayato et al. 2003) and consistent with the cell-context-dependent effects of *otd* (Johnston et al. 2011; McDonald et al. 2010; Mishra et al. 2010). These data further imply the existence of additional differentially expressed regulators that tune the effect of *otd* in a photoreceptor subtype-specific manner.

Consistent with this, other gene activities so far identified to affect the Rh5 versus Rh6 transcription decision are differentially expressed transcription factors (Fig. 5d, e). This includes the *spalt* (*sal*) complex zinc finger genes (*spalt major* and *spalt related*; Domingos et al. 2004a; Mollereau et al. 2001), and the orphan nuclear receptor zinc finger transcription factor gene *seven up* (*svp*; Mlodzik et al. 1990). *Sal* is specifically activated in the founder photoreceptors whereas *svp* is expressed in the peripheral photoreceptors (Sprecher et al. 2007). This mutually exclusive pattern is in part established through the repressive effect of *svp* on *sal*. In support of this, *sal* is expressed in peripheral BO photoreceptors when *svp* is genetically eliminated from these cells. Correlated with this, the expression of Rh5 is expanded into peripheral photoreceptors (Sprecher et al. 2007). Further, ectopic *svp* suppresses *sal* and Rh5 expression in the founder cells. *Svp* is therefore essential and sufficient for peripheral cell differentiation in the BO.

The inconsequential effect of ectopic expression of *sal* in the peripheral photoreceptors revealed that *sal* acts as an essential facilitator for Rh5 expression in the founder photoreceptors but, unlike *svp*, has no instructive effect for implementing the founder photoreceptor fate. As stated before, the mechanisms by which the differential expression of *svp* and *sal* is established remain to be explored. The presumably foundational role of *ato* is difficult to study because of the early termination of BO development in *ato*-deficient embryos (Suzuki and Saigo 2000). Taken together, however, the early specification of the founder cell state and the repression of *sal* in the secondary peripheral cell state suggest that the founder cell fate is the default state of the developing BO photoreceptors which must be modified by *svp*.

Also, the exact mechanism by which the expression of Rh5 and Rh6 are activated and controlled in the BO remain to be specifically tested and explored (Fig. 6). Based on the adult eye paradigm, it is reasonable to assume the direct involvement of *Pph13* and *otd* (Mishra et al. 2010; Tahayato et al. 2003).

Postembryonic Development

Early studies of the changes in the *Drosophila* visual system during metamorphosis found evidence that the BOs were subjected to cell death and autophagocytosis during the transition from larval to adult stage (Tix et al. 1989b). Remarkably, however, the BO is not completely decomposed: rather, only a subpopulation is removed and the remaining photoreceptors relocate towards the adult optic neuropiles (Fig. 7). Key to unraveling this process was the use of reporter gene expression to track the final fate of the BO, which led to the discovery of their relationship to a small bundle of extraretinal photoreceptors known as the Hofbauer-Buchner (HB) eyelets in the adult optic neuropiles (Helfrich-Förster et al. 2002; Hofbauer and Buchner 1989; Yasuyama and Meinertzhagen 1999). Subsequent analyses confirmed that the HB eyelets are the adult derivatives of the BO (Sprecher and Desplan 2008). During this transformation, the larval BO is reduced to its four to five founder cells (Fig. 7). The peripheral photoreceptors are removed via programmed cell death

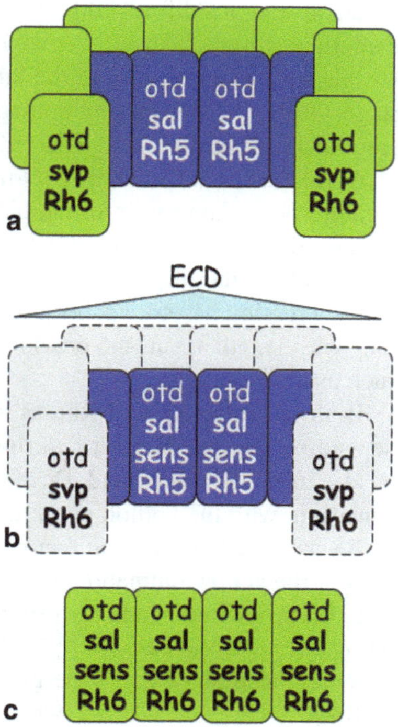

Fig. 7 Extracellular signals and differential gene expression during postembryonic Bolwig's organ development. *Bold font*: differentially expressed genes. **a** Gene expression in the Bolwig's organ during the first two larval instars. **b** Initiation of cell death in the peripheral Bolwig's organ photoreceptors (*gray background, dashed outline*) in the late third larval instar. **c** Initiation of Rh6 expression in the Bolwig's organ founder photoreceptors.

(Helfrich-Förster et al. 2002; Sprecher and Desplan 2008), consistent with the earlier notion of cell death associated with the metamorphic BO (Tix et al. 1989b). The founder cell bundle-turned-HB eyelet is relocated to a position underneath the posterior retina, and acquiring a new morphology, which is characterized by the elaboration of rhabdomeres (Helfrich-Förster et al. 2002; Hofbauer and Buchner 1989; Sprecher and Desplan 2008; Yasuyama and Meinertzhagen 1999). At this point, and throughout adulthood, the HB eyelets project into the accessory medulla, a small neuropile anterior to the medulla proper. In addition, the founder cells relinquish Rh5 expression and initiate Rh6 expression instead (Fig. 7b, c). Both the apoptotic removal of the peripheral cells and the opsin switch in the founder cells are contingent on the activation of Ecdysone (Ecd) signaling in the BO cells (Helfrich-Förster et al. 2002; Sprecher and Desplan 2008). This identifies the modification of the postembryonic BO as part of the Ecdysone-induced early onset of metamorphosis (Truman and Riddiford 2002).

Of note, the redeployment of the larval eyes as extraretinal photoreceptors has been well known in other holometabolous insect orders (Fleissner et al. 1993; Hagberg 1986; Ichikawa 1991; Sokoloff 1972), representing a defining character state of holometabolous insects (Friedrich et al. 2006; Kristensen 1999; Nuesch 1987). Moreover, the postembryonic "dedifferentiation" (Sprecher and Desplan 2008) of the BO, represents one of many facets of the general theme of reinitiated development of larval structures during metamorphosis in holometabolous insects

(Truman and Riddiford 2002). Interestingly, there are examples of convergent evolution of nymph-specific eye morphologies in direct-developing insects, which are likewise maintained into the adult but remain peripheral visual organs at this stage. One such case is the nymph-specific eye of advanced scale insects (Coccoidea: Hemiptera; Buschbeck and Hauser 2009).

At the functional level, the BO-turned-HB eyelet continues to play a role in collecting visual daytime information for the biological clock with the redundantly organized zeitgeber system of the adult fly (Helfrich-Förster et al. 2002; Veleri et al. 2007). While the physiological significance of the correlated switch from Rh5 to Rh6 expression has not yet been investigated, it is a reasonable assumption that the long wavelength opsin Rh6 provides higher sensitivity to deep tissue photoreceptors such as the HB eyelets.

Both the expression of Rh6 as well as founder cell survival and differentiation depend on the zinc finger transcription factor gene *senseless* (*sens*; Fig. 7c; Nolo et al. 2000; Sprecher and Desplan 2008). Interestingly, these events are also consistent with the homology of the BO founder cells to the adult R8 cells. *Sens* is an R8-specific transcription factor in the adult eye and Rh6 is expressed in the R8 cells of the yellow ommatidia (Fig. 3; Frankfort et al. 2001, 2004; Xie et al. 2007). Both the specification and the differentiation of the adult R8 photoreceptors depend on *sens* (Frankfort et al. 2001, 2004; Morey et al. 2008; Xie et al. 2007). Moreover, *sens* has been shown to synergistically activate Rh6 expression in combination with *otd* (Xie et al. 2007), which continues to be expressed in the BO during metamorphosis (Fig. 7). It is thus reasonable to assume that *sens* and *otd* operate in the same way in the adult R8 and the postembryonic BO founder cells.

Gene Regulatory Synapomorphies of BO and Adult Eye Development

At many levels, the BO differs dramatically from the development and organization of the *Drosophila* compound eye. It is therefore striking to note the numerous similarities at the structural, functional, and genetic level, documenting the shared evolutionary origins of larval and adult eye (Table 1). Stringent comparative evidence of a close evolutionary relationship between species, structures, or traits rests on shared derived character states (synapomorphies) as opposed to evolutionarily older traits that characterize larger, more inclusive groups of species or structures (symplesiomorphies). Therefore, demonstrating the uniquely close relationship of the BO and adult *Drosophila* eye relative to other visual organs such as the *Drosophila* ocelli, requires definition of features that are synapomorphies of the two organs rather than features that are shared with other visual organs and hence symplesiomorphies (for detailed discussion, see Friedrich 2006a).

For instance, the involvement of *eya, so, hh,* and *dpp* in primordium specification is likely an ancient characteristic of visual organ development, predating the origin of the insect compound eye (Friedrich 2006a). The same can be assumed regarding the

Table 1 Commonalities in *Drosophila* adult and larval eye development. The first four rows indicate candidate ancestral processes (symplesiomorphies). Further rows indicate candidate shared derived processes synapomorphies. (For additional discussion, see Friedrich 2006a)

Regulatory process	References
Activation of *eya* and *hh* in primordial cells by Dpp signaling	Curtiss and Mlodzik 2000; Chang et al. 2001
Positive regulation of *ato* initiation by Hh signaling	Suzuki and Saigo 2000; Dominguez and Hafen 1997; Dominguez 1999; Greenwood and Struhl 1999
Selective activation of *ato* in retinal founder cells	Dokucu et al. 1996
Founder cell induction and survival independent of EGFR signaling activation	Dominguez et al. 1998
Peripheral photoreceptor survival by EGFR signaling	Halfar et al. 2001
Founder cell specific deployment of *sal*	Domingos et al. 2004a; Mollereau et al. 2001; Sprecher et al. 2007
Peripheral cell specific deployment of *svp*	Domingos et al. 2004b; Heberlein et al. 1991; Mlodzik et al. 1990; Sprecher et al. 2007
Founder cell specific deployment of *sens*	Sprecher and Desplan 2008
Selective expression of Rh5 in founder cells	Sprecher et al. 2007
Expression of Rh1 or Rh6 long wavelength opsin in the peripheral photoreceptors	Tahayato et al. 2003; Sprecher et al. 2007
Founder cell specific deployment of *otd* in Rh5 and Rh6 regulation	Tahayato et al. 2003

involvement of *ato*, *hh*, and EGFR-signaling during early visual organ differentiation. Some of the most compelling synapomorphies of the BO and adult eye concern mechanisms which regulate the specification and differentiation of the photoreceptor subtypes. This includes the involvement of *svp* and *sal* in inner versus outer photoreceptor subtype specification (Domingos et al. 2004b; Heberlein et al. 1991; Mlodzik et al. 1990; Sprecher et al. 2007), the deployment of *otd* in Rh5 induction and Rh6 repression (Tahayato et al. 2003), and the role of *sens* in founder/R8 photoreceptor differentiation (Frankfort et al. 2001, 2004; Morey et al. 2008; Xie et al. 2007).

Head Patterning Related Differences Between BO and Adult Eye Development

The commonalities in BO and adult eye development are contrasted by major differences in the genetic regulation of precursor tissue specification. In the adult eye, the activation of *so* and *eya* depends on the expression of *toy* and *ey* in the posterior eye-disc (Czerny et al. 1999; Halder et al. 1998; 1995; Kenyon et al. 2003; Niimi et al. 1999; Ostrin et al. 2006; Punzo et al. 2002). The dispensability of the Pax6 genes *ey* and *toy* for BO development and the apparent Pax6-independent induction of the *so* and *eya* in the visual anlage thus mark profound differences in the genetic networks that regulate the early precursor specification of the two organs (Czerny et al. 1999; Finkelstein et al. 1990; Quiring et al. 1994; Suzuki and Saigo 2000).

However, the early expression of *toy* in the visual primordium (Czerny et al. 1999) implies that the BO precursor cells experience a phase of transient *toy* expression. Further, for reasons still unknown, *toy* seems to act as an upstream regulator of *ey* in the eye antennal imaginal disc but not in the visual anlage of the early blastoderm embryo. Instead *toy*, as well as *ey*, are essential for the development of the mushroom bodies in the embryo (Furukubo-Tokunaga et al. 2009; Kurusu et al. 2000).

Of note, the BO also seems to develop independently of *dachshund* (*dac*), another transcription factor gene which is an important component of primordium determination in the adult eye (Mardon et al. 1994). This is indicated by the lack of *dac* expression in the visual anlage of the embryonic blastoderm (Kumar and Moses 2001). *Dac* is expressed in optic lobe placode in the midway *Drosophila* embryo but apparently not in the BO (Anderson et al. 2006).

Looking at the correlation between eye size reduction and Pax6 independence of the BO, it has been proposed that Pax6 is essential for the development of larger visual organs (Friedrich 2006a). Alternatively, *ey* and *toy* may primarily function as region-wide commitment regulators instead of organ-specific determination genes. The dramatic reorganization and reduction of the larval head in *Drosophila* may have relinquished the need for Pax6-mediated patterning of an ocular region. This hypothesis is supported by the Pax6 sensitivity of larval eye development in non-dipteran insect species with adult-like larval head capsules (Yang et al. 2009).

In addition, one has to keep in mind that the differences in the genetic specification mechanisms of BO and adult eye may reflect ancestral differences between the embryonic and postembryonic phase of compound eye development in directly developing insects (Fig. 8; see also Chap. 11 in this volume for more details and Friedrich 2006b). The diversity of factors playing into the gene regulatory divergence between BO and adult eye development is thus considerable. Notwithstanding this, the accumulating insights into the genetic basis of BO development offers novel inroads to elucidate the developmental evolution of this visual organ in detail.

R7 Cell Reduction-Related Differences Between BO and Adult Eye Development

Compared to the gene regulatory differences in BO and adult eye specification, it is more straightforward to explain differences at the stage of photoreceptor specification and differentiation. A common denominator of the shared genetic interactions in BO and adult eye development is their restriction to the earliest phases of photoreceptor specification in the adult eye, that is, the specification of R8 and the peripheral photoreceptors (see Table 1). This finding is explained by the regressive evolution of the BO from an ancestral compound eye-like state (Melzer and Paulus 1989), which resulted in the loss retinal accessory cells and the R7 photoreceptor (Fig. 3 and 2).

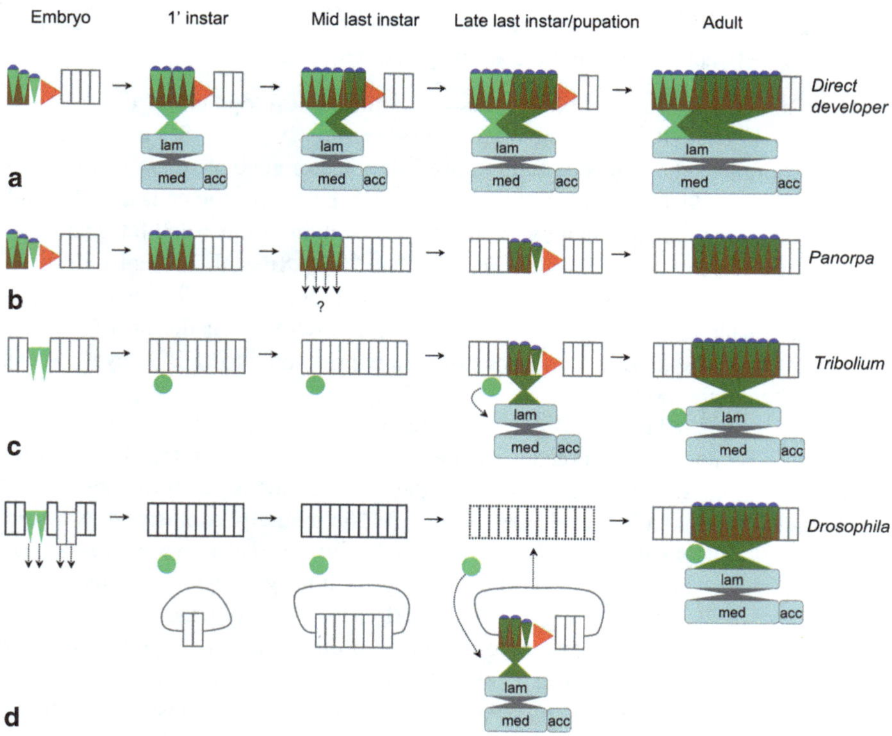

Fig. 8 Hallmarks of *Drosophila* visual system development in comparison to non-dipteran species. **a–d** Schematic alignment of corresponding stages of embryonic and postembryonic visual system development. **a** In direct-developing species, photoreceptors of the adult eye are generated during embryogenesis (*light green*) and postembryogenesis (*dark green*) by persisting differentiation at the anterior margin of the eye (*red arrow head*). The retinal photoreceptors project in the lamina (lam), forming the first optic chiasma. Axonal projections in the first optic chiasma from photoreceptors born during embryonic and postembryonic development are indicated by light and dark green, respectively. The lamina and medulla (med) neuropiles are connected through the second optic chiasma (*gray*). The accessory medulla (acc) neuropile is already present in juvenile instars receiving afferent input (not shown) from the medulla. **b** In scorpion flies of the genus *Panorpa*, a compound eye-like larval retina is formed during embryogenesis. These larval eyes do not become part of the adult retina but their postembryonic fate is not documented. **c** In *Tribolium*, highly reduced larval eyes (*circle*) are formed that lack accessory cells. During metamorphosis, the larval eyes are withdrawn in direction of the brain, where they become associated with the lamina neuropile in the adult form. *Tribolium* also develops an accessory medulla that is situated at the anterior margin of the medulla (Dreyer et al. 2010). **d** The basic course of larval eye development in *Drosophila* corresponds well with *Tribolium*. Differences include the final position of the larval eyes or Bolwig's organs as HB eyelets between retina and lamina, and the postembryonic initiation of pigment granule expression during metamorphosis. The *Drosophila* HB eyelets project into the accessory medulla (not shown). Unlike in *Tribolium*, the development of the larval eyes is integrated with the development of the adult retina from the eye-antennal imaginal disc (*light gray circular structure*) and the larval ectoderm (*black*) is replaced (*dashed black outlines*) during metamorphosis.

This dramatic reduction in cellular complexity explains the following genetic differences between the BO and the adult eye (Friedrich 2008):

1. Gene activities, which regulate accessory cell fate development or differentiate the R7 and R8 cell fates in the adult retina, are missing from BO development. One example is the homeodomain transcription factor gene *prospero* (*pros*; Chu-Lagraff et al. 1991). In the adult eye, *pros* is expressed in cone cells and R7, and is involved in processes that concern the specification of these cell fates and R7-specific differentiation (Charlton-Perkins et al. 2011; Xu et al. 2000). In addition, *pros* prevents the expression of the R8 photoreceptor cell-specific opsin paralogs Rh5 and Rh6 (Cook et al. 2003). None of these steps occur during BO development, consistent with the absence of *pros* expression during BO development (Sprecher et al. 2007).
2. Gene activities, which coordinate the differential expression of opsin paralogs in the adult R7 and R8 photoreceptors, are similarly missing from the BO. The differential expression of opsin paralogs in the inner R7 and R8 cells defines yellow and pale ommatidia in the adult eye (Fig. 3). The mutually exclusive expression of Rh5 or Rh6 in the R8 cells, coordinated with the respective expression of Rh3 or Rh4 in the R7 cells, is coordinated by complex gene network interactions that involve the genes *spineless* (*ss*), a bHLH-PAS family transcription factor, *melted* (*melt*), a pleckstrin homology domain protein, and *warts* (*wts*), a protein kinase (Mikeladze-Dvali et al. 2005). The developing BO has been probed for the expression of all of these genes (Sprecher et al. 2007). Consistent with the reduction of R7 and the associated cell fate decisions, neither *wts* nor *melt* nor *ss* are expressed in the BO.

Larval Eye Development in Other Insect Species

As previously pointed out (Friedrich 2008), there are three lines of evidence that document the evolution of insect larval eyes from ancestral compound eye ommatidia: (1) the shared gene activities in the larval and adult eyes of *Drosophila*, (2) similarities in the structure and morphogenesis of the larval eyes in holometabolous insects and the nymphal eyes of direct-developing species, and (3) the morphological conservation of ommatidial cell architecture in the larval eyes of ancestrally organized holometabolous species. Understanding the last two aspects requires taking a look at the development of the larval eye in other insect species.

Larval eyes composed of a small number of ommatidia-like units are typical for the majority of holometabolous insects, including Hymenoptera and Lepidoptera (Gilbert 1994; see also Chap. 11 in this volume). The most important key group in this context is the scorpion fly family Panorpidae. Scorpion flies (Mecoptera) represent an insect order closely related but distinctly different from the Diptera (Wiegmann et al. 2009). The eucephalic and predatory larvae of *Panorpa* species possess compound eye-like eyes (Chen et al. 2012; Gilbert 1994; Paulus 1979; Steiner 1930). This organization documents most directly the relationship of the holometabolous insect larval eye to the compound eye in the nymph of direct-developing insects (compare Fig. 8a, b; for detailed review see Friedrich 2006b). However, as in the case of other holometabolous insect larvae, the compound eye-like eyes of the *Panorpa* larva do not expand in size during larval development and are fully replaced by the newly differentiating adult compound eye during metamorphosis (Fig. 8b; Paulus 1989).

The second informative system for reconstructing the evolutionary origins of insect larval eyes is the red flour beetle *Tribolium castaneum* (Fig. 8c). At the morphological level, the visual organs of the eucephalic *Tribolium* larva represent similarly but independently reduced larval eyes that lack accessory cells (see also Chap. 11). In contrast to *Drosophila*, the *Tribolium* larval eye photoreceptors form regular rhabdomeres and a pigment cup through the endogenous expression of pigment granules (Liu and Friedrich 2004). Moreover, the *Tribolium* larval eyes are organized into two clusters on each side of the larval head (see Fig. 10 in Chap. 11). Interestingly, each of the two lateral pairs forms from an initial number of five precursor photoreceptor clusters through the differential fusion of clusters during morphogenesis (Liu and Friedrich 2004). This process is consistent with the proposal that larval eyes with noncanonical numbers of photoreceptors (>8) evolved through the fusion of ancestral ommatidial units (Paulus 1986). Of note, this aspect of the early development of the *Tribolium* larval eyes differs from the focused origin of multiple founder cells during *Drosophila* BO development. The latter could be explained by evolutionary changes facilitating the specification of several *ato*-expressing founders after the evolutionary reduction of the larval eye to a single ommatidial unit.

The independent evolutionary trajectories to the larval eyes of *Tribolium* and *Drosophila* await to be further elucidated. However, the striking correspondence between *Tribolium* and *Drosophila* with regards to the postembryonic relocation of the larval eyes into the adult optic neuropile compartments represents another piece of key evidence of the homology of larval eyes across holometabolous insect species and their relationship to part of the compound eye in direct-developing insects (Fig. 7c, d).

The Evolutionary Roots of the Bifunctional *Drosophila* Larval Visual System

The connections of the BO to the circadian network in the *Drosophila* larval brain shed further light on its evolutionary history. In many direct-developing insects, including cockroaches and grasshoppers, the most important pacemaker neurons are located in a neuropile of the visual system that is associated with the medulla and hence referred to as the accessory medulla (for review see Helfrich-Förster 2004). The latter contains several different interneurons some of which express the neuropeptide pigment dispersing factor (PDF), thus corresponding to the LN pacemaker neurons in the *Drosophila* visual system (Homberg et al. 2003; Homberg and Würden 1997;

Reischig and Stengl 2003). This system may be extremely old: PDF-expressing components have also been discovered in crustacean species (Harzsch et al. 2009; Strauß et al. 2011).

In both direct-developing and holometabolous insects alike, the accessory medulla is already present and active in the visual system during the juvenile stages. The persistence into the adult stage in direct-developing insects can be associated with extensive restructuring and expansion of connectivities in the circadian system (Vafopoulou and Steel 2012), marking a parallel to the postembryonic development of the *Drosophila* BO. However, unlike in *Drosophila*, the accessory medulla in hemimetabolous species appears to be receiving light input through intermediate neurons from the lamina, medulla, or both, as opposed to directly from retinal photoreceptors (Homberg and Würden 1997). As far as has been reported, this situation seems to be generally different in the larvae of holometabolous species. In the mosquito *Chaoborus crystallinus*, for example, the long-fibered larval photoreceptors have been shown to project directly into the accessory medulla while a population of short-fibered photoreceptors project into a larval lamina neuropile (Melzer 2009).

Taken together, these data reveal that the larval visual system of Drosophila represents an extreme compaction of the ancestral visual system that is still conserved in the nymphs of direct-developing insects, which connects the compound eye retina to both the phototactic components in lamina and medulla as well as to the circadian components in the accessory medulla.

The *Drosophila* BO as Paradigm of Regressive Evolution

Taken together, the detailed insights into the genetic regulation of BO development have corroborated earlier conclusions that this organ is paralogous to the adult compound eye, linking its origin to an ancestral postembryonic system with compound eye-like organization. While the conservation of a residual visual organ in juvenile instars is explained by the continued need for phototactic and circadian information, the question remains which adaptive or nonadaptive changes have led to the extreme reduction of the *Drosophila* larval eye. A key lead is the fact that most larval eyes in the Holometabola have been reduced to small assemblies of ommatidia-like structures (Gilbert 1994). The question therefore becomes one regarding the causes underlying this broad shared evolutionary trend in the Holometabola versus directly developing insects.

The organization of the entire body plan of holometabolous larvae exhibits a general trend towards reduction. In addition to the visual system, this is also visible regarding walking appendages, flight appendages and other sensory structures such as the antennae. In combination, these data suggest that the visual system regressed in response to factors that reshaped the entire postembryonic body plan.

A candidate factor is the adaptive acceleration of postembryonic growth, which is enhanced by efficient caloric uptake and metabolism, but also by the reduction of energy-expensive organs (Friedrich 2011). The effect on the visual system in particular is explained through the energy savings associated with the reduction of energy-expensive neuronal cells (for detailed reviews, see Niven and Laughlin 2008).

Regardless of the proximate cause of larval eye reduction in the Holometabola, it is clear that the larval eyes of *Drosophila* and higher Diptera can serve as a model case for studying mechanisms of visual system regression (Friedrich 2011). This will be greatly aided by the extensive comparative analyses, which have highlighted key steps such as accessory cell reduction and ommatidial fusion along the evolutionary path leading to the BO of schizophoran Diptera (Fig. 2; Melzer and Paulus 1989). There are thus exciting opportunities for mutually elucidating studies on the development, neurobiology, and evolution of the *Drosophila* BO. Recent work revealed the capacity of *Drosophila* larvae to detect the movement of other larvae, suggesting contrast-resolving vision (Justice et al. 2012) and raising the possibility of another component of conserved evolutionary heritage of compound eye characteristics.

Acknowledgements I am grateful to Amit Singh for the opportunity to provide this book chapter, Rewaa Yas for proofreading of the manuscript, Andrew Zelhof for sharing unpublished information and Tiffany Cook and Andrew Zelhof for valuable comments. Research in the Friedrich lab has been supported by NSF awards IOS 0951886, IOS 0091926, and EF-0334948.

References

Anderson J, Salzer CL, Kumar JP (2006) Regulation of the retinal determination gene *dachshund* in the embryonic head and developing eye of *Drosophila*. Dev Biol 297:536–549

Bao R, Friedrich M (2009) Molecular evolution of the *Drosophila* retinome: exceptional gene gain in the higher Diptera. Mol Biol Evol 26:1273–1287

Bolwig N (1946) Senses and sense organs of the anterior end of the house fly larvae. Vidensk Med Dansk Naturh Foren 109:81–217

Bonini NM, Leiserson WM, Benzer S (1993) The *eyes absent* gene: genetic control of cell survival and differentiation in the developing *Drosophila* eye. Cell 72:379–395

Bonini NM, Bui QT, Grayboard GL, Warrick JM (1997) The *Drosophila eyes absent* gene directs ectopic eye formation in a pathway conserved between flies and vertebrates. Development 124:4819–4826

Borod ER, Heberlein U (1998) Mutual regulation of *decapentaplegic* and *hedgehog* during the initiation of differentiation in the *Drosophila* retina. Dev Biol 197:187–197

Buschbeck EK, Hauser M (2009) The visual system of male scale insects. Naturwissenschaften 96:365–374

Caravas J, Friedrich M (2013) Shaking the Diptera tree of life: performance analysis of nuclear and mitochondrial sequence data partitions. Syst Entomol 38(1):93–103.

Chang T, Mazotta J, Dumstrei K, Dumitrescu A, Hartenstein V (2001) Dpp and Hh signaling in the *Drosophila* embryonic eye field. Development 128:4691–4704

Chang T, Younossi-Hartenstein A, Hartenstein V (2003) Development of neural lineages derived from the *sine oculis* positive eye field of *Drosophila*. Arthropod Struct Dev 32:303–317

Charlton-Perkins M, Whitaker SL, Fei Y, Xie B, Li-Kroeger D, Gebelein B, Cook T (2011) Prospero and Pax2 combinatorially control neural cell fate decisions by modulating Ras- and Notch-dependent signaling. Neural Develop 6:20

Chen Q, Li T, Hua B (2012) Ultrastructure of the larval eye of the scorpionfly *Panorpa dubia* (Mecoptera: Panorpidae) with implications for the evolutionary origin of holometabolous larvae. J Morphol 273:561–571

Cheyette BN, Green PJ, Martin K, Garren H, Hartenstein V, Zipursky SL (1994) The *Drosophila sine oculis* locus encodes a homeodomain-containing protein required for the development of the entire visual system. Neuron 12:977–996

Chu-Lagraff Q, Wright DM, McNeil LK, Doe CQ (1991) The *prospero* gene encodes a divergent homeodomain protein that controls neuronal identity in *Drosophila*. Development Suppl 2:79–85

Cook T, Pichaud F, Sonneville R, Papatsenko D, Desplan C (2003) Distinction between color photoreceptor cell fates is controlled by Prospero in *Drosophila*. Dev Cell 4:853–864

Curtiss J, Mlodzik M (2000) Morphogenetic furrow initiation and progression during eye development in *Drosophila*: the roles of *decapentaplegic*, *hedgehog* and *eyes absent*. Development 127:1325–1336

Czerny T, Halder G, Kloter U, Souabni A, Gehring WJ, Busslinger M (1999) *Twin of eyeless*, a second Pax-6 gene of *Drosophila*, acts upstream of *eyeless* in the control of eye development. Mol Cell 3:297–307

Daniel A, Dumstrei K, Lengyel JA, Hartenstein V (1999) The control of cell fate in the embryonic visual system by *atonal*, *tailless* and EGFR signaling. Development 126:2945–2954

Dobzhansky T (1973) Nothing in biology makes sense except in the light of evolution. Am Biol Teach 35:125–129

Dokucu ME, Zipursky SL, Cagan RL (1996) *Atonal*, *rough* and the resolution of proneural clusters in the developing *Drosophila* retina. Development 122:4139–4147

Domingos PM, Brown S, Barrio R, Ratnakumar K, Frankfort BJ, Mardon G, Steller H, Mollereau B (2004a) Regulation of R7 and R8 differentiation by the *spalt* genes. Dev Biol 273:121–133

Domingos PM, Mlodzik M, Mendes CS, Brown S, Steller H, Mollereau B (2004b) *Spalt* transcription factors are required for R3/R4 specification and establishment of planar cell polarity in the *Drosophila* eye. Development 131:5695–5702

Dominguez M (1999) Dual role for *hedgehog* in the regulation of the proneural gene *atonal* during ommatidia development. Development 126:2345–2353

Dominguez M, Hafen E (1997) *Hedgehog* directly controls initiation and propagation of retinal differentiation in the *Drosophila* eye. Genes Dev 11:3254–3264

Dreyer D, Vitt H, Dippel S, Goetz B, El Jundi B, Kollmann M, Huetteroth W, Schachtner J (2010) 3D standard brain of the red flour beetle *Tribolium castaneum*: a tool to study metamorphic development and adult plasticity. Front Syst Neurosci 4:3

Fichelson P, Brigui A, Pichaud F (2012) *Orthodenticle* and *Kruppel homolog 1* regulate *Drosophila* photoreceptor maturation. Proc Natl Acad Sci U S A 109:7893–7898

Finkelstein R, Smouse D, Capaci TM, Spradling AC, Perrimon N (1990) The *orthodenticle* gene encodes a novel homeo domain protein involved in the development of the *Drosophila* nervous system and ocellar visual structures. Genes Dev 4:1516–1527

Fleissner G, Fleissner G, Frisch B (1993) A new type of putative non-visual photoreceptors in the optic lobe of beetles. Cell Tissue Res 273:435–445

Frankfort BJ, Nolo R, Zhang Z, Bellen H, Mardon G (2001) *Senseless* repression of *rough* is required for R8 photoreceptor differentiation in the developing *Drosophila* eye. Neuron 32:403–414

Frankfort BJ, Pepple KL, Mamlouk M, Rose MF, Mardon G (2004) *Senseless* is required for pupal retinal development in *Drosophila*. Genesis 38:182–194

Friedrich M (2006a) Ancient mechanisms of visual sense organ development based on comparison of the gene networks controlling larval eye, ocellus, and compound eye specification in *Drosophila*. Arth Struc Dev 35:357–378

Friedrich M (2006b) Continuity versus split and reconstitution: exploring the molecular developmental corollaries of insect eye primordium evolution. Dev Biol 299:310–329

Friedrich M (2008) Opsins and cell fate in the *Drosophila* Bolwig organ: tricky lessons in homology inference. Bioessays 30:980–993

Friedrich M (2011) *Drosophila* as developmental paradigm of regressive brain evolution: proof of principle in the visual system. Brain Behav Evol 78(3):199–215.

Friedrich M, Dong Y, Jackowska M (2006) Insect interordinal relationships: insights from the visual system. Arthropod Syst Phyl 64:133–148

Friedrich M, Wood EJ, Wu M (2011) Developmental evolution of the insect retina: insights from standardized numbering of homologous photoreceptors. J Exp Zool B Mol Dev Evol 316:484–499

Furukubo-Tokunaga K, Adachi Y, Kurusu M, Walldorf U (2009) Brain patterning defects caused by mutations of the *twin of eyeless* gene in *Drosophila melanogaster*. Fly (Austin) 3:263–269

Gilbert C (1994) Form and function of stemmata in larvae of holometabolous insects. Annu Rev Entomol 39:323–349

Goriely A, Mollereau B, Coffinier C, Desplan C (1999) *Munster*, a novel paired-class homeobox gene specifically expressed in the *Drosophila* larval eye. Mech Dev 88:107–110

Green P, Hartenstein AY, Hartenstein V (1993) The embryonic development of the *Drosophila* visual system. Cell Tissue Res 273:583–598

Greenwood S, Struhl G (1999) Progression of the morphogenetic furrow in the *Drosophila* eye: the roles of *hedgehog*, *decapentaplegic* and the *raf* pathway. Development 126:5795–5808

Hagberg M (1986) Ultrastructure and central projections of extraocular photoreceptors in caddies-flies (Insecta, Trichoptera). Cell Tissue Res 245:643–648

Halder G, Callaerts P, Gehring WJ (1995) Induction of ectopic eyes by targeted expression of the *eyeless* gene in *Drosophila*. Science 267:1788–1792

Halder G, Callaerts P, Flister S, Walldorf U, Kloter U, Gehring WJ (1998) Eyeless initiates the expression of both *sine oculis* and *eyes absent* during *Drosophila* compound eye development. Development 125:2181–2191

Halfar K, Rommel C, Stocker H, Hafen E (2001) Ras controls growth, survival and differentiation in the *Drosophila* eye by different thresholds of MAP kinase activity. Development 128:1687–1696

Hartenstein AY 1993. Atlas of *Drosophila* development. Cold Spring Harbor Laboratory Press, USA, pp 32–34

Harzsch S, Dircksen H, Beltz B (2009) Development of pigment-dispersing hormone-immunoreactive neurons in the American lobster: homology to the insect circadian pacemaker system? Cell Tissue Res 335:417–429

Heberlein U, Mlodzik M, Rubin GM (1991) Cell-fate determination in the developing *Drosophila* eye—role of the *rough* gene. Development 112:703–712

Heberlein U, Wolff T, Rubin GM (1993) The TGF beta homolog *dpp* and the segment polarity gene *hedgehog* are required for propagation of a morphogenetic wave in the *Drosophila* retina. Cell 75:913–926

Heberlein U, Singh CM, Luk AY, Donohoe TJ (1995) Growth and differentiation in the *Drosophila* eye coordinated by *hedgehog*. Nature 373:709–711

Helfrich-Förster C (2004) The circadian clock in the brain: a structural and functional comparison between mammals and insects. J Comp Physiol A 190:601–613

Helfrich-Förster C, Edwards T, Yasuyama K, Wisotzki B, Schneuwly S, Stanewsky R, Meinertzhagen IA, Hofbauer A (2002) The extraretinal eyelet of *Drosophila*: development, ultrastructure, and putative circadian function. J Neurosci 22:9255–9266

Hinnemann A, Niederegger S, Hanslik U, Heinzel HG, Spiess R (2010) See the light: electrophysiological characterization of the Bolwig organ's light response of Calliphora vicina 3rd instar larvae. J Insect Physiol 56:1651-1658

Hofbauer A, Buchner E (1989) Does *Drosophila* have seven eyes? Naturwissenschaften 76:335–336

Holmes AL, Raper RN, Heilig JS (1998) Genetic analysis of *Drosophila* larval optic nerve development. Genetics 148:1189–1201

Homberg U, Würden S (1997) Movement-sensitive, polarization-sensitive, and light-sensitive neurons of the medulla and accessory medulla of the locust, *Schistocerca gregaria*. J Comp Neurol 386:329–346

Homberg U, Reischig T, Stengl M (2003) Neural organization of the circadian system of the cockroach *Leucophaea maderae*. Chronobiol Int 20:577–591

Ichikawa T (1991) Brain photoreceptors in the pupal and adult butterfly: fate of the larval ocelli. Zool Sci 8:471–476

Iyengar BG, Chou CJ, Sharma A, Atwood HL (2006) Modular neuropile organization in the *Drosophila* larval brain facilitates identification and mapping of central neurons. J Comp Neurol 499:583–602

Jarman AP, Grau Y, Jan LY, Jan YN (1993) *Atonal* is a proneural gene that directs chordotonal organ formation in the *Drosophila* peripheral nervous system. Cell 73:1307–1321

Jarman AP, Grell EH, Ackerman L, Jan LY, Jan YN (1994) *Atonal* is the proneural gene for *Drosophila* photoreceptors. Nature 369:398–400

Jekely G, Colombelli J, Hausen H, Guy K, Stelzer E, Nedelec F, Arendt D (2008) Mechanism of phototaxis in marine zooplankton. Nature 456:395–399

Johnston RJ Jr, Robert J, Otake Y, Sood P, Vogt N, Behnia R, Vasiliauskas D, McDonald E, Xie B, Koenig S, Wolf R, Cook T, Gebelein B, Kussell E, Nakagoshi H, Desplan C (2011) Interlocked feedforward loops control cell-type-specific Rhodopsin expression in the *Drosophila* eye. Cell 145:956–968

Jurgens G (1987) Segmental organization of the tail region in the embryo of *Drosophila* melanogaster. Roux's Arch Dev Biol 196:141–157

Jurgens G, Hartenstein V (1993) The terminal regions of the body pattern. In: Bate M, Martinez Arias A (eds) The development of *Drosophila melanogaster*. Cold Spring Harbour Laboratory Press, New York, pp 687–746

Jurgens G, Lehmann R, Schardin M, Nusslein-Volhard C (1986) Segmental organization of the head in the embryo of *Drosophila melanogaster*. Roux's Arch Dev Biol 195:359–377

Justice ED, Macedonia NJ, Hamilton C, Condron B (2012) The simple fly larval visual system can process complex images. Nat Commun 3:1156

Kaneko M, Hall JC (2000) Neuroanatomy of cells expressing clock genes in *Drosophila*: transgenic manipulation of the *period* and *timeless* genes to mark the perikarya of circadian pacemaker neurons and their projections. J Comp Neurol 422:66–94

Keene AC, Mazzoni EO, Zhen J, Younger MA, Yamaguchi S, Blau J, Desplan C, Sprecher SG (2011) Distinct visual pathways mediate *Drosophila* larval light avoidance and circadian clock entrainment. J Neurosci 31:6527–6534

Keene AC, Sprecher SG (2011) Seeing the light: photobehavior in fruit fly larvae. Trends Neurosci 35:104–110

Kenyon KL, Ranade SS, Curtiss J, Mlodzik M, Pignoni F (2003) Coordinating proliferation and tissue specification to promote regional identity in the *Drosophila* head. Dev Cell 5:403–414

Kristensen N (1999) Phylogeny of endopterygote insects, the most successful lineage of living organisms. Eur J Entomol 96:237–253

Kumar JP, Moses K (2001) Expression of evolutionarily conserved eye specification genes during *Drosophila* embryogenesis. Dev Genes Evol 211:406–414

Kurusu M, Nagao T, Walldorf U, Flister S, Gehring WJ, Furukubo-Tokunaga K (2000) Genetic control of development of the mushroom bodies, the associative learning centers in the *Drosophila* brain, by the *eyeless*, *twin of eyeless*, and *dachshund* genes. Proc Nat Acad Sci U S A 97:2140–2144

Land MF, Nilsson DE (2002) Animal eyes. Oxford University Press, Oxford

Lilly M, Carlson J (1990) smellblind: a gene required for *Drosophila* olfaction. Genetics 124:293

Liu Z, Friedrich M (2004) The *Tribolium* homologue of *glass* and the evolution of insect larval eyes. Dev Biol 269:36–54

Liu Z, Yang X, Dong Y, Friedrich M (2006) Tracking down the "head blob": comparative analysis of *wingless* expression in the embryonic insect procephalon reveals progressive reduction of ocular segment patterning in higher insects. Arthropod Struct Dev 35:341–356

Ma CY, Zhou Y, Beachy PA, Moses K (1993) The segment polarity gene *hedgehog* is required for progression of the morphogenetic furrow in the developing *Drosophila* eye. Cell 75:927–938

Malpel S, Klarsfeld A, Rouyer F (2002) Larval optic nerve and adult extra-retinal photoreceptors sequentially associate with clock neurons during *Drosophila* brain development. Development 129:1443–1453

Mardon G, Solomon NM, Rubin GM (1994) *dachshund* encodes a nuclear protein required for normal eye and leg development in *Drosophila*. Development 120:3473–3486

Mazzoni EO, Desplan C, Blau J (2005) Circadian pacemaker neurons transmit and modulate visual information to control a rapid behavioral response. Neuron 45:293–300

McDonald EC, Xie B, Workman M, Charlton-Perkins M, Terrell DA, Reischl J, Wimmer EA, Gebelein BA, Cook TA (2010) Separable transcriptional regulatory domains within Otd control photoreceptor terminal differentiation events. Dev Biol 347:122–132

Melzer RR (2009) Persisting stemma neuropils in *Chaoborus crystallinus* (Diptera: Chaoboridae): development and evolution of a bipartite visual system. J Morphol 270:1524–1530

Melzer RR, Paulus HF (1989) Evolutionswege zum Larvalauge der Insekten—Die Stemmata der höheren Dipteren und ihre Abwandlung zum Bolwig-Organ. Z Zool Syst Evol 27:200–245

Mikeladze-Dvali T, Wernet MF, Pistillo D, Mazzoni EO, Teleman AA, Chen YW, Cohen S, Desplan C (2005) The growth regulators *warts/lats* and *melted* interact in a bistable loop to specify opposite fates in *Drosophila* R8 photoreceptors. Cell 122:775–787

Mishra M, Oke A, Lebel C, McDonald EC, Plummer Z, Cook TA, Zelhof AC (2010) Pph13 and Orthodenticle define a dual regulatory pathway for photoreceptor cell morphogenesis and function. Development 137:2895–2904

Mlodzik M, Hiromi Y, Weber U, Goodman CS, Rubin GM (1990) The *Drosophila seven-up* gene, a member of the steroid receptor gene superfamily, controls photoreceptor cell fates. Cell 60:211–224

Mollereau B, Dominguez M, Webel R, Colley NJ, Keung B, Celis JF de, Desplan C (2001) Two-step process for photoreceptor formation in *Drosophila*. Nature 412:911–913

Morey M, Yee, SK, Herman, T, Nern A, Blanco E, Zipursky SL (2008) Coordinate control of synaptic-layer specificity and rhodopsins in photoreceptor neurons. Nature 456:795–799

Moses K, Ellis MC, Rubin GM (1989) The *glass* gene encodes a zinc-finger protein required by *Drosophila* photoreceptor cells. Nature 340:531–536

Moses K, Rubin GM (1991) *Glass* encodes a site-specific DNA binding protein that is regulated in response to positional signals in the developing *Drosophila* eye. Genes Dev 5:583–593

Mukhopadhyay M, Campos AR (1995) The larval optic nerve is required for the development of an identified serotonergic arborization in *Drosophila melanogaster*. Dev Biol 169:629–643

Namba R, Minden JS (1999) Fate mapping of *Drosophila* embryonic mitotic domain 20 reveals that the larval visual system is derived from a subdomain of a few cells. Dev Biol 212:465–476

Niimi T, Seimiya M, Kloter U, Flister S, Gehring WJ (1999) Direct regulatory interaction of the eyeless protein with an eye-specific enhancer in the *sine oculis* gene during eye induction in *Drosophila*. Development 126:2253–2260

Niven JE, Laughlin SB (2008) Energy limitation as a selective pressure on the evolution of sensory systems. J Exp Biol 211:1792–1804

Nolo R, Abbott LA, Bellen HJ (2000) Senseless, a Zn finger transcription factor, is necessary and sufficient for sensory organ development in *Drosophila*. Cell 102:349–362

Nuesch H (1987) Metamorphose bei Insekten. Direkte und indirekte Entwicklung bei Apterygoten und Exopterygoten. Zool Jb Anat 115:453–487

Ostrin EJ, Li Y, Hoffman K, Liu J, Wang K, Zhang L, Mardon G, Chen R (2006) Genome-wide identification of direct targets of the *Drosophila* retinal determination protein Eyeless. Genome Res 16:466–476

Paulus HF (1979) Eye structure and the monophyly of the Arthropoda. In: Gupta AP (ed) Arthropod phylogeny. Van Nostrand Reinhold Company, New York, pp 299–371

Paulus HF (1986) Evolutionswege zum Larvalauge der Insekten—ein Modell für die Entstehung und die Ableitung der ozellären Lateralaugen der Myriapoda von Fazettenaugen. Zool Jb Syst 113:353–371

Paulus HF (1989) Das Homologisieren in der Feinstrukturforschung: Das Bolwig-Organ der hoeheren Dipteren und seine Homologisierung mit Stemmata und Ommatidien eines ursprunglichen Facettenauges der Mandibulata. Zool Beitr 32:437–478

Punzo C, Seimiya M, Flister S, Gehring WJ, Plaza S (2002) Differential interactions of *eyeless* and *twin of eyeless* with the *sine oculis* enhancer. Development 129:625–634

Quiring R, Walldorf U, Kloter U, Gehring WJ (1994) Homology of the *eyeless* gene of *Drosophila* to the *Small eye* gene in mice and *Aniridia* in humans. Science 265:785–789

Ranade SS, Yang-Zhou D, Kong SW, McDonald EC, Cook TA, Pignoni F (2008) Analysis of the Otd-dependent transcriptome supports the evolutionary conservation of CRX/OTX/OTD functions in flies and vertebrates. Dev Biol 315:521–534

Reinke R, Krantz DE, Yen D, Zipursky SL (1988) Chaoptin, a cell surface glycoprotein required for *Drosophila* photoreceptor cell morphogenesis, contains a repeat motif found in yeast and human. Cell 52:291–301

Reischig T, Stengl M (2003) Ectopic transplantation of the accessory medulla restores circadian locomotor rhythms in arrhythmic cockroaches (*Leucophaea maderae*). J Exp Biol 206:1877–1886

Rodriguez Moncalvo VG, Campos AR (2005) Genetic dissection of trophic interactions in the larval optic neuropil of *Drosophila melanogaster*. Dev Biol 286:549–558

Sawin ME, Sokolowski MB, Campos AR (1995) Characterization and genetic analysis of *Drosophila melanogaster* photobehavior during larval development. J Neurogenet 10:119–135

Schmucker D, Taubert H, Jackle H (1992) Formation of the *Drosophila* larval photoreceptor organ and its neuronal differentiation require continuous *Kruppel* gene activity. Neuron 9:1025–1039

Schmucker D, Su AL, Beermann A, Jackle H, Jay DG (1994) Chromophore-assisted laser inactivation of patched protein switches cell fate in the larval visual system of *Drosophila*. Proc Natl Acad Sci U S A 91:2664–2668

Schmucker D, Jackle H, Gaul U (1997) Genetic analysis of the larval optic nerve projection in *Drosophila*. Development 124:937–948

Schmucker D, Vorbruggen G, Yeghiayan P, Fan HQ, Jackle H, Gaul U (2000) The *Drosophila* gene abstrakt, required for visual system development, encodes a putative RNA helicase of the DEAD box protein family. Mech Dev 91:189–196

Serikaku MA, O'Tousa JE (1994) Sine oculis is a homeobox gene required for *Drosophila* visual system development. Genetics 138:1137–1150

Sokoloff A (1972) The biology of *Tribolium*. Clarendon Press, Oxford

Sprecher SG, Desplan C (2008) Switch of rhodopsin expression in terminally differentiated *Drosophila* sensory neurons. Nature 454:533–537

Sprecher SG, Pichaud F, Desplan C (2007) Adult and larval photoreceptors use different mechanisms to specify the same Rhodopsin fates. Genes Dev 21:2182–2195

Sprecher SG, Cardona A, Hartenstein V (2011) The *Drosophila* larval visual system: high-resolution analysis of a simple visual neuropil. Dev Biol 358:33–43

Steiner P (1930) Studien an *Panorpa communis*. I. Zur Biologie von *Panorpa communis* L. Z Morphol Oekol Tiere 17:1–25

Steller H, Fischbach KF, Rubin GM (1987) *Disconnected*: a locus required for neuronal pathway formation in the visual system of *Drosophila*. Cell 50:1139–1153

Strauß J, Zhang Q, Verleyen P, Huybrechts J, Neupert S, Predel R, Pauwels K, Dircksen H (2011) Pigment-dispersing hormone in *Daphnia* interneurons, one type homologous to insect clock neurons displaying circadian rhythmicity. CMLS Cell Mol Life Sci 68:3403–3423

Suzuki T, Saigo K (2000) Transcriptional regulation of *atonal* required for *Drosophila* larval eye development by concerted action of *eyes absent*, *sine oculis* and *hedgehog* signaling independent of *fused kinase* and *cubitus interruptus*. Development 127:1531–1540

Tahayato A, Sonneville R, Pichaud F, Wernet MF, Papatsenko D, Beaufils P, Cook T, Desplan C (2003) Otd/Crx, a dual regulator for the specification of ommatidia subtypes in the *Drosophila* retina. Dev Cell 5:391–402

Tix S, Bate M, Technau GM (1989a) Pre-existing neuronal pathways in the leg imaginal disks of *Drosophila*. Development 107:855–862

Tix S, Minden JS, Technau GM (1989b) Pre-existing neuronal pathways in the developing optic lobes of *Drosophila*. Development 105:739–746

Truman JW, Riddiford LM (2002) Endocrine insights into the evolution of metamorphosis in insects. Annu Rev Entomol 47:467–500

Vafopoulou X, Steel CGH (2012) Metamorphosis of a clock: remodeling of the circadian timing system in the brain of *Rhodnius prolixus* (Hemiptera) during larval-adult development. J Comp Neurol 520:1146–1164

Van Vactor D Jr, Krantz DE, Reinke R, Zipursky SL (1988) Analysis of mutants in chaoptin, a photoreceptor cell-specific glycoprotein in *Drosophila*, reveals its role in cellular morphogenesis. Cell 52:281–290

Vandendries ER, Johnson D, Reinke R (1996) *orthodenticle* is required for photoreceptor cell development in the *Drosophila* eye. Dev Biol 173:243–255

Veleri S, Rieger D, Helfrich-Förster C, Stanewsky R (2007) Hofbauer-Buchner eyelet affects circadian photosensitivity and coordinates TIM and PER expression in *Drosophila* clock neurons. J Biol Rhythms 22:29–42

Wiegmann BM, Trautwein MD, Kim JW, Cassel BK, Bertone MA, Winterton SL, Yeates DK (2009) Single-copy nuclear genes resolve the phylogeny of the holometabolous insects. BMC Biol 7:34

Wiegmann BM, Trautwein MD, Winkler IS, Barrab NW, Kima J-W, Lambkin C, Bertone MA, Cassela BK, Peterson KJ, Wheeler BJ, Pape T, Sinclair BJ, Skevington JH, Blagoderovir V, Caravas J, Kutty S, Borkent A, Schmidt-Ott U, Kampmeier GE, Thompson FC, Grimaldi DA, Beckenbach AT, Courtney GW, Friedrich M, Meier R, Yeates DK (2011) Episodic radiations in the fly tree of life. Proc Natl Acad Sci U S A 108:5690–5695

Xie B, Charlton-Perkins M, McDonald EC, Gebelein B, Cook T (2007) *Senseless* functions as a molecular switch for color photoreceptor differentiation in *Drosophila*. Development 134:4243–4253

Xu C, Kauffmann RC, Zhang J, Kladny S, Carthew RS (2000) Overlapping activators and repressors delimit transcriptional response to receptor tyrosine kinase signals in the *Drosophila* eye. Cell 103:87–97.

Yang X, Weber M, ZarinKamar N, Wigand B, Posnien G, Friedrich R, Beutel R, Damen W, Bucher G, Klingler M, Friedrich M (2009) Probing the *Drosophila* retinal determination gene network in *Tribolium* (II): the Pax6 genes *eyeless* and *twin of eyeless*. Dev Biol 333:215–227

Yasuyama K, Meinertzhagen IA (1999) Extraretinal photoreceptors at the compound eye's posterior margin in *Drosophila melanogaster*. J Comp Neurol 412:193–202

Younossi-Hartenstein A, Tepass U, Hartenstein V 1993. Embryonic origin of the imaginal discs of the head of *Drosophila melanogaster*. Roux's Arch Dev Biol 203:60–73

Zelhof AC, Hardy RW, Becker A, Zuker CS (2006) Transforming the architecture of compound eyes. Nature 443:696–699

Index

A. Singh, M. Kango-Singh (eds.), *Molecular Genetics of Axial Patterning, Growth and Disease in the Drosophila Eye,* DOI 10.1007/978-1-4614-8232-1,

Kolibri 360 GmbH
Ferdinand-Jühlke-Straße 7
99095 Erfurt, Deutschland
produktsicherheit@kolibri360.de